Kant, Science, and Human Nature

Kant, Science, and Human Nature

Robert Hanna

CLARENDON PRESS · OXFORD

OXFORD
UNIVERSITY PRESS

Great Clarendon Street, Oxford OX2 6DP

Oxford University Press is a department of the University of Oxford.
It furthers the University's objective of excellence in research, scholarship,
and education by publishing worldwide in

Oxford New York

Auckland Cape Town Dar es Salaam Hong Kong Karachi
Kuala Lumpur Madrid Melbourne Mexico City Nairobi
New Delhi Shanghai Taipei Toronto

With offices in

Argentina Austria Brazil Chile Czech Republic France Greece
Guatemala Hungary Italy Japan Poland Portugal Singapore
South Korea Switzerland Thailand Turkey Ukraine Vietnam

Oxford is a registered trade mark of Oxford University Press
in the UK and in certain other countries

Published in the United States
by Oxford University Press Inc., New York

© Robert Hanna 2006

The moral rights of the author have been asserted
Database right Oxford University Press (maker)

First published 2006

All rights reserved. No part of this publication may be reproduced,
stored in a retrieval system, or transmitted, in any form or by any means,
without the prior permission in writing of Oxford University Press,
or as expressly permitted by law, or under terms agreed with the appropriate
reprographics rights organization. Enquiries concerning reproduction
outside the scope of the above should be sent to the Rights Department,
Oxford University Press, at the address above

You must not circulate this book in any other binding or cover
and you must impose the same condition on any acquirer

British Library Cataloguing in Publication Data
Data available

Library of Congress Cataloging in Publication Data
Data available

Typeset by Laserwords Private Limited, Chennai, India
Printed in Great Britain
on acid-free paper by
Biddles Ltd, King's Lynn, Norfolk

ISBN 0–19–928554–3   978–0–19–928554–9

10 9 8 7 6 5 4 3 2 1

To MTH and ETH
The fundamental things still apply

Preface and Acknowledgments

> The man bent over his guitar,
> A shearsman of sorts. The day was green.
>
> They said, "You have a blue guitar,
> You do not play things as they are."
>
> The man replied, "Things as they are
> Are changed upon the guitar."
>
> And they said then, "But play, you must,
> A tune beyond us, yet ourselves,
>
> A tune upon the blue guitar
> Of things exactly as they are."
>
> Wallace Stevens
> The Man with the Blue Guitar, verse I

In the mid-1990s I wrote an 800-or 900-page manuscript that was intended to be the definitive critical study of the immensely complex relationship between Immanuel Kant's eighteenth-century Critical Philosophy and the historical and conceptual foundations of twentieth-century analytic philosophy. It was one of those loose baggy monsters, and certainly too long for a single book. So in the end the first half of that manuscript became *Kant and the Foundations of Analytic Philosophy*, published in 2001, and most of the second half has become the core of *Kant, Science, and Human Nature*. Needless to say my intention to produce the definitive study on this topic, whether in one or many volumes, was quixotic. But it *was* my idea of a good time.

So what have I been hammering away at for the past ten years? The main aim of *Kant and the Foundations of Analytic Philosophy* was to show that Kant was essentially right about the analytic-synthetic distinction, and in particular about the nature and existence of the synthetic a priori, and that the mainstream analytic tradition from Frege to Quine was mostly wrong about both. In short, we should accept the thesis that there are two irreducibly different types of necessary a priori truth (analytic or logical, and synthetic or non-logical), or *Kantian modal dualism*. In turn, the main aim of *Kant, Science, and Human Nature* is to show that Kant was essentially right about the unknowability and methodological eliminability of a microphysical noumenal world hiding behind the directly perceivable manifest macrophysical world, and also about the priority of practical reason over theoretical reason, and that the mainstream analytic tradition from Sellars to Kripke has been mostly wrong about both. In short, we should accept the Kantian theses of *empirical realism* and *the practical foundations of the exact sciences*.

The over arching goal of the pair of books is meta-philosophical. It is to show that the tacit fundamental project of the twentieth- and twenty-first-century analytic tradition lies in its substantively reconnecting with the eighteenth and nineteenth century Kantian tradition that contains its origins, so that the two traditions can jointly become the single project of *rational anthropology in the broadest possible sense*. By this I mean the philosophical study of human persons—as embodied minds embedded in the larger natural world, as individual intentional agents capable of autonomy or self-legislation, and as participants in shared practices and society—insofar as their cognitions, volitions, emotions, and actions are all inherently open to governance and evaluation by necessary a priori principles. Or in other words, I am saying that Kant is the man with the blue guitar.

I am very grateful to successive generations of my undergraduate and graduate students at the University of Colorado at Boulder—especially Jason Potter and Bryan Hall—for listening very intelligently and patiently to my rantings about Kantian ideas, and for offering many good critical insights and thoughts in response to Kant's texts and to my rantings alike; to Bryan Hall again, for his groundbreaking PhD dissertation work on Kant's much-neglected *Opus postumum*, which forced me to take the concept of a universal causal-dynamic aether seriously and to rethink my views about Kant's theories of causation and matter; to various scholarly audiences for their highly helpful critical comments on presentations of various parts of the material in the book; to several anonymous referees at several journals for their similarly helpful critical comments on various preliminary versions of the published articles that are the bases of seven of the eight chapters; to the anonymous referees at OUP for their similarly helpful critical comments on the various preliminary versions of the book manuscript; to

A. W. Moore, Eric Watkins, and Kenneth Westphal for publishing, just in the nick of time, three first-rate books on Kant—*Noble in Reason, Infinite in Faculty* (Routledge 2003), *Kant and the Metaphysics of Causality* (Cambridge 2005), and *Kant's Transcendental Proof of Realism* (Cambridge 2004)—that helped shape my work on the final version of manuscript and especially on the last chapter; to Karl Ameriks, Paul Guyer, and Peter Strawson, for their generous encouragement and support over the years; to Bryan Hall yet again, for his help as my research assistant during the summer of 2002; to Catherine Berry, Rupert Cousens, and Peter Momtchiloff, my editors at OUP, for their efficiency and encouragement throughout the production and publication process; and finally, to all those who are acknowledged and thanked in the acknowledgments of the published articles listed below.

I am also very grateful to the Department of Philosophy at the University of Colorado for granting me research leave during the Spring semesters of 2000 and 2001, and sabbatical leave during the academic year 2003–2004; to Fitzwilliam College, University of Cambridge, for generously providing me with visiting fellowships in Lent term 2000, Lent term 2001, and the academical year 2003–2004; and to the following journals and their editors for giving their permission to include material from the following published articles, in these chapters: chapter 1: R. Hanna, "The Inner and the Outer: Kant's 'Refutation' Reconstructed," *Ratio* 13 (2000): 146–74; chapter 2: R. Hanna, "Kant and Nonconceptual Content," *European Journal of Philosophy* 13 (2005): 247–90; chapter 3: R. Hanna, "A Kantian Critique of Scientific Essentialism," *Philosophy and Phenomenological Research* 58 (1998): 497–528; chapter 4: R. Hanna, "Why Gold is Necessarily a Yellow Metal," *Kantian Review* 4 (2000): 1–47; chapter 5: R. Hanna, "Kant, Truth, and Human Nature," *British Journal for the History of Philosophy* 8 (2000): 225–50; chapter 6: R. Hanna, "Mathematics for Humans: Kant's Philosophy of Arithmetic Revisited," *European Journal of Philosophy* 10 (2002): 328–53; chapter 7: R. Hanna, "How Do We Know Necessary Truths? Kant's Answer," *European Journal of Philosophy* 6 (1998): 115–45.

Finally, there are two acknowledgments that belong in a sharply separate category. I love you so very much, Martha T. Hanna. And you too, Elizabeth T. Hanna.

R. H.

1 January 2006
Boulder, Colorado

Contents

A Note on References to Kant's Works — xv

Introduction — 1

    0.0. Kant, the Analytic Tradition, and the Exact Sciences — 1
    0.1. The Two Images Problem and Scientific Naturalism — 8
    0.2. The Critical Philosophy: A Working Idea — 17
    0.3. The Primacy of Human Nature — 29

Part I. Empirical Realism and Scientific Realism — 35

1. Direct Perceptual Realism I: The Refutation of Idealism — 37

    1.0. Introduction — 37
    1.1. Interpretation and Criticism of the Refutation — 50
    1.2. Space, Content Externalism, and Human Embodiment — 68
    1.3. The Reality of the Embodied Mind in Space — 79

2. Direct Perceptual Realism II: Non-Conceptual Content — 81

    2.0. Introduction — 81
    2.1. Kant, Non-Conceptualism, and Conceptualism — 84
    2.2. Kant's Arguments for Non-Conceptualism — 94
    2.3. The Forms of Intuition and Non-Conceptual Content — 115
    2.4. The Role of Spatiotemporal Structure in Non-Conceptual Content — 126
    2.5. Direct Perceptual Realism Again — 132

3. Manifest Realism I: A Critique of Scientific Essentialism — 140

    3.0. Introduction — 140
    3.1. What is Scientific Essentialism? — 152
    3.2. Objection I: The Empirical Inaccessibility of the Microphysical World — 160

3.3. Objection II: Why There is No Necessary A Posteriori	171
3.4. Objection III: The Antinomy of Essentialism	177
3.5. Objection IV: The Logical Contingency of the Laws of Nature	181
3.6. Concluding Anti-Scientific-Essentialist Postscript	187

4. Manifest Realism II: Why Gold is Necessarily a Yellow Metal — 191

4.0. Introduction	191
4.1. Kant's Theory of Natural Kind Terms	203
4.2. Scientific Realism in the Manifest Image	220
4.3. Kant's Other Joke	242

Part II. The Practical Foundations of the Exact Sciences — 247

5. Truth and Human Nature — 249

5.0. Introduction	249
5.1. The Definition of Truth	255
5.2. The Criterion of Truth	275
5.3. The Criteria of Truth	277
5.4. The Sense of Truth: Truth and Truthfulness	282

6. Mathematics for Humans — 287

6.0. Introduction	287
6.1. Mathematics and Transcendental Idealism	289
6.2. Why Arithmetic is Synthetic A Priori	312
6.3. The Meaning of the Concept NUMBER	320
6.4. Construction as Construal	331

7. How Do We Know Necessary Truths? — 341

7.0. Introduction	341
7.1. Epistemic Necessity and Epistemic Apriority, Kant-Style	350
7.2. Insight in Conceptual Analysis	362
7.3. The Role of Insight in A Priori Knowledge in General	379
7.4. Theoretical Technique: A Priori Knowledge in Action	383

8. Where There's a Will There's a Way: Causation and Freedom — 386

8.0. Introduction	386
8.1. Kant's Metaphysics of Causation: Three Analogies and an Antinomy	388

8.2.	The Problem of Free Will and Kant's Embodied Agency Theory	416
8.3.	Freedom and Nature	433
8.4.	Freedom, Causation, and Time's Arrow	450

Bibliography   453
Index   475

RP	*What Real Progress has Metaphysics made in Germany since the Time of Leibniz and Wolff?*, trans. T. Humphrey (New York: Arabis, 1983)
UNH	*Universal Natural History and Theory of the Heavens*, trans. W. Hastie (Ann Arbor, MI: University of Michigan Press, 1969)
VL	"The Vienna Logic," in *Immanuel Kant: Lectures on Logic*, pp. 249–377

Introduction

Kant's joke. Kant wanted to prove in a way that would dumbfound the common man that the common man was right: that was the secret joke of this soul.

Friedrich Nietzsche[1]

Appearances are not held to be a clue to the truth. But we seem to have no other.

Ivy Compton-Burnett[2]

The truly apocalyptic view of the world is that things do not repeat themselves. It isn't absurd, e.g., to believe that the age of science and technology is the beginning of the end for humanity; that the idea of great progress is a delusion, along with the idea that the truth will ultimately be known; that there is nothing good or desirable about scientific knowledge and that mankind, in seeking it, is falling into a trap. It is by no means obvious that this is not how things are.

Ludwig Wittgenstein[3]

0.0. Kant, the Analytic Tradition, and the Exact Sciences

This book is a study of Immanuel Kant's theory of the epistemological, metaphysical, and practical foundations of "the exact sciences": pure mathematics and fundamental physics. On Kant's view, pure mathematics and fundamental physics are *a priori* sciences, which is to say that the

---

[1] Nietzsche, *The Portable Nietzsche*, p. 96.
[2] Compton-Burnett, *Manservant and Maidservant*. Compton-Burnett could be read as saying either (a) that we have no clue to the truth *except* appearances, or (b) that appearances *are* the truth. I am reading her both ways.
[3] Wittgenstein, *Culture and Value*, p. 56e.

truths they disclose are necessary and knowable independently of all sense experience. But mathematics and physics, although a priori sciences, are specifically directed to *objects in the world*—as opposed to pure general logic, which is an absolutely universal a priori science that deals with human understanding and reason itself, and its underlying form or structure, and is insensitive to ontological furniture (*CPR* Bxviii–xiv, A52–3/B76–7). So, while pure general logic is presupposed by the exact sciences, it is not itself an exact science.[4] Kant's logical theory is notorious for its formal confusions and limitations.[5] Correspondingly, for most of the twentieth century, Kant's philosophies of mathematics and physics were relegated to the dustbin of the history of philosophy as uninterestingly anachronistic, on the dual grounds (i) that his account of mathematics is based on a "terrifyingly narrow and mathematically trivial, conception of the province of logic,"[6] and (ii) that his account of physics is hopelessly classical and Newtonian. But Kant's theory of the exact sciences has been recently rediscovered, re-evaluated, and resuscitated, and is now without a doubt one of the most active and fruitful areas in recent and contemporary scholarship on Kant's theoretical philosophy.[7] So in this sense *Kant, Science, and Human Nature* fits neatly into an important trend in Kant studies.

At the same time, however, this book is also a sequel to my first book on Kant's Critical Philosophy, *Kant and the Foundations of Analytic Philosophy*. There I analyzed some of the deep connections between Kant's

---

[4] According to Kant, pure general logic is in fact also *a priori moral science* (*CPR* A54–5/B78–9). Interestingly enough, Frege shares the same conception of the status of pure logic, although of course his account of the structure and content of pure logic is very different from Kant's. See Hanna, *Rationality and Logic*, ch. 7.

[5] This familiar criticism is, I think, largely misguided. The pure logic developed by Frege and Russell–Whitehead is extensional (i.e., object oriented and compositional), anti-psychologistic to a fault, focused on rough-grained (i.e., extensionally structured) propositions, and expressly designed for the reduction of mathematics to logic (logicism). Kant's logic by contrast is intensional (i.e., mode-of-presentation oriented and generative), centered on cognitive acts of judgment and fine-grained (i.e., intensionally–structured) propositions, and expressly designed to support his thesis that mathematics is synthetic a priori and irreducible to logic (anti-logicism). So Kant's logic is just *different from*, not *worse than*, Frege-Russell logic. See Hanna, *Kant and the Foundations of Analytic Philosophy*, chs. 2–3; and Hanna, "Kant's Theory of Judgment."

[6] Hazen, "Logic and Analyticity," p. 92.

[7] See, e.g., Brittan, *Kant's Theory of Science*; Buchdahl, *Kant and the Dynamics of Reason*; Butts (ed.), *Kant's Philosophy of Physical Science*; Edwards, *Substance, Force, and the Possibility of Knowledge: On Kant's Philosophy of Material Nature*; Friedman, *Kant and the Exact Sciences*; Plaass, *Kant's Theory of Natural Science*; Posy (ed.), *Kant's Philosophy of Mathematics*; Warren, *Reality and Impenetrability in Kant's Philosophy of Nature*; and Watkins (ed.), *Kant and the Sciences*.

Critique of Pure Reason and the historical and conceptual foundations of the European and Anglo-American tradition of analytic philosophy, from Gottlob Frege's *Foundations of Arithmetic* (1884) to W. V. O. Quine's "Two Dogmas of Empiricism" (1951). In order to do this, I sketched and defended what I called a "cognitive-semantic" interpretation of Kant's theoretical philosophy. The leading idea of the cognitive-semantic interpretation is that the basic goal of Kant's theoretical philosophy is to give an account of how conscious objective mental representations—and especially those that are a priori—are possible. Against that backdrop, I then argued for three basic claims: (1) that the analytic tradition emerged by struggling with some of the central doctrines of the *Critique of Pure Reason*; (2) that a careful examination of this foundational debate shows that Kant's doctrines were never refuted but rather, for various reasons, rejected; and (3) that ironically enough it is the foundations of *analytic* philosophy, not the Critical Philosophy, that are inherently shaky.

Kant, Science, and Human Nature further extends this same general line of argument by focusing on the equally deep connections between the Critical Philosophy and analytic philosophy from 1950 to the end of the twentieth century. The central topics or obsessions of the analytic tradition prior to 1950 were *meaning and necessity*, with special emphases on: (a) pure logic as the universal and necessary essence of thought; (b) language as the basic means of expressing thoughts and describing the world; (c) the sense vs. reference distinction; (d) the conceptual truth vs. factual truth distinction; (e) the necessary truth vs. contingent truth distinction; (f) the a priori truth vs. a posteriori truth distinction; and (g) the analytic vs. synthetic distinction.[8] A common and deep thread running through all of these sub-themes is the following rough-and-ready multiple necessary equivalence: logical truth ↔ linguistic truth ↔ sense-determined truth ↔ conceptual truth ↔ necessary truth ↔ a priori truth ↔ analyticity. So a very useful way of characterizing analytic philosophy from Frege to mid-century Quine is to say that it consists essentially in the rise and fall of the concept of analyticity. By vivid contrast, however, the central obsession of analytic philosophy after 1950 was and is *scientific naturalism*,[9] which

---

[8] See, e.g., Soames, *Philosophical Analysis in the Twentieth Century*, vol. 1, esp. parts 2, 3, and 5.
[9] Ibid. vol. 2, esp. parts 5 and 7.

holds—to use Wilfrid Sellars's apt phrase—that "science is the measure of all things."[10]

Here Sellars's term-of-art "science" clearly refers to the exact sciences, where the class of exact sciences has been implicitly expanded beyond Kant's own conception of the exact sciences so as to include, in particular, biology and chemistry. Kant was a non-reductivist about biology and chemistry.[11] By sharp contrast, on the standard model of contemporary scientific reduction,[12] both biology and chemistry have a fully mathematically describable and microphysical basis in fundamental physical entities, properties, and facts, and thereby are both firmly grounded in fundamental physics. So, to put it in a nutshell, scientific naturalism holds, first, that the nature of knowledge and reality are ultimately disclosed by pure mathematics, fundamental physics, and whatever other reducible natural sciences there actually are or may turn out to be, and second, that this is the *only* way of disclosing the nature of knowledge and reality. Or to put it in an even smaller nutshell, scientific naturalism is *reductive* naturalism. (But it can get to be a little cramped inside a nutshell. So see section 0.1 for an elaboration of this characterization.)

Scientific or reductive naturalism, like analytic philosophy itelf, emerged in the struggle of nineteenth- and twentieth-century European and Anglo-American philosophers with the central doctrines and implications of the Critical Philosophy. Indeed, as Michael Friedman has persuasively argued, scientific naturalism can be traced directly back to its seminal beginnings in turn-of-the-twentieth-century German neo-Kantian philosophy, as a result of the devastating *anti*-neo-Kantian critiques of early Russell, early Carnap, and other Vienna Circle logical empiricists or positivists.[13] Not altogether coincidentally of course, Russell, Carnap, and many of the leading members

---

[10] Sellars, "Empiricism and the Philosophy of Mind," p. 173.

[11] See Grene and Depew, *The Philosophy of Biology*, esp. ch. 4; and Watkins (ed.), *Kant and the Sciences*, part IV. One of the hottest issues in current philosophy of science is the question of the explanatory significance and ontological implications of the concept of *the living organism*. Somewhat ironically, but perhaps also not too surprisingly, Kant's reflections on the concepts of a "natural purpose" (*Naturzweck*) and natural teleology in the second half of the *Critique of the Power of Judgment* are right at the center of this current debate. A closely related question is whether there is, or is not, a "strong continuity" between biological life and conscious mind. See, e.g., Godfrey-Smith, *Complexity and the Function of Mind in Nature*. I will have something to say about this tangle of issues in ch. 8.

[12] See Fodor, "Special Sciences, or The Disunity of Science as a Working Hypothesis"; and Kim, "Multiple Realization and the Metaphysics of Reduction."

[13] See, e.g., Friedman, "Philosophical Naturalism"; and Friedman, *Reconsidering Logical Positivism*.

of the Circle had actually begun their philosophical lives as neo-Kantians.[14] So scientific naturalism is to an important extent a radically revisionist neo-Kantianism, created by renegade former neo-Kantians. Friedman also argues for a contemporary, "post-analytic" return to a broadly neo-Kantian approach to epistemology and metaphysics, suitably fine-tuned by sophisticated relativism, and deeply informed by a detailed knowledge of the exact sciences.[15] Fair enough. But my aim, on the contrary, is to challenge the standard, neo-Kantian interpretation of Kant's epistemology and metaphysics, and thereby reconsider and rethink the foundations of middle-to-late twentieth century analytic philosophy—and, by implication, the foundations of early twenty-first-century analytic philosophy too[16]—by thoroughly reconsidering and rethinking Kant's theory of the foundations of the exact sciences.

In *Kant and the Foundations of Analytic Philosophy*, as I have said, I argued for the adoption of a *cognitive-semantic* interpretation of the first *Critique*. The cognitive-semantic approach is sharply distinct from the classical or *metaphysical* interpretation of transcendental idealism as a theory of the necessary a priori conditions of the possibility of the mind-dependent phenomenal world in opposition to its causal source, the independently-existing noumenal world of Really Real objects and moral agents or persons. This is because the cognitive-semantic interpretation rejects the substance dualism of the classical or metaphysical interpretation, and says that transcendental idealism requires the existence of only *one world*, the world of phenomena or real empirical things, but *two sets of concepts or properties*, phenomenal concepts or properties (i.e., the actually-instantiated concepts or properties of real empirical things) and noumenal concepts or properties (i.e., non-empirical concepts or properties that are logically possibly instantiable but cannot be known by us to be instantiated).[17] Even

---

[14] See, e.g., Carnap, *Der Raum*; Russell, *An Essay on the Foundations of Geometry*; and Reichenbach, *Theory of Relativity and A Priori Knowledge*. See also Friedman, *A Parting of the Ways: Carnap, Cassirer, and Heidegger*.

[15] See, e.g., Friedman, *Dynamics of Reason*; Friedman, "Philosophical Naturalism"; and Friedman, "Transcendental Philosophy and A Priori Knowledge: A Neo-Kantian Perspective."

[16] See Hanna, "Kant in Twentieth-Century Philosophy"; and Hanna, "Kant, Wittgenstein, and the Fate of Analysis."

[17] This is what I call the Two Concept or Two Property Theory of the phenomenon-noumenon distinction, as opposed to the two standard interpretations—namely, the Two Object or Two World Theory, and the Two Aspect or Two Standpoint Theory. See section 8.2; and Hanna, *Kant and the Foundations of Analytic Philosophy*, section 2.4.

more importantly in the present context, however, the cognitive-semantic approach is also sharply distinct from the familiar and still predominant neo-Kantian or *epistemological* interpretation of transcendental idealism as a theory of the necessary a priori conditions of the possibility of justified true beliefs about the phenomenal world. The epistemological interpretation is closely tied to Norman Kemp Smith's unintentionally unfortunate translation of *Erkenntnis* as "knowledge",[18] and even more closely tied to the strong influence of C. I. Lewis's epistemological writings on Anglo-American professional philosophy in the period prior to World War II.[19] Lewis's epistemology, moreover, except for its characteristically American pragmatic orientation, is essentially the same as the *analytic phenomenalism* or *constructive empiricism* deriving from Russell's *Our Knowledge of the External World*, Carnap's *Logische Aufbau der Welt*, and Vienna Circle logical positivism in the 1930s more generally, all of which are explicitly and aggressively science-oriented, and thoroughly naturalistic in character.[20] It follows that the neo-Kantian interpretation of Kant *presupposes* the correctness of some or another version of scientific or reductive naturalism. By sharp contrast, the aim of *Kant, Science, and Human Nature* is to explore

---

[18] In middle-to-late twentieth- and twenty-first century philosophical English, "knowledge" means "justified true belief" or perhaps "justified true belief plus whatever is sufficient to close the analytical gap between justification and knowledge that was opened up by the Gettier problem." But for Kant, *Erkenntnis* means "cognition" or *the conscious mental representation of objects* (CPR A320/B376), which, when that cognition turns out to be specifically an empirically meaningful cognition and also a judgment, can be either true or false (CPR Bxxvi n., A50–1/B74–6, A58–9/B83), and certainly need not be justified. Indeed, Kant explicitly reserves a distinct technical term for knowledge-as-sufficiently-justified-true-belief, namely, *Wissen* (CPR A820–31/B848–59), which I translate as "scientific knowing." One striking consequence of these seemingly anodyne terminological points is that to the considerable extent that the first *Critique* is all about the nature, scope, and limits of human *Erkenntnis*, then it is fundamentally a treatise in cognitive semantics, and *not* fundamentally a treatise in epistemology. It should also be noted however that Kemp Smith's mistranslation is entirely forgivable, since he himself was using *Erkenntnis* and "knowledge" in ways that were much closer to the Brentano-Husserl-Meinong intentionality tradition, than they were to the C. I. Lewis-H. A. Prichard-H. H. Price epistemological tradition.

[19] See, e.g., Lewis, "A Pragmatic Conception of the A Priori"; and Lewis, *Mind and the World Order*. Lewis's influence on Kant studies in particular was directly and widely felt in North America via the writings of Lewis White Beck and Sellars. Beck and Sellars were both Lewis's PhD students at Harvard. On the other side of the Atlantic, in 1936 *Mind and the World Order* was the first contemporary philosophical text ever to be taught at Oxford, in a seminar run by J. L. Austin and Isaiah Berlin. Sellars in fact attended this Oxford seminar, started a D.Phil. dissertation on Kant with T. D. Weldon the same year, and then later transferred to Harvard.

[20] See Richardson, *Carnap's Construction of the World*.

Kant's theory of the foundations of the exact sciences from an explicitly cognitive-semantic and therefore *non*-epistemological, *non*-phenomenalist, *non*-logical positivist, *non*-scientific naturalist point of view, in order (i) to develop a Kantian critique of scientific naturalism, and also even more importantly and positively (ii) to work out the elements of a fundamentally anthropocentric or humanistic, realistic, and non-reductive Kantian theory of the exact sciences.

Abstracting away now for a moment from the domestic differences between the various approaches to Kant-interpretation, and also from the question of the truth or falsity of scientific naturalism, it should be noted that the leading interpretive assumption of *Kant, Science, and Human Nature*, like that of *Kant and the Foundations of Analytic Philosophy*, is that the history of philosophy is a genre of philosophy, pure and simple. It is purely and simply the genre in which the fundamental questions of philosophy are addressed through the explication and critical analysis of great (by which I mean the most brilliant, groundbreaking, mindchanging, and trendsetting) old books, and in which the theses, arguments, and theories found in those great old books are directly related to contemporary philosophical debates. As applied specifically to Kant, however, for me this implies two methodological principles and an overarching dictum. The two methodological principles are as follows. First, charitably attribute to Kant the best philosophical view consistent with all the texts on a given topic. Second, in cases of conflicting texts on a given topic, charitably attribute to Kant the best philosophical view consistent with at least some of his texts, and bracket the texts in which he seems confused or mistaken. Like all philosophers, Kant sometimes errs, or anyhow nods. But we respect him most by critically noting and then setting aside his slips, and by promoting his deepest and most powerful doctrines. So the overarching dictum is this: Kant's Critical Philosophy is fully worth studying, critically analyzing, charitably explicating, defending, and then independently developing in a contemporary context. This is because, in my opinion, more than any other single-authored body of work in modern philosophy the Critical Philosophy most doggedly pursues and most profoundly captures some non-trivial fragment of the honest-to-goodness truth about rational human animals and the larger natural world that surrounds them.

0.1. The Two Images Problem and Scientific Naturalism

It is plausibly arguable (and has indeed been compellingly argued by, for example, Hilary Putnam, and John McDowell[21]) that the basic problem of European and Anglo-American analytic philosophy after 1950—and perhaps also *the* fundamental problem of modern Philosophy—is how it is possible to reconcile two sharply different, seemingly incommensurable, and apparently even mutually exclusive metaphysical conceptions, or "pictures," of the world. On the one hand, there is the objective, non-phenomenal, perspectiveless, mechanistic,[22] value-neutral, impersonal, and amoral metaphysical picture of the world delivered by pure mathematics and the fundamental natural sciences. And on the other hand, there is the subjective, phenomenal, perspectival, teleological, value-laden, person-oriented, and moral metaphysical picture of the world yielded by the conscious experience of rational human beings. The deep worry about the titanic clash between these two world-pictures was first outed in the mid-1930s by Edmund Husserl, in his *Crisis of European Sciences*. Similar ideas were expressed by Wittgenstein in the thirties and forties. In 1963, having been significantly influenced by Kant, Husserl, and Wittgenstein alike, Sellars famously dubbed these two sharply opposed world-conceptions "the scientific image" and "the manifest image."[23] So I will call the profound difficulty raised by their mutual incommensurability and inconsistency *the Two Images Problem*.

It should be emphasized that the Two Images Problem expresses not merely a clash between two sharply opposed conceptions of the world. This is because human beings *also* belong to the world. So the Two Images Problem also expresses a clash between two sharply opposed conceptions

---

[21] See Putnam, *The Many Faces of Realism*; and McDowell, *Mind and World*.

[22] By *natural mechanism* I mean the thesis that the entire world—including living organisms, animals, human beings, and persons—operates according to non-teleological, mathematico-physical principles alone. Or, more precisely, natural mechanism is the disjunctive thesis consisting of *either* determinism, the view that all later events, including the intentional acts of persons, are metaphysically necessitated by the settled actual facts about the past and present according to universal natural laws *or* indeterminism, the view that the entire world, including all the intentional acts of persons, causally operates according to probabilistic or stochastic principles alone. See section 8.2.

[23] Sellars, "Philosophy and the Scientific Image of Man." For some of the Kantian influences on Sellars, see n.19. Sellars also studied both Kant's and Husserl's writings with Marvin Farber at SUNY–Buffalo in the early thirties.

of human nature. We can ask: are human beings nothing but physical objects, or are they instead essentially non-physical subjects? If human beings are nothing but physical objects, then why does it seem so obvious that *human persons* are infinitely more than *mere things*? And assuming that human persons are *also* human animals, then why does it seem so obvious that despite being animals, human persons could not possibly be *mere hunks of matter*, because mere hunks of matter are essentially inert and subject to mechanistic causal principles, while human animals are living sentient organisms and thus neither inert nor mechanical but instead spontaneous and goal-directed? But on the other hand, if human beings are essentially non-physical subjects, then precisely how are human persons related to the physical world in general, and to their own living animal bodies in particular?

As Sellars correctly notes, the Two Images Problem has its origins in early modern or seventeenth- and eighteenth-century philosophy (the period of the revolutionary rise of the natural sciences, of the predominance of Enlightenment ideas at the level of high culture,[24] and of the sometimes violent emergence of the modern political state), and especially in the writings of Hobbes, Descartes, Locke, Leibniz, Berkeley, and Hume. There we find the Problem initially surfacing in such doctrines as Hobbes's materialist rationalism and his naturalistic contractarian politics; Descartes's substance dualism and his theistic occasionalism; Locke's property-based libertarian politics and his dichotomous epistemic and metaphysical distinctions (following on from Galileo and Robert Boyle) between primary qualities and secondary qualities, and between real essences and nominal essences; Leibniz's panpsychist monadological physics and his rational theodicy; Berkeley's theologically driven arguments for the impossibility of the material world and against Locke's distinctions; and Hume's non-cognitivist proto-utilitarian ethics and his psychologically driven skeptical critiques of causation and personal identity. Each of these doctrines can be plausibly regarded as either a direct response to, or a direct result of, the intense pressures exerted by the two images on each other. It is only a

---

[24] See, e.g., Gay, *The Enlightenment*. But it does not follow that Enlightenment high culture went all the way down: indeed, the "low" or popular culture of the time was in various interesting ways opposed to and critical of Enlightenment epistemology, metaphysics, ethics, and political theory. See, e.g., Darnton, *The Great Cat Massacre*; and Ginzburg, *The Cheese and the Worms*. My point is that the standard Enlightenment conceptual dichotomies and puzzles have always had their philosophical critics.

truism, but still utterly true and worth re-emphasizing, that the repeated attempts to resolve or at least mitigate these theoretical and practical conflicts have effectively determined the development of modern philosophy.

Since the 1950s, however, a possible complete solution to the Two Images Problem has gradually emerged, in the form of scientific or reductive naturalism.[25] As I already mentioned, scientific naturalism is the product of mid- to late-nineteenth-century German neo-Kantian philosophy, and the early twentieth-century reaction against it. But now it is also important to see that scientific naturalism is at the confluence of three interestingly different philosophical sub-traditions: (i) the German neo-Kantian tradition and its renegades, the early analytic philosophers; (ii) the positivist tradition in England, France, Germany, and Austria; and (iii) British empiricism in the tradition of Locke and Hume, particularly as developed by J. S. Mill.[26] By the middle of the twentieth century, however, these three different sub-traditions had successfully achieved a stable fusion (mainly brought about by the diaspora of leading Vienna Circle logical positivists from Austria and Germany to the leading North American philosophy departments, together with the personal influence and writings of fellow travelers like C. I. Lewis and his students at Harvard) with the home—grown pragmatic, neo-Kantian, and neo-Hegelian traditions in the USA.[27] This complex synthesis—of (i) C. I. Lewis's neo-neo-Kantianism; (ii) a well-informed familiarity with the major developments of physics in the twentieth century (especially including Einstein's special and general relativity theories, the Rutherford-Bohr conception of atomic matter, and quantum mechanics); (iii) Fregean and Russellian logicism; (iv) Carnap's logical positivism and phenomenalism; (v) Mill's empiricism; (vi) C. S. Peirce's pragmatism; and (vii) Josiah Royce's neo-Hegelianism—is epitomized by the major writings of Quine and Sellars, but most especially by Sellars's writings.[28] Scientific or reductive naturalism includes four basic elements: (1) anti-supernaturalism;

---

[25] See, e.g., Danto, "Naturalism"; Friedman, "Philosophical Naturalism"; Maddy, "Naturalism and the A Priori"; McDowell, "Two Sorts of Naturalism"; Papineau, *Philosophical Naturalism*; and Stroud, "The Charm of Naturalism."

[26] See, e.g., Hatfield, *The Natural and the Normative*; and Köhnke, *The Rise of Neo-Kantianism*.

[27] See Kuklick, *The Rise of American Philosophy*.

[28] See Quine, *From a Logical Point of View*; Quine, *From Stimulus to Science*; Quine, *Ontological Relativity* (esp. "Epistemology Naturalized"); Quine, *The Ways of Paradox*; Quine, *Word and Object*; and Sellars, *Science, Perception, and Reality*. It is passing strange that Quine is almost universally taken to be the founding father of contemporary scientific or reductive naturalism, since there are some importantly anti-naturalistic strands to be found in his work. See, e.g., Fogelin, "Quine's Limited Naturalism"; and

(2) epistemological scientism; (3) physicalist metaphysics; and (4) radical empiricist epistemology. To provide a better sense of the overall scientific naturalist world-view, I will now briefly spell out each of these elements in turn.

(1) *Anti-supernaturalism.* This is the rejection of any theoretical appeal to non-physical, non-material, or non-spatiotemporal entities, properties, and causes, for example, ectoplasmic or spiritual agencies, platonic universals, God, and so on. The motivating thought here is that only what is either specifically material, or more generally a physical part of the spatiotemporal and causal order of things, can be truly real. This in turn amounts to what is sometimes called "the token-identity thesis": all particular things or events are identical with corresponding particular physical things or physical events.

(2) *Epistemological Scientism.* This says that the exact sciences are the leading sources of knowledge about the world, the leading models of rational method, and collectively the basic constraint on all other sciences and on the acquisition and justification of all genuine knowledge. In other words, nothing in the rational or the real world falls outside the theoretical purview of pure mathematics (or, for the logicist, pure mathematical logic[29]) and fundamental

---

Johnsen, "How to Read 'Epistemology Naturalized'". So in my opinion it is Sellars, and not Quine, who is the true father of contemporary scientific naturalism.

[29] The basic connections between (i) pure mathematics, (ii) pure mathematical logic as developed by Frege and Russell-Whitehead, and (iii) fundamental physics, are highly complex. But here is a super-compressed version of an account of those connections. Part of the Cartesian dream of early twentieth-century scientific philosophy was to provide an explanatory and ontological reduction of pure mathematics to pure mathematical logic. This dream was shattered by Russell's paradox of classes, the Liar Paradox, and Gödel's incompleteness theorems. Pure logic was rescued from the collapse of logicism by Tarski's retreat to elementary logic, his hierarchy of meta-languages, and his model-theoretic semantics of truth. Then at approximately mid-century the reductionist dream was reinstated by appealing to mathematical physics (containing unreduced pure mathematics) as the reductive base. In this way, *logicism* was sublimated into *physicalism*. Nevertheless, the presence of unreduced pure mathematics within physics is an explanatory and ontological embarrassment for scientific naturalists. So the more recent attempts by mathematical fictionalists to have "science without numbers," and by mathematical naturalists to naturalize away the apparent platonic commitments of pure mathematics, are basically attempts to reduce everything *including* pure mathematics to the non-mathematical parts of fundamental physics. But the enterprise of reduction itself is not a part of physics and presupposes unreduced pure logic. So current scientific naturalism proposes to reduce everything including pure mathematics to the non-mathematical parts of physics together with unreduced pure logic. But unreduced pure logic is an explanatory and ontological embarrassment for scientific naturalists, since unreduced pure logic is presupposed by the explanatory or ontological reduction of pure logic to anything else. This last problem, which seems to me an insuperable one for scientific or reductive

physics. Epistemological scientism does not mean that every form of theoretical cognition or discourse is actually scientific; nor does it mean that there cannot be "special sciences", or (except in the very extreme form of eliminativism) that non-scientific cognition or discourse is literally cognitively worthless or meaningless. But it does imply that ultimately all forms of cognition and discourse are in principle reducible to exact scientific cognition and discourse, and fully explicable, except for some non-cognitive or affect-based "noise", by means of the exact sciences.

(3) *Physicalist Metaphysics.* This says that the physical facts strictly determine all the facts. Let the term "the physical facts" stand for every fact in the world about the instantiation of physical properties. There are then two types of physical facts, and correspondingly two types of physical properties. First, there are the fundamental or first-order physical facts, that is, facts about the instantiation of the intrinsic non-relational or relational properties of microphysical entities, processes, and forces, which in turn are the proper objects of fundamental physics. And, second, there are second-order physical facts, or facts about the instantiation of second-order physical properties that specify how first-order physical facts are causally configured or patterned in relation to one another (that is, causal-functional organizations, or causal roles). And these second-order physical facts are themselves strictly determined by the first-order physical facts. So, otherwise put, according to the thesis of physicalist metaphysics, all facts or properties are *either identical to or logically strongly supervenient on*[30] the fundamental physical facts or properties.

---

naturalism, is a version of what is also known as "the logocentric predicament." See Hanna, *Rationality and Logic*, chs. 1 and 3.

[30] More technically: A-facts (or higher-level facts) about the instantiation of A-properties are strongly supervenient on B-facts (or lower-level facts) about the instantiation of B-properties if and only if (1) necessarily if anything has a B-property, then it also has an A-property ("upwards determination"), and (2) necessarily there can be no change in anything's A-properties without a corresponding change in its B-properties ("necessary covariation"). Strong supervenience, which is a cross-possible-world modal dependency relation, can be distinguished from weak supervenience, which is a one-world relation. Also sometimes supervenience is defined strictly in terms of cross-world necessary co-variation without upwards dependency, which might be called "moderate supervenience." Other types of supervenience include versions that track differences in the type of necessity (logical supervenience vs. natural or nomological or physical supervenience) or in the scope of the supervenience base (local supervenience vs. regional supervenience vs. global supervenience). A standard formulation of minimal materialism or minimal physicalism calls for the token-identity of all particular things or events with particular physical

(4) *Radical Empiricist Epistemology*. This says that all knowledge whatsoever originates in individual sensory experience, derives its significant content from sensory experiential sources, and is ultimately verified and justified by empirical means and methods alone. In other words, all epistemic facts or properties are strictly determined by (that is, they are either identical to or logically strongly supervenient on) the sensory experiential facts or properties.

To summarize, then, scientific or reductive naturalism says (a) that reality is ultimately whatever the exact sciences tell us it is; (b) that all properties and facts in the real world are ultimately nothing over and above fundamental or first-order physical properties and facts; and (c) that all knowledge is at bottom empirical. If scientific or reductive naturalism is true, then its implications are stark and profound: nothing is ultimately or irreducibly mental, first-personal, or subjective; nothing is ultimately or irreducibly semantic; nothing is ultimately or irreducibly abstract or universal; nothing is ultimately or irreducibly modal; nothing is ultimately or irreducibly logical; nothing is ultimately or irreducibly a priori; nothing is ultimately or irreducibly normative; nothing is ultimately or irreducibly free or autonomous; and nothing is ultimately or irreducibly moral.

Believe it if you can. What I mean is that the philosophical price of scientific or reductive naturalism is precisely the "disenchantment of nature",[31] and the disenchantment of *human* nature too, as artistically imagined, for example, at the turn of the nineteenth century by Heinrich Kleist in his gothic horror story *Über das Marionettentheater*, and again at the turn of the twentieth century by Robert Musil in the guise of his ironically presented and profoundly alienated anti-hero Ulrich, "the Man Without Qualities."[32] Essentially the same worry is vividly expressed by Wittgenstein[33] when he considers the possibility that in creating a world culture based on the search for exact scientific knowledge of the physical world, humanity is regressing

---

things or events, together with the strong logical global supervenience of all properties and facts on fundamental physical properties and facts. For details, see Kim, *Supervenience and Mind*.

[31] See McDowell, *Mind and World*, lecture IV.
[32] See Musil, *The Man without Qualities*. Musil wrote a thesis on Ernst Mach's epistemology at the University of Berlin.
[33] Like Musil, Wittgenstein was an Austrian and a highly aesthetically sensitive, morally serious, and philosophically sophisticated inhabitant of *fin de siècle* Vienna, where most of the important intellectual trends of the twentieth century were initially drive-tested. See, e.g., Schorske, *Fin de Siècle Vienna*; and Toulmin and Janik, *Wittgenstein's Vienna*.

towards cognitive and ethical suicide. In any case, it is entirely clear that if scientific or reductive naturalism is true, then the manifest image of human beings and their world is both explanatorily and ontologically reducible to the scientific image. Or, more starkly put, it is entirely clear that if scientific or reductive naturalism is true, *then we are nothing but naturally mechanized puppets epiphenomenally dreaming that we are real persons.* And that seems to me, as it seemed to Kleist, Musil, and Wittgenstein, philosophically tragic. It also seemed philosophically tragic to Kant:

> Now suppose that morality necessarily presupposes freedom (in the strictest sense) as a property of our will, citing *a priori*, as **data** for this freedom certain original practical principles lying in our reason, which would be absolutely impossible without the presupposition of freedom, yet that speculative reason had proved that freedom cannot be thought at all, then that presupposition, namely the moral one, would necessarily have to yield to the other one, whose opposite contains an obvious contradiction; consequently **freedom** and with it morality… would have to give way to the **mechanism of nature**. (*CPR* Bxxix)

In direct response to this profound worry, Kant developed a sharply different and highly original solution to the Two Images Problem: he proposed *to explain the scientific image in terms of the manifest image*.[34] Kant's radical explanatory proposal, however, is not in any sense reductive. This is because his manifest image of human beings and their world—the world of phenomena or "appearances" (*Erscheinungen*)—is nothing more and nothing less than the everyday common-sense world of objectively real macroscopic material entities and their intrinsic properties, *including* all the rational human animals and their intrinsic but also physically irreducible, non-empirical mental and moral properties. For Kant, we must accept that the richly structured and multidimensional natural world of appearances is the real world. This is because, as Ivy Compton-Burnett aptly puts it, "we seem to have no other." More precisely, for Kant, any world made up of noumenal objects or "things-in-themselves" (*Dinge an sich selbst*)—that

---

[34] See also Ameriks, "Kant on Science and Common Knowledge"; and Ameriks, *Kant and the Fate of Autonomy*, pp. 43–5. Ameriks says that Kant "sought to determine a positive and balanced philosophical relationship between the distinct frameworks of our manifest and scientific images" (*Kant and the Fate of Autonomy*, p. 43). I completely agree. But as I see it, Kant's way of achieving that goal is far more radical than Ameriks's formulation suggests. For Kant, the scientific image is ultimately philosophically acceptable *only if* it is fully reinterpreted, in the framework of transcendental idealism, as a proper part of the manifest image.

is, mind-independent, supersensible, unperceivable Really Real objects whose essences are constituted by intrinsic non-relational properties (*CPR* A42–9/B59–72, B306–7)—is utterly unknowable. And even though such a world *is* logically possible, and therefore barely conceivable and thinkable (*CPR* Bxxvi n., A254/B310), we must ignore it for the purposes of legitimate or objectively valid epistemology, philosophy of mind, and metaphysics. We cannot empirically meaningfully assert either that things-in-themselves exist or that they do not exist: so we must be *systematically agnostic* about them. This is what I will later call Kant's *methodological eliminativism about things-in-themselves*.

Moreover, and perhaps even more importantly, the one real world of appearances is *our* world. Thus for Kant, as for Aristotle, the natural world intrinsically includes and does not alienate purposiveness and intentional action. But, unlike Aristotle, Kant recognizes that the natural world *also* intrinsically includes pure mathematics and deterministic mathematical physics. So Kant's way of accommodating both the Aristotelian and Newtonian world pictures alike—both natural teleology and natural mechanism—is to ground both in the necessary possibility of rational human nature. According to Kant, the natural world is an objectively real material world in which human persons actually do exist, and consequently in which human persons *must also be possible*. This is not the best of all possible worlds, as in Leibniz's notorious (and according to Voltaire in *Candide*, tragically risible) theodicy. But then on the other hand it is not the worst world either. And that is because for Kant it is a world in which practical freedom of the will is both necessarily possible and also sometimes actual. In any case, it is the only real world we have got. So we had better get used to it, and give up our self-stultifying metaphysical longing for things-in-themselves (*CPR* Avii–xii, A235–9/B294–9).

It should be evident now that the philosophical (and also more broadly cultural and ideological) stakes at play in the Two Images Problem are very high indeed. In fact I cannot imagine a game with higher stakes. And I have directly opposed the Kantian and the scientific-naturalist solutions to the Two Images Problem. But, before we go on, here are two important points of clarification that may pre-empt some possible misunderstandings.

First, a point about the concept of naturalism. There are of course many different varieties of naturalism, only some of which are specifically philosophical. There are naturalistic schools of painting (for example, seventeenth-century Dutch painting), naturalistic schools of literature (for example, the nineteenth-century novels of Balzac and Zola), and so on. In turn, there are many different varieties of philosophical naturalism, only some of which are scientific or reductive. Indeed, some versions of naturalism are explicitly *non*-reductive, and to that extent will be consistent with much of what Kant has to say. This is because Kant is in fact a *liberal naturalist*,[35] who thinks that everything that really exists and is knowable and do-able must be part of material nature, but also that material nature itself is both directly accessible through our sense perception and contains some intrinsic relational properties that are physically irreducible and non-empirical, mental, strictly modal, and categorically normative. This liberal naturalism follows directly from Kant's *transcendental idealism*. By sharp contrast, scientific or reductive naturalism is definitely the most illiberal or extreme form of philosophical naturalism: all irreducibly mental, modal, or normative properties or facts are strictly prohibited—*strengt Verboten*. To be sure, most other forms of philosophical naturalism are liberal or permissive in *some* respects. For example, Humean naturalism allows for irreducible mental facts, conditional modal facts, and instrumental normative facts, although it bans irreducible strict modality and categorical normativity. Nevertheless, most forms of philosophical naturalism are also committed to some or another version of physicalist metaphysics (and in particular to the thesis that all properties and facts strongly supervene on fundamental physical properties and facts), as well as to some or another version of epistemological empiricism. Hence it would be philosophically important and instructive enough, if it were possible to show that Kant's liberal naturalistic theory of the epistemological, metaphysical, and practical foundations of the exact sciences provides both a significant critique of, and also a coherent and defensible alternative to, scientific or reductive naturalism. So that is my particular project in this book.

And, second, a point about the prevalence of scientific naturalism in recent and contemporary analytic philosophy. I certainly do not mean to say that every analytic philosopher from 1950 to the present has either covertly or overtly been a scientific or reductive naturalist. Far from it! Roderick Chisholm, for example, vigorously defended intentional realism,

---

[35] Liberal naturalism says that there are no non-spatiotemporal entities, and that everything has intrinsic physical properties, but that everything *also* has intrinsic mental properties *and* intrinsic non-empirical properties. See, e.g., Rosenberg, *A Place for Consciousness*, pp. 8–10.

mentalism, and agent causation against the scientific naturalists.[36] And Peter Strawson equally vigorously defended intensionalism, Kantian epistemology, personalism, and platonism.[37] And many contemporary analytic philosophers would probably reject the scientific naturalist label. I mean only to say (i) that many of the leading analytic philosophers of the period from 1950 to the present (and, in particular, Quine and Sellars) have either explicitly defended scientific naturalism or else have been heavily influenced by scientific naturalism; (ii) that scientific naturalism, in one version or another, has generally framed the seminal philosophical debates of the period (for example, about naturalized epistemology, naturalized semantics, naturalized ethics, scientific essentialism and the necessary a posteriori, the mind–body problem, the problem of free will vs. determinism, and so on); and (iii) that scientific naturalism effectively captures and expresses the philosophical core of a fundamental cultural and ideological tendency of the post-World War II period. I doubt that there could be serious disagreement with any of these claims.

So much for the disclaimers. In order to understand the nature of Kant's liberal naturalistic theory of the epistemic, metaphysical, and practical foundations of the exact sciences, we must now put in front of ourselves a preliminary sketch of the basics of the Critical Philosophy.

0.2. The Critical Philosophy: A Working Idea

So what *is* the Critical Philosophy? In a word, *it's all about us*. Less telegraphically put however, we can say that Kant's Critical Philosophy is a comprehensive doctrine of human nature, carried out by means of detailed analyses of human "cognition" (*Erkenntnis*), human conative volition or "desire" (*Begehren*), and human "reason" (*Vernunft*). Cognition, desire, and reason are all "faculties" (*Vermögen*), which in turn are innate, spontaneous mental "capacities" (*Fähigkeiten*) or "powers" (*Kräfte*). The innateness of a mental capacity[38] means that the capacity is intrinsic to the mind, and not the

---

[36] See, e.g., Chisholm, "Human Freedom and the Self"; and Chisholm, *Perceiving*.
[37] See, e.g., Strawson, *The Bounds of Sense*; Strawson, *Individuals*; and Strawson, *Introduction to Logical Theory*.
[38] Kant firmly rejects the existence of innate ideas or *content innateness*, but also fully accepts the existence of innate psychological capacities or *faculty innateness*. See Hanna, *Kant and the Foundations of Analytic Philosophy*, section 1.3.

acquired result of experiences, habituation, or learning. Correspondingly, the spontaneity of a mental capacity implies that the acts or operations of the capacity are:

(i) causally and temporally unprecedented, in that (ia) those specific sorts of act or operation have never actually happened before, and (ib) antecedent events do not provide fully sufficient conditions for the existence or effects of those acts or operations;
(ii) underdetermined by external sensory informational inputs and also by prior desires, even though it may have been triggered by those very inputs or motivated by those very desires;
(iii) creative in the sense of being recursively constructive, or able to generate infinitely complex outputs from finite resources; and also
(iv) self-guiding. (*CPR* A51/B75, B130, B132, B152, A445–7/B473–5)

Cognition is a faculty for the conscious mental representation of objects of any sort (*CPR* A320/376–7). (It should be noted parenthetically, however, that in the B edition Kant sometimes uses "cognition" or *Erkenntnis* much more narrowly to mean *the empirically meaningful or objectively valid conscious judgmental mental representation of empirical objects* (*CPR* Bxxvi n., A50–1/B74–6, A58–9/B83).) Unfortunately, this narrower usage has contributed to a long-standing interpretive confusion about Kant's view on the relationship between "intuitions" or *Anschauungen* and "concepts" or *Begriffe*. See chapter 2 for the complete story.) Conative volition, or the faculty of desire, is a capacity for causing actions by means of the "power of choice" (*Willkür*) (*MM* 6: 213). And reason is a faculty for cognizing or choosing according to "principles" (*Principien*) (*CPR* A405, A836/B864) (*CPrR* 5: 32), which are necessary and strictly normative rules of human thought or human action, and constitute either theoretical laws or practical laws. Theoretical reason is human thinking that is aimed at the *truth of judgments*, according to necessary and strictly normative rules of logic, and in particular according to the Law of Non-Contradiction, stating that only those propositions that are not both true and false, can be true. Practical reason by contrast is human choosing that is aimed at either *the instrumental good of actions* or *the non-instrumental good of actions*. The latter arises according to strictly normative rules of morality, and in particular according to the unconditional universal moral law or Categorical Imperative, which in its specific Formula of Universal Law says that all and only those chosen acts whose act-intentions (which can then be expressed as self-conscious policies or "maxims")—when generalized to every possible rational agent and to every possible context of intentional action, are both internally logically consistent and more broadly meaningfully coherent with the lawful willing of all possible rational agents—are morally permissible (*GMM* 4: 421).

What makes the Critical Philosophy a specifically *critical* philosophy, however, is Kant's striking and substantive thesis, which amounts to a mitigated form of rationalism, to the effect that the human faculty of reason, whether theoretical or practical, is inherently constrained by the brute fact of human finitude, or our *animality*. Otherwise put, Kantian critique is the philosophical story of how our reason, which initially misguidedly aims to occupy the standpoint of God through theoretical speculation and practical aspiration alone, rationally reconciles itself to finite cognitive and moral life in an imperfect material world. More precisely, this is to say that our faculty of reason finds itself to be inherently constrained by the special contingent conditions of all human animal embodiment: "sensation" (*Empfindung*), the "feeling of pleasure and displeasure" (*Gefühl der Lust und Unlust*), and "drive" (*Trieb*). In what ways constrained? The answer is that sensation inherently limits our theoretical reason to the cognition of sensory appearances or phenomena, and that the feeling of pleasure-and-displeasure and drive together inherently limit our practical reason to choices that are bound up with the bodily well-being or "happiness" (*Glückseligkeit*) of the human animal. So rational beings like us are nevertheless *inherently human*, and indeed *all-too-human*:

Human reason has this peculiar fate in one species of its cognition that it is burdened with questions which it cannot dismiss, since they are given to it as problems by the nature of reason itself, but which it also cannot answer, since they transcend every capacity of human reason. (*CPR* Avii)

Out of the crooked timber of humanity, nothing straight can ever be made (*IUH* 8: 23).

By sharp contrast, the theoretical reason of a divine cognizer, or "intellectual intuition" (*CPR* B72), is (barely) conceivable by us; and such a being would know noumenal objects or things-in-themselves directly and infallibly by thinking alone. Similarly, the practical reason of a divine agent or "holy will" (*GMM* 4: 439)—which is a *noumenal subject or person*, not a *noumenal object or thing*—is also (again, barely) conceivable by us, and such a being would do the right thing directly and infallibly by intending alone. Kant thinks that we cannot help being able to conceive such beings, and that a necessary feature of our rational intellectual and moral make-up is the fact that we are finite, minded, embodied, sensing beings who burden ourselves with invidious comparative thoughts about those non-finite, minded, non-embodied, supersensible beings. We crave a transcendent, superhuman ground and justification for our precious little thoughts and actions. So, for Kant, to be human is not only *to be* finite, minded, embodied, and sensuous, and also *to know* that we are finite and minded and embodied and sensuous, but most importantly of all, *to wish that we weren't*.

On the theoretical side of the rational human condition, this inherent anthropocentric limitation specifically means that human cognition is sharply constrained by three special conditions of sensibility: two formal conditions, namely, the necessary a priori representations of space and time (*CPR* A38–9/B55–6); and one material condition, namely, affection, or the triggering of cognitive processes by the direct givenness of something existing outside the human cognitive faculty (*CPR* A19/B33). The basic consequence of these constraints is *transcendental idealism*. Transcendental idealism, as the name obviously suggests, is the conjunction of two sub-theses: (1) *the transcendentalism thesis*; and (2) *the idealism thesis*.

(1) The transcendentalism thesis says that all the representational contents of cognition are strictly determined in their underlying forms or structures by the "synthesizing," or generative-and-productive, activities of a set of primitive or underived universal a priori innate spontaneous human cognitive capacities or powers, also known as "cognitive faculties" (*Erkenntnisvermögen*), insofar as those faculties are applied to original perceptual inputs from the world. These cognitive faculties include (i) the "sensibility" (*Sinnlichkeit*), or the capacity for spatial and temporal representation via intuition (*Anschauung*) (*CPR* A22/B36); (ii) the "understanding" (*Verstand*), or the capacity for conceptualization or "thinking" (*Denken*) (*CPR* A51/B75); (iii) the power of "imagination" (*Einbildungskraft*), which on the one hand comprehends the specific powers of "memory" (*Gedächtnis, Erinnerung*) (*A7*: 182–5), "imaging" (*Bildung*), and "schematizing" (*CPR* A137–42/B176–81), but also on the other hand includes the synthesizing or mental-processing power of the mind more generally

(*CPR* A78/B103); (iv) reason, which as we have already seen is the capacity for logical inference and practical decision-making; and finally (v) the capacity for "self-consciousness" (*Selbstbewußtsein*) (*CPR* B132) or apperception, which is the ground of unity for all conceptualizing, judging, and reasoning (*CPR* B406). The whole system of cognitive capacities is constrained in its operations by both "pure general logic," the topic-neutral or ontically uncommitted a priori universal and categorically normative science of the laws of thought, and also by "transcendental logic" in its truth-oriented guise as "transcendental analytic," which is pure general logic that is semantically and modally restricted by an explicit ontic commitment to the proper objects of human cognition (*CPR* A50–7/B74–82).

(2) The idealism thesis says that all the proper objects of human cognition are nothing but objects of our sensory experience—appearances or phenomena—and not things-in-themselves or noumenal objects, owing to the fact that space and time are nothing but necessary a priori subjective forms of sensory intuition (Kant calls this "the ideality of space and time"), together with the assumption (which I will later call *the intrinsicness of space and time*) that space and time are intrinsic relational properties of all appearances (*CPR* A19–49/B33–73, A369) (*P* 4: 293). Appearances, in turn, are token-identical with the intersubjectively communicable contents of sensory or experiential representations (*PC* 11: 314). Correspondingly, the essential forms or structures of the appearances are type-identical with all and only those representational forms or structures that are generated by our universal a priori mental faculties: "objects must conform (*richten*) to our cognition" (*CPR* Bxvi), and "the object (as an object of the senses) conforms to the constitution of our faculty of intuition" (*CPR* Bxvii).

The transcendentalism thesis and the idealism thesis jointly lead to a new Kantian conception of rational knowledge, as the reflective awareness of just those formal elements of representational content that express the spontaneous transcendental activity of the subject in synthesizing that content: "reason has insight only into what it itself produces according to its own design" (*CPR* Bxiii). So rational knowledge is rational *self-knowledge*. More specifically, Kantian rational insight includes elements of conceptual "decomposition" (*Zergliederung*), of pure "formal intuition" (*formale Anschauung*), and also of the "figurative synthesis" or "transcendental synthesis of the imagination" or "*synthesis speciosa*" (*CPR* A5/B9, B151, B160 n.). At the same time, however, it is crucial to acknowledge Kant's fallibilistic thesis to the effect that rational insight yields at best only a subjective aspect of a priori knowledge, or "conviction" (*Überzeugung*), but not, in and of itself, objective "certainty" (*Gewißheit*) (*CPR* A820–822/B848–850). The world must independently contribute a "given" element, the manifold of sensory content, in order for knowledge to be possible (*CPR* B145). So for Kant rational knowledge is not only rational *self*-knowledge but also *mitigated* rational self-knowledge, that is, rational self-knowledge under *anthropocentric constraints*.

In any case, putting transcendentalism and idealism together, we now have the complex conjunctive Kantian thesis of transcendental idealism:

Human beings can cognize and know only either sensory appearances or the forms or structures of those appearances—such that sensory appearances are token-identical with the semantic contents of our objective sensory cognitions, and such that the essential forms and structures of the appearances are type-identical with the representational forms or structures generated by our own cognitive faculties, especially the intuitional representations of space and time—and therefore we can neither cognize, nor scientifically know,[39] nor ever empirically meaningfully assert or deny, anything about things-in-themselves (*CPR* A369, B310–11).

But what is the *point* of transcendental idealism? Kant's immensely brilliant answer, worked out in rich (and occasionally stupefying) detail in the *Critique of Pure Reason*, is that transcendental idealism alone adequately explains how synthetic a priori propositions—that is, non-logically necessary, substantively meaningful, experience-independent truths—are semantically possible or objectively valid (*CPR* B19), and also how human freedom of the will is both logically and metaphysically (or really) possible (*CPR* Bxxv–xxx, A530–58/559–86). His two-part thought in a nutshell is this:

(1) *The synthetic apriority thesis*, which says that all and only empirically meaningful synthetic a priori propositions express one or more of the transcendental conditions for the possibility of our human experience of objective appearances.

---

[39] See n.18. See also Hanna, *Kant and the Foundations of Analytic Philosophy*, pp. 18 and 30.

(2) *The transcendental freedom thesis*, which says that the synthetic a priori proposition (call it "*F*") which says that human noumenal (a.k.a. "transcendental") freedom of the will exists, *cannot be scientifically known to be true*, yet (i) *F* is logically consistent with the true synthetic a priori proposition (call it "*G*") which says that the total mechanical system of inert macrophysical material bodies in phenomenal nature—bodies that are ultimately constituted by fundamental attractive and repulsive forces under natural laws—have deterministic temporally antecedent nomologically sufficient causes; (ii) the actual truth of *G* underdetermines the truth value of *F*; (iii) that the *truth* of *F* is a presupposition of human morality; and (iv) that the *necessary real possibility* of *F* is a presupposition of *G*.

If the synthetic apriority thesis is true, it follows that there are two irreducibly different kinds of necessary truth, namely, analytic or *logical* a priori necessities and synthetic or *non-logical* a priori necessities (which, as I mentioned in the Preface, I call *Kant's modal dualism*), and that the first principles of metaphysics are among those synthetic or non-logical a priori necessities. It also follows that the set of first principles of metaphysics, and the set of truths about how our transcendental cognitive faculties make a priori contributions to the formal structures of sensory representations, are one and the same. And if the transcendental freedom thesis is true, it follows that the law-governed mechanism of nature is not only *consistent* with human freedom of the will, but also implies the *necessary real possibility* of human freedom in nature. This shows us that the ultimate upshot of Kant's metaphysics is thoroughly anthropocentric and practical. So, perhaps surprisingly, the key to Kant's metaphysics is his ethics.

And that brings us to the practical side of the rational human condition. Here the inherent constraints on human volition are in certain ways highly analogous, but also in other ways sharply disanalogous, to the inherent constraints on human cognition. Like human cognition, whose proper objects are restricted to sensory appearances, the proper objects of human volition are *desiderated appearances, ends,* or *things that seem desirable or good to the rational human animal* and thus are bound up with its individual and social well-being or happiness. For Kant, this also directly implies that rational human animals are *radically evil* (*Rel* 6: 32–3). Despite our being *fallible* creatures, however, this does *not* mean that rational human beings are *fallen* creatures, whether in the theological (Christian) sense of original sin or in the secular (Rousseauian) sense of our inevitable corruption by socialization.⁴⁰ Radical evil implies our ability to act with *transcendental* freedom of the will, but also egoistically and wrongly, hence without occurrent *practical* freedom of the will—although it must also be added that both the capacity for and also the occurrent realization of transcendental freedom entail our possession of the *capacity* for practical freedom (*CPR* A533–4/B561–2). Perhaps even more surprisingly, radical evil also implies our ability to act freely on the basis of highly perverse non-egoistic desires, as when someone decides to loot during a natural disaster or riot even though he might very well be shot on sight for doing so, hence *whatever the consequences*. Like Hume, Kant does not regard it as contrary to reason for me to prefer the destruction of the world to the scratching of my finger. But for Kant it would be superlatively *wicked* for me to do so, since it would be a direct violation of the Categorical Imperative in its Formula of Humanity as an End-in-Itself (*GMM* 4: 429): I would thereby be treating everyone else in the world as *mere things*, worth less than my momentary mild pain.

Radical evil for Kant is actually universal in rational human beings, yet still *contingent*, and thus neither partially nor wholly constitutive of our rational human nature.⁴¹ We constitutively and necessarily desire the good, and are contingently although universally tempted by the bad. So what the thesis of radical evil means is not motivational Manicheanism, but instead the much more prosaic fact that in the developmental order of time rational humans are simply *human animals* long before they are able to actualize their capacity for reason (*Rel* 6: 26–7), and are therefore subject to the thousand natural shocks that flesh is heir to. And even when human beings have become mature adults and actualized their rational capacity, nevertheless in order to survive and flourish as animals in an often unfriendly and dangerous world that they did not create and did not ask to be born into, they must as a matter of contingent fact universally and *almost* inevitably think and act prudentially. Therefore rational humans will—not necessarily but as a matter of contingent fact universally and *almost* inevitably, given their profound vulnerability to sheer luck and their partially constitutive

---

⁴⁰ But see Wood, *Kant's Ethical Thought*, p. 288.
⁴¹ See Kain, "Kant's Conception of Human Moral Status."

animality—freely choose some things, by virtue of the power of choice inherent in their animal nature (*MM* 6: 213), in violation of the moral law (*CPrR* 5: 100).⁴²

But what is the moral law? The moral law, or Categorical Imperative—that is, an unconditional universal rational command—is our duty or strict obligation as rational animal beings or persons, and says that we ought to do only those acts whose "maxims" (i.e., act-intentions when expressed as self-conscious policies):

(i) can be consistently universalized (the Formula of Universal Law or FUL):

Act only on that maxim by which you can at the same time will that it should become a universal law (*GMM* 4: 421);

(ii) always involve treating other persons as "ends-in-themselves," or as having intrinsic value or *dignity*, and never either merely as means to the satisfaction of our own desires or as mere *things* with only an economic value or *price* (the Formula of Humanity as an End-in-Itself or FHE):

So act that you use humanity, whether in your own person or in the person of any other, always at the same time as an end, never merely as a means (*GMM* 4: 429);

(iii) inherently express our pure rational volitional nature as transcendentally free or causally spontaneous and also autonomous or self-legislating agents (the Formula of Autonomy or FA):

The supreme condition of the will's harmony with universal practical reason is *the Idea of the will of every rational being as a will that legislates universal law*. (*GMM* 4: 431); and

(iv) directly imply our belonging to an indefinitely large ideal community of persons and autonomous agents, "the Kingdom of Ends," the card-carrying members of which can self-legislate the moral law only insofar as they *also* legislate for every other member of the self-same ideal moral community (the formula of the Kingdom of Ends or FKE):

---

⁴² See Grimm, "Kant's Argument for Radical Evil."

> Never…perform any action except one whose maxim could also be a universal law, and thus…act only on a maxim *through which the will could regard itself at the same time as enacting universal law* (GMM: 433).

Now, according to Kant, the Categorical Imperative provides a universal unconditional non-instrumental reason for human action (as specified in particular unconditional non-instrumental reasons, and constructed by us in particular act-contexts) in the form of a rational command, and is to be sharply contrasted with *hypothetical* imperatives either of "skill" or "prudence," which are conditional instrumental rational commands and thereby provide conditional instrumental reasons for human action. Two examples of hypothetical imperatives would be that I ought to flip the light switch in order to turn on the light (an imperative of skill), or that I ought to bring the glass up to my lips if want to drink my tot of Buffalo Trace Kentucky Straight Bourbon Whiskey (an imperative of prudence).

But the Categorical Imperative, in and of itself, does not tell us *which* maxims or self-conscious act-intentions to form but rather only *how* we must form maxims in order to be morally good. So the Categorical Imperative is a purely *procedural* or *constructive* principle of human volition and action (exactly analogous to the Law of Non-Contradiction in relation to the truth of theoretical judgments and the soundness of arguments), and not a *substantive* or material principle that in and of itself yields maxims.[43] This means that practical reasoning must always begin with actual human desires and hypothetical imperatives as inputs or data, and then, if the moral law is to be heeded, suitably constrain intentional animal action by choosing to act only on those maxims that satisfy the four formulations of the Categorical Imperative.

In turn, recognition of the Categorical Imperative as an overriding non-instrumental reason for action causally triggers an innate higher-order emotional disposition in persons—namely, "respect" (*Achtung*)—to want our power of choice, which is the faculty of effective first-order desires (that is, desires that would, will, or actually do move us all the way to action),[44] to be moved by non-egoistic effective first-order desires in accordance with the moral law (*CPR* A802/B830) (*CPrR* 5: 97) (*MM* 6: 213). This recognition is what Kant calls *the fact of reason* (*Faktum der*

---

[43] See O'Neill (Nell), *Acting on Principle*; and O'Neill, *Constructions of Reason*.
[44] See Frankfurt, "Freedom of the Will and the Concept of a Person," pp. 14–22.

Vernunft) (*CPrR* 5: 31). What determines the will in moral agency is not pure practical reason alone, but rather pure practical reason *plus* the capacity for respect, and therefore the Categorical Imperative has both motivational and action-guiding force.

So, unlike empirically meaningful human cognition, which can never even in principle transcend the bounds of sensibility (*CPR* A42–3/B59–60), the rational human animal really *does* possess the capacity for practical freedom or autonomy, which is a spontaneous capacity for self-legislative choosing in a way which is strictly underdetermined by all alien causes (whether external to the animal or internal to it, as in the case of insanity or neurophysiological compulsion) and prudential concerns, and in self-conscious conformity with the Categorical Imperative (*CPrR* 5: 28–33). Autonomy is rational freedom of the higher-order faculty of volition, or the "will" (*Wille*) (*MM* 6: 213–14). No matter how infrequent such choices are, to exercise rational freedom of the will is to realize the rational practical aspect of human nature, and, to that extent, transcend the intrinsic constraints on ordinary human volition.

And, in this way, the fact of reason yields *compelling empirical evidence* for the actual existence of transcendental freedom—the ability of our will to cause events in an absolutely spontaneous way—and practical freedom or autonomy alike: "practical freedom can be proved through experience" (*CPR* A802/B830). More precisely, given the fact of reason, we are immediately conscious that practical freedom or autonomy actually exists, because:

(a) recognizing the Categorical Imperative as an overriding non-instrumental reason for action causally triggers the innate higher-order emotional disposition of respect in all persons to desire to have non-egoistic and morally appropriate effective first-order desires;

(b) transcendental freedom of the will is (as an online capacity) a necessary and (as an occurrent or realized capacity) sufficient condition of moral responsibility; and

(c) rational human beings are, as a psychological and social matter of fact, held morally responsible for their right and wrong intentional actions alike (*GMM* 4: 446–63).

Or more succinctly put: the *ought* and *ought not* of morality alike entail the *can* of rational volitional intentional human agency, that is, the existence of our innate capacities for both transcendental freedom and practical freedom or autonomy.

This power of autonomy is to be sharply contrasted with the power of choice that is shared by human animals and non-human animals alike, but which is *not necessarily* autonomous and therefore has no sufficient connection with moral responsibility. Yet, whenever the power of choice is realized in a rational animal, this implies at least a transcendentally free phenomenally efficacious causal responsibility and thus it is a necessary condition of autonomy and a sufficient condition of moral responsibility. No rational human animal can be practically free and operate independently of all alien causes and sensuous motivations unless it can *also*, just like any other conscious animal, move itself by means of its desires in a way that is strictly underdetermined by the universal mechanism of the causal laws of inert material nature—or, in other words, move itself *animately, purposefully, and freely*. And in this way, according to Kant, the *inertial* causal dynamics of mechanical matter are radically extended by the *vital* causal dynamics of embodied moral persons.

Finally, the transcendentally free power of choice possessed by rational human animals must also be distinguished from another necessary but not sufficient condition of autonomy or practical freedom, which is *the spontaneity of consciousness and desire*,[45] or *psychological freedom of the will*. This is manifest as a distinctive subjective experience of being constrained, uncompelled, and unmanipulated by external and internal alien causes, both in our ordinary instrumental (or "Humean") practical reasoning, whereby we are able to choose and do what we want, and also in our cognitive ability, via sensory apprehension, to order conscious perceptual sequences in an "arbitrary" (*beliebig*) and even reversible way (*CPR* A192–3/B237–8). This psychologically free power of choice is only proto-rational and proto-autonomous, however, and is shared with all conscious animals, whether human or non-human: it is therefore necessary but not sufficient for the transcendental freedom of our power of choice, just as psychological freedom and transcendental freedom alike are necessary but not sufficient for our autonomy or practical freedom. In other words, psychological freedom on its own is not enough to satisfy the demands of our pure

---

[45] See Hanna and Thompson, "Neurophenomenology and the Spontaneity of Consciousness." The spontaneity of consciousness and desire plays a crucial role in the Second and Third Analogies of Experience; see section 8.1.

practical reason (*CPrR* 5: 97). But psychological freedom remains essential to our empirical personhood, because the consciousness of being helplessly constrained, compelled, manipulated, overwhelmed, or violated by inner or outer forces utterly undermines our sense of our own rational human agency. *So don't leave home without it.*

0.3. The Primacy of Human Nature

Against the backdrop of the Critical Philosophy as I have just sketched it, Kant's overall solution to the Two Images Problem can be encapsulated in two basic theses:

1. *The empirical realism thesis.* Every self-conscious human cognizer has direct veridical perceptual or observational access to some actual macrophysical dynamic[46] material individual substances, natural kinds, events, processes, and forces in objectively real physical space and time; and she thereby has direct veridical perceptual or observational access to the essential macrophysical properties of some dynamic material things, which in turn are *intrinsic structural properties*[47] of those things (*CPR* A41–9/B59–73, B274–9, A366–80).

---

[46] For Kant's dynamics, see sections 1.0, 4.2, and 8.1.

[47] Intrinsic structural properties are necessary relational properties based on spatiotemporal form, e.g., the right-handedness of my right hand. For more details on the fundamental role of intrinsic structural properties in Kant's epistemology, metaphysics, cognitive semantics, and theory of freedom, see section 1.0 (on the general role of intrinsic structural properties in empirical realism); section 1.2 (on the orientability of space as an intrinsic structural property of the human body); section 2.4 (on the the representations of space and time as intrinsic structural properties of all non-conceptual content); section 3.5 (on laws of nature as intrinsic structural properties of all natural causes and effects); section 4.2 (on attractive and repulsive forces as intrinsic structural properties of all material substances); section 6.3 (on the representation of time as an intrinsic structural property of the concept of a number, and on numbers as instances of intrinsic structural properties of certain logico-mathematical systems); sections 8.2–8.3 (on the spontaneity of the will as an intrinsic structural property of the bodies of rational animals); and section 8.4 (on the asymmetry of time as an intrinsic structural property of naturally mechanized causation). It is sometimes argued, in a Leibnizian way, that every intrinsic property of things must be non-relational: see, e.g., Langton, *Kantian Humility*; and Langton and Lewis, "Defining 'Intrinsic'". But while Kant certainly does allow for the possibility of intrinsic non-relational properties of things—indeed, as Langton correctly points out, all *positively noumenal* properties, or properties of *things-in-themselves*, are intrinsic, non-relational, non-sensory, and mind-independent properties of things—for him (i) not every intrinsic property of things is non-relational, (ii) every intrinsic property of *empirical* things is relational and based on spatiotemporal form, and (iii) some *negatively noumenal* properties of things are intrinsic structural. For more on negatively noumenal properties, see sections 4.0 and 8.2. And for a contemporary approach to intrinsic properties that supports Kant's view in opposition to the Leibniz/Lewis/Langton view, see Humberstone, "Intrinsic/Extrinsic."

2. *The practical foundations of the exact sciences thesis.* Practical reason has both explanatory and ontological priority over theoretical reason (*CPrR* 5: 120–1).[48]

The conjunction of these two basic theses is what I will call *the primacy of human nature thesis*. What is the upshot of this thesis?

Here is one way of putting it. From a Kantian point of view scientific objectivity and human subjectivity are in fundamental conflict *only if* we make the assumption that, as Sellars puts it, "science is the measure of all things." Let us call this deeply ingrained assumption, *the primacy of the exact sciences thesis*, which of course is merely another way of expressing the doctrine of scientific naturalism. According to Kant, however, we must reject the primacy of exact sciences: "I have therefore found it necessary to deny **scientific knowing** (*Wissen*) in order to make room for **belief** (*Glauben*)" (*CPR* Bxxx). This fundamental commitment to **belief**—which I will continue to boldface in order to indicate that it is functioning here for Kant as a quasi-technical term, whose meaning is broad enough to capture not only cognitive attitudes like propositional or scientific belief, but also non-cognitive attitudes like self-resolution, confidence, and faith—entails the primacy of human nature. For Kant, **belief** is a *non-inferential* attitude that occurs only under conditions of the non-sufficiency of objective reasons: "If taking something to to be true is only subjectively sufficient and is at the same time held to be objectively insufficient, then it is called **belief**" (*CPR* A822/B850). More precisely, **belief** is either (1) a cognitive attitude in which a conscious thinking subject non-inferentially and unequivocally asserts a proposition,[49] or (2) a non-cognitive attitude in which a conscious willing subject non-inferentially and unequivocally desires something or chooses a certain course of action: "I call ... contingent beliefs, which however ground the actual use of the means to certain actions,

---

[48] See chs. 5–8. What Kant explicitly says in this crucial text is that pure practical reason has primacy over *speculative* reason. The extension of that primacy relation to practical reason more generally and to theoretical reason more specifically (including primacy over mathematics, the knowledge of necessary truths both inside and outside of mathematics, and natural science or fundamental physics), is what I hope to demonstrate. But for similar developments of Kant's thought about the primacy of practical reason over theoretical reason, see also: Neiman, *The Unity of Reason*; O'Neill, *Constructions of Reason*; O'Neill, "Vindicating Reason"; and Rescher, *Kant and the Reach of Reason*.

[49] In this respect there is an important parallel between Kant's notion of **belief** or *Glauben* and Reid's notion of an "axiom," or a non-inferential belief that is "the immediate effect of our constitution"; see Reid, *Essays on the Intellectual Powers of Man*, pp. 294–7.

pragmatic beliefs" (*CPR* A822–31/B850–9). The unequivocal nature of the attitude of **belief** for Kant in both cases captures the shared sense of a person's full *commitment* to judging or acting in a certain way, even when she lacks objective reasons. The later Wittgenstein seems to have rediscovered Kant's deep thought here in the context of trying to answer the question "how do I know how to follow a rule?":

> How can he *know* (*wissen*) how he is to continue a pattern by himself—whatever instruction you give him?—Well, how do I know?—If that means, "Have I reasons?" the answer is: my reasons will soon give out. And then I shall act, without reasons.[50]

But for Kant there are also some important differences between the two types of **belief**. Indeed the same semantic duality between the two types of **belief** can be observed in English in the important ordinary-language difference between certain "believes-that" constructions and certain "believes-in" constructions. Here we need only compare and contrast these fundamental judgments of common sense:

(1) I believe that this is my right hand and this is my left hand.
I believe that I have a body.
I believe that humans do not grow on trees.
I believe that the earth existed for a long time before I was born.[51]
And so on.

with these fundamental practical avowals or "commissives":[52]

(2) I believe in you.
I believe in what I'm doing.
I believe in universal human rights.
I believe in God's existence.
And so on.

So Kant's notion of **belief** has both a theoretical sense and a practical sense. In its theoretical sense **belief** is the non-inferential fundamental commonsensical epistemic and metaphysical commitment to the existence

---

[50] Wittgenstein, *Philosophical Investigations*, para. 211, p. 84e.
[51] Kant's notion of **belief** also has important parallels with the notion of "certainty" in later G. E. Moore and the post-*Investigations* Wittgenstein. See Moore, "A Defense of Common Sense"; Moore, "Proof of an External World"; and Wittgenstein, *On Certainty*.
[52] See Austin, *How to Do Things with Words*, pp. 151–62.

of material things in actual space and time, along with their macrophysical intrinsic structural essential properties, as the truly real and directly perceivable objects of cognition. This is equivalent to the thesis of empirical realism. And in its practical sense **belief** is the non-inferential fundamental affective and volitional commitment to the irreducible facts of conditional or unconditional practical obligation—*oughts* and *ought nots*—according either to everyday human sensuous instrumental interests and their necessary corollary, the faculty of human volition or the power of choice, or else to the non-instrumental universal moral law, the Categorical Imperative, and its necessary corollary, the faculty of autonomy or rational freedom of the will. These practical facts occur in a natural world that is, at least to the extent that it is a totality of inert matter inherently subject to attractive and repulsive forces, otherwise governed by strict deterministic mechanistic laws of mathematical physics. So practical **belief** is equivalent to the thesis of the practical foundations of the exact sciences.

In this way, Kant's radical thought is that if we fully acknowledge the overriding need for **belief** in both of these senses, then exact science and rational conscious human experience alike are both subsumable under a single coherent overarching conception of *a scientifically knowable world in which human value, human action, and human morality must be really possible, precisely because human persons are actual and so must be really possible*: and this is precisely what is asserted by the primacy of human nature thesis. For Kant, both the complete explanation and the fundamental ontology of the natural world alike presuppose persons in the moral sense: not only are persons categorically different from things, but there cannot be things without the real possibility of persons. It should be particularly noted for the purposes of later discussion that the primacy of human nature thesis does *not* entail that things cannot actually exist without existing human persons, *nor* does it entail that if all the human persons went out of existence, all the things would too. Things can exist without existing human persons, and in fact did so for millions of years before we came along. But things could not have existed unless it were really possible for us to come along. This is what I will later call Kant's *weak transcendental idealism*.

The primacy of human nature thesis has two crucial philosophical consequences. The first consequence is that the directly observed or observable and necessarily causally-ordered macrophysical natural world in space and time, the proper object of natural science, is *objectively real* but also is guaranteed to be *user-friendly* in the dual sense that it is at least in-principle knowable by creatures minded like us, and also necessarily and appropriately adjusted to the causal-dynamic powers, scale, and scope of the human body (*CPR* A213/B260). And the second consequence is that to the extent that human beings belong to the world (i) as *animals*, or sentient living organisms; (ii) as conscious, cognizing, desiring, willing, and intentionally acting *human individuals* (that is, empirical or phenomenal selves); and also and most importantly (iii) as *human moral persons* (non-empirical, negatively noumenal[53] practical agents and autonomous selves), they are, essentially and irreducibly, complex integrated phenomenal-and-noumenal (so as it were, "phenoumenal"[54]) beings or *rational human animals*.

These two revolutionary Kantian ideas—first, identifying the objectively real world with the directly observed or observable macrophysical causally-ordered material world, and, second, transforming the explanatory and ontological priority of exact science over everything into the explanatory and ontological priority of practical reason over theoretical reason—can be fruitfully and sharply contrasted with contemporary scientific realism (including scientific essentialism) on the one hand, and with anti-realist (including idealist, phenomenalist, and the weaker variety which merely reduces truth-conditions to assertibility-conditions) approaches to the philosophy of mathematics[55] and to the philosophy of science[56] on the other. Kant offers us an anthropocentric scientific realism. This goes beyond both contemporary scientific realism and scientific anti-realism alike, and at the same time effectively splits the difference between the two.

But the ultimate pay-off of Kant's thesis of the primacy of human nature is that it provides a serious philosophical alternative to scientific naturalism. Kant does this however not by *rejecting* the claims of exact science, nor by *reducing* them to phenomenal-language statements about subjective mere appearances, but instead by developing a humanistic, liberal naturalistic, and empirically realistic conception of the exact sciences that also preserves their core meaning and core truth. On this Kantian conception, what the exact sciences assert is objectively valid and in conformity with the actual

---

[53] The distinction between negative noumena and positive noumenal objects (or things-in-themselves) is crucial for understanding Kant's theory of freedom. See sections 4.0 and 8.2.
[54] Thanks to Jake LaPenter for suggesting this amusing but apt neologism.
[55] See, e.g., Wittgenstein, *Remarks on the Foundations of Mathematics*.
[56] See, e.g., Bloor, *Knowledge and Social Imagery*; Feyerabend, *Against Method*; Latour, *Science in Action*; and Van Fraassen, *The Scientific Image*.

empirical facts, yet the sciences also have an ultimately practical orientation: nothing can be meaningful, true, or knowable in the exact sciences that ultimately contradicts the real possibility of human persons and their capacity for autonomy. In this sense the final and perhaps most surprising consequence of Kant's famous "Copernican Revolution" in philosophy (*CPR* Bxvi) is to give fundamental physics and pure mathematics an explanatory and ontological grounding in rational human nature. In the late nineteenth century, 25 years before the appearance of Einstein's special theory of relativity and Russell's *Principles of Mathematics*, Nietzsche saw this as Kant's "secret joke." But from the standpoint of the twenty-first century, and in the bright light of the Two Images Problem, it looks instead like Kant's deepest insight.

PART I
Empirical Realism and Scientific Realism

1

Direct Perceptual Realism I: The Refutation of Idealism

> How could it be denied that these hands or this whole body are mine? Unless perhaps I were to liken myself to madmen, whose brains are so damaged by the persistent vapors of melancholia that they firmly maintain they are kings when they are paupers, or say they are dressed in purple when they are naked, or that their heads are made of earthenware, or that they are pumpkins, or made of glass. But such people are insane, and I would be thought equally mad if I took anything from them as a model for myself.
>
> *René Descartes*[1]

> I can prove now, for instance, that two human hands exist. How? By holding up my two hands, and saying, as I make a certain gesture with the right hand, 'Here is one hand,' and adding, as I make a certain gesture with the left, 'and here is another.' ... But did I prove just now that two human hands were in existence? I do want to insist that I did; that the proof which I gave was a perfectly rigorous one; and that it is perhaps impossible to give a better or more rigorous proof of anything whatever.
>
> *G. E. Moore*[2]

1.0. Introduction

Here is an old philosophical problem. The realist believes that there is an external world—a world outside her own mind, containing all material

---

[1] Descartes, "Meditations on First Philosophy," p. 13.
[2] Moore, "Proof of an External World," p. 166.

objects—and that she can know some things about that world. But the skeptical idealist points out (1) that he is directly aware only of the contents of his own consciousness, and (2) that for all he knows right now, he might be dreaming. Both claims seem true, and then his thesis apparently follows automatically:

(*) Possibly nothing exists outside my own conscious mental states.

Now (*) is a deeply troubling thesis, for from (*) we can easily derive not only individualism about mental content, but also both solipsism and external-world skepticism. First, if possibly nothing exists outside my own conscious mental states, then not only is it the case that those states are individuated without reference to anything existing outside themselves (individualism), it is also possible that *only* my own conscious mental states exist (solipsism). And second, if possibly nothing exists outside my own conscious mental states, then possibly every belief I have about the external world is false, and since knowledge plausibly requires the removal of all relevant doubts, it follows that I know nothing about the world (external-world skepticism). Even if the realist opts for contextualism about justification,[3] and insists that the possibility of massive deception about the external world is not a *relevant* doubt that she has to remove in order to have empirical knowledge, that still leaves the threat of solipsism to be reckoned with, not to mention individualism. Furthermore the skeptical idealist might raise the worry that contextualism is nothing but an ad hoc solution to the skeptical problem. Or else he might concede contextualism for the purposes of argument, but then insist that in the particular context of *highly demanding epistemology*—a context shared by most philosophers and many scientists—the possibility of massive deception about the external world is a perfectly relevant doubt. So, in either case, that would reinstate external-world skepticism.

For obvious reasons, then, many philosophers have tried to refute (*), including Descartes, Locke, Thomas Reid, Kant, G. E. Moore, and, more recently, Hilary Putnam and Tyler Burge.[4] But it is a well-known

---

[3] See, e.g., De Rose, "Solving the Skeptical Problem."
[4] See Burge, "Cartesian Error and the Objectivity of Perception," pp. 117–36; Descartes, "Meditations on First Philosophy," pp. 50–5; Locke, *Essay Concerning Human Understanding*, pp. 630–9; Reid, *Essays on the Intellectual Powers of Man*, pp. 625–7; Moore, "The Refutation of Idealism"; Moore, "A Defense of Common Sense"; and Moore, "Proof of an External World"; Putnam, *Reason, Truth, and History*, pp. 1–21.

philosophical "scandal" that all of these attempted refutations have failed.⁵ Indeed, Kant famously says of his predecessors that:

it always remains a scandal of philosophy and universal human reason that the existence of things outside us (from which we after all get the whole matter for our cognitions, even for our inner sense) should have to be assumed **on** [the basis of] **belief** (*auf Glauben*), and that if it occurs to anyone to doubt it, we should be unable to answer him with a satisfactory proof. (*CPR* Bxxxix n.)

Why do these purported refutations of skeptical idealism all fail? Kant's diagnostic insight, which he works out at length in the Paralogisms of Pure Reason (*CPR* A341–405/B399–432), is that the purported refutations all to some extent presuppose a Cartesian model of our mind that effectively generates the very worries the refuters are trying so hard to refute.⁶ According to this Cartesian model, the inner world of conscious experiences and the outer world of material objects are at once:

(i) essentially different and thereby metaphysically distinct from one another, in that their basic natures are incompatible (because the inner or mental is intrinsically immaterial and non-spatial, whereas the outer or physical is intrinsically material and extended in space), so it is metaphysically possible for one to exist without the other; and also

(ii) epistemically mutually independent of one another, in that from the veridical perception⁷ or knowledge of the one, no veridical perception or knowledge of the other can ever be directly accessed or immediately inferred.

---

⁵ As I will indicate directly, Kant correctly isolates the underlying flaw in Descartes's, Locke's, and Reid's anti-skeptical arguments: a Cartesian model of the mind. In turn, Moore argues that Kant's argument fails (see "Proof of an External World," pp. 147–59), yet ultimately remains unconvinced by his own argument: see "Certainty," pp. 171–96. More recently Anthony Brueckner has made it his business to refute both Putnam and Burge; see his "Brains in a Vat," and "Transcendental Arguments from Content Externalism."

⁶ See esp. the fourth Paralogism (*CPR* A253–344), and also Schwyzer, "Subjectivity in Descartes and Kant."

⁷ For the purposes of later discussion (and esp. ch. 2), I systematically distinguish between *correct sense perception* and *veridical sense perception*. Correct sense perception entails both (i) the actual existence of the object perceived, and (ii) an accurate representation of the object perceived. By contrast, veridical perception entails only the actual existence of the object perceived, but not necessarily an accurate representation of it. See also n.19 below.

Hence the anti-skeptic is driven by a sort of philosophical despair to rely upon a rationally undemonstrated **belief** either in the existence of a non-deceiving God or in the dictates of common sense.[8]

Taken either way, I fully agree with Kant's diagnosis. But I also think that Kant himself, according to one interpretation of the Refutation, as well as the other would-be refuters of idealism I mentioned, even including Putnam and Burge (who are explicitly anti-Cartesian), fall prey to a slightly different but equally fatal error: that of trying to prove *the wrong sort of thing*: namely, that we have some correct perceptions of distal material objects in space. Descartes, I think, was much closer to being correct when he noted that it was "insane" to doubt the existence of his own body; and Moore was perhaps closest of all to being correct when he selected his own right and left hands as the targets of his proof.

Nevertheless I think that Kant also grasped the same fundamental point. So I will eventually argue in section 1.3 that Kant's Refutation is, on at least one charitable reconstruction of it, a *sound* demonstration of a denial of (*), as follows:

(∼*) Necessarily for every creature having a cognitive constitution like mine, something exists outside its own conscious mental states.[9]

---

[8] The relationship between Kant's Critical Philosophy and common-sense philosophy is a subtle one. See, e.g., Guyer, "Kant on Common Sense and Skepticism." As I argued in the Introduction, Kant's decision "to deny scientific knowing in order to make room for **belief**" is a fundamental move in the Critical Philosophy. And Kant's appeal to **belief** closely parallels both Reid's common-sense appeal to "axioms" or beliefs that are "the immediate effect of our constitution," as well as the Moore-Wittgenstein notion of "certainty." Nevertheless Kant is correct, I think, in holding that, while **beliefs** supply the primitive data—or explanatory and ontological starting points—that must be held fixed in order to explain other beliefs, concepts, properties, or facts (*CPR* A731/B759 n.), **beliefs** nevertheless are *not* effective against all forms of skepticism, precisely because they have not been demonstrated from premises that the skeptic himself accepts. In other words, the rational constraints on good explanations (which include the constraint that they must start from *some* beliefs, concepts, properties, or facts taken as primitive) are not the same as the rational constraints on sound anti-skeptical arguments (which include the constraint that the conclusions of such arguments cannot themselves be **beliefs**, primitive concepts, primitive properties, or primitive facts). So Kant's objections to common-sense philosophy are directed only against its epistemological strategy, *not* against the idea that common sense is explanatorily and ontologically fundamental (*P* 4: 259–60). Indeed, the idea that common sense is explanatorily and ontologically fundamental is precisely Kant's *empirical realism*. And the basic philosophical achievement of Kant's empirical realism is that it provides the first part of a general solution to the Two Images Problem. At the same time, however, it would be a mistake to think that this implies that Kant has little or no interest in refuting skepticism; see Ameriks, *Kant and the Fate of Autonomy*, p. 43. In fact, as I will argue in this chapter, not only are empirical realism and the refutation of external world skepticism perfectly consistent, but also the latter (when charitably interpreted) is a necessary condition of the former.

[9] The "necessarily" in (∼*) expresses *synthetic* necessity, not analytic necessity. For Kant, a proposition is synthetically necessary if and only if it is true in all and only the worlds of possible human experience.

My charitable reconstruction will involve criticizing and rejecting two of Kant's own formulations of the conclusion of the Refutation, and replacing them with (∼*), which I charitably take to be the formulation that he was driving at all along. In the grand scheme of things, philosophically it does not matter very much if Kant makes a few false claims (indeed, it is highly instructive to work out what precisely makes those false Kantian claims false), as long as *other* claims he makes are arguably true.

In order to demonstrate the soundness of my reconstruction of Kant's Refutation, however, I must also show that my reconstruction satisfies a certain antecedent condition of adequacy. More precisely, it is necessary to avoid or at least de-fang Barry Stroud's influential criticism of the Refutation to the effect that Kant defeats skeptical idealism only by invoking his own questionable brand of idealism, transcendental idealism:

> I have tried to show how the Kantian view would block skepticism and supplant that traditional conception, but only by giving us a "transcendental" theory which, if we can understand it at all, seems no more satisfactory than the idealism it is meant to replace.[10]

As I noted in the Introduction, Kant's transcendental idealism is the conjunction of these two theses:

(1) All the representational contents of cognition are strictly determined in their underlying forms or structures by the "synthesizing," or generative-and-productive, activities of the a priori cognitive faculties of the human mind, insofar as those faculties are applied to original worldly perceptual inputs (the transcendentalism thesis).

(2) All the proper objects of human cognition are nothing but sensory appearances or phenomena, and not things-in-themselves or noumena (the idealism thesis).

See Hanna, *Kant and the Foundations of Analytic Philosophy*, chs. 4–5. There are, of course, other possible ways of interpreting (*), hence other possible ways of denying (*). The description "every creature having a cognitive constitution like mine" is supposed to range over all and only actual or possible human beings. By a "human being" I mean—and so I think does Kant—any animal that is *minded like us*: any animal that is cognitively, affectively, and volitionally functionally equivalent to us. Taken this way, the property of being human is multiply realizable in the sense of being "compositionally plastic." This implies that it is not in any way necessary that a human being in Kant's special sense be a member of the natural kind *Homo sapiens*, as he clearly recognizes in his speculations about extra-terrestrial embodied persons in the third part of the *Universal Natural History* (1755).

[10] See Stroud, *The Significance of Philosophical Skepticism*, p. 274. Similar charges are also made in his "Kant and Skepticism" and "Kantian Argument, Conceptual Capacities and Invulnerability."

These two theses in turn jointly imply that for Kant empirical objects are token-identical with the contents of sensory representations and also type-identical with the a priori forms or structures that are innately specified in our cognitive faculties. Furthermore, Kant thinks that both theses (1) and (2) are directly entailed by his thesis that:

> (3) Space and time are neither things-in-themselves nor ontologically dependent on things-in-themselves, but instead are nothing but a priori necessary subjective forms of all empirical intuitions of appearances (*CPR* A26–8/B42–4, A32–6/B49–53, A369),

together with one other premise that I will mention in a moment. Let us call the famous argument in the Transcendental Aesthetic that is supposed to demonstrate (3), *the Three Alternatives Argument,* in view of its basic premise to the effect that space and time are space and time are either (i) things-in-themselves; (ii) ontologically dependent on things-in-themselves (either as intrinsic non-relational properties of them or as extrinsic relations between them); or else (iii) nothing but a priori necessary subjective forms of human empirical intuition, *and there are no other alternatives.* If one assumes the soundness of the Three Alternatives Argument, only one more premise is needed to secure the truth of transcendental idealism, namely, the assumption, which I will call *the intrinsicness of space and time,* to the effect that spatiotemporal properties are intrinsic relational properties of all appearances (*CPR* B66–7).[11] For if space and time are nothing but subjective forms of human intuition, and every objective appearance or empirical object is intrinsically structured by space and time, then necessarily every empirical object is nothing but a subjective, humanly mind-dependent entity. There are two or perhaps three other arguments for transcendental idealism in the Critical Philosophy—see, for example, (*CPR* A490–7/518–25) and (*P* 4: 377)[12]—but clearly the

---

[11] For more on intrinsic structural properties, see the list of references given in the Introduction, n.47.

[12] The texts cited here contain explicit Kantian arguments for transcendental idealism from (1) the critical diagnosis of the Antinomies and the positive solution of the third Antinomy, and (2) from the solution to the problem of synthetic a priori propositions. For an interpretation and defense of the second argument, see Hanna, *Kant and the Foundations of Analytic Philosophy,* esp. chs. 1, 2, and 5. The second argument however also suggests a third but somewhat less explicit Kantian argument, according to which transcendental idealism is required in order to guarantee that objectively valid synthetic judgments, whether a priori or a posteriori, are *in fact* true. This third argument is, I think, unsound. See Hanna, "The Trouble with Truth in Kant's Theory of Meaning."

Three Alternatives Argument (plus the intrinsicness of space and time) is the primary argument, in that if *it* fails, then all the *other* arguments for transcendental idealism will fail too, since all those other arguments in fact include its conclusion, the transcendental ideality of space and time, as one of their basic premises.

Transcendental idealism is directly opposed to what Kant calls "transcendental realism" (*CPR* A369), but which I shall call "noumenal realism," so as not to confuse two different senses of 'transcendental': the one defined in the transcendentalism thesis (see also *CPR* A11–12/B25, A56/B80–1), and the one that means roughly the same as *transcending the human mind and essentially concerned with things-in-themselves or noumena in the positive sense* (*CPR* A238–9/B298). Correspondingly, noumenal realism is the two-part thesis that all knowable things (especially including all knowable things in space and time) are:

(a) metaphysically constituted by a set of intrinsic non-relational properties (either non-dispositional or dispositional,[13] but also called "categorical properties" when they are non-dispositional); and also
(b) transcendent, in the triple sense that (bi) possibly the knowable things exist even if human minds do not exist or cannot exist; (bii) possibly the knowable things exist even if all human cognizers do not know or cannot know them; and (biii) necessarily the knowable things are not directly humanly perceivable or observable.

Or, in other words, noumenal realism is the thesis that all knowable things are things-in-themselves or positive noumena: mind-independent, non-sensory, unobservable Really Real entities.

Now I do not mean to argue that Kant is *not* a transcendental idealist: of course he *is* a transcendental idealist. But what I will eventually argue is that my charitably reconstructed Kantian argument against (*) does not actually require transcendental idealism in the strong sense just spelled out as the conjunction of theses (1) and (2) above, since it rests on the logically independent grounds of a *weak* transcendental idealism (which, for the record, says that the existence of space and time requires only the necessary possibility of minds capable of adequately representing space and

---

[13] Dispositional properties have causal powers that can be triggered into activity by various situations or the presence of other objects; non-dispositional properties do not have causal powers.

44 EMPIRICAL REALISM AND SCIENTIFIC REALISM

time by means of pure or formal intuition, and does *not* say that space and time are *nothing but* necessary a priori subjective forms of human sensory intuition) that can be elicited from a careful critical analysis of Kant's argument for transcendental idealism from the ideality of space and time. I will work out the basics of weak transcendental idealism in chapter 6, in the context of reconstructing Kant's argument for the possibility of mathematics. In this and the next four chapters, I will not argue explicitly against strong transcendental idealism: but also at the same time I will never substantively appeal to the truth of any Kantian premise that presupposes more than weak transcendental idealism. So for the purposes of this chapter and indeed the rest of part I as well, what is alone crucial is my thesis that—strange as it may initially seem—on at least one charitable reading, Kant's Refutation of Idealism is perfectly consistent, first, with either the truth or the falsity of strong transcendental idealism, second, with the truth of content externalism,[14] and also above all, third, with the truth of "empirical realism":

[The] empirical realist grants to matter, as appearance, a reality which need not be inferred, but is immediately perceived (*unmittelbar wahrgenommen*). (CPR A371)

Every outer perception... immediately proves (*beweiset unmittelbar*) something real in space, or rather [what is represented through outer perception] is itself the real; to that extent, empirical realism is beyond doubt, i.e., to our outer intuitions there corresponds something real in space. (CPR A375)

Kant's empirical realism, as I will understand it from these and many other relevant texts,[15] is the conjunction of two sub-theses: (1) *direct perceptual realism*, and (2) *manifest realism*.[16]

---

[14] Content externalism is the view that the representational contents of at least some of our mental states are at least partially individuated by their direct reference, or some other direct relation, to something existing outside those states, in the worldly, historical, or social environment. See Burge, "Individualism and the Mental"; and McGinn, *Mental Content*, pp. 1–119.

[15] The question of the correct interpretation of Kant's empirical realism is highly controversial and has been much discussed in the recent Kant literature: see, e.g., Abela, *Kant's Empirical Realism*, esp. ch. 1; Collins, *Possible Experience: Understanding Kant's Critique of Pure Reason*, esp. ch. 14; Langton, *Kantian Humility*, chs. 7–10; Van Cleve, *Problems from Kant*, ch. 13; and Westphal, *Kant's Transcendental Proof of Realism*, pp. 250–68. In turn, different interpretations of Kant's empirical realism will depend heavily on the interpreter's views about non-conceptual content and transcendental idealism. On my account, Kant is both a *non-conceptualist* and a *weak transcendental idealist*. See chs. 2 and 6.

[16] For similar views in a non-Kantian context, see Johnston, "Manifest Kinds"; and Johnston, *The Manifest*, chs. 1, 5, and 7.

Direct perceptual realism says that every self-conscious human cognizer has non-epistemic,[17] non-conceptual,[18] and otherwise unmediated veridical[19] perceptual or observational[20] access to some macrophysical dynamic material objects in objectively real space and time. This thesis requires some further elaboration.

According to Kant, "matter is the movable in space" (*MFNS* 4: 480). Material objects are *dynamic* insofar as they involve change or motion, and thereby involve active *forces* (*Kräfte*). From Kant's earliest work on the philosophy of nature at the very beginning of the pre-Critical period, his *Thoughts on the True Estimation of Living Forces* of 1747, and also right throughout the pre-Critical and Critical periods as well, there are two fundamentally different kinds of active force. Active forces either essentially have their sources inside or internally to substances (hence are called "living," spontaneous, or *non-inertial* forces), or else essentially exist outside substances and are merely externally and relationally applied to those substances (hence are called "dead," mechanical, or *inertial* forces).[21] Living forces and dead forces, in turn, correspond closely to the Scholastic distinction between "immanent" causation and "transeunt" causation. Immanent causes are substantial causal *sources*, or agents; by contrast, transeunt causes are successive worldly causal *happenings*, or spatiotemporal

---

[17] A cognition is non-epistemic if and only if its representational content does not require either the cognitive presence or the truth of any belief. See Dretske, *Seeing and Knowing*, ch. II.

[18] A cognition is non-conceptual if and only if its representational content does not require either the cognitive presence of a concept or the correct application of a concept. More generally, a cognition is non-conceptual if and only if its representational content is not fully determined by our conceptual capacities. Since beliefs intrinsically contain concepts, necessarily if a cognition is non-epistemic then it is also non-conceptual. See ch. 2 below for Kant's theory of non-conceptual content.

[19] As mentioned in n. 7, I am using the term "veridical perception" in a precisified way to mean *sense perception that requires the actual existence of the object perceived but not necessarily an accurate representation of it*. For example, I inaccurately and thus incorrectly, but still veridically, see that actual rose as a tulip. By contrast, I am using the term "correct perception" in a similarly precisified way to mean *sense perception that requires both the actual existence of its object and also an accurate representation of it*. For example, I accurately and thus correctly see that actual rose as a rose. Correct perception entails veridical perception, but not the converse.

[20] A "perception" (*Wahrnehmung*) is a conscious objective sensory representation (*CPR* B160). An "observation" (*Beobachtung*), as I am using that notion, is a perceptual judgment that is also non-epistemically and non-conceptually accessible via memory or projective imagination. Hence I have observed, can observe, and will observe, many things that I am not currently perceiving or thinking about. The existence of unobserved observables is a crucial feature of Kant's empirical realism.

[21] See Schönfeld, *The Philosophy of the Young Kant: The Precritical Project*, chs. 1–2; and Watkins, *Kant and the Metaphysics of Causality*, p. 108.

events. So direct empirical realism says that self-conscious human cognizers have direct veridical perceptual or observational access to some material objects in objectively real space and time that either change and move themselves spontaneously or immanently (agents or persons), or else change and move mechanically and transeuntly by virtue of the inertial forces that externally and relationally govern them (things).[22]

After 1755, Kant defended the (heavily Newton-inspired) view that natural science or fundamental physics is exclusively focused on the latter, mechanical kind of material objects, *material things*, and their transeunt or event-based causality, insofar as material things are literally constituted by, and insofar as their causality is also fully determined by, fundamental attractive and repulsive forces (*CPR* A49/B66–7, A265/B321) (*MFNS* 4: 524–5, 533–4) (*OP* 21: 215–33, 22: 239–42).[23] Furthermore, as we shall see in section 4.2, this causal-dynamic Newtonian approach to material nature remains fixed throughout Kant's philosophical life, even despite the various domestic changes and evolutions in his philosophy of physics—the most important of which are:

(a) the move from his postulating a "force-shell atom" theory of matter in the *Physical Monadology* of 1756; through
(b) his definite rejection of atomism in the *Metaphysical Foundations of Natural Science* of 1786; to
(c) his final definite acceptance of the existence of a fundamental physical fluid aether in the unfinished *Transition from the Metaphysical*

---

[22] It is often thought that agent causation and event causation are mutually inconsistent with one another—see, e.g., Chisholm, "Human Freedom and the Self," and Watkins, *Kant and the Metaphysics of Causality*. But while it is indeed true that for Kant agent causation and event causation are logically distinct, since not every event-cause is also an agent-cause, nevertheless they are also *logically consistent*, since every human or non-human animal that is an agent-cause by virtue of having the power of choice or *Willkür*, is also an event-cause. The key to understanding this is the recognition that, for Kant, all individual material substances are complex events such that: (i) all their proper parts are temporally successive observable momentary states of the totality of matter; (ii) these states are intrinsically linked according to a necessary causal law; and (iii) each individual substance also stands in simultaneous dynamic interaction with all other existing material substances (*CPR* A191–211/B236–56). See section 8.1. Then an intentional agent or person is nothing more and nothing less than a living, conscious, willing, free, rational causal-dynamic material substance. See sections 8.2–8.3. See also Hanna, "Kant, Causation, and Freedom: A Critical Notice of *Kant and the Metaphysics of Causality*."

[23] See Laywine, *Kant's Early Metaphysics and the Origins of the Critical Philosophy*, ch. 3; Schönfeld, *The Philosophy of the Young Kant: The Precritical Project*, chs. 3–7; and Watkins, *Kant and the Metaphysics of Causality*, ch. 2.

Foundations of Natural Science to Physics of the 1790s, later published in the *Opus postumum*.

In any case, Kant's direct perceptual realism is a form of *epistemic* realism, because it is a thesis about the objective representational character of perceptual cognition. Correspondingly, then, his manifest realism is a form of *metaphysical* realism. As the Refutation of Idealism shows, all metaphysical realists hold minimally that some knowable things exist in objectively real physical space and not merely in one's phenomenal consciousness or inner sense. In addition to this minimal metaphysical realism, manifest realism says that the essential properties of dynamic material individual substances, natural kinds, events, processes, and forces in objectively real physical space and time are nothing but their directly humanly perceivable or observable *intrinsic structural* (by which I mean necessary, relational, and spatiotemporally-based) *macrophysical properties*:

A permanent appearance in space (impenetrable extension) contains mere relations and nothing absolutely intrinsic, and nevertheless can be the primary substratum of all outer perception. (*CPR* A284/B340)

Whatever we cognize only in matter is pure relations (that which we call their intrinsic determinations is only comparatively intrinsic); but there are among these some self-sufficient and permanent ones, through which a determinate object is given to us. (*CPR* A285/B341)

The Kantian notion of an intrinsic structural property[24] is particularly crucial here, since it gets relevantly between the notion of an *intrinsic non-relational (non-dispositional or dispositional) property* on the one hand, and the notion of an *extrinsic (non-dispositional or dispositional*[25]*) relational property* on the other hand. And this fact has several profound epistemic and metaphysical implications. According to Leibniz and contemporary scientific realists, the knowable essential properties of matter are all intrinsic non-relational properties of microphysical entities, namely (in traditional terms), *primary qualities of real internal constitutions*. According to Empiricists and contemporary anti-realists (whether empirical idealists, phenomenalists, or the more cautious clan that merely reduces truth-conditions to assertibility-conditions), there

---

[24] See the Introduction, n.47, and esp. Humberstone, "Intrinsic/Extrinsic."
[25] The distinction between dispositional and non-dispositional properties thus cuts across the distinction between relational and non-relational properties.

are no knowable essential properties of matter precisely because all directly humanly perceivable or observable properties, including causal powers, are merely extrinsic relational properties, namely (again in traditional terms), *secondary or tertiary qualities*. Kant, by sharp contrast to both, wants to say that:

(1) Intrinsic non-relational properties of material objects would be definitive of their being (if they actually existed) *unknowable things-in-themselves*.
(2) Extrinsic relational properties of material objects are definitive of their being *mere subjective appearances*.
(3) The objectively real essential properties of material objects, including their causal powers, are definitive of their being *authentic appearances*.
(4) The objectively real essential properties of material objects, including their causal powers, are knowable by means of direct human perception and observation *only if* those properties are intrinsic structural properties.

The seminal role of intrinsic structural properties in Kant's empirical realism is a central theme of this and the next three chapters.

In this connection it is not a trivial feature of Kant's manifest realism that for him all material things are dynamic: indeed, as I have just mentioned in passing, he holds that matter is *essentially* causal-dynamic. Hence for him the manifest essential properties of material objects are not only causal-dynamic properties, but *also* intrinsic structural properties. More precisely, Kant defends what I will call (with apologies for its being a terminological monstrosity) *a causal-dynamic structuralist theory of matter*.[26]

In this chapter and the next, I will work out and defend Kant's argument for direct perceptual realism. Chapters 3 and 4 will be devoted to working out and defending his argument for manifest realism. Above all, however, in these four chapters, I want to emphasize the sharp contrast between Kant's empirical realism and contemporary scientific realism.[27]

---

[26] See sections 3.2, 4.2, and 8.1–8.2. See also: Edwards, *Substance, Force, and the Possibility of Knowledge*; Warren, "Kant's Dynamics"; and Warren, *Reality and Impenetrability in Kant's Philosophy of Nature*.

[27] See, e.g., Churchland, *Scientific Realism and the Plasticity of Mind*, ch. 1; Kripke, *Naming and Necessity*; Leplin (ed.), *Scientific Realism*; Putnam, *The Many Faces of Realism*; and Smart, *Philosophy and Scientific Realism*.

Contemporary scientific realism in most of its forms is committed to *microphysical* noumenal realism. Therefore contemporary scientific realism in most of its forms is also committed to some form of *indirect* epistemic realism,[28] and to some form of *non-manifest* metaphysical realism, according to which the essential properties of individual material substances, natural kinds, events, processes, and forces are all unobservable, precisely because those properties are reducible to the microphysical, intrinsic, non-relational (non-dispositional or dispositional), supersensible, and mind-independent properties of metaphysically ultimate objects or *things-in-themselves*.[29] My goal in these four chapters is to show that Kant's eighteenth-century *empirical* scientific realism provides an intelligible and at least prima facie plausible alternative to contemporary noumenal scientific realism. Thus the first half of this book is in effect a direct reply to J. J. C. Smart, who writes in *Philosophy and Scientific Realism* that

Kant's so-called Copernican Revolution was really an anti-Copernican counter-revolution. Just when man was being taken away from the center of things by the astronomers, and when he was soon to be put in his biological place by the theory of evolution, Kant was, by means of his metaphysics, putting him back in the center again. It is a major theme of this book to oppose this Kantian tendency, and try to show that philosophical clarity helps us, just as scientific knowledge does, to see the world in a truly objective way and to see that man is in no sense at the center of things.[30]

Kant, as I have said in the Introduction, is an anthropocentic scientific realist. But it is certainly *not* Kant's considered view that human beings are literally at the center of things, if this implies either a divine creationist teleology or a solipsistically idealistic world-making (let us call this *strong anthropocentrism*). The world was not made for us by God, nor did we make it ourselves. It is instead, far more modestly, Kant's considered view that truth, objectivity, scientific knowledge, and the natural world itself are

---

[28] Indirect epistemic realism says that the relation between cognition and its objects is always and necessarily mediated, and never immediate—whether the means of mediation are theories, inferences, beliefs, propositions, concepts, intentional objects, non-representational sense data, non-representational qualia, or purely physical causal intermediaries of some sort.

[29] To this extent contemporary scientific realism is profoundly Leibnizian, in that it also accepts the reducibility of all relations to the intrinsic non-relational dispositional or non-dispositional properties of fundamental perceptually and observationally "hidden" microphysical entities. See Langton, *Kantian Humility*.

[30] Smart, *Philosophy and Scientific Realism*, p. 151.

impossible without the necessary real possibility of rational human animals or persons (let us call this *weak anthropocentrism*). Given the brute contingent fact that we actually exist as rational human animals, or persons, the real world could not have been such that creatures minded like us were *not* really possible: for better or worse, the objectively real world is *our* world. Or, otherwise put, Kant's empirical realism directly entails the thesis *that the objectively real world contains some human persons, and as a consequence of that brute contingent fact together with the nature of persons, the objectively real world is necessarily user-friendly*. To say that a world is "user-friendly" is just to say that it is in-principle knowable by means of our cognitive faculties and that it is causal-dynamically pre-configured in scale and scope to the causal-dynamic potentials of our living animal bodies. Sadly, however, it neither guarantees epistemic infallibility, nor does it prevent things from being fairly frequently FUBAR.

The intitial phase in my overall presentation of Kant's empirical realism is my charitable reconstruction of the Refutation of Idealism. But first of all, in order to motivate that reconstruction, I want to go through it carefully step-by-step, and then criticize the "unreconstructed" version of it.

1.1. Interpretation and Criticism of the Refutation

The nerve of Kant's Refutation can be found at B275–6 in the "Postulates of Empirical Thought" section in the first *Critique*; but I will also take into account the three "Notes" that immediately follow it in the text at B276–9, as well as a crucial footnote that Kant added at the last minute to the second Preface (*CPR* Bxxxix–xli).[31] For each step I will offer a short commentary. Then, as I mentioned, I will develop several criticisms of the argument as it stands, with the ultimate aim of eliciting a defensible version of it.

The view that Kant is aiming to refute is what he variously calls "dreaming idealism" (*P* 4: 293), "empirical idealism," "skeptical idealism," or "problematic idealism," and which directly corresponds to Descartes's methodological skepticism in the first two *Meditations*:

---

[31] There is also a wealth of relevant material in the *Reflexionen*, e.g., at (R 5653–4, 5709, 5984, 6311–16, 6323; 18: 306–13, 332, 416, 610–23, 643), and a brief treatment in the *Prolegomena* (*PFM* 5: 336–7).

[T]he empirical idealism of Descartes...was only an insoluble problem, owing to which he thought everyone at liberty to deny the existence of the corporeal world because it could never be proved satisfactorily. (*P* 4: 293; see also *CPR* A367–70)

[T]he **skeptical idealist** [is] one who **doubts** [the existence of matter], because he holds [matter and its existence] to be unprovable. (*CPR* A377)

Problematic idealism...professes only our incapacity for proving an existence outside us from our own [conscious existence] by means of immediate experience. (*CPR* B275)

Skeptical or problematic idealism says that *possibly* the external world does not exist. This is to be contrasted with what Kant calls "dogmatic idealism" or "visionary idealism," which corresponds to Berkeley's metaphysics, and takes the modally stronger position that the external world "is false and impossible," that is, that the external world both actually and necessarily does not exist (*CPR* B274; see also *P* 4: 294). Both skeptical or problematic idealism and dogmatic or visionary idealism are of course directly opposed to noumenal realism, which in turn is directly opposed to empirical realism; and by virtue of his diagnostic insight that skeptical idealism, dogmatic idealism, and noumenal realism all share a false Cartesian model of the mind, Kant then takes this relation of direct opposition to be transitive: so problematic and dogmatic idealism are also both directly opposed to empirical realism. Since Kant takes on the modally weaker form of skepticism, he has given himself a heavier burden of proof than would be required to refute Berkeley, since it is always—I take it—harder to show that something is impossible than to show merely that its denial is actual or possible. But, on the other hand, if Kant can show that *problematic* idealism is false, then that will also suffice to show that *dogmatic* idealism is false, and more generally that "**material** idealism"—which is the inclusive disjunction of problematic and dogmatic idealism (B274)[32]—is false. So if the Refutation is sound, it will kill three skeptical birds with one argumentative stone.

Assuming now that Kant is pursuing this interestingly strenuous strategy, here is his argument against problematic idealism:

(1) "I am conscious of my existence as determined in time" (*CPR* B276).

---

[32] For some reason, however, Kant identifies material idealism with Cartesian or problematic idealism in the *Prolegomena* (*P* 4: 337).

Step 1. Kant begins with what he elsewhere in the first *Critique* calls "empirical apperception" (*CPR* A107). Empirical apperception is empirical self-consciousness, or empirical reflective consciousness. So what he is saying here is that I have an empirical reflective consciousness *of* myself, as I consciously exist in "inner sense." Inner sense for Kant is the subject's intuitional awareness of a temporal succession of representational or sensory contents (*CPR* A22/B37, A107, B152–5, A357–9, A361–3, B420, B422–3 n.). Intuitional awareness, in turn, is (i) immediate or directly referential, (ii) sense-related, (iii) singular, (iv) object-dependent, and (v) logically prior to thought or non-conceptual (*CPR* A19/B33, A51/B75, B132, B146–7, A320/B377) (*JL* 9: 91) (*P* 4: 281–2).[33] Occasionally in the first *Critique*, Kant seems to confuse inner sense and empirical apperception by calling them both "consciousness." But when he is being careful, we can see that he invokes a distinction between (i) a first-order unreflective reflexive consciousness[34] of the phenomenal contents (whether objectively representational or merely sensory) of one's own mental state, and (ii) a second-order reflective reflexive consciousness of one's own first-order consciousness. In one of the *Reflexionen* and in the *Prolegomena*, he says this of inner sense:

(The inner sense) Consciousness is the intuition of its self. (*R* 5049; 18: 72)

[The ego] is nothing more than the feeling of an existence without the slightest concept and is only the representation of that to which all thinking stands in relation. (*P* 4: 334 n.)

By contrast, he says of empirical apperception that it is "**one** consciousness of myself" through which "I can say of all perceptions that I am conscious of them" (*CPR* A122). And in the *Anthropology* he distinguishes usefully between what he calls "taking notice of oneself" (*das Bemerken*), that is, an unreflective reflexive consciousness of oneself in inner sense at a given time, as opposed to "observing oneself" (*Beobachten*) (*A* 7: 132), that is, the introspective function of empirical apperception, which is repeatable over time and directly accessible via memory.

---

[33] See also Hanna, *Kant and the Foundations of Analytic Philosophy*, pp. 194–211, and ch. 2 below.
[34] Frankfurt very aptly calls this sort of first-order consciousness "immanent reflexivity." See his "Identification and Wholeheartedness," p. 162.

This difference between two levels of consciousness is crucial to Kant's argument against problematic idealism. To use some non-Kantian terminology borrowed from William James and Thomas Nagel, inner sense is both a "stream of consciousness"[35] and also captures "what it is like to be, for an organism":[36] inner sense is a constantly changing succession of unreflectively reflexive egocentric phenomenal states in a human or non-human animal cognizer. In other words, inner sense is the *phenomenal consciousness* of an animal cognizer. Empirical apperception, by contrast, is a second-order judgmental consciousness of myself as a singular or individuated first-order stream of unreflectively reflexive representations. The propositional element in empirical apperception makes it imperative that we further distinguish it from what Kant variously calls "pure apperception," "transcendental apperception," or "the original synthetic unity of apperception." This is an a priori or empirically underdetermined, universal, innate, spontaneous capacity for anonymous content-unification and for propositional and conceptual self-representation in general: more precisely, it is a capacity for attaching the cognitive prefix "I think" to any concept-involving representational content of the mind whatsoever (*CPR* B131–9, A341–8/B399–406). Empirical apperception, which presupposes transcendental apperception, is perhaps best regarded as the realization of that latter capacity under concrete, particular, empirical conditions. So through empirical apperception, in a context, by carrying out an introspective judgment, I become conscious of my own first-order consciousness *as* constituting a determinate conscious human individual: "**I**, as a thinking being, am an object of inner sense, and am called 'soul'" (*CPR* A342/B400).

Kant's idea in this first step, then, is that even the most refractory skeptic would have to allow for the patent fact of empirical introspection. To deny it would entail either (i) that we are always unconscious, or (ii) that even if we are sometimes conscious, then we are never conscious of our own consciousness ("meta-conscious"), or (iii) that even if we are sometimes meta-conscious, then we are never able to make first-person psychological reports. There may well be living creatures that are always unconscious (for example, humans in persistent vegetative states), or

---

[35] See James, *Principles of Psychology*, vol. I, pp. 224–90.
[36] See Nagel, "What is it Like to be a Bat?," p. 160.

animals that have consciousness without meta-consciousness (for example, newborn human infants and many non-human animals), or animals that have both consciousness and meta-consciousness without the capacity for introspective judgment (for example, pre-linguistic toddlers and Great apes): but these are not creatures sharing the complete set of our online rational human cognitive capacities.

(2) "All determination in time presupposes something **persistent** in perception" (*CPR* B276).

Step 2. For Kant, to "determine" (*bestimmen*) something X, is either (necessarily or contingently) to ascribe or apply some definite attribute (i.e., a monadic property) to X, or to show how X enters (necessarily or contingently) as a relatum into some definite relation, and thereby takes on the attribute of belonging to that relation, or to show how X (necessarily or contingently) grounds or supports some definite relational property. That all time-determination presupposes "that which persists," is a direct consequence of the arguments given by Kant in support of the first Analogy of Experience, the "principle of the persistence of substance" (*CPR* A182–9/B224–32). In the first Analogy, Kant asserts that:

that which persists, in relation to which alone all temporal relations of appearances can be determined, is substance in the appearance, i.e., the real in the appearance, which as the substratum of all change always remains the same. (*CPR* B225)

The rationale behind this is the plausible thought that every change of attributes or relations in time requires something which remains unchanging and self-identical over time and throughout those changes. Now, when we apply Kant's reasoning to strictly psychological phenomena, it grounds the conclusion that every determinate sequence of successive changes of conscious mental contents in time requires some or another unchanging self-identical substratum (something which persists) to which those changes are directly ascribed or applied. We need not, for my purposes in this chapter, accept Kant's questionable further argument in the first Analogy—which seems to reflect a quantifier shift fallacy—to the effect that therefore there exists one and only one unchanging self-identical material substratum in nature to which every change of attributes or relations whatsoever is ascribed or applied, in order to buy into his original point. Nor need we, for my present purposes, accept his insufficiently argued assumption that the unchanging self-identical material substratum is either absolutely or even sempiternally persistent, rather than only relatively or temporarily persistent.³⁷

In any case, the crucial point Kant is driving at in step 2 has to do with psychological "determination in time." This phrase could be read as referring merely to the application of temporal predicates to my various experiences. But I think that by using this phrase Kant is instead invoking something slightly stronger than this, namely, the *individuating determination* of my stream of experiences.³⁸ This seems to be clearly implied by his use of the unusual phrase "my existence (*meines Daseins*) as determined in time"—as opposed to, say, "my various experiences as determined in time"—and by his telling remark in the B Preface footnote to the effect that:

this consciousness of my existence in time is thus <u>bound up identically</u> (*identisch verbunden*) with the consciousness of a relation to something outside of me. (*CPR* Bxl, underlining added)

Kant's idea is that if I am to exist in inner sense as a constantly changing yet individuated stream of consciousness, and as an object of empirical apperception, then that stream must be essentially discriminable or uniquely

---

³⁷ This is *not* to say, however, that Kant does *not* hold that there is one and only one unchanging sempiternal material substratum in nature. In fact he *does* hold this: the single unchanging sempiternal material substratum is the fluid aether, which in turn is the total relational dynamical complex, and total relational dynamical structure, of attractive and repulsive moving forces in nature. My point here is only that the argument of the first Analogy is insufficient to prove the existence of the aether. Kant himself explicitly recognizes this in the *Metaphysical Foundations of Natural Science*, and then later offers a new argument for it in the unfinished *Transition from the Metaphysical Foundations of Natural Science to Physics* in the *Opus postumum*. The other crucial point to recognize in this context is that for Kant the notion of a single unchanging sempiternal substratum is not in any way intended by him to exclude the possibility of a real plurality of temporarily existing individual substances. And there are two reasons for this. First, the third Analogy explicitly requires a real plurality of simultaneous causally-dynamically interacting individual substances. And, second, since the aether is nothing but a *complex totality* and *structure* of moving forces, then it follows that the many real individual substances are simply *positions* in that structure, or *integral parts* of that structure, and in this way they do not in any way compromise the unity or singularity of the aether. I further develop this "causal-dynamical structuralist" interpretation of Kant's theory of matter in sections 3.2, 4.2, and 8.1.

³⁸ Some determinations are non-individuating by virtue of their form, e.g., "N.N. is a such-and-such" (an indefinite description). But other determinations are designed for individuation: e.g., "N.N. is here right now" (an indexical predication) or "N.N. is the such-and-such" (a definite description). Kant is committed to the view that only intuition (i.e., a mental representation that is essentially indexical) can ground a genuine individuating determination: "Since only individual things, or individuals, are thoroughly determinate, there can be thoroughly determinate cognitions only as *intuitions*, but not as *concepts*" (*JL* 9: 99). This is of course in sharp opposition to the Leibniz-Russell tradition.

identifiable, in the sense that it is distinguishable from any other such flow. To individuate my stream of consciousness is to confer upon all the separate and various experiential elements of that stream—sensations, conceptions, images, judgments, etc.—a contingent yet *particular* ordering. This ordering activity occurs primarily through what Kant calls "the synthesis of reproduction" (*CPR* A100–2), which I think is best construed as our cognitive capacity for (short-term, long-term, semantic, episodic, and procedural) memory.[39] In any case, what reproductive synthesis does is to convert that otherwise undifferentiated stream of mental contents into a single personal history or autobiography, whereby my inner life takes on a definite psychological-temporal shape, profile, or structure.[40] Now, according to Kant, the individuating determination (through reproductive synthesis) of any such flow of changing mental contents requires a relatively fixed underpinning or matrix, that is, a psychologically self-identical and persistent factor which "as the substratum of all [psychological] change always remains the same."

It is hard to know *precisely* what Kant means by this, but I think that an analogy taken from physical geography, the example of a *river*, is very illuminating precisely because it involves essentially dynamic facts and processes while abstracting away from psychological factors.[41] It is also an aptly contrastive analogy, because Hume used the same example in the *Treatise* in support of a radically different conclusion—a thoroughgoing skepticism about the reality of identity over time.[42] In any case, here is the analogy. A given river over historical time can be individuated only in relation to a spatially fixed material underpinning or matrix that includes its banks and riverbed, its beginnings and its terminus, and more generally the total path or locus it follows in getting from one end of the river to the other. Let us call this total path or locus its "geophysical route." The Mississippi, for example, flows south along a certain geophysical route from northern Minnesota to the Gulf of Mexico, and could not be the self-same

---

[39] See, e.g., Sternberg, *Cognitive Psychology*, chs. 7–8.

[40] See Campbell, *Past, Space and Self*, chs. 4 and 7; and Campbell, "The Structure of Time in Autobiographical Memory."

[41] Kant lectured on physical geography throughout his career, and closely connected it with his studies in anthropology. See Caygill, *A Kant Dictionary*, pp. 214–15. More generally, Kant seems to have been fascinated by dynamical natural phenomena of all kinds, and in this way his philosophy of nature significantly anticipates contemporary Dynamical Systems Theory. See sections 8.1–8.4.

[42] Hume, *Treatise of Human Nature*, book I, part IV, section vi, p. 258.

river unless it did so. Since the actual quantity of water in the Mississippi at any given time is always changing and running off into the Gulf of Mexico, the Mississippi would then seem to be uniquely defined by three physical factors: first, the fact that it is always water that is flowing in it, and not (say) beer or gasoline; second, the actual history of all the water that has already flowed through it over the years; and, third, its geophysical route. This geophysical route can of course vary slightly within certain parameters, due to flooding or erosion, but those defining parameters continue to exist in a fixed way all the same. Now, like water in the Mississippi, which is always changing and running off into the Gulf, the contents of my stream of consciousness are always changing and running off into the past. So, analogously, my own individual psychological life would seem to be uniquely defined and distinguished from all other such "streams of consciousness," or conscious psychological processes in other minds, by three analogous psychological factors: first, the fact that only conscious human sensations, memories, concepts, etc., are flowing in it, and not (say) either non-sensory or "intellectual" intuitions or completely alien sorts of sense perceptions (*CPR* B71–2); second, the actual history of the various conscious mental contents that have already occurred in my psychological life; and third, its "psychological route": a fixed underpinning or matrix that remains invariant in relation to the constantly changing flow of my sense-qualities and representations in time. All the psychological changes in my inner life must be changes *of*, or changes ascribed or applied *to*, this particular fixed or unchanging something, which in turn functions as a source of unity for my otherwise ever-changing stream of consciousness. But just as a river cannot be individuated without its geophysical route (its underlying self-identical geophysical substratum or defining parameters—that which geophysically persists in relation to it), so too the individuation of my stream of consciousness requires a psychological route (its underlying self-identical psychological substratum or defining parameters—that which psychologically persists in relation to it). And also by analogy we can predict, as in the case of the river's geophysical route, that small variations within my individuating self-identical psychological determining substratum will also be permissible, so long as they always remain within certain fixed parameters. Here, as elsewhere, we can see that for Kant there is a deep and intimate connection between the notion of *a substance* and the notion of *an intrinsic dynamic material structure*.

(3) "But this persisting element cannot be an intuition in me. For all the determining grounds of my existence that can be encountered in me are representations, and as such they themselves need something persisting distinct from them, in relation to which their change, and thus my existence in the time in which they change, can be determined" (*CPR* Bxxxix n.).

Step 3. This is the first of the two most crucial steps of the proof. From step 2 we know that every changing conscious individuating determination of myself in time presupposes something that self-identically persists, in relation to which I can uniquely determine and individuate the conscious stream of contents in my inner sense. But this self-identical persistent thing must be outside my own conscious mental states, and not merely inside me. For if it were merely inside me, it would then *belong* to the ever-changing stream of consciousness, and so could not provide a uniquely determining substratum for the mental modifications I experience directly. Hence it must fall outside the proper domain of my inner sense, that is, outside the series of first-order phenomenally conscious representational states that I am directly aware of via my second-order introspective consciousness or empirical apperception.

Right at this point it might well occur to us that something else in inner sense might plausibly function as the self-identically persisting substratum, namely, the *form* of inner sense, as opposed to its contents. Indeed, according to Kant the form of inner sense always remains the same, since it is invariantly presupposed by any actual or possible inner experience (*CPR* A22–3/B37, A31/B46). But the form of inner sense is nothing other than the representation of *time*. If the Transcendental Aesthetic is sound, then either the mere representation of time or time itself (if we accept the transcendental ideality of time) is a necessary formal precondition for the series of changes in my stream of consciousness. Now, it is incoherent to suggest that either the mere representation of time or time itself could be a self-identically persisting or enduring thing *in* time. To hold that the mere representation of time occurs in time, would be to confuse properties of the psychological *vehicle* of a representation (which does indeed occur in inner sense, hence in time) with semantic properties of its representational *content*. And, correspondingly, it would plainly be conceptually incoherent to hold that time itself occurs in time. So neither the mere representation of time nor time itself could also function as an enduring self-identical substance or substratum *to which* my changing conscious representational states are ascribed or applied. Hence nothing in either the content or the form of inner sense can function as the self-identically persistent element or substratum that is required for the individuation or unique determination of my stream of consciousness.

(4) "Thus the perception of this persistent thing is possible only through a **thing** outside me and not through the mere **representation** of a thing outside me. Consequently the determination of my existence in time is possible only by means of the existence of actual things that I perceive outside myself" (*CPR* B275–6).

Step 4. This is the second crucial step. In order uniquely to determine or individuate my own successive existence in time, I must presuppose the existence of outer things perceptually represented by me, and not merely the existence of my internally flowing conscious representations of those outer things. The radical nature of what Kant is saying here cannot be overemphasized. He is saying that any individuating temporally determinate introspective awareness of myself is necessarily also a direct non-conceptual veridical representation of some real material thing existing *outside* my stream of conscious experiences and *at a distance from me in space*. The latter factor is especially to be noted. In the Transcendental Aesthetic, Kant argues that "in order for certain sensations to be referred to something outside of me" they must be referred to "something in another place in space from that in which I find myself" (*CPR* A23/B38). My unique individuality at the level of first-order phenomenal consciousness is therefore *inherited* from the world of distal physical objects. In this way, despite the fact that via empirical apperception in a loose and everyday sense we introspect "an object of inner sense [which is] called 'soul'," there is strictly speaking for Kant no independent "inner object" of inner sense:

inner sense, by means of which the mind intuits itself, or its inner state, gives, to be sure, no intuition of the soul itself, as an object. (*CPR* A22/B37)

That is, the individual empirical self or person for Kant cannot be reified: it is neither a noumenal inner thing nor a phenomenal inner thing, precisely because it is essentially a well-ordered complex of conscious contents. The empirical self or person is a psychological-temporal *intrinsic dynamic structure* of conscious contents, not a mere thing.[43] Moreover, as he puts it in the

---

[43] This point yields a result very similar to Shoemaker's attack on what he calls the "perceptual model of self-knowledge" in his seminal paper "Self-Reference and Self-Awareness," pp. 562–4, and

first Note concerning the Refutation, "inner experience itself is ... only mediate and possible only through outer experience" (*CPR* B277). So insofar as I am aware of myself in empirical apperception as a uniquely determined psychological being, an empirical self or person, then I must directly and non-conceptually ascribe or apply the changing contents of my mental states to the objective furniture of the distal material world.

This crucial point needs further emphasis. Far from having the problem of escaping from a "Cartesian box" into the outer world, *Kant's* problem in the first *Critique* is instead that of distinguishing himself from various surrounding material objects in the outer world!⁴⁴ This problem comes out clearly if we put it in non-Kantian terminology, this time borrowed from G. E. Moore and Jean-Paul Sartre. Kant's view of inner sense in the Refutation comes very close to an amazing doctrine defended by Moore in his 1903 essay, "The Refutation of Idealism," a doctrine which he calls the "transparency of consciousness":

[W]hen we refer to introspection and try to discover what the sensation of blue is, it is very easy to suppose that we have before us only a single term. The term 'blue' is easy enough to distinguish, but that other element which I have called 'consciousness' ... is extremely difficult to fix. That many people fail to distinguish it at all is sufficiently shown by the fact that there are materialists. And, in general, that which makes the sensation of blue a mental fact seems to escape us: it seems, if I may use a metaphor, to be transparent—we look through it and see nothing but the blue. We may be convinced that there *is something* but *what* it is no philosopher, I think, has yet clearly recognized.⁴⁵

Here consciousness is again *not an inner mere thing*, and in spades: instead it is nothing but a noetic searchlight on outer things. Later, in the

---

also the "object perception model" of self-knowledge in his 1993 Royce Lectures, "Self-Knowledge and 'Inner Sense'," pp. 249–69. The later Wittgenstein makes a closely related point when he says that a conscious sensation "is not a *something*, but not a *nothing* either!" (*Philosophical Investigations*, para. 304, p. 102e).

⁴⁴ Kant of course attempts to solve this problem in the second Analogy of Experience by sharply distinguishing between arbitrary (subjective) and necessary (objective) orderings of conscious sensory representations in time (*CPR* A189–211/B232–56). Peter Strawson and Guyer have convincingly tied this argument both to the B Deduction and to the Refutation; see the former's *Bounds of Sense*, pp. 72–152, and the latter's *Kant and the Claims of Knowledge*, chs. 3–5, 10, and 12–14. Unfortunately Kant's solution fails, I think, because he does not notice that *both* waking experiences and dreams can include either arbitrary or necessary orderings of perceptions. See Hanna, "The Trouble with Truth in Kant's Theory of Meaning," pp. 13–17.

⁴⁵ Moore, "The Refutation of Idealism," p. 37.

1930s (but presumably without having read Moore), Sartre pushes this idea of transparency one step further and describes something he calls "the transcendence of the ego."[46] Sartre's idea is that the ego receives its first-order unreflective reflexive subjective unity solely and directly from the outer things it is transparently conscious of. So this is not merely *content*-externalism: it is also *phenomenal consciousness*-externalism. The conscious mind is as much "out there in the world" as it is "in here." *Ego*centricity is representational *ec*centricity. Phenomenal consciousness is nothing but *consciousness-of* or intentionality. And essentially the same view is currently held by defenders of the "first-order representational theory of consciousness."[47] All of these later affinities show how radical and philosophically prescient Kant's doctrine really is. Add the Sartrean transcendence of the ego and the representationalist theory of intentionality to the Moorean transparency of consciousness, and you have, in effect and in essence, Kant's doctrine in step 4.

(5) Now consciousness [of my existence] in time is necessarily bound up with consciousness of the [condition of the] possibility of this time-determination. Therefore it is also necessarily combined with the existence of the things outside me, as the condition of time-determination. (*CPR* B276)

Step 5. This step is fairly straightforward. Insofar as I am empirically self-aware, and individuate myself in time, I must also be directly consciously aware of this act of time-determination. Hence I must also be directly consciously aware of the existence of a distal persistent thing outside me that individuates me, since this is the necessary condition of time-determination.

(6) "I.e., the consciousness of my existence is at the same time (*zugleich*) an immediate consciousness of the existence of other things outside me" (*CPR* B276).

Step 6. This adds a crucial factor to step 5. The "immediate consciousness of" something is a direct veridical consciousness of that thing. So Kant is saying that for any particular empirical apperception of myself as uniquely determined in inner sense, I am also *simultaneously* directly veridically perceptually aware, via outer intuition, of some existing or actual distal material object in space, as the individuating substratum to which I ascribe

---

[46] See Sartre, *The Transcendence of the Ego: An Existentialist Theory of Consciousness.*
[47] See Carruthers, "Natural Theories of Consciousness." Contemporary defenders of the first-order representational theory of consciousness include Dretske and Michael Tye.

or apply the changing conscious representational contents of my mind. So to sum up the whole Refutation: Necessarily, if I am determinately aware of myself in empirical apperception, then I am also thereby at that very same moment directly veridically perceptually aware of some actual distal material object in space.

I now move on to critical objections. It seems to me that both steps 1 and 3 are acceptable, assuming the correctness of both Kant's philosophical psychology (of inner sense, outer sense, and apperception) and also the "identity-oriented" reading of the first Analogy that I proposed.[48] Yet critics of the Refutation often hold that the fundamental gap in the proof is to be found in step 3.[49] Why, such critics ask, is it necessarily the case that the intuition of what is self-identically persistent be an intuition of something outside me? Why couldn't it instead be an intuition of some self-identically persistent thing *inside* me—that is, of some "thinking thing"? This option immediately fails, however, when we remember just what sort of intuition an inner intuition is:

the determination of my existence can occur only in conformity with the form of inner sense, according to the particular way in which the manifold that I combine is given in inner intuition, and therefore I have **no cognition** of myself **as I am** but merely as I **appear** to myself. (*CPR* B157–8)

In other words, in empirical self-consciousness I am not directly aware of myself as a Cartesian ego-in-itself. That would require an "intellectual intuition" (*CPR* B72) of myself. But as a finite human cognizer who is not merely rational but also an animal, my intuition is strictly sensory and not intellectual: in inner sense, I am directly aware only of the phenomenal flotsam and jetsam of consciousness. That is, in inner sense, I am directly aware only of *my phenomenally conscious states and their phenomenal contents (whether objectively representational or not), as occurring in a certain psychological-temporal intrinsic dynamic structure*, not of some deeper noumenal substratum of those phenomenally conscious states.

---

[48] I work out a more detailed reading of the first Analogy in section 8.1. For the time being, I will only note that it seems to me quite illuminating to regard the first Analogy as, among other things, a direct reply to Hume's skepticism about both personal and object identity over time in the *Treatise of Human Nature*, book I, part iv, section vi.

[49] See, e.g., Vogel, "The Problem of Self-Knowledge in Kant's 'Refutation of Idealism': Two Recent Views," pp. 875–87.

But even if steps 1 to 3 hold up tolerably well, nevertheless, in my opinion, steps 4 to 6 as they stand are highly questionable. Here is a worry about step 4. Even granting that my empirical self-consciousness of my stream of consciousness in inner sense requires an outer intuition of something persistent, nevertheless it does not seem to follow that inner intuition in general requires any outer intuition of actually existing distal material objects in space. For so long as *space alone, as an object*, can be represented by means of a "pure intuition" or "formal intuition," as Kant explicitly argues in the Transcendental Aesthetic and again later in the B edition's Transcendental Deduction of the Pure Concepts of the Understanding (*CPR* B160 n.), then *that* seems sufficient to meet the requirement that there be a single self-identically persistent thing over against me, to which I must intuitionally refer and ascribe my ever-changing conscious inner states. And the pure or formal intuition of space does not logically require the existence of any distal material objects *in* space. Kant says explicitly that "one can never represent that there is no space, although one can very well think that there are no objects to be encountered in it" (*CPR* A24/B38–9).

What is the pure or formal intuition of space? Five features are at least individually necessary for it. First, the pure or formal intuition of space is *a non-empirical presupposition of all empirical intuitions of objects in space*: "[this representation of space] is a necessary representation, a priori, which is the ground of all outer intuitions" (*CPR* A24/B39).[50] Second, the pure or formal intuition of space is *non-conceptual*: "[this representation of] space is not a discursive or, as is said, general concept of relations of things in general, but a pure intuition" (*CPR* A25/B39). Third, the pure or formal intuition of space represents space as a *unique object*: "one can represent only a single space" (*CPR* A25/B39). Fourth, the pure or formal intuition of space represents space as a *unified structured manifold*:

Space and time are represented a priori not merely as **forms** of sensible intuition, but as **intuitions** themselves (which contain a manifold), and thus with the determination of the **unity** of this manifold... Space, represented as **object** (as

---

[50] There is, however, an interpretive subtlety here: the pure or formal intuition of space is *a* presupposition of all empirical intuitions of *fully-determined objects of experience* in space, but it is not *the* presupposition of all empirical intuitions of *any sort of apparent object* in space: only the bare intuitional representation of space, the form of our outer intuition, is uniquely presupposed by all empirical intuitions of apparent objects. The formal intuition of space entails the form of outer intuition, but the form of our outer intuition does not entail the formal intuition of space. For the important distinction between *formal intuitions* and *forms of intuition*, see *CPR* B 160 n. and section 2.3.

is really required in geometry), contains more than the mere form of intuition, namely the **putting-together** (*Zusammenfassung*) of the manifold. (*CPR* B160, text and note combined)

Fifth, and finally, the pure or formal intuition of space represents space as an *infinite totality*: "space is represented as a given infinite magnitude" (*CPR* A25/B40). For the moment, I need not unpack Kant's extremely interesting doctrine of pure spatial representation any further.[51] My point right now is only (a) that the pure or formal intuition of space is a necessary a priori non-conceptual representation having a referent—space itself—which is represented as a unique unified structured manifold and an infinite totality, and (b) that this unique unified structured manifold and infinite totality has not been ruled out as the self-identically persisting element Kant needs in order to meet the requirement of step 2. It is incoherent to suppose that time itself might function as a persisting entity in time, but *not* incoherent to suppose that space itself might function as a self-identically persisting entity in time. And if space itself can meet that requirement, then, since Kant explicitly says that space can be represented as empty of all material objects (*CPR* A24/B38–9, A291/B347), it follows that Kant has not ruled out the possibility that I ascribe my changing mental states directly to empty space itself.[52]

Just in case my objection to step 4 is not convincing, however, here is another objection, this time to step 6. Even if we grant what I think we should *not* grant—namely, that my self-conscious awareness of my stream of consciousness in inner sense entails that I have some direct outer experiences of actual distal material objects in space—it does not seem to follow from that, that on every occasion of self-awareness I must be *simultaneously* directly correctly perceptually aware of a distal material external object. What about dreams and hallucinations? In Note 3 of the Refutation, Kant himself admits that:

---

[51] In section 2.3, I will argue that the pure or formal intuition of space (and of time) is in fact a higher-order self-conscious or apperceptive and concept-involving representation of the first-order representation of space (and of time), which is in turn the form of outer (and of inner) sense. But this subtlety, and the correspondingly subtle distinction between "formal intuitions" and "forms of intuition" (see n.50), which is crucial in the context of a discussion of Kant's theory of non-conceptual content (see ch. 2), and later is equally crucial in the context of discussing the semantic and epistemic roles of the pure intuition of time in arithmetic (see ch. 6), are not essential in this context.

[52] Later in this chapter I will argue that Kant *is* ultimately able to rule out the possibility that empty space is the permanent element needed for step 2. But this cannot be done without bringing in some resources from other Kantian texts.

from the fact that the existence of outer objects is required for the possibility of a determinate consciousness of our self it does not follow that every intuitive mental representation of outer things includes <u>at the same time</u> (*zugleich*) their existence, for that may very well be the mere effect of the imagination (in dreams as well as in delusions). (*CPR* B278, underlining added)

So Kant is certainly aware of the dream problem, and he must then implicitly grant that step 6 as it stands, with the simultaneity condition, is false.

Where does this leave us? By virtue of his admission of the dream problem, Kant has implicitly admitted that not every self-conscious awareness of my own uniquely determined conscious existence in time entails a simultaneous direct correct perception of a distal external object. So, since Kant is certainly no fool, it seems to me that his concluding step 6 is most charitably and plausibly interpreted as saying the same as these three alternative formulations of the conclusion of the Refutation:

The proof that is demanded must therefore establish that we have **experience** and not merely **imagination** of outer things, which cannot be accomplished unless one can prove that even our **inner experience**, undoubted by Descartes, is possible only under the presupposition of outer experience. (*CPR* B275)

The mere, but empirically determined, consciousness of my own existence proves the existence of objects in space outside me. (*CPR* B275)

By means of external experience I am conscious of the existence of bodies as external appearances in space, in the same manner as by means of internal experience I am conscious of the existence of my soul in time. (*P* 4: 336)

Taken together, these formulations say that my having a self-conscious experience of my individuated stream of inner consciousness entails my also having some direct correct perceptions of distal material objects in space. Even so, there is no necessity that I have a direct correct outer perception of a distal material object at the very *same* time that I am in one of these self-conscious states, so long as I also have some direct correct outer perceptions of distal spatial objects at *other* times. The simultaneity condition can be dropped.[53]

---

[53] Nevertheless, as we shall see later, Kant really was driving at something—albeit somewhat obscurely—by including the simultaneity condition: the necessary simultaneous dynamical compresence of a self-locating perceiver's living human body and at least one other material object that is directly referred to by that perception.

This charitable interpretation is backed up by a footnote appended to the Refutation, which says that even when we are dreaming or hallucinating, and merely *imagining* space, it is presupposed that we already have an outer sense through which we do sometimes get direct correct perceptual access to outer material things:

> In order for us even to imagine something as external, i.e., to exhibit it to sense in intuition, we must already have an outer sense, and by this means immediately distinguish the mere receptivity of an outer intuition from the spontaneity that characterizes every imagining. For even merely to imagine an outer sense would annihilate the faculty of intuition, which is to be determined through the power of imagination. (*CPR* B276–7 n.)

In other words, space cannot be even imagined without our already having a capacity, sometimes actually realized, for directly and correctly perceiving or empirically intuiting distal material objects in space. And this reading is in turn backed up by two other texts. First, in the *Anthropology*, Kant notes that imagination "cannot bring forth a representation that was *never* given to the power of sense; we can always trace the material of its representations" (*A*7: 168). And, second, in one of the *Reflexionen* he is even more explicit:

> Dreams can represent to us things as outer, which are not there; however, we would never be able to represent something as outer in dreams, if these forms were not given to us through outer things. (*R* 5399; 18:172)

So it seems to be Kant's view that even our capacity for "imagination of something as external" is parasitic upon some direct correct outer sense perceptions of distal material objects, at some time or another.[54] If he is right, and if we interpret step 6 in such a way as not to commit Kant to the implausible thesis that every individuating act of empirical self-consciousness requires a simultaneous direct correct perceptual awareness of a distal material object, then he in fact avoids the dream problem.

But even so, *is* he right? Well, it seems likely that it is empirically true, as a fact in cognitive psychology, that normal image-construction

---

[54] It is unclear from the texts whether Kant thinks that the dependence of imagination on correct perception of outer objects implies that the relevant correct perception must be in the past, so that in order to imagine something at a certain time, I must *already* have correctly perceived some outer things, or whether my first correct perception could in fact occur *after* I have imagined some things. This apparently trivial point actually makes a subtle but important difference in some of the *Matrix*-type thought-experiments I will describe.

and manipulation is originally funded by direct correct sense-perception of distal material objects.[55] But is it *necessarily true* for creatures minded like us? Surely we can conceive of a possible human being in this world whose empirical imagination-content is entirely funded by some source other than direct correct sense-perception of distal material objects. Or, to put it another way, if a creature had been born with, or had developed, a capacity for imagining external things that was entirely empirically funded in some non-standard way which was systematically insulated from direct correct perceptual contact with the distal outer world—suppose, for example, that someone was fitted from birth with a microscopically thin computer-driven "virtual reality suit" covering her entire body, or that (as in *The Matrix*) she was born hooked up to an all-encompassing computational system, so that again all her perceptions were in fact false digital images—would she thereby fail to be *one of us*? I think not.[56] Such a human cognizer, cocooned inside her all-encompassing perceptual prosthetic, or unconsciously supplied with a massively complex and detailed but still phoney digital image of her actual surrounding world, would certainly be odd, and perhaps somehow slightly cognitively handicapped (or perhaps not even slightly handicapped, in light of the actual empirical fact of "neural plasticity," as manifest in the effective neural and behavioral adaptation of actual human cognizers to inverting lense goggles, Tactile-Visual Substitution Systems, etc.[57]). But she would certainly nevertheless, I think, still fully share our human cognitive constitution.[58] So Kant's thesis of the dependency of imagination on correct perception is false, and the Refutation as it stands is therefore unsound.

While I am considering *Matrix*-type scenarios as counterexamples to Kant's thesis of the dependency of imagination on correct perception,

---

[55] See, e.g., Kosslyn, *Image and Brain*, esp. ch. 9.
[56] And then I disappeared. (Old Cartesian joke.)
[57] See Hurley and Noë, "Neural Plasticity and Consciousness."
[58] Admittedly, this is controversial territory. Putnam's famous argument in *Reason, Truth, and History*, pp. 1–21, against the "brain-in-a-vat" hypothesis seems to show that even if the Cartesian hypothesis is conceivable, nevertheless it is metaphysically impossible. Three things can be briefly said in reply. First, Putnam's argument is probably unsound: see, e.g., Brueckner, "Brains in a Vat." Second, even if Putnam's argument is sound, it holds only for the entire community of cognizers/speakers, not for an isolated individual. And, third, it seems to me that neither the possibility of a human-person-in-a-virtual-reality-suit-for-life nor the *Matrix*-scenario is a "Cartesian hypothesis" at all, but rather just a slightly unusual and overtly science-fictional but still entirely logically and metaphysically possible "plain possibility" in Thompson Clarke's sense. See Clarke, "The Legacy of Skepticism," pp. 754–69; and Chalmers, "*The Matrix* as Metaphysics."

however, there is also one important subtlety that needs to to mentioned. Kant *might* reply to my counterexamples by saying that the dependency of imagination on correct perception does not require that the imaginer have *already* correctly perceived some outer objects: all it requires is that *sooner or later* she does. And even in *The Matrix*, we are primarily shown the point of view of characters who have come to learn the difference between mere digital appearances and sordid reality. But consider again the person who is hooked up to the Matrix from birth. Supposing that she *never* is unhooked from the Matrix and in fact dies without learning the awful truth that her apparent external world is nothing but a computer image: would she thereby fail to be a human cognizer? No. She would still be a fully human cognizer, albeit a systematically deluded one. The importance of this subtlety is that it teaches us not to confuse semantic conditions for the possibility of human *cognition*, including both objectively valid perception and objectively valid judgment, with epistemic conditions for the possibility of human *knowledge in the scientific sense, or justified true belief*. Perceptual *knowledge in the scientific sense* generally requires the correct perception of some outer objects: but perceptual *cognition* does not. Perceptual cognition requires only the *veridical* perception of some outer objects. I will come back to this crucial point later.

But in any case my preliminary conclusion is that if we take together as a cumulative package my objections to steps 4 and 6, then Kant's proof as it stands is unsound.

1.2. Space, Content Externalism, and Human Embodiment

I would like now to shift philosophical gears, and move from the quite negative evaluation of the Refutation given at the end of the last section, toward a more positive line of analysis. Indeed I think that Kant's Refutation implicitly contains something of real and even fundamental philosophical significance. Suppose, now, that steps 4 and 6 are indeed fallacious as they stand. Nevertheless it seems to me the case that Kant *has* indeed proved *this* weaker thesis:

[I]nner experience in general is possible only through outer experience in general. (*CPR* B278–9)

My reasoning is this. Crucial to this thesis are two phrases: "inner experience in general" and "outer experience in general". I want to read "inner experience in general" as meaning "to have a self-conscious awareness of myself in inner sense," and I want to read "outer experience in general" as meaning "to have an actual outer sense." That is, I want to read the thesis as saying:

To have a self-conscious awareness of myself in inner sense is possible only through my also having an actual outer sense,

not:

Each and every inner self-conscious experience of a given mental state of my own is possible only through some direct correct outer perception of an actual distal material object in space.

That my proposed reading is at least arguably Kant's or anyhow recognizably Kantian, on the charitable assumption that he was intending to say something true and not something false, is also well-supported by a passage in the *Reflexionen*:

The question, whether something is outside of me, is just the same as to ask, whether I represent to myself an actual space. For this is outside of me. (R 5400; 18: 172)

Otherwise put, I want to distinguish quite sharply between three distinct meanings of the phrase "*X* is outside my own conscious mental state":

(1) *X* is a mind-independent substance.
(2) *X* is a material object in another part of space from that in which I am located = *X* is a distal material object.
(3) *X* is necessarily spatial in character.

What I want to argue on Kant's behalf is that in order to refute (*), or skeptical idealism, it is necessary only to prove that *I myself* satisfy (3), not to prove that *something else* satisfies (1) or (2). The issue on the table right now is whether a dreamer or hallucinator could have a capacity for imagining external things without having an actual outer sense. Again, I think not. That is, I want to argue that a capacity for imagining external things, even in dreams or hallucinations, is not possible without an actual outer sensibility. This becomes evident when we recognize that having an actual outer sense is at a bare minimum equivalent to having a capacity to uniquely represent space in intuition. For Kant, every representation of space whatsoever either simply is, or else necessarily requires,[59] the singular intuition of space. Indeed, even every *concept* of space presupposes the singular pure intuition of space: "in respect to [the general concepts of spaces] an a priori intuition [of space] (which is not empirical) grounds all concepts of [space]" (*CPR* A25/B39). Further, every imaginary representation of space presupposes the singular intuition of space, because the imagination is necessarily constrained by our forms of intuition (*CPR* B151).

Now, as we saw above, the singular pure intuitional representation of space functions as a directly referential term in Kant's cognitive semantics. So, even in dreams and hallucinations, the pure intuitional representation of space picks out *space* if it picks out anything at all. If it failed to pick out space in some context or in some possible set of circumstances, then it would be vacuous: it would not represent *anything*. In dreams or hallucinations I can of course have illusory images of material objects and their spatial locations, and I can also certainly be mistaken about *which* particular part of actual physical space I am in right now, or at other times; but there is no such thing as a "fake representation of space per se." Even visual illusions—say, Escher's famous drawings of impossible figures—require the genuine singular representation of space in order to fool the eye. So our capacity to represent space remains unaffected by incorrect perceptions of spatial objects. Or, more positively put, our capacity to represent space is veridical.[60] Only if space were a thing-in-itself would an illusory representation of space per se be possible. But space, as Kant argues in the Transcendental Aesthetic, is *empirically real* in that the representation of space functions as a necessary a priori form of outer sensibility, regardless of whether the empirical representations of that outer sensibility are correct or not (*CPR* A22–5/B37–40). And this would be impossible if space were a thing-in-itself, because then the representation of space would be cognitive-semantically "empty" (*leer*) in Kant's sense,

---

[59] What I mean is that the pure or formal intuition of space requires a less richly structured and non-conceptual singular intuitional representation of space, i.e., the form of outer sense. See notes 50–1.

[60] In ch. 2, I will argue that this is a consequence of the dual fact that our capacity for representing space and time is essentially non-conceptual, and also more cognitively basic than concept-involving cognition.

that is, not objectively valid or empirically meaningful. Therefore, even in dreams or hallucinations of space, or of spatial objects, the genuine singular representation of space must be invoked.

Because the Kantian thesis that I am interested in says that having a self-conscious experience of my determinate or individuated empirical self in inner sense is possible only through my also having an actual outer sense, we need not assume that every or even any act of introspection maps directly, perceptually, and correctly onto an actual material object at some distance from the conscious subject in space. By contrast, on this reading, Kant is only saying that we could not have the introspective experiences that we in fact have, without also at the very least actually representing space, that is, without having a genuine singular representation of space, even if that representation is only of empty space. And that is a pretty interesting conclusion, since it yields content externalism. Content externalism, as I am construing it, is the view that the representational content of at least some of our mental states is at least partially individuated by direct reference or direct relation (be it causal or otherwise physical, social, or historical) to something existing outside the human mind, in the worldly environment. Kant has shown, I think, that the empirical self-consciousness of oneself as an individuated stream of consciousness in inner sense presupposes at the very least a direct veridical singular intuitional representation of actual space as the self-identical substratum or persistent element to which I ascribe those states. So Kant is a content externalist. Now granting only Kant's content externalism, and, as I proposed above, explicitly bracketing his strong transcendental idealism, it also seems to me plausible, consistently with some other views that Kant expresses elsewhere, that a sound argument can be teased out of the Refutation.

In order to do this, I will rely heavily on a short essay published by Kant in 1768, "Concerning the Ultimate Ground of the Differentiation of Directions in Space." Despite its brevity, and despite its being a pre-Critical writing, this essay is seminal for the Critical Philosophy as a whole because it effectively prepares the ground for the theory of space, time, and sensibility that Kant first worked out in his Inaugural Dissertation in 1769. Indeed, as Kant remarks in one of the best-known *Reflexionen*: "the year '69 gave me great light" (*R* 5037; 18: 69). For my purposes, the most important passage in the "Directions in Space" essay is this one:

72 EMPIRICAL REALISM AND SCIENTIFIC REALISM

Because of its three dimensions, physical space can be thought of as having three planes, which all intersect each other at right angles. Concerning the things which exist outside ourselves: it is only in so far as they stand in relations to ourselves that we have cognition of them by means of the senses at all. It is, therefore, not surprising that the ultimate ground on the basis of which we form our concept of directions in space, derives from the relation of these intersecting planes to our bodies... Even our judgments relating to the cardinal points of the compass are, in so far as they are determined to the sides of our body, subject to the concept we have of directions in general. Independently of this fundamental concept, all we know of relations in heaven or on earth is simply the positions of objects relative to each other. No matter how well I may know the order of the compass points, I can only determine directions by reference to them if I know whether this order runs from right to left, or from left to right.... The same thing holds of geographical and, indeed, of our most ordinary knowledge of the position of places. Such knowledge would be of no use to us unless we could also orientate the things thus ordered, along with the entire system of their reciprocal positions, by referring them to the sides of our body. (DS 2: 378–9)

As I read this text, Kant is arguing as follows. First, every space represented by creatures like us in sensory intuition contains *directions*, namely, special topological features that partially determine not only the qualitative or extrinsic non-relational properties of material objects and their intrinsic structural properties but also the extrinsic relative positioning of material objects, yet which cannot be determined merely by any intrinsic non-relational properties that material objects might possess. Indeed, as Kant argues two pages later than the above text, the fact of "incongruent counterparts" or enantiomorphic objects—for example, 3-D physical objects such as my right and left hands, which (ideally speaking) are isomorphic mirror images of one another and yet cannot be made to coincide by any continuous translation of the figures in 3-D space—shows the falsity of the relational (or Leibnizian) theory of space, which says that spatial relations are logically strongly supervenient on the intrinsic non-relational properties of noumenal objects or monads, whether these are taken to be mental monads (as Leibniz did) or physical monads (as Kant himself did in the pre-Critical *Physical Monadology*): for if this were true, then the right and left hands would be congruent.[61] Second, directions in space

---

[61] See (DS 2: 381–3). Kant famously uses the very same argument from incongruent counterparts in 1783 in the *Prolegomena* (P 4: 285–6), in order to prove the transcendental ideality of space. But, as

THE REFUTATION OF IDEALISM 73

are unintelligible unless there exists a fixed frame of reference for spatially orienting distal objects. Third, any fixed frame of reference for distal spatial orientation is necessarily centered on an egocentric origin-point or "egocentric space"[62] consisting of the 3-D rectilinear axes of the cognizing subject's own *body*: up/down, right/left, in front/behind. Fourth, therefore, necessarily for all creatures with cognitive constitutions like mine, if any one of those creatures represents space, then its own body must also exist in that space.

Now, can my own body, construed as an egocentric, indexically-fixed material reference-frame for any possible representation of an intrinsically directionally-structured space, function as the persisting substratum in the original version of the Refutation? If so, then it seems to me that the Refutation can be charitably reconstructed as really arguing that necessarily for every creature cognitively constituted like me, a self-conscious awareness of its own uniquely determined stream of consciousness in inner sense requires the existence of its own body in space. Here is a more explicit version of that argument:

(1) I represent myself through empirical self-consciousness as an individuated stream of consciousness in inner sense.

(2) In order to represent myself in empirical self-consciousness as an individuated stream of consciousness in inner sense, I must represent myself, at the very least, in a direct relation to actual space itself, which in turn functions as a self-identical substratum for temporally determining my own stream of consciousness.

[The rationale for this crucial step is given by Kant's argument for steps 1 to 3 in the original version of the Refutation, plus my criticism of steps 4 and 6.]

(3) In order to represent actual space itself in some empirical context, I must represent it as having directions.

---

many have noted, since Kant uses an argument from incongruent counterparts in "Directions in Space" to prove *Newton's* theory of space, which is a noumenal realism about space (CPR A39/B56), then an appeal to incongruent counterparts cannot alone establish idealism. As I see it, however, the fact that Kant uses the case of enantiomorphs to argue in different contexts for spatial noumenal realism and spatial idealism alike shows that the incongruent counterparts argument is *neutral* as between realism and idealism, and equally consistent with both.

[62] See Evans, *The Varieties of Reference*, pp. 153–4.

(4) In order to represent directions in space, I must interpret those directions by means of a 3-D egocentric frame of reference centered on my own body.

(5) Therefore, necessarily, if I represent myself in empirical self-consciousness as an individuated stream of consciousness in inner sense, then my body must also exist in space.

By using modus ponens on (1) and (5), and then generalizing the result, we easily obtain:

(∼*) Necessarily for every creature having a cognitive constitution like mine, something exists outside its own conscious mental states.

Reconstructed in this way, the Refutation conforms very smoothly to Kant's leading idea that all conscious changes in inner sense are necessarily immediately ascribed to an actual spatially existing self-identical persisting thing or substratum. For we can now see that the most natural way of reading this is as saying that necessarily the contents of my own consciousness literally belong to my own living human body [63] or *Leib*.[64] More specifically, the big problem with steps 4 and 6 in the original version of the Refutation was the assumption that the external substratum in question was *distal*, not *proximal*, in relation to the self-conscious subject. But suppose instead that the external substratum Kant is talking about is strictly proximal: suppose that the external substratum is none other than my own living human body in actual space. Then Kant is saying (i) that my conscious mind is necessarily an *embodied mind*,[65] and (ii) that in order

---

[63] See also Cassam, "Inner Sense, Body Sense, and Kant's Refutation of Idealism." This necessary ascription of conscious states to the body should not, I think, be construed as a Davidson-style token-token identity thesis, whereby mental events or states literally are physical events or states, although Meerbote takes it this way: see his "Kant's Refutation of Problematic Material Idealism." Instead, I think that Kant anticipates the subtly but crucially different view that the human mind is an "embodied mind." See n.65.

[64] For Kant, there is therefore a strong continuity between the psychological life of my mind and the biological life of my own body. As he puts it in *Prolegomena* and in the *Critique of the Power of Judgment*: "Life is the subjective condition of all our possible experience" (*P* 4: 335) and "The mind for itself is entirely life (the principle of life itself)" (*CPJ* 5: 278). Since biological life can occur in beings lacking consciousness, it follows that for Kant biological life is non-conscious mind; that sentient animal life is conscious mind; and that sapient or rational animal life is self-conscious mind. Furthermore the feelings of pleasure and pain, and of the bodily affects more generally, constitute "the feeling of life" (*CPJ* 5: 204, 278). So consciousness entails embodied life: conscious beings are necessarily also living organisms. These are crucial points that I will come back to in sections 8.2–8.3.

[65] According to the embodied mind view, all conscious states or cognitive states necessarily occur in and throughout living animal bodies. See, e.g., Bermúdez et al. (eds.), *The Body and the Self*; Clark,

to individuate myself psychologically and as a unique member of my own species,⁶⁶ then I must ascribe each of my mental states directly to my own living human body in actual space.⁶⁷ Or, in other words, the ascription of my mental states to my own living human body individuates my mental states by *locating them uniquely*.⁶⁸

In the important 1786 essay, "What is Orientation in Thinking?," Kant calls this capacity of a rational human animal for bodily self-location "self-orientation":

> To self-*orient* (*sich orientiren*) in the proper sense of the word, means to use a given direction—and when we divide the horizon into four of these—in order to find the others, and in particular that of *sunrise*. If I see the sun in the sky and know that it is now midday, I know how to find south, west, north, and east. For this purpose, however, I must necessarily be able to feel a difference within my own *subject*, namely that between my right and left hands. I call this a *feeling*, because these two sides display no perceptible difference as far as external intuition is concerned...I can now extend this geographical concept of the process of orientation to signify any kind of orientation within a given space, i.e., orientation

---

Being There: Putting Brain, Body, and World Together Again; Clark, "Embodiment and the Philosophy of Mind"; Damasio, *The Feeling of What Happens: Body and Emotion in the Making of Consciousness*; Hanna and Thompson, "The Mind–Body–Body Problem"; Merleau-Ponty, *Phenomenology of Perception*, part one, ch. 3. pp. 98–147; Rockwell, *Neither Brain nor Ghost: A Nondualist Alternative to the Mind–Brain Identity Theory*; Thompson, *Mind in Life*; Thompson and Varela "Radical Embodiment"; and Varela, Thompson, and Rosch, *The Embodied Mind*.

⁶⁶ Or, in other words, in order to determine *my empirical personal identity over time*, then I must ascribe each of my mental states directly to my own human body in actual space. If I am right about the Refutation, then Kant is in fact an "animalist" and, to that extent, an Aristotelian, about personal identity. See Kain, "Kant's Conception of Human Moral Status"; and see Olson, *The Human Animal*. It should also be noted, however, that Kant's concept of the empirical human person is not the same as his concept of the rational human person. The identity of the empirical person is a necessary but not sufficient condition of the existence of the rational human person.

⁶⁷ How does the doctrine that my changing mental contents are directly ascribed to my living human body in space, square with Kant's claim that "inner sense, by means of which the mind intuits itself, or its inner state, gives, to be sure, no intuition of the soul itself, as an object" (CPR A22/B37)? One possibility is that for Kant the contents of inner sense are not the intrinsic non-relational qualitative phenomenal mental properties or property-instances (i.e., phenomenal qualia or sense data) of either a Cartesian soul or a mere hunk of inert matter, but rather are nothing but irreducible intrinsic structural phenomenal mental properties, or "lived experiences" (*Erlebnisse*), of an embodied mind. See n.65.

⁶⁸ This should be compared and contrasted with Strawson's famous argument in *Individuals*, ch. 2, that one can in principle individuate oneself purely in terms of a system of auditory experiences alone, as an isomorphic analogue of spatial representation. If correct, Strawson's argument would suffice to show that psychological individuation does *not* require that I ascribe my mental states directly to my own living human body in space. But Strawson's sound-system cannot distinguish between incongruent counterparts, hence cannot distinguish between the very different experiences of the same sound heard as "coming from my right" and as "coming from my left."

in a purely *mathematical* sense. In the darkness, I can orientate myself in a familiar room so long as I can touch any one object whose position I remember. But it is obvious that the only thing which assists me here is an ability to define the positions of the objects by means of a *subjective* distinction: for I cannot see the objects whose position I am supposed to find; and if, for a joke, someone had shifted all the objects round in such a way that their relative positions remained the same but what was previously on the right was now on the left, I would be quite unable to find my way about a room whose walls were in other respects identical. But in fact, I can soon self-orient simply by the feeling of difference between my two sides, my right and my left. (*OT* 8: 134–5)

There are obvious parallels between this text and the passage from the "Directions in Space" essay quoted earlier. But the most important thing is this: when we take the idea of self-orientation from "What is Orientation in Thinking?," and add it to the original Refutation and to the "Directions in Space" essay, we can derive a profound Kantian doctrine to the effect that to be self-consciously aware of my own uniquely determined psychological life is automatically also to be intuitionally aware of the unique location—and also, I think, of the unique locus of movement,[69] or motility—of my own living human body in space. All human empirical apperception is thus "orienting apperception": necessarily, I become aware of myself as *myself* only by way of establishing my own living human body, which is the unique location and kinesthetic locus of all my mental states, as a 3-D egocentric frame of reference in a directionally-structured encompassing total singular space. Or to use the entirely apt and now canonical language introduced by Nagel: necessarily, when I introspectively find myself as a subject enjoying phenomenal consciousness, which is "what it is like to be, for an organism," I also find myself essentially having a "single point of view."[70]

This result, happily enough, also entails direct perceptual realism, if we add one further assumption. For it is obvious, I think, that *accurate* self-orientation—that is, knowing where one is relatively to other real things in

---

[69] See Brewer, "Self-Location and Agency." Brewer argues that perceptual self-location is based on a more basic perceptual grasp of determinate possibilities of spatial bodily self-movement. The text I quoted from "What is Orientation in Thinking?" can be used to support Brewer's conclusion, although in both "Directions in Space" and in the first *Critique* Kant places more emphasis on perceiving the motions of other bodies than on the movement of one's own body. Also, some qualifications will be needed to handle tricky cases in which self-movement is, as a matter of contingent fact, drastically reduced or altogether ruled out. See n.78.

[70] See Nagel, "What is it Like to be a Bat?," pp. 160 and 167; and Hurley, *Consciousness in Action*, ch. 2.

actual space—requires perceptual access to at least one distal material object in space: in order to orient myself accurately, I need to get a genuine "fix" on at least one actually existing material object that is "out there." And this is where we can begin to make some positive sense of Kant's including a simultaneity condition in the conclusion of the original formulation of the Refutation. What gives it its positive sense is the Third Analogy of Experience. I can accurately orient myself in space only if the distal material object I am getting a fix on is in perceptual "dynamic community" with me—that is, only if that material object *actually exists now, is simultaneously dynamically interacting with me, and is being perceptually effectively tracked by me.*[71] Objects presented in dreams, hallucinations, other non-veridical illusions,[72] or time-lag perceptions,[73] will not suffice. Indeed, Kant argues that by accurately uniquely locating one's own animate human body in space, one thereby determines the relative locations of *all* other simultaneously existing material things in a macrophysical causal-dynamically structured spatial system of reciprocally interacting bodies:

> From our experiences it is easy to notice that only continuous influence in all places in space can lead our sense from one object to another, that the light that plays between our eyes and the heavenly bodies effects a mediate community between us and the latter and thereby proves the simultaneity of the latter, and that we cannot empirically alter any place (perceive this alteration) without matter everywhere making the perception of our position possible: and only by means of its reciprocal influence can it establish their simultaneity and thereby the coexistence of even the most distant objects (though only mediately). (*CPR* A213/B260)[74]

Furthermore, this accurate self-orienting fix on at least one actual material object that is in dynamic community with a living human body,

---

[71] Also I can empirically intuit very distant actual material objects, remotely past objects, and even futural actual material objects, providing such objects have a "connection with some actual perception in accordance with the analogies of experience" (*CPR* A225–6/B273–4); see also Hanna, *Kant and the Foundations of Analytic Philosophy*, p. 210.

[72] Non-veridical illusions are phenomenal representations without any existing objects, and can vary radically in content from context to context and from perceiver to perceiver. By sharp contrast, veridical illusions—e.g., the straight stick in water that appears to be bent—imply the actual existence of the object perceived, and how we represent them remains essentially the same across contexts and perceivers. Also Kant holds that perceivers can stand in non-epistemic and non-conceptual dynamical community with the objects of veridical illusions—see section 2.2.

[73] Sailors using an astrolabe or sextant, e.g., navigate by referring to the contemporary apparent stars in the night sky, not by referring to the much earlier objects or events that are the causal sources of those light signals.

[74] See also Longuenesse, *Kant and the Capacity to Judge*, pp. 391–3.

and therefore also with my my veridical perception of it, is essentially intuitional, and thereby is a *direct* (that is, a non-epistemic, non-conceptual, and otherwise unmediated) perception. In this way my reconstructed version of the Refutation yields not only externalism and the necessity of minded human embodiment, but also, granting one further assumption, direct perceptual realism: creatures minded like us necessarily have direct veridical perceptual access to macrophysical dynamic material objects in phenomenal space and time *if* they are empirically self-conscious—which necessarily guarantees their self-orienting embodiment—and *if* that self-orienting embodiment is also an accurate one.

Nevertheless, it should be noted that since the direct veridical perceptual access at issue need not *in fact* be either locationally accurate or epistemically correct—because of the lingering dream problem—Kant's direct perceptual realism will not satisfy the *traditional* refuters of idealism. For such refuters want not only a direct veridical perceptual access to external objects and to the worldly environment at the very same time that I am empirically self-conscious: they also want a representationally accurate or correct access to distal material objects that *justifies* my perceptual judgments. To repeat, however, it takes only a *non-epistemic or belief-independent, non-conceptual, and object-dependent* use of one's perceptual capacities to be able to carry out an accurate self-locating act of empirical self-consciousness. That is, as long as I can "feel" the primitive difference between one's right and left sides, and can accurately intuitionally locate myself in space relative to some other actual material things, whatever those things turn out to be, then neither true beliefs nor correctly applied concepts about the specific details of my actual outer situation are required.[75] Nevertheless, it still follows that my immediate, non-inferential, and therefore "blind"[76] judgments about my own living human body and (assuming an accurate self-orientation) also about some material objects in my general vicinity *are* indeed *fully*

---

[75] See Eilan "Objectivity and the Perspective of Consciousness," pp. 247–8, for a defense of the opposing position. Eilan's view implausibly entails, however, that if I wake up suddenly in total darkness and cannot for the life of me remember where I am, then I am neither perceptually conscious nor empirically self-conscious.

[76] Here there is a profound connection between (1) Kant's notion of a "blind" intuition (*CPR* A51/B75), (2) Kant's notion of an "aesthetically certain" perceptual judgment (*JL* 9: 39), and (3) the later Wittgenstein's deep idea that we follow rules "blindly" (*Philosophical Investigations*, para. 219, p. 85e) and thus we can be *defeasibly warranted* or *legitimated* in making judgments without being *infallibly inferentially justified* in making them. See section 2.5.

warranted by the charitably reconstructed Refutation of Idealism, even if those perceptual judgments are not in fact true.

1.3. The Reality of the Embodied Mind in Space

If I am right, then, although Kant's Refutation of Idealism fails to establish the conceptual correctness and judgmental truth of our outer perceptions of distal material objects, it does nevertheless establish the truth of content externalism. And when it is supplemented by some points from "Directions in Space" and "What is Orientation in Thinking?", it also refutes skeptical idealism and establishes direct perceptual realism by means of the quite substantive thesis of the necessity of self-orienting embodiment for all self-conscious creatures sharing our cognitive constitution, as well as the non-inferential warrant of perceptual judgments about some proximal and distal material objects—more specifically, judgments about my own living human body and also about any other material object in my general vicinity that is needed to orient myself accurately in space.

It must be re-emphasized, however, that even assuming the soundness of this reconstructed Kantian argument for the necessity of self-orienting embodiment, it certainly does not follow that I must be able to know *what* my own particular living human body *is like*. I could be wrong that I actually have limbs, as "phantom limb" cases show. Or, for all I know, my living human body might be nothing but a brain floating listlessly in a vat. Still, for creatures like us there is no such thing as having a "phantom body." I could not be utterly disembodied and still be *me*. Without a real living human body of some sort, I would necessarily lack self-orientation in space, and without self-orientation in space I would necessarily fail to individuate my stream of changing conscious mental states.[77] Moreover,

---

[77] Strawson, by contrast, argues that although persons are normally and paradigmatically embodied, it is at least conceptually possible for them to be disembodied; see *Individuals*, pp. 115–16. So a Strawsonian critic of my account might grant Kant's point about the necessary connection between psychological individuation and unique spatial location, and also claim that one can in principle be uniquely located in egocentric space, yet disembodied. For example, it might seem that someone could have a disembodied purely visual awareness of directions and of various purely visual objects in a purely visual phenomenal space. But that seems to me absurd. It would be to say that I could somehow be uniquely located in egocentric space without literally occupying any space. If I were an extensionless point, I certainly could not *look around*.

the necessity of self-orientation also carries with it the further implications that my living human body enters into dynamic interaction with other material objects and is (at least in principle[78]) motile. So, necessarily, if I am self-consciously aware of myself as an individuated stream of consciousness in inner sense, then my living human body also exists as a uniquely located, dynamic, and (in principle) motile 3-D egocentric material reference-frame for directions in space.

Seen in this light, ironically enough, Moore's notorious proof of an external world is *almost* sound, but not at all for the reasons Moore thought. Moore was quite wrong about being certain that he had a right *hand* and a left *hand*: but he was absolutely correct about being certain that he had *a right side of his body* and *a left side of his body*. And, even more ironically, we can see that on this one point at least Descartes was absolutely correct when, for a brief moment in the first Meditation, he saw that it was utter madness to doubt that he had a body.

---

[78] As a matter of contingent fact I might have a body that is tied down, paralyzed, limbless, or listlessly floating around in a vat. But in principle even such a body could be moved about, either by myself or by some intrusive external agency, hence my body is not *intrinsically* static or inert, precisely because it is *a sentient living organism or animal* and also the locus of my rational human embodied first-order volition or "power of choice" (*Willkür*). See sections 8.2–8.3.

2

Direct Perceptual Realism II: Non-Conceptual Content

> To make a concept, by means of an intuition, into a cognition of an object, is indeed the work of judgment; but the reference of an intuition in general is not.
>
> PC 11: 310–11

> The informational states which a subject acquires through perception are non-conceptual, or non-conceptualized. Judgements based upon such states necessarily involve conceptualization: in moving from a perceptual experience to a judgement about the world (usually expressible in some verbal form), one will be exercising basic conceptual skills. But this formulation (in terms of moving from an experience to a judgement) must not be allowed to obscure the general picture. Although the judgments are *based upon* his experience (i.e., upon the unconceptualized information available to him), his judgements are not *about* the informational state. The process of conceptualization or judgment takes the subject from his being in one kind of informational state (with a content of a certain kind, namely, non-conceptual content) to his being in another kind of cognitive state (with a content of a different kind, namely, conceptual content).
>
> Gareth Evans[1]

2.0. Introduction

In the last chapter I developed a charitable reconstruction of the Refutation of Idealism. According to this reconstruction, even though Kant fails to prove

---

[1] Evans, *The Varieties of Reference*, p. 227.

that necessarily whenever a human cognizer is empirically self-conscious then some distal material objects in space also exist at the same time, nevertheless he *does* give a sound argument for the thesis that necessarily if a human cognizer is empirically self-conscious then something exists outside her own conscious mental states—namely, her own living (and in principle motile) human body. This conclusion follows because in order to determine her own individual mental life in inner sense, a human cognizer necessarily has a self-orientating embodiment in objectively real or actual space. The further assumption of an *accurate* self-orientating embodiment then entails the thesis of direct perceptual realism: necessarily if a human cognizer is empirically self-conscious and able accurately to locate herself uniquely in space, then she has direct (that is, non-epistemic, non-conceptual, and otherwise unmediated) veridical[2] perceptual or observational access to some macrophysical dynamic material things in objectively real space and time. And direct perceptual realism constitutes the first half of Kant's empirical realism.

In this chapter I want specifically to explore the "direct" aspect of direct perceptual realism, and in particular Kant's theory of non-conceptual perceptual content.[3] A cognition is direct in the Kantian sense if and only if it refers "immediately" (*unmittelbar*) to an object, and in turn, a cognition refers immediately to an object if and only if it is non-epistemic (belief-independent), non-conceptual (concept-independent), and otherwise unmediated (in the sense that it does not, or at least need not, refer by

---

[2] It should be remembered from ch. 1, n. 7 that I am systematically distinguishing between correct sense perception and veridical sense perception. Again, correct sense perception necessarily involves both (i) an accurate representation of the object represented and (ii) the actual existence of the object. By contrast, veridical perception necessarily involves only the actual existence of the object represented, but not necessarily an accurate representation of it. Among other things, this implies that there can be "veridical illusions": see section 2.2.

[3] This indicates a central difference between my interpretation of Kant's empirical realism and Abela's. Abela's basic account of empirical realism is three-part: (1) all objective cognition is judgment-dependent; (2) objective cognition pertains exclusively to empirical nature, the causally ordered world of actual and possible human experience; and (3) something is real if and only if it is the recognition-independent truth-maker of a true judgment (Abela, *Kant's Empirical Realism*, pp. 287–8). I fully agree with (2), but disagree with (1) and (3): (1*) in fact Kant holds that only *some* objective cognition is judgment-dependent, and (3*) being a recognition-independent truth-maker of a true judgment is a sufficient but not a necessary condition of something's being real. For earlier defenses of these claims, see Hanna, *Kant and the Foundations of Analytic Philosophy*, chs. 1–2 and 4. In any case, since Abela is committed to the judgment-dependence of objective cognition and even more specifically to its concept-dependence, he is also explicitly committed to the rejection of the existence and representational significance of non-conconceptual content (Abela, *Kant's Empirical Realism*, pp. 102–6). So, if my argument in this chapter is correct, then it entails another denial of Abela's claim (1).

means of any other sort of representational faculty, representational content, psychological intermediary, or physical intermediary). Since all beliefs intrinsically contain concepts, then non-conceptuality is both necessary and sufficient for a cognition's being non-epistemic. And, as I will argue in section 2.5, non-conceptuality is also both necessary and sufficient for a *perception*'s being otherwise unmediated. So non-conceptuality is both necessary and sufficient for the directness of perception.

The issue of non-conceptuality may seem recondite. But in fact the issue of whether non-conceptual content exists and is representationally significant is not only a fundamental issue in the theory of mental content—for reasons I will rehearse in the next section—but also a fundamental point of contention between Kant's empirical realism and contemporary scientific realism. Because contemporary scientific realism[4]—with the sole exception of eliminativist scientific realism—is committed to *indirect* epistemic realism (see section 2.5 and 3.0), it is also thereby committed to the explicit rejection of non-conceptual content. And while it is quite true that eliminativist scientific realists like Paul Churchland, Richard Rorty, and Sellars[5]—who assert the truth of physicalism together with the non-existence of all mental substances, properties, and facts—are fully committed both to the existence of a fundamentally real microphysical world and to the fact of our ability to refer to it directly in language and cognition, nevertheless an essential element of their doctrine is the overdetermination of all linguistic and cognitive reference by background theories and conceptual frameworks. The thesis of semantic overdetermination says that linguistic and cognitive reference are radically *underdetermined* by worldly inputs to our background theories and conceptual frameworks, and also necessarily *relativized* to those theories and frameworks. So the eliminativist scientific realists reject the Myth of the Given, and replace it with what we might call *the Dogma of Semantic Overdetermination*. But the Dogma *also* directly implies the rejection of non-conceptual content. So, by contraposition, Kant's affirmation of non-conceptual content entails the denial of all forms of contemporary scientific realism. And by now the very fact of Sellars's fourfold commitment to scientific realism, eliminativism, the Dogma of Semantic Overdetermination, and the rejection of non-conceptual content

---

[4] See ch. 1, n.27.
[5] See Churchland, *Scientific Realism and the Plasticity of Mind*; Rorty, *Philosophy and the Mirror of Nature*, ch. 2; and Sellars, "Empiricism and the Philosophy of Mind."

should already strongly indicate that there is a set of deep philosophical connections that indissolubly link together non-conceptual content, Kant's theory of cognition, and the Two Images Problem. I will explore this threefold linkage further in the next section.

2.1. Kant, Non-conceptualism, and Conceptualism

For the purposes of this chapter, by the notion of "cognitive content" I mean conscious representational content, whether object-directed (intentionality) or self-directed (reflexivity). And for every type of cognitive content there is a corresponding cognitive capacity by means of which a minded creature generates, possesses, and deploys that content. We could allow for non-conscious cognitions. Indeed—as I will mention in passing below—Kant holds that some cognitions are non-conscious; and it is also the case that some contemporary philosophers of cognition have claimed that non-conceptual content is non-conscious sub-personal information-content.[6] But the topic of non-conscious mind raises controversial and subtle issues beyond the scope of this chapter.[7] So, to keep things fairly simple, I will focus almost exclusively on conscious cognitions.

Assuming that caveat, we can then formulate the thesis of *non-conceptualism* about cognitive content. Non-conceptualism holds that non-conceptual content exists and is representationally significant—that is, meaningful in the "semantic" sense of describing or referring to states-of-affairs, properties, or individuals of some sort. More precisely, non-conceptualism says: (a) that there are cognitive capacities which are not determined (or at least not *fully* determined) by conceptual capacities; and (b) that the cognitive

---

[6] See Stalnaker, "What Might Nonconceptual Content Be?" Stalnaker argues that non-conceptual content reduces to purely physical information. But virtually all contemporary non-conceptualists and conceptualists are non-reductivists about cognitive content, however—so Stalnaker's view is oddly orthogonal to the contemporary debate.

[7] In particular, since Kant identifies mind with life, and since biological life can occur in beings lacking consciousness, it follows that for him biological life is non-conscious mind, and that human consciousness entails biological life. See ch. 1, n.64. Kant's commitment to the strong continuity of biological life and mind has crucial implications for his theory of the will. See section 8.3. To be sure, this Kantian line of reasoning is in direct opposition to many contemporary theorists of non-conscious cognition, who construe such cognition mechanistically (and, more specifically, computationally) and physicalistically; see, e.g., Jackendoff, *Consciousness and the Computational Mind*. But, on the other hand, some contemporary "embodied mind" theorists of cognition follow the Kantian line. See Clark, *Being There: Putting Brain, Body, and World Together Again*; and Clark, *Mindware*, chs. 6–7.

capacities which outstrip conceptual capacities can be possessed by rational and non-rational animals alike, whether human or non-human.

In my view, contemporary non-conceptualism (defended and developed by, for example, José Bermúdez, Tim Crane, Fred Dretske, Richard Heck, Susan Hurley, Sean Kelly, M. G. F. Martin, Christoper Peacocke, Michael Tye, and others[8]) can be traced directly back from Evans's *Varieties of Reference*[9] to the first *Critique* in a three-linked chain of philosophical influences, via (i) Russell's notion of "acquaintance," (ii) the Brentano-Husserl-Meinong notion of an intentional "presentation," and finally (iii) Kant's notion of intuition. I will not attempt to rehearse the blow-by-blow details of this story here, except just to note that Russell's tendency to extend the notion of acquaintance to an infallible grasp of sense data[10] is in fact a wrongheaded *anomaly* within this tradition, which is otherwise focused on how conscious minds achieve direct perceptual contact with real worldly objects, and that there are some very deep philosophical affinities, recorded in the two epigraphs, between the terminal points of this historical chain: *Kant's* theory of cognition and *Evans's* theory of cognition.

So my first claim is that non-conceptual cognitive content in the contemporary sense is, for all philosophical intents and purposes, identical to *intuitional* cognitive content in Kant's sense. Indeed, in my opinion the contemporary distinction between non-conceptual cognitions and their content, and conceptual cognitions and their content, is essentially the same as Kant's distinction between intuitions and concepts. Correspondingly, if I am correct, then the contemporary distinction between non-conceptual capacities and conceptual capacities is also essentially the same as Kant's cognitively seminal distinction between the sensibility and the understanding.

As I noted in the Introduction, both the sensibility and the understanding are innate spontaneous mental capacities, or *faculties*. For Kant, the faculty

---

[8] Bermúdez, "Nonconceptual Mental Content"; Bermúdez, *The Paradox of Self-Consciousness*; Bermúdez and Macpherson, "Nonconceptual Content and the Nature of Perceptual Experience"; Crane, "The Nonconceptual Content of Experience"; Dretske, "Conscious Experience"; Dretske, *Knowledge and the Flow of Information*; Dretske, *Seeing and Knowing*, ch. II; Heck, "Nonconceptual Content and the 'Space of Reasons'"; Hurley, *Consciousness in Action*, ch. 4; Kelly, "The Nonconceptual Content of Perceptual Experience: Situation Dependence and Fineness of Grain"; Kelly, "What Makes Perceptual Content Non-conceptual?"; Martin, "Perception, Concepts, and Memory"; Peacocke, "Does Perception Have a Nonconceptual Content?"; Peacocke, "Nonconceptual Content Defended"; Peacocke, *A Study of Concepts*, ch. 3; Tye, *Consciousness, Color, and Content*; Tye, *Ten Problems of Consciousness*; and Gunther (ed.), *Essays on Nonconceptual Content*.
[9] Evans, *The Varieties of Reference*, chs. 2–7.
[10] See, e.g., Russell, *The Problems of Philosophy*, ch. 5.

of sensibility vs. faculty of understanding distinction is cognitively seminal precisely because it exhausts the "fundamental sources of the mind." Now the sensibility is the perceptual, imaginational, affective (in the broad sense of "feelings," not the narrower sense of desires or volitions, which for Kant belong to the faculty of desire or the will) capacity of the mind, which produces intuitions as outputs, given external "affections"—causal-informational triggerings—as inputs. By contrast, the understanding is the logical and discursive capacity of the mind, which produces concepts as outputs, given intuitions as inputs. Intuitions and concepts together "constitute the elements of all our cognition," in the sense that intuitions and concepts are combined together by the non-basic "faculty of judging" (*Vermögen zu urteilen*) (*CPR* A69/B94) in order to form judgments, which are the central cognitive acts of the rational personal mind.[11] And for Kant there are no other basic *content-producing* faculties over and above the intuition-producing faculty and the concept-producing faculty (*CPR* A50/B74).

Here it should be noted that I am construing the sensibility as only *relatively* passive, but not *entirely* passive (as, for example, in Locke's account of sensibility), by virtue of its expressing a mental power for spontaneous synthesis, or mental processing. This mental power is the "power of imagination" (*Einbildungskraft*), and it is delivered in two distinct basic stages or moments: (i) a **synopsis** of the manifold *a priori* through sense" or "synthesis of apprehension"; and (ii) a **synthesis** of this manifold through the imagination" or "synthesis of reproduction in imagination" (*CPR* A94, A98–102). In the B edition of the first *Critique* these two basic stages of mental processing are said to have a single shared innate psychological ground in the "transcendental" or "productive" imagination, which carries out the operation of "figurative synthesis" or *synthesis speciosa* (*CPR* B151), whose precise cognitive function it is to produce representations of static or dynamic spatiotemporal forms, patterns, or shapes. Kant's general thought here can be expressed as the thesis that "imagination is a necessary ingredient of perception itself" (*CPR* A120 n.).[12]

Otherwise put, Kant's sensibility vs. understanding distinction captures the difference between the *sub-rational* or *lower-level* spontaneous cognitive

---

[11] See Hanna, "Kant's Theory of Judgment."
[12] See Hanna, *Kant and the Foundations of Analytic Philosophy*, ch. 1. And, for an earlier acknowledgment of Kant's deep idea that perception is pervaded by the activity of the imagination, see Strawson, "Imagination and Perception."

powers of the human or otherwise animal mind, and the *rational* or *higher-level* spontaneous cognitive powers of the human or otherwise animal mind. On this Kantian picture of our cognitive capacities, it is not to be assumed that rational animals do not *also* have the sub-rational or lower-level cognitive powers: on the contrary, for Kant all rational animals also have sub-rational or lower-level cognitive powers that they share with non-rational animals, whether human or non-human. In this connection, Dretske very relevantly remarks in *Seeing and Knowing* that:

[v]isual differentiation, as I am employing this phrase, is a pre-intellectual, pre-discursive sort of capacity which a wide variety of beings possess [and it] is an endowment which is largely immune to the caprices of our intellectual life.[13]

The crucial point grasped by Kant, Dretske, and Evans alike, I think, is that non-conceptual cognitive capacities are "sub-rational" or "non-rational" capacities only in the sense that they are necessary but not sufficient for our rational cognitive capacities, not in the sense that they are irrational or arational. So non-conceptual content does not *exclude* rationality: on the contrary, on the Kant-Dretske-Evans picture, non-conceptual cognition and its content constitute the *proto*-rationality of conscious animals.

This brings me to the thesis of *conceptualism* about cognitive content. Conceptualism hold that non-conceptual content neither exists nor is representationally significant. More precisely, conceptualism says: (a) that all cognitive capacities are fully determined by conceptual capacities, and (b) that none of the cognitive capacities of rational human animals can also be possessed by non-rational animals, whether human or non-human.

It should be noted, for completeness's sake, that there are also at least three weakened versions of conceptualism. The first weakened version says that non-conceptual content indeed exists but is *not* representationally significant, because such content is nothing but the intrinsic non-relational qualitative content of sensations: namely, phenomenal qualia, whether qualia are taken to be sensory types or sensory tokens. In other words, this sort of conceptualism is prepared to admit non-conceptual content, but only if it is *pure sensory* content. Oddly this sort of conceptualism could also, with a little squinting and tweaking, be regarded as a super-weak version of *non*-conceptualism: a "pure sensationalist non-conceptualism." And this

---

[13] Dretske, *Seeing and Knowing*, p. 29.

sort of faux-non-conceptualism is a favorite target of those who also want to accuse non-conceptualists of investing in the Myth of the Given. But contemporary Evans-influenced non-conceptualists and Kantian non-conceptualists alike will simply not go in for this, because it is crucial to their view that non-conceptual content is *representationally significant* and *world-oriented*, not phenomenalistic.

By contrast, the second weakened version of conceptualism says that while there are non-conceptual *cognitions*, there are nevertheless no non-conceptual *contents*: the contents of non-conceptual cognitions are themselves conceptually fully determined.[14] But contemporary Evansian non-conceptualists and Kantian non-conceptualists will not accept this formulation either: their view is not merely about cognitive acts, processes, or states—it is about the *semantics* of those acts, processes, or states.

Finally, the third weakened version of conceptualism says that (a) is true but denies (b): some non-rational human or non-human animals also have primitive or "proto" conceptual capacities. Not surprisingly, this sort of conceptualism is favored by some of those interested in non-human animal cognition.[15] The problem with this move, however, from the standpoint of non-conceptualism is that it forces its conceptualist defender to posit "simple concepts," or more generally some sort of pre-logical and pre-linguistic "proto-concepts" that are possessed by both rational humans and non-rational human or non-human animals. But at this point in the discussion it becomes almost impossibly difficult to tell the difference between concepts and non-concepts: what distinguishes between a *proto*-conceptual content and a *non*-conceptual content? That problem is closely connected with what, two paragraphs on, I will call "the concept problem" in the contemporary debate about non-conceptual content. In any case, in what follows I will focus exclusively on the full-strength version of conceptualism.

This is because the most influential version of contemporary conceptualism is in fact full-strength conceptualism, as defended by John McDowell in *Mind and World*.[16] The crucial point for our present purposes, however, is that McDowell not only frequently cites Kant in support of his

[14] This point is also made by Heck in "Non-Conceptual Content and the 'Space of Reasons'."
[15] See, e.g., Griffin, *Animal Minds*, ch. 7. It is however possible to defend the thesis that non-human animals can think, and *also* be a non-conceptualist. See Bermúdez, *Thinking without Words*.
[16] McDowell, *Mind and World*, lecture III, and Afterword, part II. See also See Brewer, *Perception and Reason*, ch. 5; and Sedivy, "Must Conceptually Informed Perceptual Experience Involve Non-conceptual Content?"

conceptualism, but even takes himself to be working out a *Kantian* theory of cognition against *Evans's* theory of cognition, via his—I mean McDowell's—interpretations of various selected texts by Donald Davidson and especially Sellars. In turn, McDowell's conceptualist interpretation of Kant has also been further developed by Paul Abela, in the context of the realist/anti-realist debate.[17]

Now, Davidson and Sellars are both clearly thoroughgoing conceptualists *avant la lettre*. Indeed, and more generally, twentieth-century conceptualism—whether in its early linguistic guise as "Russell-Frege semantics" (a.k.a. *descriptivism*), or in its 1970s linguistic reincarnation (spearheaded by the early McDowell) as the semantics of "*de re* senses,"[18] or in the guise of the semantics of theoretical overdetermination favored by Sellars, or finally in the explicitly cognitive version of conceptualism developed by the later McDowell in *Mind and World*—is one of the basic commitments of mainstream analytic philosophy.[19] Michael Dummett even claims in *The Origins of Analytical Philosophy* that, insofar as Evans was a non-conceptualist, "he was no longer an analytical philosopher."[20] So the debate about non-conceptual content is seminal. And, in a way that is fully within the classical nineteenth- and twentieth-century tradition of neo-Kantian philosophical polemics,[21] McDowell has drafted Kant into service in support of the conceptualist/descriptivist cause, without acknowledging even so much as the possibility of a non-conceptualist reading of Kant's theory of cognition.[22]

McDowell's positive case for conceptualism is heavily based on his strong endorsement of Sellars's attack on the Myth of the Given,[23] on Sellars's theory of intentionality,[24] and above all on Sellars's controversial

[17] See n. 3.
[18] See McDowell, "On the Sense and Reference of a Proper Name." The notion of a *de re* sense is McDowell's version of Evans's revisionist reading of Frege's theory of reference in *The Varieties of Reference*.
[19] See Hanna, *Kant and the Foundations of Analytic Philosophy*, esp. ch. 4.
[20] Dummett, *Origins of Analytical Philosophy*, p. 4.
[21] See Hanna, "Kant in Twentieth-Century Philosophy."
[22] McDowell's conceptualist interpretation of Kant is by no means uncontroversial. Indeed, other cognitively oriented Kantians have adequately acknowledged Kant's theory of non-conceptual cognition: see, e.g., Brook, *Kant and the Mind*, p. 125; and Kitcher, *Kant's Transcendental Psychology*, p. 161. And it has been quite plausibly argued that McDowell's Kant is more Hegelian than Kantian: see Sedgwick, "McDowell's Hegelianism."
[23] See McDowell, *Mind and World*, lectures I–III, and Afterword, part I; and Sellars, "Empiricism and the Philosophy of Mind."
[24] See McDowell, "Having the World in View: Sellars, Kant, and Intentionality."

reading of Kant.[25] Sellars in turn had explicitly deployed his attack on the Myth, his theory of intentionality, and his reading of Kant in tandem as necessary parts of his scientific realist and eliminativist resolution of the Two Images Problem.[26] As I formulated it in the Introduction, this resolution proposes to reduce the manifest image to the scientific image by means of scientific or reductive naturalism. And Sellars's conceptualism is absolutely necessary for his scientific naturalism and realism, via the Dogma of Semantic Overdetermination: according to him, our cognitive access to the "really real" microphysical world of basic entities, processes, forces, and fundamental physical properties is necessarily mediated and in fact overdetermined by exact scientific theories, by judgments, and above all by language, and so finally by concepts, which are the basic semantic constituents of language and thought. Therefore, if Kant's non-conceptualism is correct, then not only is Sellars's scientific naturalism false insofar as it depends on conceptualism: also his naturalistic resolution of the Two Images Problem is unsound. Correspondingly, it follows that every *other* argument for scientific or reductive naturalism which employs the thesis of conceptualism as either an explicit or implicit premise, is similarly unsound. And, in section 3.2, I will offer some reasons for thinking that every possible version of microphysical scientific realism must also be committed to some form of conceptualism.

Now there is at least one other significant fact concerning the seminal debate about non-conceptual content that also needs to be pointed up: just as McDowell has not acknowledged Kant's non-conceptualism, oddly enough contemporary non-conceptualists have not acknowledged their debt to Kant *either*. Indeed, as far as I can tell, non-conceptualists have made no attempt whatsoever to trace the historical sources of their doctrine back beyond Evans's writings—although surely the line of influence runs like the M40 directly through Peter Strawson's *Bounds of Sense* to the first *Critique*.[27] In any case, add the non-conceptualists' odd non-acknowledgment to the intimate and inextricable entanglement of McDowell's and Sellars's writings with Kantian doctrines and themes, and the result, I think, is the recognition *that Kant's theory of intuition* is *the hidden historical origin*

---

[25] See Sellars, *Kant and Pre-Kantian Themes*; Sellars, *Kant's Transcendental Metaphysics*; and Sellars, *Science and Metaphysics: Variations on Kantian Themes*.

[26] See Sellars, "Philosophy and the Scientific Image of Man."

[27] Evans and McDowell were both students and then later younger colleagues of Strawson at Oxford in the 1960s and 1970s.

of **both** *sides of the debate between conceptualists and non-conceptualists.* But if Kant's theory of intuition covertly sponsors both conceptualism and non-conceptualism, then revisiting his theory of intuition can surely teach us something new and important about the issue of non-conceptual content.

This last claim is closely connected with three salient problems in the contemporary debate about non-conceptual content: (1) the lack of a suitably fine-grained classification of different types of non-conceptual content (*the classification problem*); (2) the lack of a corresponding account of the nature of concepts (*the concept problem*); and (3) the worry that there may in fact be no unitary phenomenon of non-conceptual content to be explained (*the unity problem*).[28] It seems to me, however, that at least the first two of these basic problems can be, if not actually solved, then at least pre-emptively mitigated, so that I can concentrate for the rest of this chapter on the third problem.

Here then is a pre-emptive response to the classification problem: for me, non-conceptual content is cognitive content that: (i) absolutely lacks concepts either globally or locally[29] (*very strongly* non-conceptual content); or (ii) does not require the *correct application* of concepts even if it requires concepts (*fairly strongly* non-conceptual content); or (iii) does not require concepts even if it happens to include concepts that correctly apply (*moderately* non-conceptual content); or else (iv) requires both concepts and also their correct application but does not require the possession or self-conscious rational grasp of those concepts by the user of those concepts (*weakly* non-conceptual content).[30] The primary rationale here is classifying by inverse proportionality to the degree of involvement of conceptual capacities in cognition: the less they are involved, the greater the degree of non-conceptuality.

And here is a pre-emptive response to the concept problem. For me, concepts are (1) abstract structured semantic items with cross-possible-worlds

---

[28] See Bermúdez, "Nonconceptual Mental Content," p. 7; and Speaks, "Is There a Problem about Nonconceptual Content?"

[29] When I say that a certain cognitive content lacks concepts "globally," I mean that the relevant cognizer (owing to temporary or permanent cognitive disruption or selective breakdown, e.g., agnosias) either lacks conceptual capacities altogether or else has no *online* conceptual capacities. By contrast, when I say that a certain cognitive content lacks concepts "locally," I mean that the relevant cognizer (again, owing to temporary or permanent cognitive disruption or selective breakdown) either lacks a specific conceptual capacity altogether or else lacks a specific online conceptual capacity, *in relation to that content*, even though it otherwise possesses some conceptual capacities, some of which are online.

[30] In a similar way, Speaks distinguishes between *absolute* non-conceptual content and *relative* non-conceptual content; see "Is There a Problem about Non-Conceptual Content?", section 1.

extensions (fine-grained intensional entities), and *also* (2) psychological items in the triple sense that they are: (2a) tokened in some particular conscious mental states; (2b) express subjective modes of presentation in affect or emotion, perception, judgment, thought, and intentional action; and (2c) entail the existence of psychological capacities for generating, possessing, and applying concepts. That is, for me concepts are *intensionally-structured mental representation types*. Furthermore, I think that this conception of concepts is perfectly consistent with Kant's theory of concepts.[31] More generally, concepts for me and for Kant are at once the basic objects of conceptual analysis, psychological rules for classifying and identifying perceptual objects, and the basic elements of cognitive rationality. It also seems to me, as I think it would seem to Kant, that concepts will satisfy Evans's "generality constraint": the subsumption of an object under a concept implies a dual pair of cognitive capacities for applying that same concept to distinct objects and for applying different concepts to the same object.[32] The primary rationale for this overall approach to concepts is smoothly reconciling the semantics of concepts and the psychology of concepts. It should also be noted, however, that because I am postulating a fairly high-powered and overtly Kantian conception of concepts, it will tend to rule out by fiat those overly concessive forms of conceptualism that illegitimately respond to non-conceptualist arguments by merely "downsizing" their notion of a concept in order to accommodate various sorts of good evidence in favor of the existence and representational significance of non-conceptual content.

I hasten also to re-emphasize that my responses to the classification and concept problems are merely pre-emptive and not in any way decisive. Each would no doubt require its own book to justify it adequately. My claim is only that each of the responses is prima facie *somewhat plausible*. But, if that is so, then it reduces our philosophical task load significantly and leaves us with the unity problem as *the* central problem about non-conceptual content. Simply put, the worry about unity is that the phenomenon of non-conceptual content is nothing but a jumbled collection of apparently similar but ultimately heterogeneous cognitive facts, without a single underlying structure or nature. But then conceptualism would turn out to be true about cognitive content by default—just by virtue of being the only game in town.

---

[31] See Hanna, *Kant and the Foundations of Analytic Philosophy*, chs. 1 and 3.
[32] See Evans, *Varieties of Reference*, pp. 75 and 100–5.

By way of offering a Kantian response to the unity problem, my line of argument will be as follows. In section 2.2, I will show that Kant gives various arguments for the existence and representational significance of very strongly, fairly strongly, moderately, and weakly non-conceptual content in inner sense and outer sense (*innere Sinn, äussere Sinn*), feeling or affect (*Gefühl*), imagination (*Einbildungskraft*), sense perception (*Wahrnehmung*), and judging (*Urteilen*). Even more importantly, however, as I will show in section 2.3, in the Transcendental Aesthetic he traces back the very possibility of non-conceptual content to our representations of space and time, which in turn are necessary and non-empirical or a priori conditions of every mental representation generated by means of human sensibility. As we saw in chapter 1, these are what Kant calls "the forms of intuition."

Kant also famously claims in the Transcendental Aesthetic that we can have direct non-conceptual representations of the forms of intuition as unique non-empirical objects, and he calls these representations "pure intuitions" and sometimes also "formal intuitions." Pure or formal intuition in turn is taken by Kant to be the semantic and epistemic foundation of mathematics, or more precisely, of arithmetic and geometry. In chapter 6, I will try to show how Kant's strange-sounding claim that the pure intuition of time is the semantic and epistemic foundation of arithmetic can make sense and even be defensible. But what I want to highlight in this chapter are *the forms of intuition*, not *formal intuitions*. Even so, the subtle distinction between forms of intuition and formal intuitions is directly relevant to Kant's theory of non-conceptual content, and will be briefly discussed in section 2.3.

Nevertheless, the crucial point is that if Kant is right, then there are two and only two forms of intuition, our a priori representations of space and time; and these representations of space and time are not only presupposed by all non-conceptual content but also account for the existence, cognitive significance ("objective validity"), and psychological coherence ("subjective validity") of every type of non-conceptual content. In this way, the forms of intuition provide a *fundamental explanation* of non-conceptual content. Moreover, I also believe that Kant's fundamental explanation can be directly transferred to the seminal debate about non-conceptual content, and in fact significantly advance it (section 2.4). Finally, and perhaps most importantly for the purposes of this book, assuming the essential connection between non-conceptual content and the nature of sense perception, then

Kant's fundamental explanation of non-conceptual content in terms of the forms of intuition can also be used to provide a deeper understanding of his direct perceptual realism (section 2.5).

2.2. Kant's Arguments for Non-Conceptualism

According to Kant, the central fact about the mind is its capacity to represent (*vorstellen*), which is to say that the mind has something "to put before" (*stellen ... vor*) it, and this something is a mental "representation" (*Vorstellung*) (*CPR* A320/B376–7). Representations, in turn, can be either conscious or non-conscious (*CPR* A78/B103).[33] The primary cognitive role of consciousness (*Bewußtsein*) is to contribute subjective integrity, or a well-focused egocentric organization, to a representation (*CPR* B139).

In turn, every conscious representation has both (i) a "form" (*Form*) and (ii) a "matter" (*Materie*) or "content" (*Inhalt*) (*CPR* A6/B9) (*JL* 9: 33). *Materie* is qualitative sensory content (more on this in the next paragraph). *Inhalt* by contrast is intensional content: what Kant calls a conscious representation's "sense" or *Sinn* and also its "meaning" or *Bedeutung* (*CPR* A239–40/B298–9). The form of a conscious representation in Kant's sense is somewhat similar to what Descartes called the "formal reality" of an idea, and the intensional content of a conscious representation in Kant's sense is somewhat similar to what Descartes called the "objective reality" of an idea. More precisely, however, for Kant the form of a conscious representation is what for lack of a better name I will call its *representational character*, by analogy with the "phenomenal character" of phenomenal consciousness. Representational character includes: (a) the difference between clarity and unclarity, and between distinctness and indistinctness; (b) different subjective attitudes of all sorts, or what Locke called "postures of the mind," including

---

[33] Kant says that the synthesis (i.e., the mental processing) of the imagination is a "blind though indispensable function of the soul ... of which we are rarely even conscious (*selten nur einmal bewußt*)," in that the mental operations that are applied to inputs typically occur without conscious implementation (*CPR* A78/B103). Still, these mental operations can often also be consciously implemented; and, even when non-conscious, they can to some extent be indirectly consciously recovered by acts of higher-order "reflection" (*Reflexion, Überlegung*) on our faculties (*CPR* B2, A260–263/B316–319). So the difference between conscious and non-conscious cognition for Kant is always only a matter of degree, not of kind. This is sharply different from most contemporary conceptions of the non-conscious mind, which have been heavily influenced by computational theory or by Freud. See, e.g., Jackendoff, "Unconscious Information in Language and Psychodynamics." But for a contemporary view fairly similar to Kant's on this particular issue, see Searle, *Rediscovery of the Mind*, ch. 7.

but not restricted to propositional attitudes; and (c) our direct conscious awareness of and ability to distinguish between and generalize over types of mental acts or mental operations of all different sorts (for example, analysis, synthesis, memory, imagination, thought, judgment, etc.), which Kant calls "reflection" (*Überlegung*) (*CPR* A260/B316) and which is somewhat similar to Locke's "ideas of reflection." By contrast to representational character, the intensional content of a conscious representation is *what it is about*, or its *topic*: more precisely, it is *a package of information about something, an X*. The intensional content of a conscious representation can be held fixed while varying its representational character (say, from unclearly seeing *A* to seeing *A* clearly; or from asserting that *P* to doubting that *P* to denying that *P*); and the representational character can also be held fixed while varying its intensional content (say, from being a memory of *A* to being a memory of *B*). But an individual representation is uniquely determined by its intensional content and not by its representational character.

Conscious representations can be either subjective or objective, but in either case they are necessarily accompanied by "sensations" (*Empfindungen*). The "matter" or qualitative phenomenal content of sensations—what we would now call "qualia"—are intrinsic non-relational phenomenal properties of conscious representations. More precisely, sensation is "the effect of an object on the capacity for representation, insofar as we are affected by it" (*CPR* A19–20/B34), or, in other words, a sensory content is nothing but how the subject directly responds to either endogenously caused changes or exogenously caused changes in its own state. Endogenously caused sensations are "subjective sensations," or feelings (*CPJ* 5: 206); and exogenously-caused sensations are "objective sensations," such as the sensations that accompany the perception of external objects (*CPJ* 5: 206). Whether subjective or objective, however, for Kant sensations are always *cognitively transparent* features of the mental states in which conscious representations occur,[34] in the sense that they interpose neither an intensional content nor an intentional object[35] between the conscious subject and its representations: "sensation in itself is

---

[34] For a very different approach to Kant's theory of sensations, which takes them to be direct cognitive relations to sense data (and, in effect, confuses them with intuitions), see, e.g., Falkenstein, "Kant's Account of Sensation"; and Van Cleve, *Problems from Kant*. This interpretation is characteristic of phenomenalist, or C. I. Lewis-inspired, approaches to Kant's transcendental idealism.

[35] Intentional objects are part-and-parcel of the Brentano-Husserl-Meinong theory of intentionality: they are the direct objects of every act of intentionality; they are uniquely determined by the representational contents of intentionality (i.e., they are whatever we take them to be); and they have independent ontic status ("ideality" or "subsistence") even if those representational contents are not

not an objective representation" (*CPR* B208). This is because sensations refer *only* to the conscious subject's direct response to changes in its inner or outer world: "a perception (*Perception*) that refers to the subject as a modification of its state is a **sensation** (*sensatio*)" (*CPR* A320/B376). To borrow a relevant formulation from the later Wittgenstein,[36] for Kant a sensation is not a *something*—it is neither an intensional content nor an intentional object—but not a *nothing* either. A sensation is simply a *direct response* of the cognitive subject to outer or inner stimuli, hence a genuine "lived experience" or *Erlebnis*.

Because sensations are cognitively transparent *Erlebnisse*, they must be distinguished from both subjective conscious representations and objective conscious representations alike. Subjective conscious representations are conscious awarenesses of "mere appearances" (*bloße Erscheinungen*) (*CPR* A46/B63), or the flotsam and jetsam of representational life, such as the phenomenal mental images (*Bilder*) that are constantly being generated in the course of conscious psychological processes by the empirical imagination (*CPR* A141/B181), but might not have any coherence or representational significance. So, in other words, a subjective conscious representation is a loosely organized and relatively unstructured conscious state, the mere result of what Hume called "the association of ideas," and what Kant in the A edition of the first *Critique* calls the "empirical synthesis of reproduction" (*CPR* A101). By sharp contrast, however, an objective conscious representation, or cognition (*Erkenntnis*), is always either outwardly directed to some object or another and thereby has "intentionality" in the

---

satisfied by any actually existing objects. The early Moore-early Russell notion of a "sense datum" is essentially the same as that of an intentional object of sense perception. Some Kant interpreters have suggested that for Kant every cognition includes an intentional object: see, e.g., Aquila, *Representational Mind*. From a strictly philosophical point of view, the main problem with the intentional object theory is that it is ontologically highly extravagant: not only is an extra intentional object generated for every actually existing object that is correctly represented (the reduplication problem), but also for every distinct mode of perceptual presentation of a real world object—e.g., something's looking like a cylinder from sideways on, but also looking like a circle from directly above—there is another intentional object that is generated as well (the fine-grainedness problem). From a textual point of view, the intentional object theory also does not square with Kant's remarks about sensation, which, as I have just shown, imply cognitive transparency. Nor does it square with Kant's remarks about the "transcendental object = X," which are best interpreted in terms of the notion of a generic object of mental representation—a notion that is *internal* to representational content; see Hanna, *Kant and the Foundations of Analytic Philosophy*, pp. 107–8, 110–12.

[36] See Wittgenstein, *Philosophical Investigations*, p. 102e, §304.

Brentano-Husserl-Meinong sense (aboutness, object-directedness),[37] or else it is self-directed and reflexive. Self-directed and reflexive cognition for Kant is meta-cognition, or a cognizer's *objective* conscious representation of itself as the *subject* of conscious representation. Now cognitions—conscious mental states with intentionality or reflexivity—are of two distinct kinds: (1) intuitions, and (2) concepts (*CPR* A320/B376–7). So far, so good. But here is where things get fairly tricky.

That is because Kant defines intuitions and concepts in such a way that they are logically independent of one another, yet he also explicitly asserts that they are cognitively complementary and semantically interdependent. And this brings us to perhaps the most famous and widely quoted—but I think also the most generally misunderstood—text in Kant's writings, which expresses what I will call *the togetherness* (of intuitions and concepts) *principle*:

Intuition and concepts...constitute the elements of all our cognition, so that neither concepts without intuition corresponding to them in some way nor intuition without concepts can yield a cognition.

Thoughts without [intensional] content (*Inhalt*) are empty (*leer*), intuitions without concepts are blind (*blind*). It is, therefore, just as necessary to make the mind's concepts sensible—that is, to add an object to them in intuition—as to make our intuitions understandable—that is, to bring them under concepts. These two powers, or capacities, cannot exchange their functions. The understanding can intuit nothing, the senses can think nothing. Only from their unification can cognition arise. (*CPR* A50–1/B74–6)

What does the togetherness principle mean, and how does Kant argue for it? Well, "thoughts" for Kant are mental acts that essentially involve concepts. Although a concept can be entertained on its own in a "mere" thought, the only "use" (*Gebrauch*) or application of a concept is to judge by means of it (*CPR* A68/B93); hence every application of a concept involves a

---

[37] Unlike the Brentano-Husserl-Meinong tradition, however, Kant does *not* include intentional objects in his theory of objective mental representation. More precisely, Kant avoids the reduplication and fine-grainedness problems by allowing for two sharply distinct kinds of cognitions (concepts and intuitions), and, correspondingly, two sharply distinct kinds of representational content (conceptual and non-conceptual/intuitional). All and only directly referential intuitions pick out individual actual objects; and all and only indirectly referential concepts capture descriptive modes of presentation of individual actual objects.

corresponding judgment. Judgments are higher-order self-consciously unified complex representations (*CPR* A69/B94, B140–2) that are systematically composed of empirical concepts, intuitions (as singular terms in "atomic" singular categorical judgments, out of which all other judgments are logically constructed[38]), and logical forms—the latter of which Kant calls "functions of unity in judgments" or "pure concepts of the understanding." In the Metaphysical Deduction, he stresses that the pure concepts of the understanding are also necessarily applied to the semantic contents of the intuitions that occur in judgments:

> The same function that gives unity to the different representations *in a judgment* also gives unity to the mere synthesis of different representations *in an intuition*, which, expressed generally, is called the pure concept of the understanding. (*CPR* A79/B104–5)

The semantic content of a judgment is a "proposition" (*Satz*), and a proposition takes a truth-value if and only if it has "objective validity" (*objektive Gültigkeit*) (*CPR* A58/B83, B142, A155–6/B194–5), that is, cognitive significance or anthropocentric empirical meaningfulness. The subjective validity of a representation, by contrast, is its mere psychological coherence under the laws of association (*CPR* B142). We have already seen how cognitions in general are objective conscious representations, and that both concepts and intuitions are cognitions. In the B edition of the first *Critique* however Kant also highlights a much narrower notion of "cognition" that means *objectively valid judgment* (*CPR* Bxxvi n., B146), and this is in fact how he is using it in the famous texts at A50–1/B74–6. This narrow conception of cognition as objectively valid judgment, in turn, plays a fundamental role in the B edition version of the Transcendental Deduction of the Pure Concepts of the Understanding (*CPR* B129–69). Kant also says at A111 that "intuition without thought [is] never cognition, and would therefore be as nothing to us," and there are similar remarks at A112 and A120. And, finally, the togetherness principle is also explicitly supported by at least one other text:

> The understanding cognizes everything only through concepts; consequently, however far it goes in its divisions [of lower concepts] it never cognizes through mere intuition but always yet again through lower concepts. (*CPR* A656/B684).

---

[38] See Hanna, *Kant and the Foundations of Analytic Philosophy*, section 1.4; and Hanna, "Kant's Theory of Judgment," section 2.

These texts have led many readers and interpreters of Kant—and, in particular, Sellars and McDowell—to deny the cognitive and semantic independence of intuitions and concepts. I fully accept the truth of the togetherness principle as Kant states it, and I also accept his arguments in support of it. But I think that the Sellars–McDowell *interpretation* of the togetherness principle, despite its being widely held, is wrong. It is wrong not only because it does not conform to what Kant actually says, but also because it pays insufficient attention to the fine-grained details of Kant's cognitive semantics. As I have argued in detail elsewhere,[39] and will not repeat here, what Kant is actually saying in the famous text at A50–1/B74–6 is that intuitions and concepts are indeed cognitively complementary and semantically interdependent, *but only for the specific purpose of constituting objectively valid judgments*. From this it does not follow that there cannot be "empty" concepts or "blind" intuitions *outside the special context of empirically meaningful judgments*. Therefore "empty concept" for Kant does not mean either "bogus concept" or "meaningless concept": rather it means "concept that is not objectively valid," and there can be very different sorts of concepts that are not objectively valid. Some concepts that are not objectively valid are indeed bogus or meaningless (or at least necessarily uninstantiated) in the sense of being either nonsensical or conceptually absurd, for example, the concepts of a furiously sleeping colorless green idea or of a round square. But according to Kant there can also be concepts that are not objectively valid yet still fully intelligible, for example, concepts of things-in-themselves or positive noumena (*CPR* B148–9, A238/B293, B307). Similarly, "blind intuition" for Kant does not mean either "bogus intuition" or "meaningless intuition": rather it means *objectively valid non-conceptual intuition*. So Kant's term-of-art "blind intuition" no more implies the denial of intuitional cognition than our contemporary psychological term-of-art "blindsight" implies the denial of visual cognition: "blindsight" is veridical visual cognition without visual qualia, and "blind intuition" is veridical intuitional cognition without concepts.[40]

In this way, my first basic point about intuitions and concepts for Kant is that despite its being true, according to the togetherness principle, that they must be combined with one another in order to generate objectively valid judgments, nevertheless intuitions can also occur independently of concepts

---

[39] See Hanna, *Kant and the Foundations of Analytic Philosophy*, pp. 46–65, and 202–3.
[40] See Weiskrantz, *Blindsight*.

and remain objectively valid. And, in particular, to the extent that intuitions are cognitively and semantically independent of concepts, they are *non-conceptual cognitive contents*. So the togetherness principle is perfectly consistent with Kant's non-conceptualism. Now, as my second basic point about intuitions and concepts, I need to say more about the nature of an intuition.

Intuitions for Kant are objective cognitions that are: (i) immediate; (ii) sense related; (iii) singular; (iv) object-dependent; and (v) prior to thought. As before, I have argued in detail for this interpretation elsewhere,[41] so will also not repeat that argumentation here. The two most important things for our present purposes are that these five features are individually necessary and jointly sufficient for any objective cognition's being an intuition, and that the fifth feature is the same as the non-conceptuality of an intuition. For completeness's sake I will briefly gloss the first four features, and then zero in on the fifth feature.

(i) *Immediacy*. Kant says that an intuition "refers immediately (*bezieht sich unmittelbar*) to the object" (*CPR* A320/B377) and again more explicitly that

in whatever mode and by whatever means a cognition may refer to objects, intuition is that through which it immediately refers to them, and to which all thought is mediately directed. (*CPR* A19/B33)

I take this to be the same as the *referential directness* of an intuition, in the strong sense[42] that it picks out objects without necessarily being mediated by any sort of descriptive content (whether propositional or conceptual) or by any other sort of representational faculty, representational content, psychological intermediary, or physical intermediary. In other words: an intuition refers to its object even if there is no corresponding propositional or conceptual description of that object; an intuition refers to its object even if there *is* a corresponding description of that object but it is false of that object, or vague; an intuition refers to its object even if no other cognitive faculty apart from sensibility is involved; an intuition refers to

---

[41] Hanna, *Kant and the Foundations of Analytic Philosophy*, ch. 4.

[42] The criterion of referential directness that I am using here is somewhat stronger than the one I used in *Kant and the Foundations of Analytic Philosophy*, pp. 196–7. My rationale is that whereas the earlier formulation was designed to capture the Kantian notion of immediacy that is relevant to direct *linguistic* reference, the current formulation is designed to capture the Kantian notion of immediacy that is relevant to direct *perceptual* reference.

its object without requiring any psychological intermediary other than intuition itself; and an intuition refers to its object without requiring any physical intermediary other than what is already intrinsically involved in intuition itself—the living and (in principle) motile body of the intuiting subject (see section 1.3).

(ii) *Sense relatedness*. Kant says that "it comes along with our nature that **intuition** can never be other than **sensible**, i.e., that it contains only the way in which we are affected by objects" (*CPR* A51/B75), and again even more explicitly that

[intuition] ... takes place only in so far as the object is given to us; but this in turn, *for humans at least*, is possible only if it affects the mind in a certain way. The capacity (receptivity) to acquire representations through the way in which we are affected by objects is called *sensibility*. Objects are therefore given (*gegeben*) to us by means of sensibility, and it alone affords us intuitions. (*CPR* A19/B33)

In this way, while for Kant it is in principle logically possible for a minded being (in particular, a divine being with "intellectual intuition" (*CPR* B72)) to have intuitions that are not based on the givenness of objects and do not involve natural dynamical processes that externally trigger or affect our sensibility, nevertheless necessarily all creatures minded like us (that is, conscious human animals) have a specifically *sensible* kind of intuition.

(iii) *Singularity*. Kant says that intuition "refers immediately to the object and is singular (*einzeln*)" (*CPR* A320/B377), that "an intuition is a singular representation" (*JL* 9: 91), and that

since individual things, or individuals, are thoroughly determinate, there can be thoroughly determinate cognitions only as *intuitions*, but not as *concepts*. (*JL* 9: 99)

The singularity of intuition must not be confused with the definiteness or particularity of a definite description, because a concept, no matter how specific, can never necessarily guarantee reference to a fully determinate or concrete material individual in space and time: "a [material] thing can never be represented **through mere concepts**" (*CPR* A284/B340). Even a concept that is satisfied by one and only one thing in the actual world might have a counterpart in another possible world that shares all its intrinsic non-relational properties but is not identical with the original object, namely, *this very object right* *here and now*. This Kantian idea is sharply anti-Leibnizian. On a Leibnizian theory, object-identity is determined entirely by intrinsic non-relational properties, which in turn are picked out exclusively by concepts. But, for Kant, spatiotemporal properties are *intrinsic structural properties* of all appearances, and in particular of all real material objects; furthermore, for him, intrinsic structural spatiotemporal and causal-dynamic properties entirely determine the natures of real material objects (*CPR* A281–6/B337–42);[43] and last but not least only an intuition can representationally capture this *essentially indexical*, or irreducibly actual-world-bound and spatially or temporally context-dependent, sort of object-identity.

(iv) *Object-dependence*. Kant says that "our mode of intuition is dependent on the existence (*Dasein*) of the object" (*CPR* B72) and that "an intuition is such a representation as would immediately depend upon the presence (*Gegenwart*) of the object" (*P* 4: 281). This is the veridicality of an intuition. In other words, intuition is essentially a *relational* form of cognition, in that the existence of the object of intuition is a necessary condition of both the objective validity or cognitive significance of the intuition and also the existence of the intuition itself. If the putative object of an intuition fails to exist, then it is not only not an objectively valid intuition, it is not even authentically an intuition (*P* 4: 282) but rather only an output of our faculty of imagination (*CPR* B278). By contrast, a concept can still both exist and be objectively valid even if it is not satisfied by anything in the actual world, so long as it can be satisfied by something in some other possible world (*CPR* A239/B298).

(v) *Priority-to-thought*. Kant says that "that representation that can be given prior to all thinking is called *intuition*" (*CPR* B132), and all thoughts essentially involve concepts, so intuitions can be given prior to all concepts. Furthermore it is clear that this priority of intuition to thought is both cognitive and semantic. Thus an act of intuition can occur without any corresponding act of conceptualization, and also an intuition can be objectively valid independently of any concept:

---

[43] More specifically, for Kant all real material objects have both intrinsic relational spatiotemporal properties (shape, position, handedness, simultaneity, duration, temporal asymmetry, etc.) and also intrinsic relational causal-dynamical properties (nomologically sufficient connections to their actual and possible causes and effects, and to the surrounding simultaneous world). The thesis that all real material objects have intrinsic structural properties, in turn, is a fundamental feature of Kant's manifest realism. See the Introduction, n. 37, and chs. 3–4.

Objects can indeed appear to us without necessarily having to be related to functions of the understanding. (*CPR* A89/B122)

Appearances can certainly be given in intuition without functions of the understanding. (*CPR* A90/B122)

Appearances could after all be so constituted that the understanding would not find them in accord with the conditions of its unity.... [and] in the succession of appearances nothing would offer itself that would furnish a rule of synthesis and thus correspond to the concept of cause and effect, so that this concept would be entirely empty, nugatory, and without significance. <u>Appearances would nonetheless offer objects to our intuition, for intuition by no means requires the functions of thinking.</u> (*CPR* A90–1/B122–3, underlining added)

The manifold for intuition must already be given prior to the synthesis of the understanding and independently from it. (*CPR* B145)

In other words, the priority-to-thought of an intuition is its *non-conceptuality*. Since on my view there are four different basic types of non-conceptuality, it is also crucial to see that correspondingly an intuition can be non-conceptual in at least four different ways: (1) it is possible for a cognitive subject to intuit an object while the content of that intuition absolutely lacks concepts either globally or locally (= *very strong non-conceptuality*); (2) it is possible to intuit an object even if concepts are required but they are false of that object or under-discriminate that object (= *fairly strong non-conceptuality*); (3) it is possible to intuit an object even if there is a corresponding concept that happens to apply correctly to that object, but this very same intuition could have occurred even without that concept or even if the concept had been false of that object or had under-discriminated that object (= *moderate non-conceptuality*); and (4) it is possible to intuit an object even if there is a corresponding concept that is required for that cognition and whose correct application is also required but that concept is not self-consciously and rationally possessed by the user of that concept (= *weak non-conceptuality*).

Since intuitional cognitive content in Kant's sense and non-conceptual cognitive content are identical, I want to show now that Kant offers defensible proofs for the existence and representational significance of very strongly, fairly strongly, moderately, and weakly non-conceptual content, in inner sense and outer sense, feeling or affect, imagination, sense perception, and empirical judging. In each case, his proofs for non-conceptuality are broadly speaking *phenomenological* or ostensively psychological: he pumps our philosophical insight by appealing to introspectively or intersubjectively given self-evident facts about conscious cognitions.

(1) *Very strong non-conceptuality.* As we have seen already in chapter 1, inner sense is a temporally successive stream of phenomenal mental contents or states in time, by means of which a conscious subject directly intuits herself:

Inner sense [is that] by means of which the subject intuits itself, or its inner state. (*CPR* A22/B37)

(The inner sense) Consciousness is the intuition of its self. (*R* 5049; 18: 72)

Everything that belongs to the inner determinations is represented in relations of time. (*CPR* A23/B37).

Through inner sense, the subject is intuitively directly aware of herself as phenomenal or apparent, and never as noumenal: "inner sense ... gives ... no intuition of the soul itself as an object" (*CPR* A22/B37); and "inner sense ... presents even ourselves to consciousness only as we appear to ourselves, not as we are in ourselves" (*CPR* B152-3). Moreover inner sense contains a "subjective unity of consciousness, which is a determination of inner sense, through which [the] manifold of intuition is empirically given" (*CPR* B139). So inner sense is what we would now call *phenomenal consciousness*.

As we also saw in chapter 1, in rational human animals like us, inner sense is always accompanied by a cognitive capacity for "apperception" or self-consciousness (*Selbstbewußtsein*), which is an innate ability for forming self-directed judgments and thereby imposing a higher-order unity on all the cognitive faculties and their representational outputs. This capacity for self-consciousness—which Kant calls "transcendental apperception," and which (we know from the Refutation of Idealism) generates, under real-world psychological conditions, empirical apperceptions of myself as a single living, (in principle) motile, embodied, self-orientating empirical person—is necessary for the representation of determinate states-of-affairs, that is, individual material substances in space and time together with all their monadic and relational properties. Transcendental apperception also constitutes an "objective unity" of consciousness (*CPR* B136-B43), by virtue of its introducing conceptual and propositional logical form into the structure of every representation that is accessible to self-consciousness. In this way, the capacity for self-consciousness necessarily implies conceptual abilities, and this necessary connection is captured by the characteristic self-directed discursive representation "I think" (*CPR* B131–2). Conversely all conceptual abilities have the capacity for self-consciousness as a necessary condition (*CPR* B133–4 n.). But at the same time Kant holds that it is possible for non-rational or proto-rational animals—and in particular human infants and most non-human animals—to have inner sense without apperception (*A*7: 127–8) (*PC* 11: 52); hence consciousness without self-consciousness. Indeed, Kant explicitly notes that his contrast between inner sense and apperception sets his philosophical psychology sharply apart from earlier systems: "it is customary in the systems of psychology to treat **inner sense** as the same as the faculty of **apperception** (which we carefully distinguish)" (*CPR* B153). Therefore it is possible for non-rational or proto-rational animals to have both intuitions in inner sense and also consciousness without any concepts, or indeed without any (online) conceptual capacities at all, which directly implies the very strong non-conceptuality of intuition in inner sense and of consciousness alike.

We already know that by contrast to inner sense, outer sense is a cognitive capacity for representing objects outside the living, (in principle) motile, embodied, self-orientating empirical subject in space: "by means of outer sense ... we represent to ourselves objects as outside us, and all as in space" (*CPR* A22/B37). Now Kant holds that it is possible for non-human animals—say, an ox—to have outer sense intuitions of material objects in space—say, a barnyard gate—without any corresponding concepts and indeed without any conceptual capacities whatsoever (*FS* 2: 59). The ox sees the gate, but cannot see the gate *as* a gate. To borrow another formulation from Wittgenstein, the ox has universal "aspect blindness"[44] built right into its cognitive constitution. So, like the case of inner sense without apperception, the case of non-human animal perception implies what I call the *global* very strong non-conceptuality of intuition, since self-representing conscious states and external perceptual cognitions can both occur without any (online) conceptual capacities.

Perhaps the most interesting Kantian example of the very strong non-conceptuality of intuition, however, is this one:

---

[44] Wittgenstein, *Philosophical Investigations*, p. 213e.

If a savage (*Wilder*) sees a house from a distance for example, with whose use he is not acquainted, he admittedly has before him in his representation the very same object as someone else who is acquainted with it determinately as a dwelling established for humans. But as to form, this cognition of one and the same object is different in the two. With one it is *mere intuition*, with the other it is *intuition* and concept at the same time. (*JL* 9: 33)

Leaving aside minor worries about Enlightenment cultural condescension (which would hardly be unique to Kant in any case), here the so-called "savage" is clearly a rational animal, and would therefore obviously have concepts for describing and recognizing relatively large material objects in space, and also have concepts for describing and recognizing dwellings of some sort. Thus Kant's point is not that he lacks all (online) conceptual capacities whatsoever: he merely lacks a specific (online) capacity for conceptualizing *houses*. Unlike the ox, he is only *accidentally and partially* but not *constitutionally and wholly* aspect-blind. This is therefore a case of *local* very strong non-conceptuality, which seems virtually ubiquitous in our own everyday experience: I can see the particle-accelerator over there perfectly well without seeing it *as* a particle-accelerator and indeed without having any specific (online) capacity for conceptualizing *particle-accelerators*. In fact, the world around me is just chock full of objects I can clearly and distinctly perceive without also being able to recognize or understand. While awake, I try to keep my eyes and ears open, and I manage to get around in the world most of the time without too much hassle. But basically I am just an ignorant slob. So, insofar as I am constantly perceiving things that I do not know how to conceptualize, I am no doubt a rational "savage" many times daily.

(2) *Fairly strong non-conceptuality.*   Kant's key argument for the fairly strong non-conceptuality of intuition is from the existence of what I will call "veridical illusions," quite familiar to us now from the cognitive science literature on modularity and in particular from the evidence (for example, from the persistence of the Müller-Lyer illusion) for what Jerry Fodor aptly dubbed "encapsulation,"[45] or the resistance of a given peripheral information-processing capacity (for example, vision) to penetration by the "central" processes of conceptualizing, judgment, and inference:

---

[45] See Fodor, *The Modularity of Mind*.

The astronomer can[not] prevent the rising moon from appearing larger to him, even when he is not deceived by this illusion. (*CPR* A297/B354)

The purest form of this example requires a naïve perceiver, that is, yet another rational "savage," this one not informed about astronomy, who falsely judges that the rising moon is bigger than the ordinary moon. In cases of this sort we can see that "truth and illusion are not in the object, insofar as it is intuited, but in the judgment about it insofar as it is thought" (*CPR* A293/B350). In other words, the intuition in outer sense is *veridically* illusory[46] in the sense that our pre-theoretical, untutored, or "uncivilized" capacity for perception reliably presents an actually existing object (the moon) just as it would seem to any other untutored creature equipped with our cognitive faculties under those contextual conditions—i.e., as seeming bigger near the horizon than when it is higher in the sky—so the error or illusion lies in the corresponding concept and not in outer sense. This again directly implies the fairly strong non-conceptuality of outer sense perception, especially when taken along with the empirical fact that such illusions perceptually persist *even after the acquisition of conceptual sophistication about them.*

Perhaps most importantly, however, such cases also imply the *evidential force* of some non-conceptual contents, since for Kant false perceptual judgments based on veridical illusions are *non-inferentially warranted* even though false (*JL* 9: 38, 71). If Kant is right, then McDowell is simply wrong when he claims that non-conceptual perceptual content cannot have evidential force because it is outside the Sellarsian "space of reasons," and therefore does not justify belief but instead only "exculpates"—merely causes or triggers—belief. Kant by sharp contrast thinks that *non-conceptual intrinsic phenomenal structure carries its own form of epistemic normativity.* I will develop this crucial epistemological point further in section 2.5.

Beyond veridical illusions, Kant also offers a different proof for the fairly strong non-conceptuality of intuition, from cases of what he calls "indistinct" perception:

We glimpse a country house in the distance. If we are conscious that the intuited object is a house, then we must necessarily have a representation of the various parts of this house, the windows, doors, etc. For if we did not see the parts, we would

---

[46] As opposed to *falsidical* or non-object-dependent, non-intersubjectively-shared illusions (e.g., hallucinations), in which the apparent perceptual object fails to exist, the representation is highly idiosyncratic in character, and the material world is incorrectly represented.

not see the house itself either. But we are not conscious of this representation of the manifold of its parts, and our representation of the object indicated is thus itself an indistinct representation. (*CPJ* 5: 34)

Kant's claim is that we can intuitionally perceptually cognize objects that are under-discriminated by our concepts: I see the country house, but not *as* a country house, rather only as a big undifferentiated blob over there in the distance. Strictly speaking, there are two slightly distinct possible versions of the "country house" example: one in which the cognizer is a rational "sophisticate" who has a conceptual capacity for recognizing country houses, and one in which the cognizer is again a rational "savage" who lacks the specific capacity for conceptualizing houses, whether urban or country—who has "aspect-blindness" for seeing houses as houses. I see it now as a blob, but when I get closer, I see it *as a country house*. The rational "savage" also sees the house now as a blob: but when he gets closer, by contrast, he sees it more simply *as a slightly-bigger-than-mid-sized material object over there*, which in turn implies the existence of a lower-level, but still rational, dimension of the content of his own seeing.

From the perspective of the contemporary debate about non-conceptual content however, the crucial feature of this example is that our ability non-conceptually to perceive a manifold of phenomenal content indistinctly, directly implies that the "richness" or "fine grain" of perceptual content exceeds the reach of our conceptual capacities, since for Kant our "manifold of intuitions"—our ambient perceptual array—will always contain indistinct regions. Peacocke makes the same good point.[47] McDowell replies by constructing a demonstrative conceptual device for capturing fineness of grain, "that shade (of colour)," with suitable variations for different sorts of fine-grained perceptual content, which would allow the conceptualizer either to invent new concepts or to activate previously offline conceptual capacities.[48] Peacocke counter-replies by making another good point: that it is going to be difficult for conceptualists to give an account of concept-learning by this means without going all the way to an implausible radical nativism—like the early Fodor's—about basic perceptual concepts such as COLOR.[49] In the end, however, Heck makes what I take to be the really decisive point, which is that the reference of demonstratives is

---

[47] See n.8.   [48] See McDowell, *Mind and World*, Afterword, part II
[49] See Peacocke, "Does Perception Have a Non-Conceptual Content?"

fixed non-conceptually.⁵⁰ But all of this could have been avoided by just rereading Kant, since it is clear enough from his theory of intuition that all *essentially indexical* cognition whatsoever is *intuitional* cognition.⁵¹

(3) *Moderate non-conceptuality.* Feelings and other affects supply some very interesting examples of moderately non-conceptual content. For Kant, feelings are subjective sensations that necessarily involve either pleasure or pain (although they need not be exhausted by their pleasure/pain component). Pleasure and pain in turn are modes of "the feeling of life" (*CPJ* 5: 204), which is an immediate subjective experience of dynamic natural vitality that expresses our existence as living organisms and embodied minds: "the mind (*Gemüt*) for itself is entirely life (the principle of life itself)" (*CPJ* 5: 278). In aesthetic experience of the beautiful, according to Kant, we get a "disinterested pleasure" that expresses the harmonious and life-enhancing interaction between our various embodied cognitive faculties—and in particular between the understanding and the imagination—as they jointly operate in order to represent the phenomenal form⁵² of the beautiful object (*CPJ* 5: 217–9). On the basis of this disinterested pleasure, we non-inferentially judge that the object—say, this rose—is beautiful. But at the same time, "the judgment of taste...determines the object, independently of concepts, with regard to satisfaction and the predicate of beauty" (*CPJ* 5: 219). In other words, even though the object falls under some concept or another (we not only see the rose but also see it *as* a rose), this conceptual fact is accidental to its being beautiful, since its being beautiful consists merely in the relation between its phenomenal form and the pleasure we experience in the harmonious interplay of our cognitive faculties. So, despite the fact that the judgment of taste includes concepts and as it happens those concepts correctly apply to that object (that is, this is indeed a rose), nevertheless even if the concepts were false of the object (say, I judged that this tulip is beautiful), or even if my concepts

---

⁵⁰ See Heck, "Nonconceptual Content and the 'Space of Reasons'." See also Hanna, "Direct Reference, Direct Perception, and the Cognitive Theory of Demonstratives"; and Hanna, *Kant and the Foundations of Analytic Philosophy*, pp. 213–15. Heck does not fully elaborate his good point, but my idea is that non-conceptual spatiotemporal cognition is what determines the reference of demonstratives. The deeper Kantian transcendental cognitive-semantic explanation for this will be worked out in section 2.3.

⁵¹ See Hanna, *Kant and the Foundations of Analytic Philosophy*, ch. 4.

⁵² In the case of aesthetic objects like roses, their phenomenal form is their spatial shape; in the case of aesthetic objects like sounds, their phenomenal form is their dynamic profile in time.

under-discriminate that object (suppose that I cannot actually tell roses apart from tulips), or even if I lacked the specific concept ROSE, or even if it did not actually exist as such (suppose it is a hallucination), still the aesthetic judgment of taste has a direct object and remains valid. *This* is beautiful in any case: "I do not need [a concept] in order to find beauty in something" (*CPJ* 5: 207). This in turn directly implies the moderate non-conceptuality of feeling.

The role of the imagination in the non-conceptuality of feeling is crucial. The imagination belongs to sensibility (*CPR* B151), and is a cognitive function of intuition: "imagination ... [is a power] of intuiting even when the object is *not* present" (*A*7: 153). Superficially this formulation is inconsistent with Kant's definition of intuition, since intuition includes object-dependence as a necessary condition. So it could perhaps be read as a case of Kant nodding. But, more charitably, I think that the best overall construal of what he is saying is that imagination is essentially intuition *minus* object-dependence, in the sense that imagination can be defined as an immediate, sense-related, singular, and non-conceptual cognitive capacity for representing either existing or non-existing objects. Otherwise put, imagination is *quasi-intuition*.

That there is a cognitive function for sensibly representing objects that do not exist—or at least do not presently exist—is obvious in the case of the synthesis of reproduction (that is, memory) and mental imagery. More generally, however, as I mentioned in section 2.1, the power of imagination for Kant is not merely a capacity for reproducing past sensory representations and generating mental images, but also an all-purpose cognitive engine for representational synthesis or mental processing (*CPR* A78/B103). The operations of this engine in turn have a transcendental ground in the productive imagination (*CPR* B152) (*A* 7: 167). Amongst the characteristic outputs of productive imagination is a special class of representations called "schemata" (*CPR* A137–42/B176–81). Schemata are essentially directed to objects that are spatiotemporally formed, patterned, or shaped, because schematic representations are the direct result of figurative synthesis or *synthesis speciosa* (*CPR* B151). But they are also inherently *sortal* because they can be used to organize sensory images under concepts: "this representation of a general procedure of the imagination for providing a concept with its image is what I call the schema for this concept" (*CPR* A140/179–80). More precisely, and translated out of Kantspeak for a moment, schemata can directly encode both sensory and discursive information in a phenomenal spatiotemporal structural format—Kant's example is a monogram (*CPR* A142/B181), but a better example would be a map—and thus are mental icons, outlines, models, or templates of what they represent.[53] Schemata do not, like concepts, *describe*; rather they *depict*. These fine-grained details of Kant's theory of non-conceptual content are not just exegetical window-dressing: I will return in sections 2.3 and 2.4 to the fundamental role of spatiotemporal representation in non-conceptual content. But for the moment we need only note that insofar as schemata, as functions of the imagination, are also quasi-intuitions and thereby *not* inherently conceptual in nature,[54] it follows that the content of imaginational representation is non-conceptual.

So far, we have seen how Kant is committed to the existence of very strongly, fairly strongly, and moderately non-conceptual content in inner sense, outer sense, feeling or affect, empirical judging (that is, the aesthetic judgment of taste), and imagination. And I have also highlighted the role of spatiotemporal representation in non-conceptual content for later consideration. But there are some further implications of Kant's arguments that we should also briefly note. Since a sense perception of an object is merely a conscious outer intuition of a material object in space (*CPR* B160, B275–9), it follows directly from the very strong and fairly strong non-conceptuality of outer sense that sense perception is also very strongly and fairly strongly non-conceptual. Moreover, consider a perceptual judgment like "This bent stick in the water is three feet long," accompanied by a visual sense perception that provides good prima facie evidence for the truth of that judgment. Suppose, however, that what you are actually looking at is a (relatively) straight snake in a pond filled with gin and it is actually only two feet long. This veridical illusion guarantees the non-conceptuality of that perceptual judgment, so it follows that for Kant perceptual judgments are also fairly strongly non-conceptual. Fairly strong non-conceptuality may be inconsistent with *true* perceptual judgments,[55] but it is perfectly consistent with false ones.

---

[53] See also Blachowicz, "Analogue Representation Beyond Mental Imagery," pp. 78–83; and Johnson-Laird, *Mental Models*, pp. 2, 190, 407, and 415.

[54] Since schemata can encode discursive or conceptual information, there is also a sense in which schemata are quasi-conceptual, and thereby mediate between intuitions and concepts (*CPR* A137–42/B176–81) (R 5661; 18: 320). But, to the extent that schemata are *functions* of intuition and intrinsically intuitional in nature, they are strictly speaking only *compatible* with concepts, and not intrinsically conceptual in nature.

[55] Philosophical intuitions differ here however. For example, the well-known "Donnellan cases" in the theory of reference—i.e., referential uses of definite descriptions such as the use of 'Smith's

If sense perceptions and perceptual judgments alike are fairly strongly non-conceptual, then so are both perception-based desires and volitional intentions. This becomes obvious when we consider that I can want that bent stick in water (for my private collection of extremely interesting bent sticks, of course) and also intend to grab that bent stick in water (so that I can take it home with me). More generally, to the extent that desires and volitional intentions are all based on appearances of the good, that is, on things that *seem* good for me, it is obvious that not only can I be wrong about whether the F that I want or the F that I intend to act upon is in fact good for me, but also more generally I can be wrong about whether *this* or *that* is in fact the F, yet nevertheless non-conceptually want or intend to act upon precisely this or precisely that.

It should now be clear that Kant has solid reasons for holding that very strong, fairly strong, and moderately non-conceptual content are common and indeed pervasive in the mental lives of minded animals, including rational human animals like us. There is, however, one other kind of non-conceptuality noted by Kant that we need to look at very briefly before moving on, because it is importantly different from the other kinds.

(4) *Weak non-conceptuality.* So far, I have concentrated on cases of non-conceptuality in which for one reason or another, a human or non-human minded animal's capacity for sensibility in some way cognitively dominates over its capacity for understanding, even if concepts are required by mental content and even if those concepts happen to be correctly applied. But Kant also points up cases in which there is a cognitive dominance of sensibility over the understanding even though the cognition in question is necessarily conceptual in character and also the relevant concepts must be correctly applied. These are cases in which the sensibility-driven *use* of a concept dominates over the *possession* of that concept, or, more precisely, cases in which a concept is correctly applied by a subject even though the subject cannot self-consciously and rationally in the theoretical sense (that is, logically and analytically) grasp that concept. Kant's compelling examples

murderer' in the statement "Smith's murderer is insane," when said of someone right in front of you who isn't Smith murderer but just happens to be insane—imply that some fairly strongly non-conceptual perceptual judgments are also true. See Donnellan, "Reference and Definite Descriptions"; and Hanna, "Direct Reference, Direct Perception, and the Cognitive Theory of Demonstratives."

in this connection again trade on his notion of indistinct representations. He says:

> The difference between an indistinct and a distinct representation is merely logical and does not concern the content. Without doubt the concept of **right** that is used by a healthy understanding contains the very same things that the most subtle speculation can develop (*entwickeln*) out of it, only in common and practical use (*gemeinen und praktischen Gebrauche*) one is not conscious of these manifold representations in these thoughts. (CPR A43/B61)

> When we compare the thoughts that an author expresses about a subject, in ordinary speech as well as in writings, it is not at all unusual to find that we understand him better than he understood himself, since he may not have determined his concept sufficiently and hence sometimes spoke, or even thought, contrary to his own intention. (CPR A314/B370)

Here, even though a subject engages in the "common and practical use" of a certain concept, nevertheless he does not possess that concept because its specific content is "indistinct" (*undeutlichen*). Conceptual indistinctness—or more precisely, what Kant calls "intellectual indistinctness," because there can also be strictly aesthetic or perceptual indistinctness, as we saw in the country house case above—is a specific psychological predicate or *representational character* of conceptual content, such that the conscious subject of a certain conceptual representation C is unable either to analyze the content of C into its several necessary sub-conceptual constituents, its component intensions, or to give any other sort of account of the logical details of its conceptual microstructure (JL 9: 33–5, 61–4). This entails that the cognizing subject lacks possession of the concept RIGHT, just as he would lack possession of the concept BACHELOR if he were unable to judge that necessarily every bachelor is unmarried and male. Consider, for example, your average five- or six-year-old boy who has minimal mastery of "right" and "bachelor" in English. He is able correctly to pick out some instances of right action (perhaps because they superficially resemble other cases in which his parents gave moral approbation to some action), just as he might be able correctly to identify some bachelors (perhaps by the fact that they superficially resemble some bachelors he has seen on reruns of *Seinfeld*). But he is unable to give even a partial analysis of either the concept RIGHT or the concept BACHELOR. Of course not only *children* correctly use concepts without possessing them, and in the second of the indented texts quoted immediately above Kant specifically notes cases in which philosophers who are fully rational adults—that is, "fully rational" in the sense that they possess an undamaged and online faculty for reason, not in the sense thay they always use this faculty in an ideally successful way—also correctly deploy concepts, yet indistinctly. In other words, the correct use of concepts without their possession happens all the time. Kant's overall point is that representing subjects can fail to possess a concept even though they can correctly apply it under real-world conditions, and thus concept-use-without-concept-possession is weakly non-conceptual.

As Kantian theorists of non-conceptual content, what we need to know more precisely is just what sort of cognitive activity is actually going on in weak non-conceptuality. Unfortunately Kant does not tell us explicitly, except for an intriguing analogy between a subject's conscious awareness of the intensional content of her concepts and our cognition of maps (A9: 64). This chimes in with contemporary work in the cognitive psychology of concepts, in that it strongly suggests that much of our ordinary concept-use has little or nothing to do with conceptual analysis but in fact is largely determined by our ability to match items in the world with "stereotypes" or "prototypes,"[56] which in Kantian terms would be schemata consisting of classificational patterns of linguistic and non-linguistic imagery of perceptually salient and pragmatically important features of objects and situations. Assuming that this is pretty much what Kant has in mind, it implies that for the purposes of "common and practical use" of concepts, the human capacity of outer sense plus the schematizing function of the imagination can cognitively dominate over our self-conscious rational capacities for concept-possession, even in cognition that requires both concepts and their correct application.

At this point, if Kant's "phenomenological" proofs of non-conceptuality are prima facie rationally compelling—as I think they are—then we are in a good position to assert the existence and representational significance of non-conceptual content. But now the further question arises: what *accounts for* non-conceptual content? Kant's important answer is that all non-conceptual content can be explained in terms of our basic cognitive capacities for spatial and temporal representation: *the a priori forms of intuition*.

---

[56] See, e.g., Smith and Medin, *Categories and Concepts*.

Here I do not mean that the *qualitative or sensory content* of non-conceptual experiences will be explained by the forms of intuition, but rather only that the *representational or cognitive content* of non-conceptual experiences will be so explained. Kant's transcendental explanation of non-conceptual content has not been explored by contemporary non-conceptualists, although it certainly seems to be implicit in at least some of their accounts. For example, in my opinion, Peacocke's Evans-inspired theory of "scenario content" is essentially a tacit reworking of the Transcendental Aesthetic in contemporary terminology.[57] And no doubt the missing link between Evans, Peacocke, and Kant is the strong dual influence of Strawson's teaching and his *Bounds of Sense* on his younger Oxford colleagues. In any case, I will now unpack Kant's transcendental theory of non-conceptual content.

2.3. The Forms of Intuition and Non-Conceptual Content

In order to do this, however, we need the notion of a *transcendental deduction*. Since the publication of *Bounds of Sense*, Kant's transcendental deductions have typically been construed as special epistemic arguments, with an eye to defeating skepticism—whether Cartesian evil demon skepticism or external world skepticism, or one of the Humean brands of skepticism.[58] But, if I am correct, a transcendental deduction for Kant is really a *cognitive-semantic* argument of a special kind. In the first *Critique*, in the first section of the Transcendental Deduction of the Pure Concepts of the Understanding, Kant observes that "we have already traced the concepts of space and time to their sources by means of a transcendental deduction, and explained and determined their a priori objective validity" (*CPR* A87/B119–20). And in the *Prolegomena* Kant speaks of a "transcendental deduction of the concepts of space and time" (*P* 4: 285). More precisely, then, I take a Kantian transcendental deduction to be a demonstration of the objective validity—the empirical meaningfulness or cognitive significance—of an a priori representation R (whether that representation is an a priori concept, an a priori intuition, an a priori necessary proposition, or a systematic corpus of a priori

---

[57] See Peacocke, *A Study of Concepts*, pp. 64–98.
[58] See, e.g., Stern, (ed.), *Transcendental Arguments: Problems and Prospects*; and Stern, *Transcendental Arguments and Skepticism*.

necessary propositions), by means of demonstrating that R is the presupposition of some other representation R*, which is assumed for the purposes of the argument to be objectively valid (*CPR* A84–94/B116–27, A156/B195). What follows is a reconstruction of Kant's transcendental deduction of our a priori representations of space and time. In terms of my schema for transcendental deductions, this will mean that the representations of space and time are slotted in for R, and that empirical intuitions of appearances of ourselves in inner sense or of material objects in outer sense are slotted in for R*.

Corresponding to my cognitive-semantic reconstruction of this transcendental deduction of our representations of space and time, is also an explicitly cognitive-semantic reading of Kant's overall *theory* of space and time in the Transcendental Aesthetic, as opposed to either a narrowly metaphysical or a narrowly epistemic reading of that theory.[59] On my cognitive-semantic reading, Kant's theory of space and time is not exclusively an investigation into either "the question of the ontological status of space and time,"[60] or the question of how we obtain justified true beliefs about space and time, but instead is essentially an investigation into the basic semantic features of the a priori "concepts" or representations of space and time. This investigation, to be sure, will have metaphysical and epistemic *implications*. But Kant's turn away from metaphysics narrowly conceived (as in classical Leibnizian-Wolffian Rationalism) or epistemology narrowly conceived (as in classical Lockean-Humean Empiricism) towards cognitive semantics, via his transcendental deduction of the representations of space and time implies, among other things, that his notorious thesis of the transcendental ideality of space and time, saying that space and time are *nothing but* a priori necessary subjective forms of human sensibility (*CPR* A28/44, A36/B52), is *not* in fact a premise or a conclusion of Kant's transcendental deduction of the representations of space and time. Instead, on my reading, transcendental ideality is a logically independent thesis, supposedly proven by the Three Alternatives Argument, according to which:

(1) Space and time are either (a) things-in-themselves (Newton's theory); (b) either intrinsic non-relational properties of things-in-themselves or extrinsic relations between things-in-themselves (Leibniz's theory); or else (c) transcendentally ideal (Kant's theory).

---

[59] For a defense of the cognitive-semantic approach to Kant's transcendental idealism, see Hanna, *Kant and the Foundations of Analytic Philosophy*, chs. 1–2.
[60] Allison, *Kant's Transcendental Idealism*, p. 81.

(2) But space and time are obviously both not-(a) and not-(b).
(3) Therefore space and time are transcendentally ideal (*CPR* A23/B37–8, A39–40/B56–7).

In my opinion, for reasons that lie beyond the scope of this chapter and will be discussed in some detail in section 6.1, this equally famous and notorious argument is unsound. But I also believe, for reasons that *don't* lie beyond the scope of this chapter, that Kant's transcendental deduction of our *representations* of space and time is sound. And this means that for the purposes of this chapter, I can temporarily bracket the thorny issue of the nature and justification of Kant's idealism, and focus instead on clearly presenting his transcendental deduction of our representations of space and time.[61]

Unfortunately, however, Kant's use of the term "concept" or *Begriff* in the Aesthetic (and also in the treatise on which the Aesthetic was originally based, *On the Form and Principles of the Sensible and Intelligible World*, or the "Inaugural Dissertation") is consistently ambiguous in one respect. He explicitly discusses the "concepts" of space and time. But an intermediate conclusion of the Aesthetic is that neither the representation of space nor the representation of time is a "discursive" representation or a "general concept": rather, both are intuitions and therefore not concepts (*CPR* A24–5/B39, A31/B47). So, in order to be charitable to Kant and to avoid the absurdity of his arguing that the concepts of space and time are not concepts, I think that we must take all his references to the "concepts" of space and time—with the single exception of a special case that I will mention in the next paragraph—to invoke a broad meaning of *Begriff* that is essentially the same as that of the neutral term *Vorstellung* or "representation." This comports well with Kant's usage of *Begriff* in the pre-Critical writings and in the *Reflexionen*. It also makes sense of an otherwise unintelligible passage in the first *Critique* in which he explicitly distinguishes between "two sorts of concepts of an entirely different kind, which yet agree with each other in that they both relate to objects completely a priori manner, namely the concepts of space and time, as forms of sensibility, and the categories, as concepts of the understanding" (*CPR* A85/B118). So, in order to avoid confusion, I will consistently use "representation" where the broader sense of "concept" is clearly intended by Kant.

---

[61] For a general discussion of Kant's idealism, see also Hanna, *Kant and the Foundations of Analytic Philosophy*, ch. 2.

118 EMPIRICAL REALISM AND SCIENTIFIC REALISM

Just to make things even more complicated, however, Kant does speak in at least two places of "the general concept of spaces" (*CPR* A25/B39) and of a "general concept of space (which is common to a foot as well as an ell)" (*CPR* A25). Since these passages are juxtaposed with arguments *against* construing spatial representation as conceptual—and again on the charitable assumption that Kant is not simply contradicting himself—he must actually be arguing that despite the fact that the representation of space is not itself a concept, it is nevertheless possible for us to *construct* some sort of general mental representation, or (in the narrower sense of *Begriff*) concept, of space. But this general concept of space that is constructed by us will be parasitic on a more basic intuition of space, just as we might form the general concept BEING SOCRATIC on the basis of a direct empirical intuitional acquaintance with Socrates himself.

I will now spell out Kant's transcendental deduction of our representations of space and time. For convenience, I will abbreviate "the representation of space" as "r-space" and "the representation of time" as "r-time," display the individual steps of the argument along with supporting texts (whether from the Transcendental Aesthetic or elsewhere in the first *Critique*), and also give a brief commentary on each step.

A Step-by-Step Reconstruction of the Transcendental Deduction of R-Space and R-Time

Prove: that r-space and r-time, as the forms of intuition, are the a priori necessary subjective forms of all empirical intuitions of appearances.

(1) Empirical intuitions are singular representations of undetermined apparent or sensible objects, and those representations in turn possess both matter and form.

The undetermined object of an empirical intuition is called **appearance**. (*CPR* A20/B34)

I call that in the appearance which corresponds to sensation its **matter**, but that which allows the manifold of appearance to be intuited as ordered in certain relations I call the **form** of appearance (*Form der Erscheinung*). (*CPR* A29/B34)

Commentary: We learn later in the first *Critique* that empirical intuitions must be combined with concepts in the context of judgments in order to "determine" perceptual appearances by representing *determinate* objects of experience (*CPR* A51/B75). But empirical intuitions are, as such, very strongly non-conceptual (see *CPR* A90/B122, and section 2.2). The object of such a representation is not a determinate object of experience, but instead an undetermined or at best partially determined object of the senses, that is, an appearance. These objects, as represented, have both a material component that corresponds to our objective sensory perceptions of them, and also a formal-structural (that is, spatiotemporal) component that remains fixed across variations in the material component.

It is very important, however, to recognize that this formal-structural spatiotemporal component is *immanent*, or literally intrinsically present, in the objective sensory representations themselves. It is indeed possible, by an act of conceptual abstraction, to consider the spatiotemporal component apart from any sensory representational content (*CPR* A20–1/B34–5). The semantic residue of this act of abstraction is then what Kant calls "a mere form of sensibility" (*CPR* A21/B35) or "mere form of intuition, without substance" (*CPR* A291/B347). These *mere* forms of intuition are *not* however the same as the forms of intuition in the proper sense, which are *empirically realized* formal-structural spatiotemporal representational frameworks. So, as Kant puts it in the case of r-space, "if extended beings were not perceived, one would not be able to represent space" (*CPR* A292/B349). One crucial implication of this doctrine is that the mere form of spatial intuition is nothing but an "empty intuition without an object" or an *ens imaginarium* (*CPR* A292/B348), lacking any determinate structure:

[The mere universal form of intuition called "space"] is something so uniform and so indeterminate with respect to all specific properties that certainly no one will look for a stock of natural laws in it. (*P* 4: 321–2)

By sharp contrast, however, the empirically realized form of spatial representation—that is, the form of intuition in the proper sense—for Kant represents a three-dimensional, homogeneous rectilinear or Euclidean, egocentrically-oriented topological structure that in turn guarantees, for example, the incongruence of exact physical counterparts that are also enantiomorphs (mirror-reflected isomorphs), such as the right and left hands, two spherical triangles sharing the same base, and double helix figures like "oppositely spiralled snails" (*P* 4: 284–6). In short and more generally, whereas the mere forms of intuition are merely indeterminate or *thin* spatiotemporal representational structures—presumably picking out homogeneous rectilinear spaces or events but without any special topological features such as limited dimensionality or "handedness"—by contrast, the forms of intuition are empirically-realized, determinate or *thick* spatiotemporal representational structures.[62] I will come back to this crucial point in section 2.4, under the heading of what I will call the "designated" structures of r-space and r-time.

(2) Appearances or objects of the senses are represented in empirical intuition by means of either outer (or spatial) sense or inner (or temporal) sense. R-space and r-time are the mutually distinct and jointly exhaustive (although not mutually exclusive) forms of intuition, and also the subjective forms of outer and inner sense respectively.

> By means of outer sense (a property of our mind) we represent to ourselves objects as outside us, and all as in space. In space their shape, magnitude, and relation to one another is determined, or determinable. Inner sense, by means of which the mind intuits itself, or its inner state, gives, to be sure, no intuition of the soul itself, as an object; yet it is still a determinate form, under which the intuition of its inner state is alone possible, so that everything that belongs to the inner determinations is is represented in relations of time. (*CPR* A22–3/B37)

> Time can no more be intuited externally than space can be intuited as something in us. (*CPR* A23/37)

> [R-]space is nothing other than merely the form of all appearances of outer sense, i.e., the subjective condition of sensibility, under which alone outer intuition is possible for us. (*CPR* A26/B42)

> [R-]time is nothing other than the form of inner sense, i.e., of the intuition of our self and our inner state. (*CPR* A33/B49)

Commentary: The contrast between outer sense and inner sense is phenomenologically self-evident and primitive: roughly speaking, the outer is whatever stands in some determinate sensory relation to the living body of the subject (see also *CPR* A23/B38), and the inner is whatever is sensory and

---

[62] This subtle semantic distinction between the thick and thin intuitional representations of space also generates a corresponding subtlety in the general concept of space, since there then can be either (i) a general concept of a "thick" Euclidean structure with special topological features such as three-dimensionality or handedness (i.e., egocentric directional orientation, enantiomorphism), or else (ii) a general concept of a "thin" Euclidean structure lacking special topological features. From the general concept of a thin Euclidean structure, it would then seem to be only a short step to the formation of a *super*-general concept of space which also abstracts away from homogeneous rectilinearity or Euclidean-ness, and allows for non-homogeneity or curvature (*CPR* A220–1/B268).

non-outer. Otherwise put, inner sense is phenomenal consciousness (see sections 1.1 and 2.2). R-time, as the form of outer sense, is *the intrinsic phenomenal structure*, or "the immediate condition" (*die unmittelbare Bedingung*) (*CPR* A34/B51),[63] of inner sense; and correspondingly r-space, as the form of outer sense, is the intrinsic phenomenal structure, or immediate condition, of outer sense. Because the contrast between outer sense and inner sense is phenomenologically self-evident and primitive, and because r-space and r-time are the forms of inner and outer sense, it follows that the contrast between r-space and r-time is phenomenologically self-evident and primitive. It does not follow, however, that r-space and r-time exclude one another; on the contrary, they are strictly complementary, just as outer and inner sense are strictly complementary.[64]

(3) R-space and r-time are necessary conditions for the empirical intuition of appearances in outer and inner sense.

> [R-]space is a necessary representation, a priori, which is the ground of all outer intuitions. One can never represent that there is no space, although one can very well think that there are no objects to be encountered in it. (*CPR* A24/B38)

> [R-]time is a necessary representation that grounds all intuitions. In regard to appearances in general one cannot remove time, though one can very well take the appearances away from time. (*CPR* A31/B46)

Commentary: R-space and r-time belong to the formal constitution of the senses, so as a matter of conceptual necessity they cannot be removed from our representations of appearances; but it is conceivable and therefore

---

[63] This is a slightly different use of "immediate" or *unmittelbar* than in the case of immediate cognition, where it means the *referential directness* of a cognition. To say that r-time as a formal representation is "the immediate condition" of inner sense is, I think, to say that r-time is a necessary and immanent phenomenal structure that essentially distinguishes inner sense from other sorts of phenomenal experience. And this is what I mean by saying that that r-time is the "intrinsic phenomenal structure" of inner sense. *Mutatis mutandis*, the same goes for r-space and outer sense.

[64] Kant states explicitly that r-time is "the mediate condition of outer appearances" (*CPR* A34/B50), which is to say that the empirical intuition of objects in space also automatically implements temporal form: "all appearances in general, i.e., all objects of the senses, are in time, and necessarily stand in relations of time" (*CPR* A34/B51). Moreover, the very possibility of representing the motion of material objects in space presupposes r-time (*CPR* B48–49). Correspondingly, according to Kant, we necessarily represent our own inner mental states in relation to space. I can introspectively "find myself" only if there is "something in another place in space from that in which I find myself" (*CPR* A23/B38). And, in the Refutation of Idealism, Kant argues that "inner experience is…only mediate and possible only through outer experience" (*CPR* B277). In other words, r-time is the immediate condition of inner sense and the mediate condition of outer sense, and r-space is the immediate condition of outer sense and the mediate condition of inner sense.

possible that r-space and r-time can exist without spatial and temporal objects; hence r-space and r-time are strictly necessary for the empirical intuition of appearances, although the converse is not the case.

(4) R-space and r-time, the forms of intuition, by means of an act of self-consciousness, can also be treated as "pure intuitions," or singular non-conceptual representations of themselves as unique abstract relational totalities or formal-structural frameworks, thereby in turn representing space and time as singular infinite given wholes.

> [R-]space is not a discursive or... general concept of relations of things in general, but a pure intuition. (*CPR* A24–5/B39)
>
> Space is represented as a given infinite magnitude. (*CPR* A25/B39)
>
> [R-]time is no discursive or... general concept, but a pure form of sensible intuition. (*CPR* A31/B47)
>
> The infinitude of time signifies nothing more than that every determinate magnitude of time is only possible through limitations of a single time grounding it. The original representation, [r-]time, must therefore be given as unlimited. (*CPR* A32/B48)
>
> [R]-space and [r]-time and all their parts are **intuitions**, thus individual representations along with the manifold that they contain in themselves (see the Transcendental Aesthetic), thus they are not mere concepts by means of which the same consciousness is contained in many representations, but rather are many representations that are contained in one and in the consciousness of it; they are thus found to be composite, and consequently the unity of consciousness, as **synthetic** and yet as original, is to be found in them. This singularity of theirs is important in its application. (*CPR* B136 n.)

Commentary: R-space and r-time, by means of an act of self-consciousness, can be treated as non-conceptual singular intuitions that represent *themselves* as unique individuals—but not in any way as empirical objects, rather only as unique abstract relational totalities or formal-structural frameworks (*CPR* A291/B347). These frameworks in turn represent space and time as singular infinite given wholes because, although empirical quantities are possible only through space and time, they are also presented as intrinsically unlimited, non-denumerable, or "ideal" totalities (*CPR* A438/B466). All intuitions are singular representations (see section 2.2), but the singularity of a *pure* intuition is partially constituted by a special synthetic unity of consciousness, which directly and necessarily connects pure intuition with self-consciousness or apperception: "[t]he supreme principle of all intuition in relation to the understanding is that all the manifold of intuition stand under conditions of the original synthetic unity of apperception" (*CPR* B136). In other words, even though a pure intuition is non-conceptual, it is only *weakly* non-conceptual, because the capacity for pure intuition stands in a necessary relation to the understanding and thereby to our conceptual capacities, via the capacity for self-consciousness (again see section 2.2).

(5) R-space and r-time are a priori. (From (3), (4), and the definition of 'a priori' as absolute experience-independence, or underdetermination by all possible sets and sorts of sensory impressions. That is: to say that X is a priori is to say that X is not strongly supervenient on sensory impressions.[65])

[W]e will understand by a priori cognition not those that occur independently of this or that experience, but rather those that occur *absolutely* independently of all experience. (*CPR* B3)

Commentary: To the extent that r-space and r-time can be treated as pure intuitions via self-consciousness, they lack all sensory "matter" or sensory qualitative content by the definition of "pure" (*CPR* B3), and so automatically satisfy the definition of apriority.

(6) Since r-space and r-time are (a) mutually distinct and jointly exhaustive (although complementary) necessary forms of the empirical intuition of appearances, (b) subjective forms of outer and inner sense, and (c) able to be treated, via self-consciousness, as pure a priori non-conceptual

---

[65] See Hanna, *Kant and the Foundations of Analytic Philosophy*, p. 248. Roughly speaking, Y is strongly supervenient on X if and only if the X-features of something are sufficient for its Y-features, and there cannot be a change in anything's Y-features without a corresponding change in its X-features. For more details, see the Introduction, n.30. The basic idea behind strong supervenience is that it captures an asymmetric modal dependency relation that is weaker than identity and consistent with irreducibility. And the point of deploying the notion of strong supervenience in the present connection is that it allows us to say that a cognition is a posteriori or *dependent on sensory impressions* just in case its form or its semantic content is strongly supervenient on sensory impressions; but a cognition is a priori or *absolutely independent of all sensory impressions* just in case its form or its semantic content is not strongly supervenient on sensory impressions and is instead strongly supervenient on one or another of our innate spontaneous cognitive capacities, or faculties (*CPR* B2–3).

intuitions of themselves as unique abstract relational totalities or formal-structural frameworks, they are therefore the a priori necessary subjective forms of all empirical intuition of appearances. (From (1)-(2) and (5).)

Commentary: Step (6) establishes the objective validity of r-space and r-time and thus completes the transcendental deduction of r-space and r-time. But it is crucial to see that this conclusion invokes a basic distinction between r-space and r-time as (1) *the forms of intuition*, or the a priori necessary subjective non-conceptual forms of all empirical intuition of appearances, and (2) *pure or formal intuitions*, that is, pure self-conscious a priori non-conceptual intuitions of r-space and r-time themselves as unique abstract relational totalities, which in turn represent space and time as singular infinite given wholes. As Kant puts it in the B edition version of the Transcendental Deduction of the categories:

[R-]space, represented as an **object** (as is really required in geometry) contains more than the mere form of intuition, namely the **putting-together** (*Zusammenfassung*) of the manifold given in accordance with the form of sensibility in an **intuitive** representation, so that the **form of intuition** (*Form der Anschauung*) merely gives the manifold, but the **formal intuition** (*formale Anschauung*) gives unity of the representation. (*CPR* B160 n.)

Otherwise put, the basic distinction between r-space and r-time as "forms of intuition" on the one hand, and as "formal intuitions" on the other hand, is this. Forms of intuition require only a *subjective unity of consciousness* and do not necessarily involve a synthetic unity of self-consciousness or apperception, hence forms of intuition are very strongly non-conceptual, but by contrast formal intuitions require an *objective unity of consciousness* that is determined by the capacity for self-consciousness or apperception, hence formal intuitions are only weakly non-conceptual and thus necessarily related to our conceptual capacities.[66] Still otherwise put, the forms of intuition

---

[66] An important consequence of the forms of intuition vs. formal intuitions distinction is a sharp difference in the way in which phenomenal extensive quantities are represented by the forms of intuition alone, as opposed to the way in which phenomenal extensive quantities are represented by pure or formal intuitions. More precisely, when phenomenal spatial and temporal extensions are represented by the forms of intuition alone, as in non-conceptual perceptual experience, they have what some contemporary theorists of non-conceptual content aptly call a "unit-free" phenomenal character (see, e.g., Bermúdez, "Nonconceptual Mental Content," p. 4.). This means that phenomenal distances and phenomenal time-stretches are given in perceptual experience as, e.g., "being *that* far" or as "taking *that* long to happen," without these quantities being represented as determinately measured or measurable, hence without these quantities being brought or bringable under number-concepts.

are involved in rational cognition and *proto-rational* cognition (say, of pre-linguistic human children or non-human animals) alike, whereas formal intuition strictly requires a capacity for self-conscious *fully rational* cognition.

In this connection, it is very important not to confuse (1) the distinction between a subjective unity of consciousness and an objective unity of consciousness, with (2) the distinction (noted in section 2.2) between subjective consciousness and objective consciousness. Objective consciousness is the representational consciousness of either some intentional object or oneself (i.e., objective reflexive consciousness, or meta-consciousness). Subjective consciousness is consciousness without an underlying unity of content; hence a subjective consciousness does not even have a *subjective* unity of consciousness, and represents neither an intentional object nor oneself. Moreover, a unity of representational content available to consciousness, whether a subjective unity or objective unity, is a necessary condition of all objective consciousness. But not every objective consciousness has an *objective unity*. Hence there can be an objective consciousness that has a merely *subjective unity* of consciousness, that is, a unity provided by the forms of intuition alone.

Now, taking together step (6) with the material in section 2.2, we can immediately derive the target thesis in this section:

(6*) Since all non-conceptual content is intuitional content, and since non-conceptual intuitional content exists and is representationally significant, and since all non-conceptual content is either empirical or non-empirical, it follows that r-space and r-time, as the forms of intuition, are the a priori necessary subjective forms of all non-conceptual content. (From (6) and section 2.2.)

Translated again out of Kantspeak for a moment, what I am asserting on Kant's behalf is that our capacities for spatial and temporal representation

---

Kant's explanation for this is that if and only if the representational content of r-time is *combined with* the representational content of the logical categories of quantity, is the representation of the natural numbers possible (*CPR* A142–3/B182, A242/B300) (*P* 4: 283)—see also section 6.3. In turn, as the Axioms of Intuition show, the representation of the natural numbers then makes possible one-to-one correlations between the numbers and limited discrete parts of the spatiotemporal phenomenal manifold, thereby yielding determinate spatial or temporal magnitudes (*CPR* A162–6/B202–7). But, as Kant stresses in the famous footnote at B160, the representational contents of r-time and r-space cannot be combined with or brought under the categories of quantity (as, e.g., r-space is combined with and brought under the categories of quantity in geometry) until they have already been represented as unique abstract structural wholes or frameworks—that is, until they have already been represented through pure or formal intuition.

constitutively explain non-conceptual content: non-conceptual content is *nothing but* cognitive content that is essentially structured by our a priori representations of space and time. I reiterate however that by this thesis I do not mean that the sensory *qualitative* content of non-conceptual cognition is to be explained in this way, but rather only that the *representational* content of non-conceptual cognition is to be so explained. In particular, then, Kant is saying that what determines our cognitive reference to the uniquely individual material objects of empirical non-conceptual or intuitional representations, are the spatiotemporal features of those representations alone. To cognize *this* or *that* individual material object non-conceptually or intuitionally in inner sense, outer sense perception, feeling or affect, imagination, and empirical judgment, is simply *to locate it uniquely here-and-now or there-and-then*. Or as the real estate agents say: it's all about location.[67]

2.4. The Role of Spatiotemporal Structure in Non-Conceptual Content

We are now in a good position to see how the Kantian forms of intuition provide a fundamental explanation of non-conceptual content. Kant's way of formulating this, as we have just seen, is that the forms of intuition are the apriori conditions of the possibility of—the a priori semantic presuppositions of—all intuitional content, as guaranteed by the transcendental deduction of our representations of space and time. But once we have translated Kant's thesis about the forms of intuition out of Kantspeak and into more contemporary terms, we can recognize that he is making an intelligible, substantive, and plausible claim that significantly extends the recent debate about non-conceptual content, by solving the unity problem about non-conceptual content.

The key to recognizing the Kantian solution to the unity problem lies in the answer to the following question: is the underlying nature of cognitive content exhausted either by its functional components or by its purely logico-rational components? Those who answer "yes" to this question will

---

[67] This includes both *fixed location*, or spatiotemporal place, and also *successive-change-in-spatial-location*, or motion.

deny either the existence or at least the representational significance of non-conceptual content, whereas those who answer "no" will assert the existence and representational significance of non-conceptual content. Now Kant's fundamental explanation for non-conceptual content via the forms of intuition gives us good reason to answer "no," and here is why.

If Kant is right, then forms of intuition introduce intrinsic spatial or temporal structures directly into phenomenal cognitive content, which is to say that all sensory representations of material objects or of the individual subject herself are necessarily informed, infused, or "matted" by our representations of space or time. As Kant puts it in the Aesthetic, you *can* conceive of space or time as empty of apparent objects or subjects, but you *cannot* conceive of apparent objects or subjects without also representing space or time (*CPR* A24/B38, A31/B46). This claim has metaphysical modal force because for Kant properly constrained conceivability entails real or metaphysical possibility, and the proper constraints on conceivability are yielded by Kant's theory of objective validity.[68] But the crucial point is that our representations of space and time are *intrinsic phenomenal structures* of cognitive content.

This claim requires a side comment to avoid misunderstandings. For me an "intrinsic property" is an inherent or necessary property of something, and an "extrinsic property" is an accidental or contingent property of something. So for me the "intrinsic property vs. extrinsic property" distinction is *not* the same as the "necessary non-relational (that is, monadic or 1-place) property vs. contingent non-relational or relational (that is, polyadic or many-place) property" distinction, although some philosophers, influenced by Leibniz's views on the reducibility of all relational properties to the internal, necessary, non-relational properties of ontologically basic micro-entities, have offered definitions of "intrinsic property' and "extrinsic property" to this effect.[69] Definitions of concepts are of course perfectly acceptable philosophically. But when a stipulative usage of a *word* gains currency, it may carry with it an entirely false impression of metaphysical inevitability. And it seems to me this now-popular stipulative usage of "intrinsic" has in fact tended to carry with it the entirely false impression that it is a priori impossible for there to

---

[68] For the details of Kant's theory of objective validity, see Hanna, *Kant and the Foundations of Analytic Philosophy*, ch. 2.
[69] See, e.g., Langton and Lewis, "Defining 'Intrinsic'."

be intrinsic relational properties based on spatiotemporal form, or what I call *intrinsic structures*.

According to Kant, as I have said, our representations of space and time are intrinsic phenomenal structures of cognitive content. Now, as a direct consequence of this, since r-space and r-time are more specifically intrinsic phenomenal *structures*, then they are irreducible to phenomenal qualia or sensory qualitative content. This is because qualia are intrinsic *non-relational* features of cognitive content, whereas our representations of space and time are fully relational (B67). Moreover, since our representations of space and time are *intrinsic* phenomenal structures, they are irreducible to functional features of cognitive content. This is because functional features of cognitive content are *extrinsic* relational patterns or structures within content, that trace causal mappings from processing inputs to processing outputs in animals or machines. In a materialist representational framework, these mappings could ultimately be either behavioral, computational, otherwise mechanical, or neurobiological. The extrinsicness of functional structures consists in the fact of their ontological promiscuity or unconstrained multiple realizability: unrestrictedly many different (kinds of) things can be the causal-role players, so the functional structure has no intrinsic properties of its own, *as* structure. By contrast, our representations of space and time are also multiply instantiated but at the same time *ontologically constrained* in that they must be realized in all and only *the phenomenal states of animals minded like us*. Finally, then, since our representations of space and time are intrinsic *phenomenal* structures, they are irreducible to purely logico-rational features of cognitive content. This is because the purely logico-rational features of cognitive content, as *purely* logical and rational in character, are of course thereby also *non-phenomenal* in character.

So: Kant's thesis is that the intrinsic phenomenal structures of cognitive content that are introduced by our representations of space and time are not only required by but also immanently configure, organize, and "pre-format" all phenomenal cognitive content. And, while it is not implausible to hold that all the conceptual parts of phenomenal cognitive content *can* be accounted for (reductively, or non-reductively) in either functional or logico-rational terms alone,[70] nevertheless the non-conceptual

---

[70] See, e.g., Chalmers, *The Conscious Mind*, chs. 1 and 6–7, for an argument that all or at least massively most of the conceptual elements of cognition are functional.

spatiotemporal elements of phenomenal content necessarily *resist* functional or logico-rational reduction.

Significantly, Kant's thesis of the cognitive autonomy of non-conceptual spatiotemporal representation has also received some independent *empirical* confirmation, at least as far as spatial representation is concerned, in experiments involving commissurotomy patients. Commissurotomy is the surgical severing of the corpus callosum, which is the primary neural connection between the right and left hemispheres of the brain. Commissurotomy patients typically manifest some cognitive dissociation between types of information normally processed in the right hemisphere and types of information normally processed in the left hemisphere. In one particularly interesting experiment carried out by Colwyn Trevarthyn,[71] a commissurotomy patient was instructed to carry out a left-hand task that referred to an object on the right side of the visual field, while focusing his vision on a central point between the two sides of his visual field. What happened was that as soon as the left-hand movement (controlled by the right hemisphere) started, the visual appearance of the object in the right-hand side of the visual field (controlled by the left hemisphere) disappeared, thus vividly indicating a strong dissociation between information processed in the two hemispheres. But most importantly for our purposes, even though the visual representation of the object disappeared, *the right-hand phenomenal visual field remained both intact and continuous with the left-hand side of the visual field.* In an elaboration of these results, Trevarthyn proposed "that neo [i.e., recent] commissurotomy in man may...divide cortical vision for perception of detail and identification of objects, without producing a similar division in the perception of ambient space."[72] In other words, it is possible for commissurotomized humans to dissociate from the conceptual content of visual experiences and judgments, while still retaining the uncompromised non-conceptual grasp of visual space. So the visual representation of space seems to be both non-conceptual and also more cognitively basic than visual conceptualization.

This leads me up to my last leading question: how, more precisely, do the forms of intuition according to Kant play their constitutive role in

---

[71] See Trevarthyn, "Analysis of Cerebral Activities that Generate and Regulate Consciousness in Commissurotomy Patients."
[72] See Trevarthyn and Sperry, "Perceptual Unity of the Ambient Visual Field in Human Commissurotomy Patients," p. 547.

non-conceptual content? Remember now that the role of empirical non-conceptual or intuitional cognition, along with its non-conceptual cognitive content, is that of uniquely locating individual material objects. So Kant's answer to my last leading question, I think, is that the forms of intuition introduce a single spatiotemporal—by which I mean spatial *or* temporal, so weakly disjunctive—phenomenal framework, or a phenomenal *field*, into cognitive content. Only within the framework of such a field can real objects in the world be uniquely located by our cognitive capacities.

Absolutely essential to this uniquely locating representational function is the fact that this phenomenal field is not merely a set of spatial or temporal representational relations, but also has what I will call a "thick" or "designated" structure. This means that the set of spatial relations and the set of temporal relations found in the sensible experience of rational animals like us have further *special* constraints on them, that cannot be found in every logically possible set of spatial or temporal relations.

For example, according to Kant, the spatial part of the spatiotemporal phenomenal field is not only represented as homogeneous and rectilinear (that is, as Euclidean), as filled with points, as figural, and as extended, but also as *an egocentrically oriented three-dimensional Euclidean manifold*. Obviously the Euclidean designation sets it apart from non-Euclidean spaces; and just as obviously the 3-D designation sets it apart from higher-dimensional Euclidean spaces or higher-dimensional non-Euclidean spaces. Nevertheless the egocentrically oriented designation is particularly important. In chapter 1, we saw Kant arguing that all three-dimensional spaces represented by creatures like us necessarily have "centered" or egocentric axes for right-left, front-back, and up-down directionality (*DS* 2: 381–3) (*OT* 8: 134–5). So, according to him, it is possible for me to cognize the incongruence between my right and left hands *non-conceptually*, despite their being exact one-to-one analytical counterparts and thus indistinguishable conceptually, merely by possessing an outer sense (*P* 4: 285–6).

In turn, according to Kant, the temporal part of the spatiotemporal phenomenal field is not only represented as successive, filled with moments, and as linear or one-dimensionally extended (*CPR* A33/B50), but it is also represented as *asymmetric or irreversible in its succession* (*CPR* A191–3/B236–8). The asymmetric character of time—a.k.a. "Time's Arrow"—is of course crucial in our representation of the causal order of nature, and guarantees that our representation of material causal processes, in particular those involving motion (*CPR* B48–9), will represent such processes as always and only flowing successively forwards into the future and never backwards into the past. This important and highly prescient Kantian idea has also received some confirmation in contemporary theories of the non-linear, non-equilibrium thermodynamics of irreversible natural processes, including organismic biological processes, which Kant called "self-organizing" (*CPJ* 5: 374) processes.[73] I will come back to this connection briefly in section 8.4. But the crucial point right now is that for Kant, just as our representation of the oriented directionality of space depends solely on the constitution of outer sense and is non-conceptually cognizable, so too our representation of the irreversibility of time is an essentially "centered" or egocentric feature of time that depends solely on the constitution of inner sense and is non-conceptually cognizable.

Now both the Euclidean 3-D orientability built into the representation of phenomenal space and the Time's Arrow irreversibility built into the representation of phenomenal time are conceptually or logically *contingent* features of space and time themselves. This is apparent in non-Euclidean geometries, higher-dimensional geometries, the notion of "non-orientable" surfaces like the Möbius strip, and the conceivability of backwards causation and backwards time-flow in the linear equilibrium dynamics described by classical Newtonian mechanics. But, on the other hand, our representations of 3-D Euclidean "centered" or egocentric orientability and Time's Arrow alike seem to be built right into the metaphysics of animal minds, in the sense that for Kant the conscious states of animals, whether human or non-human, are necessarily framed by the non-conceptual spatiotemporal phenomenal field. Correspondingly, orientable space and dynamical temporal irreversibility seem to be built right into the metaphysics of the living bodies of conscious animals. Therefore, on the Kantian assumption that mind is identical to life (*CPJ* 5: 278), the "centeredness" or egocentricity of conscious cognitive content necessarily requires real spatial orientation and real temporal asymmetry in the animal whose cognitive content it is.

Not only that, but perhaps even more radically, for Kant the designated formal intuitional spatiotemporal structure of non-conceptual cognitive content *just is* its subjective or first-person character. *It is precisely an animal's unique non-conceptual spatiotemporal perspective or "point of view" that constitutes*

---

[73] See, e.g., Nicolis and Prigogine, *Self-Organization in Nonequilibrium Systems*.

the subjective character of its experience, and not the "objective unity of consciousness" in the Kantian sense of a necessarily conceptual capacity for rationally self-conscious and proposition-based unification of a phenomenal manifold of sensory or representational content.⁷⁴ As the commissurotomy cases vividly show, the unity of consciousness in this Kantian sense is a relatively sophisticated and fragile achievement of rational animals, but unnecessary for animal consciousness and conscious animal cognition in general, whether the animal is rational or non-rational, and whether the animal is human or non-human.⁷⁵

2.5. Direct Perceptual Realism Again

Kant's direct perceptual realism, I have proposed, is the first part of his empirical realism. In turn, for Kant, perception is *realistic* precisely to the extent that it is object-dependent or veridical; and it is *direct* precisely to the extent that it is non-epistemic, non-conceptual, and otherwise unmediated. What I want to do now is to explore three important epistemic implications of Kant's theory of the spatiotemporal grounding of non-conceptual content for his direct perceptual realism.

(1) *Non-Conceptual content and the justification of perceptual belief.* McDowell has influentially argued in *Mind and World* that since (1) all epistemic justification requires inferential reasons for beliefs, but (2) non-conceptual perceptual content cannot constitute an inferential reason, it follows that (3) non-conceptual perceptual content cannot have any evidential force with respect to perceptual beliefs.⁷⁶ Kant would reject McDowell's conclusion by rejecting the first premise of his argument. For Kant, a perceptual

---

⁷⁴ Nagel has of course famously associated the first-person character of phenomenal consciousness with having a "point of view" in "What is it Like to be a Bat?" But he never actually unpacks this idea in terms of its spatiotemporal character. And in fact in "Brain Bisection and the Unity of Consciousness," in *Mortal Questions*, pp. 147–64, Nagel explicitly commits himself to the thesis that a *unity of consciousness* in the broadly Kantian sense is a necessary and sufficient condition for the subjectivity of phenomenal consciousness. But it is quite possible to read the empirical evidence from the commissurotomy cases as making the essentially Kantian point that an animal's subjectivity is preserved by its non-conceptual spatiotemporal representational capacities, even when the unity of consciousness is disrupted by brain bisection.

⁷⁵ For similar points formulated in a slightly different way, see Hurley, *Consciousness in Action*, chs. 2 and 4.

⁷⁶ See McDowell, *Mind and World*, lectures I–III.

belief can be non-inferentially defeasibly warranted solely by means of what he calls the "aesthetic certainty" of a very strongly or fairly strongly non-conceptual perceptual experience (*JL* 9: 39), thus giving rise to an "empirical certainty" which is "unmediated" or "original"(*JL* 9: 71). This is taken by Kant to be precisely analogous to the way in which a purely rational belief can be non-inferentially defeasibly warranted solely by means of "a priori insight" (*Einsicht a priori*), thus giving rise to a "rational certainty" which is also "immediately certain" (*JL* 9: 71).[77]

In other words, for Kant, even though very strongly or fairly strongly non-conceptual contents cannot count as *inferential reasons* for perceptual beliefs (because, as non-propositional, non-conceptual contents cannot be premises in arguments), they can nevertheless still be *non-inferential reasons*, for either a defeasible warrant or a decisive justification—which would be the case in which the warrant for taking a proposition to be true is appropriately combined with the actual truth of that belief—of perceptual beliefs. In a precisely analogous way, there can be *insight-based* non-inferential reasons for the defeasible warrant or decisive justification of pure rational beliefs. McDowell fails to see this point because he falsely assumes that all epistemic warrant or justification must be inferential. And that is because he falsely assumes that the only alternative to an inferential justification of a belief is something that merely *causes or triggers* ("exculpates") the belief. But Kant clearly sees that warrant or justification can be *either* inferential *or* non-inferential, and that under the right cognitive conditions non-conceptual perceptual experience can itself constitute a non-inferential reason for a genuine defeasible warrant or decisive justification of perceptual beliefs.

How is a non-inferential non-conceptual perceptual warrant or justification possible? Or, otherwise put, since an epistemic reason is a fact that normatively supports beliefs, how can non-conceptual content be an epistemic reason that normatively supports perceptual beliefs? Here is what Kant says:

aesthetic certainty. This rests on what is necessary in consequence of the testimony of the senses, i.e., what is confirmed through sensation and experience. (*JL* 9: 39)

Empirical certainty is original (*originarie empirica*) insofar as I become certain of something *from my own* experience. (*JL* 9: 71)

---

[77] I work out the basic details of this Kantian theory of pure rational knowledge in ch. 7.

Kant's idea here—which can I think be viewed as his original development of a fascinating but underexploited theoretical strand in Descartes's doctrine of "clear and distinct perception"[78]—is that the intrinsic structural phenomenal (in Kantspeak, "aesthetic") character of such experiences is such that it confers an *optimal phenomenal articulation or lucidity* upon their non-conceptual perceptual content, and thereby, just in virtue of this optimally articulated or lucid content, *synthetically necessitates*[79] the perceiver's assertoric belief in the corresponding propositional content that is cognitively built right on top of that non-conceptual perceptual content. Or, as we might say in a specifically visual context, *seeing is believing*.

Indeed, the initially surprising Kantian thesis that spatiotemporal intrinsic structures within non-conceptual representational content—or what we would now call "mental models"[80]—when self-consciously associated with corresponding propositional contents, can have phenomenological synthetically necessary normative epistemic implications for propositional attitudes directed to those contents, turns out upon reflection not to be very surprising after all. This is for two reasons, one purely philosophical and the other textual.

First, the epistemic normativity associated with spatiotemporal non-conceptual perceptual structures on the Kantian account derives from the epistemic normativity of *pattern-matching activities* more generally. For example, getting a square peg into a square hole is *getting it right*, whereas failing to get a square peg into a round hole is *getting it wrong*. But once you get it right, the isomorphic matching of a representational pattern onto a corresponding real pattern in the world guarantees belief: you cannot deny that it fits. As this simple example indicates, the epistemic normativity of pattern-matching activities extends down into the proto-rational domain of human toddlers and relatively cognitively sophisticated non-human animals like monkeys and apes. But precisely the same sort of epistemically normative pattern-matching activity also occurs at a higher cognitive level in the imaginational "mental rotation" acts carried out by ordinary rational human thinkers dozens of times every day—as when I ask you to imagine taking your house-key out of your pocket, then turning the key right-side up, putting it in your front door lock, and opening the

---

[78] See Gaukroger, *Descartes: An Intellectual Biography*, pp. 115–24, and 167–72.
[79] See Hanna, *Kant and the Foundations of Analytic Philosophy*, ch. 5.
[80] See Johnson-Laird, *Mental Models*; and ch. 7.

door with it.[81] According to Kant, precisely the same sort of epistemically normative pattern-matching activity recurs yet again, this time at the cognitively highest level, in abstract mathematical reasoning in arithmetic and geometry. Here, by means of the pure productive imagination in its schematizing function, in what Kant calls mathematical "construction" (*Konstruktion*), we pattern-match our abstract propositional representations of arithmetic and spatial structures to those very mathematical structures themselves. And when this yields a consciously experienced isomorphism of the propositional representation with a real mathematical structure, then it automatically generates an a priori mathematical insight and the objective certainty of a rational "conviction" (*Überzeugung*). I will unpack these fascinating and little-studied themes in Kant's epistemology in sections 6.4 and 7.2–7.4.

Second, the epistemic normativity associated with spatiotemporal non-conceptual perceptual structures is architectonically and analogically right in line with Kant's better-known thesis that some intrinsic structures within *conceptual* content, when self-consciously associated with corresponding propositional contents, can have *analytically* necessary implications for propositional truth and apriority: indeed, the notion of an intrinsic-structural necessitation between various conceptual contents that are "contained in" or "contained under" one another provides the basic rationale for Kant's theory of analyticity.[82]

Turning back to perceptual contexts now, it is important to remember that because the phenomenological necessitation of an assertoric propositional attitude can happen under conditions of veridical illusion as well as under conditions of perceptual correctness, it is a *defeasible* attitude. Thus I can have a perfectly good non-conceptual non-inferential warrant for holding a perceptual belief, yet unfortunately still be in error. Kant's general point, again, is that non-conceptual perception can provide me with a reason sufficient for a defeasible warrant or a decisive justification *without* being an inferential reason. So for Kant not all epistemic warrant or justification is inferential, precisely because some sufficient reasons for a posteriori or a priori belief are not inferential: some defeasible warrants or justifications are

---

[81] See, e.g., Shepard, "The Mental Image"; Shepard and Chipman, "Second Order Isomorphisms of Internal Representations: Shapes of States"; Shepard and Cooper, *Mental Images and their Transformations*; and Shepard and Metzler, "Mental Rotation of Three-Dimensional Objects."

[82] See Hanna, *Kant and the Foundations of Analytic Philosophy*, ch. 3.

based on epistemic reasons that are given in the clear and distinct intrinsic structure of non-conceptual phenomenology itself. It is true that not all seeing is believing: but *some* is, and such seeing can successfully operate without conceptual constraints. For Kant, non-conceptual perceptual normativity is a necessary but not a sufficient condition of conceptual normativity. So non-conceptual perceptual epistemic reasons essentially outrun conceptual epistemic reasons, and yet also anchor conceptual normativity in the natural world of embodied human and non-human animal perception. Martin makes the same crucial point about memory.[83]

(2) *What is affection?* The second important epistemic implication of Kant's theory of the spatiotemporal basis of non-conceptual content has to do with the nature of the causal relation of "affection" (*CPR* A19/B33–34) that runs between perceived material objects and conscious perceptual states. Kant argues in the second Analogy of Experience that all alteration of material substances in time not only requires but also intrinsically structurally involves causation (*CPR* A189/B232-A211/B256). And, as the third Analogy explicitly states, perceivers can also be causally affected by material objects that are *simultaneous* with their conscious perceptual states (*CPR* A211/B257–8, A213/B260). The relative simultaneity of a perceiver and the material object she presently perceives in turn involves a worldwide causal-dynamic community of mutually interacting material substances (*CPR* A211–15/B256–62). Therefore, since all causation is either diachronic or synchronic, perceivers can be causally affected either by past or present objects. But, quite obviously, not every causal relation is a causal relation of affection. For Kant, in opposition to the common caricature of Berkeleyan idealism, any number of trees can fall down in any number of forests without anyone's actually perceiving them. So at this point the obvious question arises: just what sort of causal-dynamic relation *is* affection for Kant?

The correct answer, I think, is that for Kant affection is what I will call *a causal-dynamic interactive non-conceptual spatiotemporal tracking relation* between the perceived object and the conscious perceiver. More precisely, I am saying that for Kant an actual macrophysical material object O affects a conscious perceiver P if and only if:

---

[83] See Martin, "Perception, Concepts, and Memory."

(a) O stands in a diachronic or synchronic causal-dynamic relation with P;
(b) P is able consciously and non-conceptually to locate O uniquely in objectively real space relative to P's body;
(c) and P is also able consciously and non-conceptually to follow O through O's various local changes and local movements in objectively real space and time.

The perceiver's ability for *uniquely* locating O relative to her body, and also the perceiver's ability for following O through *local* changes and *local* movements, are crucial features of this analysis of affection. This is for the following reason. According to Kant, through empirical intuition we can also *non*-uniquely locate and *non*-locally follow the changes and motions of very distant actual material objects, remotely past actual objects, and even futural actual material objects, providing that these objects have a "connection with some actual perception in accordance with the analogies of experience" (*CPR* A225–6/B273–4). And this, I think, is best construed as the thesis that for Kant empirical intuition can under some conditions be a form of "deferred ostensive reference."[84] But although empirical intuition is *required* by perception, empirical intuition alone is not *sufficient* for perception: perception is empirical intuition *plus* a subjective unity of consciousness (*CPR* B160). Furthermore, as we saw in section 2.4, for Kant the subjective or first-person character of an animal's consciousness (its unique "point of view") is essentially guaranteed by that animal's non-conceptual capacities, and in particular by its capacities for spatiotemporal representation. So whenever a cognizer's capacity for empirical intuition is supplemented by a non-conceptual ability to locate, consciously and uniquely, a material object in space relative to her own body (which is itself the outer aspect of the unique point of view that belongs intrinsically to her inner sense or phenomenal consciousness) and also consciously to follow the various causal-dynamic local changes and local movements of the actual material object (which of course rules out direct perceptual cognitive access to very distant, remotely past, and futural material objects), then this dual combination of capacities constitutes the cognitive capacity for the direct perception of that actual macrophysical material object.

---

[84] See Hanna, *Kant and the Foundations of Analytic Philosophy*, p. 210 and n.58.

In this way, the affection relation—the causal-dynamic interactive non-conceptual spatiotemporal tracking relation—can hold between the perceiver and past actual material objects, and also between the perceiver and simultaneous actual material objects. But in neither case is the causal-dynamic interactive tracking relation fully captured or entirely *exhausted* by either diachronic causation or synchronic causation, since it also intrinsically involves the non-conceptual spatiotemporal representational capacities of the conscious perceiver, which constrain the scope of interactive tracking to the recent, local causally efficacious past and the local simultaneous sphere of causal-dynamic community. Therefore the causal-dynamic interactive tracking relation is inherently scaled to and shaped by the brute contingencies of the specific spatiotemporal configuration, specific causal and motile propensities, and also the specific cognitive parameters, of the conscious perceiver's living body.[85] So Kant's theory of perception is not merely a *causal* theory of perception:[86] more than that, it is also an *embodied* causal theory of perception.

(3) *Non-Conceptual content and direct perceptual realism.* This leads me to the third and final important epistemic implication of Kant's theory of the spatiotemporal basis of non-conceptual content. The directness of direct perception, I have said, consists in the conjunction of its non-epistemic character, its non-conceptuality, and its not being otherwise mediated. It is obvious enough, I think, given the essential occurrence of concepts in beliefs, that the non-conceptuality of a cognition is necessary and sufficient for its being non-epistemic. But it may not be similarly obvious that the non-conceptuality of a perception is sufficient for its being not in

---

[85] Assuming that Special Relativity is correct, we might wonder: why is the speed of light a physical absolute in the actual material universe? From a Kantian point of view the answer could be: because, as a sheer matter of fact, it determines the absolute outer limits of the specific spatiotemporal configuration, specific causal and motile propensities, and specific cognitive parameters of the living body of a rational human animal. The actual material universe was not *designed* to be like this: it just accidentally came out this way. But once it *had* accidentally come out this way, that places some weak anthropocentric modal constraints on anything that counts as an actual or possible part of this actual material universe. More precisely, necessarily for anything that is a part of this actual material universe, then the living human body of a rational animal with such-and-such a specific spatiotemporal configuration, specific causal and motile propensities, and specific cognitive parameters, must *also* be possible. In short, it is impossible for something to belong to actual material nature in such a way that it makes us impossible. This, I think, is the Kantian grain of truth in the controversial "Anthropic Principle" discussed by some contemporary physicists.
[86] See, e.g., Grice, "The Causal Theory of Perception."

any sense mediated, whether by beliefs or concepts, by another cognitive faculty or content, by a psychological intermediary other than perception itself, or by a physical intermediary other than the perceiver's body. Nevertheless, when we recognize that non-conceptual content has its fundamental explanation in the spatiotemporal representational capacities of the conscious perceiver, then it becomes altogether clear how for Kant non-conceptuality is essential for the non-mediated character of direct perception, and thereby also essential for the directness of direct perception itself. In other words, Kant's theory of non-conceptual content is the *core* of his direct perceptual realism.

3

Manifest Realism I: A Critique of Scientific Essentialism

I am apt to doubt that, how far soever humane industry may advance useful and *experimental* Philosophy *in physical Things*, *scientifical* will still be out of our reach: because we want perfect and adequate *Ideas* of those very Bodies, which are nearest to us and under our Command... Having no *Ideas* of the particular mechanical Affections of the minute parts of bodies, that are within our reach, we are ignorant of their Constitutions, Powers, and Operations.

John Locke[1]

The *real* essence (the nature) of any object, that is, the primary *inner* ground of all that necessarily belongs to a given thing, this is impossible for human beings to discover... [T]o recognize the real essence of matter, the primary, inner, sufficient ground of *all* that *necessarily belongs to* matter, this far exceeds the capacity of human powers. We cannot discover the [real] essence of *water, of earth*, or the [real] essence of any other empirical objects.

PC 9: 143–4

3.0. Introduction

Where are we now? My project in this book is to work out a Kantian solution to what, following Sellars's canonical formulation, I have called the Two Images Problem: How is it possible to reconcile the the objective, non-phenomenal, perspectiveless, mechanistic, value-neutral, impersonal, and amoral metaphysical picture of the world delivered by pure mathematics and the fundamental natural sciences (*the scientific image*), with the subjective, phenomenal, perspectival, teleological, value-laden, person-oriented, and

---

[1] Locke, *Essay Concerning Human Understanding*, p. 556–7, IV.iii.26.

moral metaphysical picture of the world yielded by the conscious experience of rational human beings (*the manifest image*)? In this first part of the book I have been working on the relation between Kant's empirical realism and contemporary scientific realism. And I have argued this so far: (i) that Kant's empirical realism is the conjunction of direct perceptual realism and manifest realism; (ii) that his empirical realism is logically independent of his transcendental idealism, insofar as the *strong* version of his transcendental idealism is not necessary for empirical realism; (iii) that a charitably reconstructed version of the Refutation of Idealism yields direct perceptual realism; (iv) that Kant's theory of non-conceptual content is the core of his direct perceptual realism; and (v) that by virtue of Kant's non-conceptualism his direct perceptual realism is sharply opposed to both the Dogma of Semantic Overdetermination that is presupposed by eliminativist scientific realism and also the indirect perceptual realism that is presupposed by the other contemporary versions of scientific realism.

But what is scientific realism?[2] Scientific realism in the broadest sense, or what I will call *minimal scientific realism*, is the two-part thesis that:

(1) Some knowable things (whether individual material entities, natural kinds, events, processes, or forces) exist in objectively real physical space and not merely in consciousness.
(2) These knowable individual material entities, natural kinds, events, processes, and forces, as described by natural science, have explanatory primacy in our best theory of the natural world.

Noumenal scientific realism then adds to these two theses, the further two theses that:

(3) Each knowable physical spatial thing is ontologically constituted by a set of intrinsic non-relational properties.

---

[2] In fact there are many slightly different competing conceptions of scientific realism: to take just two recent examples, see Derden, "A Different Conception of Scientific Realism: The Case for the Missing Explananda"; and Kitcher, "Real Realism: The Galilean Strategy." To allow for this diversity, I have tried to formulate a version of scientific realism that is robust enough to include every theory that satisfies Putnam's thesis of "metaphysical realism"—see, e.g., *Reason, Truth, and History*, p. 49—but otherwise neutral as between different conceptions. It should also be noted that if, as I have suggested earlier (section 1.0), we think of metaphysical realism as encompassing *any* substantive doctrine about the intrinsic properties of things, then certainly not every metaphysical realism is "metaphysical realism" in Putnam's technical sense of that phrase. Indeed, Putnam's "metaphysical realism" is explicitly a version of *noumenal* realism: see, e.g., *The Many Faces of Realism*, Lecture I. But there are also some *non*-noumenal forms of metaphysical realism, e.g., Kant's *manifest* realism.

(4) All knowable physical spatial things are *transcendent*, in the triple sense that (4i) possibly the knowable physical spatial things exist even if human minds do not or cannot exist, (4ii) possibly the knowable physical spatial things exist even if all human cognizers do not know or cannot know them, and (4iii) necessarily the knowable physical spatial things are not directly and non-conceptually humanly perceivable or observable but instead only at best either *semantically overdetermined* by background theories and concepts or else *indirectly* humanly perceivable or observable.

Finally, *maximal* scientific realism is noumenal scientific realism plus two further theses that:

(5) There is exactly one true description of the world of knowable physical spatial things, and *truth* is the correspondence-relation between thought or language and the several mind-independent facts making up the world of knowable physical spatial things.

(6) The essential properties of all knowable physical spatial things are *microphysical* properties.

We have already seen that Kant's empirical realism runs contrary to maximal scientific realism by virtue of his commitment to direct non-conceptual perceptual realism. In the next two chapters, I want to explore the opposition between Kant's empirical realism and maximal scientific realism from another angle, by developing his manifest realism, which I have defined as follows:

Manifest Realism: All the essential properties of dynamic individual material substances, natural kinds, events, processes, and forces in objectively real physical space and time are nothing but their directly humanly perceivable or observable intrinsic structural macrophysical properties.

Or in other words, for Kant, both cognitively and ontologically speaking, *nothing is hidden*. Here I am borrowing, and retro-fitting for contemporary Kantians, a famous remark of the later Wittgenstein:

If it is asked: "How do [true empirical cognitions in the natural sciences] manage to represent?"—the answer might be: "Don't you know? You certainly see it, when you [cognize] them." *For nothing is concealed.* How do [true empirical cognitions in the natual sciences] do it?—Don't you know? *For nothing is hidden.* But given this answer: "But you know how [true empirical cognitions in the natural sciences] do it, for nothing is concealed" one would like to retort, "Yes, but it all goes by so quick, and I should like to see it as it were laid open to view".[3]

The present chapter makes a negative case for Kant's manifest realism by criticizing the leading contemporary version of maximal scientific realism, namely, *scientific essentialism*, and then chapter 4 works out the positive theory lying behind this negative case. A leftover Big Issue that is specifically raised by the fifth thesis mentioned above—the question of the nature of the relation between realism and truth—will be discussed in detail in chapter 5.

From the 1960s onwards, much philosophical activity has been directed to the interpretation and explanation of true propositions[4] about natural kinds such as gold and water. In turn, the question of the nature of natural kind propositions lies at the convergence of three highly influential lines of philosophical theory within the mainstream analytic tradition after 1960: (a) the semantics of direct reference (NB. direct reference semantics holds that the semantic value of a term is its referent, and that reference is not mediated by descriptions or Fregean senses); (b) the epistemology and metaphysics of scientific realism; and (c) the modal metaphysics of essentialism (NB. essentialism holds that individuals and kinds have intrinsic properties, some of which are also constitutive of those individuals and kinds). But at the same time the whole topic has a decidedly Kantian provenance. According to Kant, the natural kind proposition.

(GYM) Gold is a yellow metal

is analytically true, necessary, and a priori:

All analytic judgments are a priori even when the concepts are empirical, as, for example, "Gold is a yellow metal": for to know this I require no experience

---

[3] Wittgenstein, *Philosophical Investigations*, §435, p. 128e, text modified slightly.
[4] For the purposes of this and the next chapter I will not discriminate carefully between different types of truth-bearers: e.g., propositions (*Sätze*), assertoric judgments (*Urteile*), beliefs, statements, or assertively uttered sentences. The Kantian arguments should apply no matter what sort of item is favored. Nevertheless for the purposes of later chapters it is useful to note some Kantian notions and distinctions. Propositions are the truth-bearing semantic contents of judgments, or "what is judged" in the act of judging (*VL* 24: 934)(*R* 3100; 14: 659–60). And the power of judgment (*Urteilskraft*) is the *central* cognitive faculty of the human mind, in the sense that all of the other cognitive faculties must operate coherently and systematically together in order to produce a judgment. More precisely, judgments are built up out of sub-acts of intuition, conceptualization, and self-conscious synthetic predicative unification under universal a priori logico-grammatical constraints. See Hanna, *Kant and the Foundations of Analytic Philosophy*, sections 1.3–1.5, and 2.1; and Hanna, "Kant's Theory of Judgment."

beyond my concept of gold, which had as its content that this body is yellow and metal. For just this constituted my concept: and I need only decompose (*zergliedern*) it, without looking beyond it elsewhere. (*P* 4: 267)

For Kant, (GYM) is analytically true and necessary because its predicate expresses part of the *conceptual microstructure* of the natural kind concept GOLD. In traditional terms, the predicate-concept is "contained in" the subject-concept.[5] Furthermore (GYM) is semantically a priori in the sense of being semantically experience-independent, because the meaning and truth of the intrinsic conceptual connection between the predicate-concepts and the subject-concept is underdetermined by—that is, is not strongly supervenient on—(GYM)'s empirical verification conditions.[6] Not only is the proposition (GYM) semantically a priori, however, it is also *known* a priori in the sense that, on the assumption that (GYM) is in fact known by some means or another, the justification of a thinker's true belief in it requires only a direct decompositional insight into GOLD's conceptual microstructure, and is thereby underdetermined by—again, is not strongly supervenient on—any outer or sensory experiences or sensory processes by means of which the concept GOLD was learned or acquired.

In sharp contrast to Kant, however, Saul Kripke and Hilary Putnam[7] have influentially argued that propositions such as (GYM) and "Water is a clear liquid" are neither analytic, nor necessary, nor a priori. As Kripke puts it:

Kant...gives as an example [of an analytic statement] 'gold is a yellow metal,' which seems to me an extraordinary one, because it's something I think that can turn out to be false.

---

[5] See Hanna, *Kant and the Foundations of Analytic Philosophy*, ch. 3.

[6] Kant's doctrine of apriority as experience-independence (*CPR* A1–2/B1–2) is both semantic and epistemic, although he does not always carefully distinguish between the two types of apriority; and to complicate matters further, he also claims that apriority is strongly equivalent with the semantic notions of necessity and "strict universality." This can all be sorted out, I think; but it takes some work. See Hanna, *Kant and the Foundations of Analytic Philosophy*, section 5.2.

[7] See Kripke, "Identity and Necessity"; and Kripke, *Naming and Necessity*. In the 1970s, pursuing the same topics more or less independently, Putnam also argued for a similar view in a series of papers, including "Is Semantics Possible?," "Explanation and Reference," "Meaning and Reference," and (especially) "The Meaning of 'Meaning'." Putnam's version of scientific essentialism is not perfectly equivalent with Kripke's, however. Moreover Putnam later changed his views on these issues in both subtle and substantive ways—see, e.g., Putnam, "Is Water Necessarily H_2O?" To keep things relatively simple, I will treat Kripke's doctrine as the paradigm of scientific essentialism, and use Putnam's equally important views only insofar as they accord with, or deepen, Kripke's account.

Note that Kant's example, 'gold is a yellow metal,' is not even *a priori*, and whatever necessity it has is established by scientific investigation; it is thus far from analytic in any sense.[8]

The Kripke-Putnam view is based fundamentally on the doctrine of scientific essentialism,[9] which Kripke glosses as follows:

> Scientific investigation generally discovers characteristics of gold which are far better than the original set. For example, it turns out that a material object is (pure) gold if and only if the only element contained therein is that with atomic number 79. Here the 'if and only if' can be taken to be *strict* (necessary). In general, science attempts, by investigating basic structural traits, to find the nature, and thus the essence (in the philosophical sense) of the kind.[10]

According to scientific essentialism, then, natural kinds are identical to[11] their underlying particulate (molecular, atomic, or sub-atomic) physical microstructures; and these microstructures are knowable solely by means of the fundamental natural sciences (in particular, fundamental physics and chemistry). From this it follows that true propositions about the micro-entities or the micro-properties making up those physical microstructures, such as:

(GE) Gold is the element with atomic number 79

and "Water is H_2O" are statements of essential identity acquired and justified solely by empirical means. (GE) and other propositions like it are therefore both (i) logically necessary, or at least metaphysically necessary in the "strict" or Leibnizian sense that they are true in all logically

---

[8] Kripke, *Naming and Necessity*, pp. 39 and 123 n.63.
[9] See also Ellis, *Scientific Essentialism*.   [10] Kripke, *Naming and Necessity*, p. 138.
[11] There is some question as to whether the proposition "Water is H_2O" really means the same as "Water is identical to H_2O" or "Necessarily something is water if and only if it is H_2O" or "Water is materially constituted by H_2O" or "Water is strongly supervenient on H_2O." This is an issue because the four interpretations are logically independent: the first interpretation entails the second, third, and fourth; the second interpretation entails the third and fourth but not the first (due to the difference between property identity and necessary coextensionality); the third interpretation entails the fourth but does not entail either the first or the second (due to the possibility of multiple realization under material constitution); and the fourth interpretation does not entail either the first, second, or third (due to the possibilities of multiple realization and material non-coincidence under supervenience). See Baker, "Why Constitution is Not Identity"; and Kim, *Supervenience and Mind*. In any case, for the purposes of this chapter and the next, I will assume that "Water is H_2O" means that water is identical to H_2O, which of course also yields necessary coextensionality, material constitution, and strong supervenience.

possible worlds,[12] and also (ii) a posteriori in the epistemic sense that their being justifiably believed (and thereby known to be true) solely depends on specific factual and sensory experiential conditions. Thus their full-strength modality is anchored firmly in the natural world; and their acquisition apparently has nothing to do with a priori investigation, whether it be via conceptual analysis, rational intuition, or transcendental philosophy.

On the other hand, according to scientific essentialists, propositions like (GYM) are not in any way about the *metaphysically deep*, non-apparent or non-phenomenological, fundamental, lower-level, or microphysical properties of the natural kinds gold and water, but rather are only about the *metaphysically superficial*, apparent or phenomenological, derivative, higher level, or macrophysical properties of those kinds. Consequently, on the assumption that microphysical and macrophysical properties of natural kinds differ categorically—since microphysical properties are intrinsic non-relational dispositional properties, and macrophysical properties are extrinsic relational properties—there will be possible worlds in which gold (that is, the microphysically-defined stuff around here) exists but is neither yellow nor metallic, and in which our microphysically-defined water exists but is neither clear nor a liquid; and there will also be possible worlds in which the yellow, metallic, ductile, etc., stuff is not our microphysically defined gold and in which the clear, drinkable, liquid, etc., stuff is not our microphysically

---

[12] This raises a tricky modal issue. Certainly Putnam construes the necessity of essentialist identity statements about natural kinds in the logical or strict metaphysical, Leibnizian way ("The Meaning of 'Meaning'," p. 233). Kripke generally speaks this way too (see, e.g., *Naming and Necessity*, pp. 38, 125, 138). But there is a more cautious characterization that he mentions in passing ("Identity and Necessity," p. 164, and *Naming and Necessity*, pp. 109–10): namely, that a statement or proposition about a certain individual or natural kind is metaphysically necessary if and only if it is true in every possible world in which that individual or kind *exists*. Far from being an accidental or optional feature of metaphysical necessity, however, it seems to me that this existential element is crucial to it. For with the exception of mathematical truths, truths about God, and a very few others—which are held by Kripke to contain singular terms for *necessarily existing* objects—most essentialist identity statements contain singular terms for *actually and contingently existing* objects only. But this in fact can be used to drive a wedge between (1) logical necessity, or Leibnizian "strict" metaphysical necessity, or what is now sometimes called "weak metaphysical necessity" (truth in every logically possible world), and (2) non-logical necessity, or what is now sometimes called "strong metaphysical necessity" (truth in all and only the members of a specially restricted class of logically possible worlds that are all in some way irreducibly dependent on the actual world). See Chalmers, *The Conscious Mind*, pp. 136–8; and Farrell, "Metaphysical Necessity is not Logical Necessity." Otherwise put, strong metaphysical necessity is *synthetic* necessity. See Hanna, *Kant and the Foundations of Analytic Philosophy*, chs. 4–5. This is of course grist for the Kantian mill, and will be seen below to have important negative implications for scientific essentialism.

defined water.¹³ The association of the familiar observable macrophysical properties with the underlying physical microstructures is an adventitious empirical fact of this actual world, and goes no deeper. Hence (GYM) and other propositions like it are strictly synthetic, contingent, and a posteriori.

What would Kant say about contemporary scientific essentialist identity propositions about natural kinds? It seems obvious enough that even had Kant been aware of them, they would not have counted for him as analytically true. For, just to focus on (GE), it is obviously not logically true by Kant's lights since its denial is logically consistent in Kant's formal logic;¹⁴ and in addition it would not involve analytic "containment" of the predicate-concept in the subject concept, because for Kant the empirical concept GOLD contains only phenomenological, "identificational" sub-concepts, and therefore would not include within its conceptual microstructure the concept ELEMENT HAVING ATOMIC NUMBER 79.¹⁵

---

¹³ It is not often noticed, I think, that if scientific essentialism is combined with physicalist "micro-reductionism" about macrophysical properties, then the familiar "Twin Earth"-style modal counterexamples to the Kantian doctrine do *not* go through. (Kripke himself is quite aware of this point, however; see *Naming and Necessity*, pp. 128, n.66, and 144–55.) For suppose that the higher level or macrophysical properties of natural kinds are identical with, or at least necessarily coextensive with, their corresponding lower level or microphysical properties. Then, since the existence of the microphysical properties is necessary and sufficient for the existence of the macrophysical properties, there will *not* be a possible world in which the kind that is the element with atomic number 79 is not the kind that is yellow, metallic, etc.; nor will there be worlds in which the yellow, metallic, etc., stuff is not gold. Therefore, only if the scientific essentialist is a *property dualist* as regards microphysical and macrophysical properties, can the Kripkean and Putnamian Twin Earth examples work. In this chapter, to keep things relatively simple, I will not explicitly consider how Kant might respond to the scientific essentialist or scientific realist who is also a micro-reductive physicalist.

¹⁴ Obviously, in some very general sense, Kant's formal logic is Aristotelian/Scholastic (*CPR* Bvi–vii). But characterizing it more precisely in modern terms is tricky. In my opinion, Kant's formal logic is a *bivalent* (= classical 2-valued) *second-order* (= quantification not only over individuals but also properties or concepts of individuals) *monadic* (= quantification into one-place properties or concepts only) *intensional* (= irreducible modal operators and fine-grained intensions) logic. See Hanna, *Kant and the Foundations of Analytic Philosophy*, pp. 69–83; and Hanna, "Kant's Theory of Judgment," sections 2.1.1 and 2.1.2.

¹⁵ In conversation, Michael Friedman reminded me that the contemporary essentialist who is a micro-reductive physicalist can also regard (GE) as analytic, even though this option is not open to the original Kripke-Putnam version of scientific essentialism. The deep issue being raised here is how correctly to construe the meaning of "gold," and, correspondingly, the content of the concept GOLD. Kripke apparently treated the meaning of "gold" and the content of GOLD as univocal. But this leads to insuperable problems in trying to explain how we can at once have the "illusion of contingency" that possibly gold is not the element with atomic number 79 and also know that necessarily gold is the element with atomic number 79. The standard neo-Kripkean strategy nowadays—officially encoded in the doctrine of "two-dimensional modal semantics" (see ch. 4, n.8, for details)—is to treat "gold" as ambiguously expressing two meanings or concepts, one (called "the primary intension") which expresses the higher level or macrophysical properties of gold, and another (called "the secondary intension") which expresses the lower level or microphysical properties of gold. The micro-reductive

148   EMPIRICAL REALISM AND SCIENTIFIC REALISM

Moreover, and perhaps very surprisingly to those steeped in the lore of twentieth-century physics, Kant's theory of matter itself necessarily rules out the truth of (GE). Kant's theory broadly resembles the physical theories of Newton and Roger Boscovich,[16] both of whom were atomists. But he also radically extends those theories. Indeed, because in his Critical period Kant holds what I call a *causal-dynamic structuralist* conception of physical matter, he vigorously rejects both the standard "corpuscularian" or "rigid atom" atomic theory of the seventeenth and eighteenth centuries, and also atomic theories more generally, even including his own earlier pre-Critical non-corpuscularian atomic theory in the *Physical Monadology* of 1756, which appeared almost contemporaneously with Boscovich's essentially similar doctrine in his *Theory of Natural Philosophy* of 1758. Strictly speaking, both Boscovich's theory of matter and Kant's physical monadology posit the existence of atoms, or fundamental indivisible micro-entities, at the basis of matter. Nevertheless Boscovichian-Kantian "atoms" are not metaphysically *real* atoms but instead only *virtual* atoms, or more precisely what Thomas Holden calls "force-shell atoms," since they are nothing but unextended point-particles, each of which projects a diffused divisible shell of force that is in turn defined entirely in terms of relational attractive and repulsive kinematic laws.[17]

As Nietzsche pointed out in the late nineteenth century, Boscovich's conception of the virtual or force-shell atom represents a fundamental conceptual shift in philosophical thinking about the nature of matter:

While Copernicus had persuaded us to believe, contrary to all the senses, that the earth did *not* stand fast, Boscovich taught us to abjure the belief in the last part of the earth that "stood fast"—the belief in "substance," in "matter," in the earth-residuum and particle-atom: it is the greatest triumph over the senses that has been gained on earth so far.[18]

Otherwise put, Boscovichian "atoms" are *purely structural entities*. So my way of describing the Boscovichian-Kantian shift in the philosophy of matter

---

physicalist who is also a two-dimensionalist can then claim that (GE) is analytically necessary according to the secondary intension of "gold." Nevertheless, I believe that the first and fourth Kantian objections to scientific essentialism in sections 3.2 and 3.5 cast serious doubts on this thesis.

[16] See, e.g., Whyte, "Boscovich, Roger Joseph."
[17] See Holden, *The Architecture of Matter: Galileo to Kant*, esp. ch. 6.
[18] Nietzsche, *Beyond Good and Evil*, pp. 19–20.

is to say that it represents a transition from metaphysical substantialism and compositionalism about matter, to *metaphysical structuralism about matter*. In turn, this radically new conception of matter can be defined within the Kantian framework as follows:

Kantian Metaphysical Structuralism about Matter. Individual material things and natural kinds are not ontologically independent things-in-themselves, each defined by a set of intrinsic non-relational, mind-independent, non-sensory, unobservable properties, but instead are essentially *determinate positions* or *determinate roles* in a maximally large relational structure or system of empirical nature as a whole. The specific properties expressing these determinate positions or determinate roles are then constitutive intrinsic structural properties of those material things and natural kinds, that is, *structural essences* of those things and kinds.

Kant's mature Critical theory of matter is fully metaphysical structuralist in character, but also overtly causal-dynamic in its direct appeal to attractive and repulsive forces acting at a distance,[19] and, above all, it is explicitly anti-atomistic and radically macrophysicalist. More precisely, then, according to Kant's causal-dynamic metaphysical structuralist theory of matter:

(1) Matter as a substantial whole or as a single totality (which the later Kant ultimately takes to be a physical *aether*) is essentially relationally constituted by a real spatiotemporally-organized structural complex of attractive and repulsive physical forces that are nomologically determined by synthetically necessary causal and interactive laws (*CPR* A49/B66–7, A265/B321) (*MFNS* 4: 524–5, 533–4) (*OP* 21: 215–33, 22: 239–42).
(2) The plurality of particular macrophysical material substances (individual objects or specific natural kinds) are nothing but positions in that total spatiotemporal structural complex, or otherwise put, are nothing but determinate dynamic causal roles, without any hidden causal role-players (*CPR* A226/B273, A265/B321).[20]

---

[19] See Carrier, "Kant's Mechanical Determinations of Matter in the *Metaphysical Foundations of Natural Science*"; and Friedman, "Matter and Motion in the *Metaphysical Foundations* and the First *Critique*: The Empirical Concept of Matter and the Categories."

[20] For a further explication and defense of Kant's causal-dynamic structuralism about matter, see section 4.2.

On this Kantian metaphysical structuralist conception, objectively real matter has no positive noumenal (that is, no intrinsic, non-relational, mind-independent, non-sensory, and unobservable) physical properties, nor is it composed of unobservable indivisible extended atomic or sub-atomic parts (indeed, the very intelligibility of this notion is rejected in the Second Antinomy (*CPR* A435–443/B463–71)), but instead material individuals and kinds are essentially identical with a certain set of manifest intrinsic structural causal-dynamic properties—their *manifest dynamic structural essences*. So Kant is not only an anti-atomist; he is a *radical anti-microphysicalist*. Hence it would quite trivially be the case that for him there exist possible worlds in which gold is *not* the element with atomic number 79: for him, the actual world is just such a possible world! So (GE) would obviously not be analytically true.

And it gets even better than that. Kant's causal-dynamic manifest structuralist metaphysics of matter entails, or at least so I will argue later in the chapter, that if Kant had stepped into a time machine, traveled forward to the present day, and *had in some sense accepted* our contemporary Bohr-Rutherford atomic theory of matter, with its combination of elementary particles of various kinds, electromagnetic forces, charged atomic nuclei, nuclear transitions, electron orbits, discontinuous energy emissions, and quantum leaps, he would *still* have insisted that there are not only logically or conceptually possible worlds but also experienceable worlds in which gold exists but is not the element with atomic number 79.[21] Furthermore, even if Kant were to have accepted that (GE) is necessary, he would have also insisted that it is thereby a priori in that its necessity directly depends upon transcendental conditions for the possibility of human experience. So, merely for the sake of argument let Kant accept, in some sense, the contemporary claim that gold is the element with atomic number 79. Nevertheless, on his view of things, the meaning-content of that proposition would depend importantly upon certain special intuitional or essentially indexical (that is, irreducibly actual-world-dependent and egocentric or subject-centered) formal or material conditions that can

---

[21] Donnellan, in "Kripke and Putnam on Natural Kind Terms," imagines a scenario in which Locke is time-transported to the present, learns about contemporary scientific discoveries, and then opposes his empiricist theory of natural kinds and natural kind terms to that of scientific essentialism. I fully agree that an updated Locke is a natural enemy of essentialism. What I want to argue here, however, is that Kantian doctrines supply an even more direct and effective criticism: that Kant is the *nemesis* of scientific essentialism.

vary across logically possible worlds.²² Therefore (GE)—and, by extension, any other true scientific essentialist identity proposition about natural kinds—must be both synthetic and not strictly metaphysically necessary.

In these ways, the conflict between Kant and scientific essentialism is a sharply defined one. If the scientific essentialist account of natural kind terms and propositions is correct, then obviously Kant must be: (a) wrong that propositions like (GYM) are analytic; (b) wrong that propositions like (GE) are synthetic and not logically or strictly metaphysically necessary; and (c) wrong that all necessary propositions are a priori. That is, natural kind propositions interpreted in the scientific essentialist way would count as crucial *counterexamples* to the Kantian theory of meaning, necessity, and apriority. As Putnam observes:

> Since Kant there has been a big split between philosophers who thought that all necessary truths are analytic and philosophers who thought that some necessary truths were synthetic *a priori*. But none of these philosophers thought that a (metaphysically) necessary truth could fail to be *a priori*: the Kantian tradition was as guilty as the empiricist tradition of equating metaphysical and epistemic necessity. In this sense Kripke's challenge to received doctrine goes far beyond the usual empiricism/Kantianism oscillation.²³

Now Putnam is certainly right about one thing: if Kripke is correct, then Empiricists and Kantians alike are wrong. But looked at the other way around, if Kant is correct, then the scientific essentialists are: (a*) wrong that essentialism provides any good reasons for holding that (GYM) and other such propositions are not analytic; (b*) wrong that (GE) and other scientific essentialist identity propositions about natural kinds are logically or strictly metaphysically necessary; and (c*) wrong that essentialism provides any genuine counterexamples to the claim that all necessary truths are a priori. So the Kant vs. scientific essentialism debate is of the utmost importance for mainstream analytic philosophy in the late twentieth and early twenty-first cenuries. And since maximal scientific realism is a central feature of scientific or reductive naturalism, then the Kant vs. scientific essentialism debate is also of the utmost importance for the Two Images Problem. Right now I will not undertake the somewhat arduous task of unpacking and defending Kant's positive theory of natural kinds, natural kind terms, and

---

²² See Hanna, *Kant and the Foundations of Analytic Philosophy*, ch. 4.
²³ Putnam, "The Meaning of 'Meaning'," p. 233.

152 EMPIRICAL REALISM AND SCIENTIFIC REALISM

natural kind propositions,[24] together with explicating and defending the background Kantian metaphysical theory of manifest realism that supports this positive theory. I will instead reserve all that for chapter 4. In the rest of this chapter I will work out only the necessary negative prolegomenon to that task, a Kantian critique of scientific essentialism.

3.1. What is Scientific Essentialism?

Since the publication and philosophical assimilation of Kripke's *Naming and Necessity* and Putnam's "The Meaning of 'Meaning'," it has been too often assumed that scientific essentialism reflects a set of fairly simple, perspicuous, and even self-evident philosophical insights about the meaning and truth of the members of certain classes of propositions—a set of insights that any careful honest thinker would readily, or at least ultimately, assent to. These insights are usually presented as logical and semantic in character. Thus scientific essentialism is held to follow directly from:

(1) A general thesis concerning the correct analysis of necessary or strongly modal statements (namely, that they are logically or strictly metaphysically necessary and "Leibnizian," or true in all logically possible worlds accessible from a designated world, our actual world).[25]

(2) A general theory of the semantics of natural kind terms (namely, that every natural kind term is a "rigid designator": it holds its actual-world reference fixed across all possible worlds in which its referent exists, and never picks out anything else otherwise).[26]

(3) A specific thesis about the modality of identity-statements involving rigid designators (namely, that if they are true, then they are necessarily true).[27]

---

[24] Some interesting initial steps towards unpacking—if not actually defending—Kant's theory of natural kind terms and natural kind propositions have already been taken by others; see Anderson, "Kant, Natural Kind Terms, and Scientific Essentialism"; and Kroon and Nola, "Kant, Kripke, and Gold."

[25] See Kripke, *Naming and Necessity*, pp. 35–6; and Kripke, "Semantical Considerations on Modal Logic." Accessibility is a between-world relation of compatibility or joint consistency.

[26] See Kripke, *Naming and Necessity*, pp. 9, 26, 48–9; see also Putnam, "The Meaning of 'Meaning'," pp. 229–35. The theory of rigid designation partially coincides with direct reference semantics.

[27] See Kripke, "Identity and Necessity," pp. 162–7; and Marcus, "Modalities and Intensional Languages."

(4) A specific linguistic claim to the effect that natural kind terms are rigid designators, based on some proposals concerning the nature of the linguistic mechanisms of reference-fixing at work in society at large and in the natural sciences in particular (namely, the "causal theory of names,"[28] and the socio-linguistic hypothesis of "linguistic division of labor"[29]).

The rational compulsion supposedly induced by the conjunction of these insights is then used directly against the alternative views, be they empiricist or Kantian.

In point of fact, however, scientific essentialism is a thoroughly *hybrid* doctrine, more or less covertly incorporating several non-trivial metaphysical and epistemological assumptions into an enabling theoretical context *in which and as a consequence of which* its basic logical and semantic claims make philosophical sense. Essentialism thereby more or less covertly presupposes a fairly complex justificatory argument-structure whose own presuppositions, particular premises, inferential connections, validity, and soundness are far from being immediately self-evident. So, as proponents of an alternative Kantian position, we had better scrutinize the basic theses of essentialism with some critical care.

Scientific essentialism is a doctrine about the logic, semantics, metaphysics, and epistemology of *natural kinds*.[30] But what are natural kinds? As I

---

[28] See Kripke, *Naming and Necessity*, pp. 91–7.
[29] See Putnam, "The Meaning of 'Meaning'," pp. 227–9.
[30] One caveat: Apart from gold and water, other familiar examples of natural kinds include cats, elms, and human beings—that is, biological species. Intuitively, biological species are indeed natural kinds. Nevertheless in this chapter and the next I will avoid discussing biological natural kinds and focus solely on inorganic natural kinds, for three reasons: (a) It is not at all obvious to me that the essentialist microphysical metaphysics designed for inorganic natural kinds carries over automatically to organic natural kinds (is "being alive" a *micro*physical or a *macro*physical property of an organic kind?); (b) Kant's doctrine of "teleological judgments" about organic or living kinds, including animals, as worked out in the First Introduction and second half of the *Critique of the Power of Judgment*, is importantly different from his treatment of propositions about inorganic or inert natural kinds. For him, teleological judgments are merely "regulative" (heuristic, hypothetical, abductive), not "constitutive" or "determining" (objectively valid, empirically meaningful, deductive); (c) In chapter 8, I will argue that some non-conceptually represented biological features of our own animate bodies play a crucial role in Kant's theory of freedom, and because the cognitive semantics of these representations is non-conceptual, their content cannot be assimilated to teleological *judgments* and their mere regulativity. So, even though teleological judgments are only regulative, non-conceptual teleological *intuitions* are still referential. But all this means that Kant's theory of biological kinds is equivocal, for it entails that while he is a biological *anti-realist* at the level of concepts and judgments, he is nevertheless a biological *realist* at the level of intuitions and perceptions.

have glossed them, natural kinds are "real classes" of material things; and I have also mentioned that familiar examples of natural kinds include gold and water. More precisely, however, according to a philosophical tradition originating with Aristotle,[31] natural kinds are themselves material substances in the robust sense that they are ontologically independent beings having an essence, constitution, or nature. Unlike individual material substances, however, natural kinds are inherently plural or multiple, and also complete within their kinds. So natural kinds are either complete collections of individual material substances made out of the same concrete stuff (referred to by "count nouns" such as "diamonds") or else complete mereological conglomerates of bits of the same concrete stuff (referred to by "mass nouns" such as "gold"). Members or bits of a given natural kind are often widely dispersed in space and time. Yet every such member or bit of a natural kind possesses a more or less definite set of shared apparent physical properties, or *macro*physical properties, by virtue of which it is identified and empirically sorted or categorized into that kind. This set of macrophysical properties is normally expressed in the dictionary definition of the general word used to designate the kind and all its members or bits.

Nevertheless, it is not the shared set of macrophysical properties that actually *makes* the various members or bits into items of the same natural kind. Instead, all the members or bits of the same natural kind are intrinsically linked by virtue of their sharing in common the same underlying material basis or stuff (which for convenience I will call the "substratum" of the kind). The relevant set of macrophysical properties is then held to *depend* on the substratum in the threefold sense that:

---

[31] Actually, the historical provenance of the contemporary philosophical picture of natural kinds is complex, since it involves not merely Aristotelian modal essentialism (= the view that all and only substantial kinds have necessary properties) together with noumenal scientific realism (= the view that mind-independent unobservable fundamental microphysical entities exist in space and are knowable) and their impact on Kripke's semantics, epistemology, and modal metaphysics, but also the mediation of Locke and Leibniz. See Ayers, "Locke vs. Aristotle on Natural Kinds." Roughly speaking, Locke artfully yet rather confusingly fuses the tradition of Aristotelian essentialism with (i) an empiricist epistemology that is vigorously skeptical about the possibility of knowledge beyond the limits of our sensory ideas and also with (ii) Robert Boyle's version of scientific realism: the view that macrophysical bodies have mind-independent microphysical "corpuscular" constitutions supporting their really real or primary qualities, which in turn cause their merely apparent or secondary qualities. The great unsolved puzzle in Locke is whether the corpuscles, their real internal constitutions, and their primary qualities are ultimately observable or unobservable. What Leibniz adds to this are (1) the crucial idea that the essential properties of natural kinds are intrinsic and non-relational, and (2) the equally crucial idea of noumenally real possible worlds.

(i) the macrophysical properties are predicated of the substratum;
(ii) the existence of the substratum is metaphysically and also causally sufficient for the existence of the relevant set of macrophysical properties; and
(iii) there cannot be a change in any of the relevant macrophysical properties of something without a corresponding change in that thing's substratum properties.

In other words, the macrophysical properties *causally strongly supervene* on the properties of the substratum.[32]

Assuming this relatively "neutral" conception of natural kinds, scientific essentialism can then be displayed as a set of further theses, premises, and conclusions taking the form of a single extended argument, as follows:

(1) *The Metaphysical Doctrine.* Natural kinds, like gold or water, are to be metaphysically identified with their substrata: in other words, the substrata are the *essences* or *constitutions* or *natures* of those kinds. The substrata in turn are entirely and ultimately composed of various distinctive intrinsic non-relational microphysical properties and their corresponding physical micro-constitutions. These micro-constitutions involve molecular, atomic, or sub-atomic particles in various chemical and electromagnetic relations and causal interactions. Because the particles and their particulate intrinsic non-relational properties, physical micro-constitutions, and causal interactions are all essentially *microphysical* in character, thus falling far below the physical scale and dynamical scope of the human

---

[32] As I mentioned in the Introduction, n.30, the strong supervenience of macro-level properties on first-order physical properties, together with token physicalism, constitutes a minimal materialism in the sense that it is consistent with reductive and non-reductive materialism alike, and thus can be tightened up to yield reductive materialism if desired. Scientific essentialists who follow Kripke and Putnam must regard the connection between the micro-level and the macro-level as both (i) causal and (ii) weaker than identity or logically necessary equivalence (to account for Twin Earth cases), and may even regard it as a causal relation that is weaker than nomological equivalence (if Twin Earth shares all our natural laws). It is worth noting, too, that even the Humean radical empiricist about kinds will preserve the form of this causal dependency relation; see, e.g., Hume's remarks in the *Enquiry Concerning Human Understanding*, pp. 17–22, on the "secret structure of parts" and the "secret powers" supposed by us to underlie the sensible qualities of natural kinds. For Hume, that the causal relation is empirically unknown or even unknowable by us is quite consistent with at least the bare supposition of its existence. And even more radically conventionalist theories of natural kinds will preserve the form of the causal dependency relation as a set of stipulative linguistic rules governing the use of referential expressions; see, e.g., Sidelle, *Necessity, Essence, and Individuation*, chs. 1–3 and 6.

body and its sensory organs, then not only the micro-entities and their intrinsic non-relational micro-properties but also their micro-constitutions, are all completely inaccessible to sense perception or ordinary empirical observation—that is, they are all completely non-apparent or "hidden."[33] On the other hand, and by sharp contrast, the macrophysical properties ascribed to natural kinds reflect only our human subjective sensory responses to causal interactions with the microphysical entities, their intrinisc non-relational microphysical properties, and the micro-constitutions that entirely and ultimately compose the natural kinds. In this way, the macrophysical properties of natural kinds are (a) strongly superveniently caused by the substrata of the kinds, and also (b) merely extrinsic relational features of those kinds.[34]

(2) *The Epistemological Doctrine.* The essential microphysical properties and physical microstructures of at least some natural kinds are actually known by contemporary natural scientists;[35] and the microphysical properties and microstructures of all the other natural kinds are in principle knowable through future scientific investigation. This deep scientific metaphysical knowledge is empirically derived or a posteriori, and therefore cannot be gained by any sort of a priori investigation. By contrast, knowledge of the macrophysical properties of natural kinds, whether by perceptual observation or by analytic a priori knowledge of the dictionary meanings of natural kind terms, belongs strictly to "folk science" or to conceptual "stereotypes" of common sense, and is of practical use only.[36]

(3) *The Semantic Doctrine.* A term is a rigid designator if and only if it is a term that determines the same referent that it has in the actual world in every logically possible world in which the referent exists, and never refers to anything else otherwise. Some, but not all, rigid

---

[33] See Putnam, "The Meaning of 'Meaning'," p. 235.

[34] See Kripke, *Naming and Necessity*, p. 133: "The property by which we identify [the natural kind] originally, that of producing such and such sensation in us, is not a necessary property but a contingent one."

[35] See Kripke, *Naming and Necessity*, p. 124: "Let us suppose that scientists have investigated the nature of gold and found that it is part of the very nature of this substance, so to speak, that it have the atomic number 79."

[36] See Putnam, "The Meaning of 'Meaning'," pp. 247–52.

designators are directly referential.[37] Directly referential terms are terms: (a) whose referents exist (or at least subsist)[38] in the actual world; (b) that do not express Fregean senses or "modes of presentation" as their semantic contents; (c) whose semantic values are their referents themselves and whose semantic contents thereby literally contain, or at least are determined by, their referents; and (d) that do not uniquely determine their referents by means of descriptions but instead fix their referents by means of various pragmatic and contextual factors (sometimes called "meta-semantic"). All rigid designators have their reference fixed by causal-historical chains of word use that originate in non-stipulative or stipulative "dubbings" of the referents themselves. By contrast, a term that is not predicable of its actual world referent in every possible world is a non-rigid designator. No non-rigid designators are directly referential: they all fail to have their actual-world reference in some other possible worlds; they express Fregean senses or modes of presentation as their semantic contents; their semantic value is something other than their referents themselves and thus their semantic content neither contains nor is determined by their referents; and they uniquely determine their reference—when they have a reference—solely by descriptive, non-pragmatic, and non-contextual means.

(4) *The Logical Doctrine.* Necessary statements are true in every logically possible world (that is, they are "strictly metaphysically necessary"). And all identity statements containing only rigid designators are necessarily true if true at all, even when the truth of such statements is discovered a posteriori.

(5) *The Crucial Premise.* Both the names of natural kinds (for example, "gold," "water," etc.) and what for lack of a better label can be called "scientific natural kind terms" (that is, definite descriptive terms expressing the physical microstructures of those kinds, such

---

[37] For an extended presentation of the theory of direct reference, see Kaplan, "Demonstratives: An Essay on the Semantics, Logic, Metaphysics, and Epistemology of Demonstratives and Other Indexicals," and "Afterthoughts." The paradigms of the directly referential term are the indexical, the proper name, and the names of natural kinds. Hence it is important to recognize that some rigid designators are *not* directly referential, e.g., rigid definite descriptions such as "the element having atomic number 79" and "H_2O".

[38] This is to allow for the possibility of direct reference to abstract objects, fictional objects, future objects, etc.

as "the element with atomic number 79" and "H₂O") are rigid designators, because they pick out the microphysical essences of the kinds themselves. Names of natural kinds are directly referential, but scientific natural kind terms are not directly referential even though they are rigid designators. Nevertheless, the word-use chains for names of natural kinds and scientific natural kind terms alike have their original dubbings determined by natural scientists, who function as linguistic experts in relation to the whole community of language-users. By contrast, what again for lack of a better term can be called "manifest natural kind terms" (that is, descriptive terms expressing macrophysical properties of natural kinds, such as "yellow metal") are non-rigid designators, and also not directly referential terms, because they do not pick out the lower-level microphysical essences of the kinds they refer to, but instead merely describe accidental higher-level macrophysical features of them. Furthermore, manifest natural kind terms express sense-like or conceptual "stereotypes" that are of practical or commonsensical use for discriminating between kinds, but do not in any way uniquely determine the cross-possible-worlds extension of the natural kind terms: indeed, stereotypes may not even determine actual-world extensions.

(6) *The First Conclusion.* Given (1)–(5), it follows that every truth about natural kinds which takes the form of a scientific natural kind proposition involving an identity-predication between two natural kind terms—one being a name of that kind, and the other being a scientific natural kind term (for example, "Gold is the element with atomic number 79")—is a logically or strictly metaphysically necessary a posteriori proposition. It is logically or strictly metaphysically necessary because it employs only rigid designators of the natural kind, and thus is a true identity statement in all logically possible worlds. But it is also epistemically a posteriori (and therefore not analytically true, which would entail epistemic apriority) because it is known through empirical science alone.

(7) *The Second Conclusion.* Given (1)–(5), it also follows that every truth about natural kinds expressed as a predication between the name of a natural kind and a manifest natural kind term (for example, "Gold is a yellow metal"), is a merely contingent a posteriori proposition. This is so because, despite its being empirically true, its manifest natural

kind term is not a rigid designator: hence the proposition says nothing essential about the natural kind but reflects only extrinsic relational apparent properties of it and our pre-theoretical, commonsensical ways of knowing it.

As presented in this skeletal form, the overall argument for scientific essentialism is still undoubtedly somewhat abstract. But at least its general thrust and upshot should be clear. In any case, what I want to do in the next four sections is to offer a direct Kantian critical rejoinder to scientific essentialism. In doing so, I will focus only on the Metaphysical and Epistemological doctrines. I do this for two reasons.

First, while the Semantic and Logical doctrines, the Crucial Premise, and the two Conclusions have been quite carefully and extensively explored in the critical literature on scientific essentialism,[39] the Metaphysical and Epistemological doctrines have been somewhat less closely or widely examined.[40] But second, and more importantly, it should be obvious enough that the Metaphysical and Epistemological doctrines are *pivotal conditions* of the whole argument: without them in place, the scientific essentialist conclusions simply do not go through. Even assuming the correctness of the theories of rigid designation and direct reference, and of the logical intuitions about modality and identity, there would be no reason whatsoever to believe that names of natural kinds and scientific natural kind terms alike actually *are* rigid designators and that manifest natural kind terms actually *are* non-rigid designators and not directly referential terms, unless the Metaphysical and Epistemological Doctrines were true. For the Metaphysical Doctrine guarantees the radical difference between the referential properties of a natural kind name or scientific natural kind term, and those of a manifest natural kind term. And the Epistemological Doctrine guarantees that the contrasting sorts of terms actually do apply

---

[39] See, e.g., Cartwright, "Some Remarks on Essentialism"; Donnellan, "Kripke and Putnam on Natural Kind Terms"; Donnellan, "Necessity and Criteria"; Quine, "Reference and Modality"; Salmon, "How *Not* to Derive Essentialism from the Theory of Reference"; Salmon, *Reference and Essence*; Wiggins, "Putnam's Doctrine of Natural Kind Words and Frege's Doctrines of Sense, Reference, and Extension: Can they Cohere?"; and Zemach, "Putnam's Theory on the Reference of Substance Terms."

[40] There are, of course, a few notable exceptions. On the Metaphysical Doctrine, e.g., see Mellor, "Natural Kinds." And on the Epistemological Doctrine, see Bealer, "The Philosophical Limits of Scientific Essentialism"; Casullo, "Kripke on the A Priori and the Necessary"; and Kitcher, "Apriority and Necessity."

(through the mediation of empirical natural science on the one hand, and via ordinary perception or common sense on the other) in the sharply contrasting ways specified. Only by invoking these guarantees, can the Crucial Premise be derived. In this way, both the semantics of rigid designation or direct reference and the logical theory of modality and identity can be logically detached from the theory of scientific essentialism: they do not *alone* entail essentialism, and can in fact be consistently combined with non-scientific essentialist metaphysical or epistemological theories.[41] But, most importantly of all, without the Metaphysical and Epistemological doctrines, the Crucial Premise simply does not hold; and without the Crucial Premise it simply will not follow that propositions like (GE) are strictly metaphysically necessary and that propositions like (GYM) are merely contingent. Therefore, in order to expose and probe the soft underbelly of scientific essentialism, Kant would need to focus only on its metaphysics and epistemology.

There are, I think, at least four distinct ways that Kant would criticize scientific essentialism. The interpretation of scientific essentialism worked out in this section strongly suggests that the most sublimated and vulnerable parts of essentialism are its epistemological and metaphysical components. So, correspondingly, two of the Kantian counterarguments will be epistemological and two will be metaphysical.

3.2. Objection I: The Empirical Inaccessibility of the Microphysical World

It is a basic assumption of scientific essentialism that microphysical properties and physical microstructures are knowable through empirical inquiry in the natural sciences. But, as Locke first pointed out in his famous doctrine of "real essences" and "real internal constitutions" in the *Essay Concerning Human Understanding*, microphysical properties and physical microstructures are in fact necessarily hidden from such empirical inquiry and are therefore totally empirically inaccessible, because they are thoroughly *non*-apparent

---

[41] The inferential gap between modal logic plus rigid designator or direct reference semantics on the one hand, and scientific essentialism on the other, has been pointed up in various ways by Putnam in "Is Water Necessarily H$_2$O?"; by Salmon in "How *Not* to derive Essentialism from the Theory of Reference"; and by Sidelle in "Rigidity, Ontology, and Semantic Structure."

and *non*-phenomenological, and thereby inherently *unobservable*.[42] The most we can know empirically about natural kinds are their "nominal essences" or the corresponding complex general sensory ideas (in Kantspeak, the "empirical concepts") that we have acquired experientially and that we use more or less conventionally for the practical purposes of identifying and naming the kinds.[43] So the real essences of natural kinds are both perceptually and conceptually closed to us.

This worry has two degrees of salience. First, the microphysical properties and physical microstructures cannot be directly perceived, or even empirically conceived, because of basic contingent limitations of our human (all-too-human!) sensory and concept-forming apparatus. More specifically, the microphysical properties and physical microstructures occur at a spatial or temporal scale that falls well beneath the spatial or temporal scale of the human body; and the causal powers of those microproperties and microstructures fall well outside the dynamical scope of our sensory organs, and therefore cannot be detected by them. As contingent, these limitations can at least in principle be overcome by the use of improved ancillary detection devices (microscopes, etc.) or by epistemically favorable human evolution. Indeed, there is no a priori reason why our perceptual capacities cannot be extended artificially or evolutionarily to the point at which we can detect all sorts of natural phenomena that are currently too small or too large or too far away for us to intuit empirically. This is what Kant calls the "empirical progress" or "empirical regress" of objects of possible experience in space, according to necessary rules or laws of nature (*CPR* A496/B524)

But, second, the Lockean doctrine suggests a further and much deeper difficulty to the effect that humans are simply not cognitively constituted so as to have direct or even an *indirect, progressive, or regressive* epistemological access to the sorts of strange and wholly uncommonsensical entities—microphysical particles, energy quanta, light waves, gravitational fields, etc.—required by the scientific image of the world. Indeed, even Sellars's root idea that the microphysical conception of the natural world constitutes a meaningful "image" is put seriously at risk here. For, in

---

[42] See Locke, *Essay Concerning Human Understanding*, pp. 554–5 (IV.iii.24) and 556–7 (IV.iii.26). As a matter of historical and textual accuracy it must be admitted, however, that Locke is somewhat inconsistent on this point: occasionally he seems to suggest that human sensory access to physical microentities, microproperties, and microstructures is at least possible—see, e.g., pp. 301–3 (II.xxiii.11–12).

[43] Locke, *An Essay Concerning Human Understanding*, pp. 439 (III.vi.2) and 452–62 (III.vi.24–35).

order for images to make sense, there must be isomorphisms, analogies, or some other sort of basic similarity of properties between the image and imaged object. But, as cognizers with our specific sort of sensibility and our specific sort of sensibility-funded conceptual capacities, we are not cognitively equipped to bring about *any* sort of significant mapping from the microphysical order into the macrophysical order. This is not therefore an issue of scale or scope, or of the empirical regress or progress of perceptual experiences in space—which are in effect merely matters of technological or evolutionary "engineering." Instead it is an issue of the impossibility of perceiving things-in-themselves:

> [T]he objects of experience are **never** given **in themselves**, but only in experience, and they do not exist at all outside it. That there could be inhabitants of the moon, even though no human being has ever perceived them, must of course be admitted; but this means only that in the possible progress of experience we could encounter them; for everything is actual that stands in one context with a perception in accordance with the laws of the empirical progression. Thus they are real when they stand in an empirical connection with my real consciousness, although they are not real in themselves, i.e., outside this progress. (*CPR* A492–3/B521)

This is a fundamental point. The microphysical constitution of the table in front of me, for example, *is not another table*, only more insubstantial, schematic, pointillistic, and energetic than this one, so that the laptop computer sitting on this table, to use Arthur Eddington's evocative description, "is poised as it were on a swarm of flies and sustained in shuttlecock fashion by a series of tiny blows from the swarm underneath."[44] Indeed Eddington's famous discussion of the "two tables" is crucially misleading because it radically understates the nature of the "explanatory gap" between the microphysical and macrophysical worlds. The truth of the matter is that there really are *no* macrophysical objects (tables, laptop computers, etc.) whatsoever in the microphysical world! The supposed microphysical counterpart of this table not only does not play the same set of causal roles as the macrophysical table—for example, unlike the table, it is not a rigid, static, continuously dense object with a continuously extended surface—but also presumably does not even count as a *single* entity from the standpoint of physical theory. *Hence there aren't two tables here, whether for Eddington or for anybody else. There is one and only one table, the macrophysical material object on*

---

[44] See Eddington, *The Nature of the Physical World*, pp. xi–xix.

which my laptop computer is sitting; and it has no genuine microphysical counterpart at all, but instead, at best, only a set of hypothetical lawlike correlations with wholly unobservable entities. So the macrophysical and microphysical worlds are not merely far apart, explanatorily and ontologically. In fact they are *explanatorily and ontologically incommensurable.*

This deeper Lockean worry dovetails perfectly with the sharp distinction that the scientific essentialist draws between the strictly semantical/metaphysical vs. the strictly epistemological sides of the language community's employment of natural kind terms. As Kripke points out, even granting the truth of (GE), there are two senses in which gold "might not have been" the element with atomic number 79:

(1) It is both logically and analytically possible that the actual world might have been other than it is (so the actual empirical fact that gold is the element with atomic number 79 might, in principle, not have obtained).

(2) It is "epistemically possible" that gold might not have been the element with atomic number 79, in the sense that given the perceptual epistemic situation of those who understand or believe that gold is the element with atomic number 79, they did not *have to* believe it:[45] that is, given the very same perceptual epistemic situation, some thinkers like us could easily fail to believe that gold has any sort of physical microstructure (for example, as in Aristotelian-Scholastic science), or could instead believe that gold has a quite different physical microstructure (for example, as in seventeenth-century corpuscularianism).

Such an epistemic possibility, that is, depends entirely on the fact that cognizers of gold have access only to the sensory appearances of gold,

---

[45] See Kripke, *Naming and Necessity*, pp. 103–5, 123–5, 141–4, 150–3; and also Putnam, "The Meaning of 'Meaning'," p. 233. Unfortunately, neither Kripke nor Putnam carefully defines epistemic possibility or epistemic necessity. As I understand it, however, a proposition *P* is epistemically possible if and only if understanding or considering *P* does not lead automatically to a belief in *P*'s denial—that is, nothing necessarily rules out a thinker's believing in *P*. Thus, if *P* is "Gold is not the element with atomic number 79," to say that *P* is epistemically possible is to say that one *doesn't absolutely have to believe that gold is the element with atomic number 79 just because one has understood or considered P: one can either believe P or the denial of P*. For *P* to be epistemically necessary is for *P* to be such that any rational understanding or consideration of *P* also suffices for belief in *P*: i.e., *P* is rationally unrevisable, so merely to understand it or consider it is also to believe it with a warrant. This warrant is still defeasible however. It is also plausible to hold that if a proposition is epistemically necessary, then whenever it is believed or known, it is believed or known a priori. So epistemic necessity yields an a priori warrant.

or to concepts that describe those sensory appearances (Locke's "nominal essences"). Now assuming:

(3) that it is an actual fact that gold is the element with atomic number 79, it then follows (by the Logical Doctrine) that gold is *logically or strictly metaphysically necessarily* the element with atomic number 79: on that assumption, there are no logically possible worlds in which gold shows up but is not possessed of precisely that physical microstructure. So, for the scientific essentialist, although it is always epistemically possible that gold is not the element with atomic number 79, it is nevertheless logically or strictly metaphysically *impossible* that gold is other than the element with atomic number 79, so long as it is an actual fact that gold is the element with precisely that atomic number.

The obvious problem here is that, given the sharp distinction between logical or strict metaphysical necessity and possibility, on the one hand, and epistemic necessity and possibility on the other, there is no guarantee whatsoever that the inherently empirical reference-fixing means by which the linguistic community attaches its natural kind term "gold" to the natural kind gold (that is, by perceptual ostension, or by the use of reference-fixing definite descriptions expressing empirical concepts or "nominal essences") will *ever* suffice to give empirical knowledge of that actual fact. On its own this skeptical worry about the limitations of empirical scientific inquiry is troublesome enough for the scientific essentialist.[46] But from a Kantian point of view it can be seen that the worry leads to an even deeper problem.

The deeper problem arises from the fact that the perceptual objects or empirical concepts of any a posteriori inquiry into natural kinds are necessarily sensory, apparent, phenomenological, or macrophysical in character. This remains true even if sensory evidence is used to ground an "inference to the best explanation" that invokes unperceived, unperceivable, and thus unobservable causally efficacious microphysical and microstructural physical entities or properties,[47] since, as we shall see below, Kant regards this as at best a useful logical fiction. And it also remains true even if, as Sellars holds, the unconceptualized sensory Given is nothing but a Myth, because qualitative sensory content is inevitably semantically

---

[46] See Zemach, "Putnam's Theory on the Reference of Substance Terms," pp. 123–4.

[47] For what (I think) is the first explicit employment of the inference-to-the-best-explanation argument in a contemporary scientific realist context, see Russell, *The Problems of Philosophy*, ch. 2.

overdetermined by background theories and concepts.⁴⁸ For the threefold Kantian fact is: (i) that sense-perception is *not* the awareness of sense data, since there are no such things as sense data, and since sensations are nothing but non-representational modes of conscious transparent access to bodily affection; (ii) that empirical inquiry is not *empirical* inquiry unless its concepts apply exclusively to perceptual objects and apparent properties; and (iii) perceptual objects and their apparent properties are originally presented to us directly and non-conceptually through empirical intuition, and *not* conceptually via judgments and theories (see chapter 2). So the physical micro-objects, microproperties, and microconstitutions investigated by the natural scientist according to the doctrine of scientific essentialism, are in principle *never* directly presented or presentable to human sense perception, and are therefore in principle *never* directly observable.

But this also entails that the microphysical world is never *in*directly observable either. This is because indirect observation requires some pre-established ground of isomorphism, analogy, or similarity between the object observed and the indirect means used to observe it. Yet the pre-establishment of such a ground implies an independent direct access to the object, which must be either some sort of direct observation (thus contradicting the assumption of indirect realism that observation is never direct), or else a neutral, decontextualized, or "sideways on" representation of the object that is neither a direct observation of it nor an indirect observation of it (thus stepping beyond perceptual representations into speculative theoretical models of the object). Hence, *either* one must be completely skeptical about the claims of natural science to know microphysical objects, microphysical properties, or physical microconstitutions *or else*, if one still assumes that a posteriori knowledge of them is possible, then one must paradoxically claim that our perceptual-empirical mode of access to them is inherently abductive, conceptual-hypothetical, or otherwise theoretically speculative. But then, avoiding skepticism and opting for the second option, the scientific essentialist must be prepared to admit that a posteriori knowledge of physical micro-constitutions is "a posteriori" only in a Pickwickian sense: that is, such knowledge is in fact *essentially rational* or *non-empirical* in character.

This critical point corresponds directly to Kant's views about the epistemic scope and limits of natural science. Kant's doctrine begins with

---

⁴⁸ See Sellars, "Empiricism and the Philosophy of Mind."

the very plausible idea that empirical concepts and empirical propositions about physical nature are applicable only to objects given in actual or possible sensory experience:

> Nature considered *materialiter* is the *totality of all objects of experience*. And with this only are we now concerned; for, besides, things which can never be objects of experience, if they were to be cognized as to their nature, would oblige us to have recourse to concepts whose meaning (*Bedeutung*) could never be given *in concreto* (by any example of possible experience). (*P* 4: 295)

This basic cognitive-semantic scruple about the meaningfulness of empirical concepts and propositions—that they are objectively valid or empirically meaningful if and only if applicable to objects of actual or possible human experience—in turn puts inherent limits upon the evidential scope of natural science. In a discussion of the nature of "real definitions" in the *Jäsche Logic*, Kant explicitly identifies the "cognition of the object according to its inner determinations," or intrinsic non-relational properties, with cognition of a Lockean "real essence" (*JL* 9: 143–4). But the attempt to comprehend completely the real essence of a material object transcends the boundaries of all possible experience. Kant expresses this view most forcefully in a letter to K. L. Reinhold, which I have already quoted as one of the epigraphs for this chapter:

> The *real* essence (the nature) of any object, that is, the primary *inner* ground of all that necessarily belongs to a given thing, this is impossible for human beings to discover... [T]o recognize the real essence of matter, the primary, inner, sufficient ground of *all* that *necessarily belongs to* matter, this far exceeds the capacity of human powers. We cannot discover the [real] essence of *water, of earth*, or the [real] essence of any other empirical objects. (*PC* 11: 36–7)

Putting the same idea in a slightly different way, what I am arguing is that a microphysical "real essence," a set of unobservable intrinsic non-relational dispositional properties of individual material things and natural kinds, is nothing but a Kantian thing-in-itself or *Ding an sich*: nothing but the positive *noumenon*, as translated into the theoretical framework of contemporary physics.[49] The "Really Real" microphysical properties and

---

[49] Russell states that "Kant's 'thing in itself' is identical *in definition* with the physical object, namely, it is the cause of sensations" (*Problems of Philosophy*, p. 48, n.1). The deep connection between Kant's notion of the thing-in-itself, the causal theory of perception, and contemporary scientific realism

physical micro-constitutions of the scientific essentialist are not, and never can be, either perceivable by the human senses or intelligibly describable by means of empirical concepts. We can concede that scientific essentialist physical micro-entities, microproperties, and micro-constitutions may, as a logical possibility, exist in *some* sort of space and time: but in fact this is only space and time *in themselves*, not in Kantian space and time, which are necessarily representable by (and also, according to Kant's official doctrine of the strong transcendental ideality of space and time, nothing but[50]) our subjective forms of intuition, which in turn are our subjective forms of outer and inner sensibility (see *CPR* A19–49/B33–73, B160 n., and section 2.3). That is, according to Kant, space and time are necessarily representable by us as non-empirical, unique, essentially relational, "thickly structured" (see section 2.4), infinite totalities that are necessarily and exclusively applicable to all actual and possible macrophysical material things and their macrophysical properties. But, for the scientific essentialist, the space-time framework of the microphysical world (which the contemporary essentialist would of course also take to be Einstein-relativistic and Bohr-Rutherford-atomic, not, like an eighteenth-century scientific realist, absolute in Newton's sense and incorporating action-at-a-distance), is strongly mind-independent.

The scientific essentialist is thus a defender of transcendental or noumenal realism:

transcendental realism... regards time and space as something given in themselves (independent of our sensibility). The transcendental realist therefore interprets outer appearances (if their reality is conceded) as things-in-themselves, which

---

was later explicitly developed by Sellars in *Science and Metaphysics*, chapters II and V, pp. 31–59 and 116–50. More recently, the same idea has been resuscitated by Wilson in "The 'Phenomenalisms' of Kant and Berkeley," and by Langton in *Kantian Humility*. Langton persuasively argues that for Kant all the properties of positive noumena in general and of noumenal material objects more specifically are unobservable intrinsic non-relational properties. For the distinction between positive and negative noumena, see Hanna, *Kant and the Foundations of Analytic Philosophy*, p. 106. Negative noumena are objects possessing some intrinsic non-sensible properties, e.g., physical objects in four-dimensional spaces. The notion of *a phenomenal thing that possesses some negatively noumenal properties* will be central later in my discussion of the nature of Kant's theory of freedom: see sections 8.2–8.3.

[50] To say that time and space are "nothing but" our subjective forms of sensibility is to say that space and time are identical with our subjective forms of sensibility (the strong transcendental ideality of space and time). But, on the other hand, to say that time and space are "necessarily representable by" our pure forms of sensibility, is only to say that actual space and time are necessarily possibly structurally isomorphic with our pure or formal intuitional representation of space and time (the weak transcendental ideality of space and time). See section 6.1.

would exist independently of us and our sensibility and thus would also be outside us according to pure concepts of the understanding. (*CPR* A369)

Given noumenal realism, it then follows that the microphysical world in the scientific essentialist's sense is epistemically independent of all actual and possible experiential knowledge: the "true correlate [of sensibility], i.e., the thing in itself, is not and cannot be cognized, through [representations of our sensibility], but is also never asked after in experience" (*CPR* A30/B45). Indeed, if the essentialist's microphysical essences are directly perceptually knowable at all, it could only be through "intellectual intuition" (*CPR* B72), that is, the sort of perceptual cognition available to a divine cognizer alone. But as non-divine finite sensory cognizers, *we* cannot make any sense of how intellectual intuition is possible:

[I]f we understand by [a noumenon in the positive sense] **an object of a non-sensible intuition**, then we assume a special kind of intuition, namely intellectual intuition, which, however, is not our own, and the possibility of which we cannot understand. (*CPR* B307)

So here, in a nutshell, is the first objection: The scientific essentialist wants to claim that empirical knowledge of microphysical essences is possible. But since by their own semantically driven metaphysical theory the essences *actually are not* and *never can be* cognized (sense perceived or empirically conceptualized), there is very good reason to doubt that natural scientists have as a matter of fact any empirical knowledge whatsoever of the noumenal microphysical and physical microstructural world. Not only that, but if in order to avoid skepticism the scientific essentialist then appeals to some abductively, conceptually, or speculatively—hence, essentially rationally—enriched sort of empirical inquiry, the Kantian can retort by pointing out that the very idea of a special mode of human so-called "a posteriori" knowledge that transcends the bounds of sense perception and empirical concepts is itself unintelligible. The scientific essentialists may indeed have some knowledge of their unperceivable and unobservable microphysical world, but it must be purely rational, or non-empirical, in character. If this line of counterargument is cogent, then obviously the Epistemic Doctrine of scientific essentialism fails.

This substantive Kantian conclusion requires just a little more elaboration in order to avoid an important possible misconception. It would be a big mistake to think that the demonstrative force of this objection to scientific essentialism rests on Kant's *strong* transcendental idealism. As I have mentioned, strong transcendental idealism is the two-part doctrine to the effect that (i) all the proper objects of cognition are nothing but phenomenal appearances, not things-in-themselves, and (ii) the constitutive structures of those phenomenal objects are strictly imposed upon them by the transcendental faculties of the cognizing subject (*CPR* A41–2/B59–60, A369). In my opinion—which I am only asserting here, but will explicitly argue for in section 6.1—Kant is indeed explicitly and officially a transcendental idealist in the strong sense, but he is also at the same time committed to a significantly weaker and correspondingly more philosophically defensible position I call *weak transcendental idealism*. Weak transcendental idealism says that by their very nature actual space and actual time properly satisfy, or are correctly represented by, our pure or formal intuitions of them. Therefore the actual or possible existence of material things and natural kinds in actual objectively real spacetime directly entails *the necessary possibility* of rational human minds. In other words, according to weak transcendental idealism, actual space and time can exist in a possible world (including of course the actual world) even if no rational human minds *actually* exist in that world—or do not actually exist at some times in that world, say, prior to the evolutionary appearance of *Homo sapiens*—provided that if there *were* rational human minds in that world, then they *could* correctly represent space and time.

Clearly, there is much more to say here about the rationale for and implications of weak transcendental idealism, which I will also reserve for section 6.1. The crucial point for my present purposes is that, while weak transcendental idealism is indeed a necessary condition of Kant's empirical realism, nevertheless his empirical realism is perfectly consistent with the denial of strong transcendental idealism. So Kant's empirical realism is *logically independent of* strong transcendental idealism. So, if I am right, and if we explicitly apply here my two methodological principles of charitably reading Kant as being committed (1) only to the best overall philosophical view consistent with all the texts, and (2) in cases of textual conflict, only to the best philosophical view supported by at least some of the texts,[51] then

---

[51] See section 0.0. It must be admitted that I haven't actually offered any argument for these two principles, but rather have only announced my resolution to use them. If I *were* pushed to defend them, however, I would do so on the grounds that these principles are canons of the method that is most likely to produce a philosophically robust history of philosophy.

Kant is best interpreted as *a weak transcendental idealist and an empirical realist*. He is thereby committed to the thesis that human cognizers have direct veridical perceptual access to some macrophysical dynamic material objects in objectively real space and time (direct perceptual realism), and also to the further thesis that the essential properties of dynamic material things and natural kinds are their directly humanly perceivable or observable intrinsic structural macrophysical properties (manifest realism), under the assumption that while this implies the necessary possibility of creatures minded like us, nevertheless material objects and natural kinds in objectively real space-time can still exist even if human cognizers do not actually exist. The main point I am making here is that it is only Kant's *empirical realism*, not his *strong transcendental idealism*, that supplies the rationale behind the first objection to scientific essentialism.

Now the following critical question naturally arises. "How can someone be an empirical realist and a *scientific* realist at the same time: aren't all sense-perceivable objects and sensory properties inherently idiosyncratically subjective and therefore illusory?" And the Kantian answer comes back immediately:

If I say: in space and time intuition represents both outer objects as well as the self-intuition of the mind as each affects our senses, i.e., as it **appears**, that is not to say that these objects would be a mere **illusion** (*Schein*). For in appearance the objects, indeed even properties that we attribute to them, are always regarded as something really given (*wirklich Gegebenes*). (CPR B69)

That is, what is "really given" to natural science are causal-dynamic macrophysical individual material things in objectively real space and time, and real classes of those things (manifest natural kinds). These macrophysical objects are not in any way merely subjective items or illusory sense data, rather they are *intersubjectively objectively real material things*:

We ordinarily distinguish quite well between that which is essentially attached to the intuition of appearances, and is valid for every human sense in general, and that which pertains to them only contingently because it is not valid for the reference of sensibility in general but only for a particular situation or organization of this or that sense. (CPR A45/B62)

Matter is *substantia phaenomenon*. What pertains to it internally I seek out in all parts of space that it occupies and in all effects that it carries out, and which can certainly always be only appearances of outer sense. (CPR A277/B333)

It should not be inferred from all this, however, that Kant *altogether* rules out transcendent or noumenal thinking in the physical sciences. But for him concepts or propositions about the "inner nature" or "real essence" of the dynamic forces, or about the "inner nature" or "real essence" of natural kinds—including both the seventeenth-century and twentieth-century atomistic microphysical theories adopted by scientific essentialists and other maximal scientific realists—are nothing but (barely) thinkable Ideas of reason, and not objectively valid concepts of the understanding. Hence they belong to the "regulative" use of reason, not to its "constitutive" use. The function of the regulative use of reason in natural science is to produce strictly heuristic speculative schemes that promote the unity or coherence of sets of scientific beliefs about the motions and interactions of macroscopic material bodies in nature (*CPR* A642–68/B670–96). We are to act in our scientific theorizing *as if* we believed in microphysical properties and microphysical essences, even though we do *not* actually believe in them, precisely because it is often useful to posit the logical fiction of unobservable microphysical grounds of otherwise seemingly disparate classes of observable natural things and kinds in order to promote the discovery of causal natural laws governing those observable things and kinds. Constitutive uses of reason, by contrast, imply the full empirical meaningfulness and truth-valuedness of the propositions expressed by that use. This strongly suggest that if, on the fictional time-travel scenario, Kant were indeed to adopt (GE) and the Bohr-Rutherford atomic theory of matter, he would ultimately regard it as only *a regulatively useful hypothesis and a convenient logical fiction*, not as a constitutive truth.

3.3. Objection II: Why There is No Necessary A Posteriori

The Kantian criticism described in the last section leads directly to another. It is a primary feature of scientific essentialism that knowledge of propositions like (GE) is thoroughly a posteriori or experience-dependent. But the essentialist thesis to the effect that there exist necessary a posteriori propositions such as (GE) runs directly counter to Kant's doctrine that all necessary propositions semantically are, and also are known, a priori:

172   EMPIRICAL REALISM AND SCIENTIFIC REALISM

[Experience] tells us, to be sure, what is, but never that it must necessarily be thus, and not otherwise ... Now such universal cognitions, which at the same time have the character of inner necessity, must be clear and certain for themselves, independently of experience; hence one calls them *a priori* cognitions. (CPR A2)

If a proposition is thought along with its **necessity,** it is an *a priori* judgment. (CPR B3)

And this provides us with the rudiments of another Kantian epistemological objection to essentialism. Specifically, the objection has to do with the extent to which the truth and modal status of (GE) and other essentialist identity propositions are in any genuine sense known a posteriori.

For the purposes of argument, let us grant this much to the essentialist: that the truth of (GE) is discovered empirically. But (GE) is not merely true and known to be true: it is necessarily true; and it is known to be necessarily true. Granting the empirical discovery of at least its truth-value, however, how does that fact affect either the specific modal status[52] of (GE) or the way in which we know that modal status? Consider, now, the following manifestly *bogus* argument:

Prove: That some analytic propositions are known only a posteriori.

(1) It is an a posteriori truth that Kant is a bachelor, and another such truth that Kant is not married. For one thing, each of the claims can be empirically confirmed; for another, it is logically and analytically possible that he, Kant, might not have been a bachelor; and thirdly, even given all available empirical evidence, it is possible that a thinker like us could believe that Kant was secretly married—so it is "epistemically possible" that Kant is married despite his actually being unmarried.

---

[52] "Specific modal status" is a useful notion employed by Casullo in "Kripke on the A Priori and the Necessary," p. 163. He distinguishes between: (a) the truth-value of a proposition—its being true or false; (b) the specific modal status of a proposition—its being necessarily true, necessarily false, contingently true, or contingently false; and (c) its "general modal status"—its being necessary or contingent. It is possible to know a proposition *P*'s general modal status (say, its being contingent) a priori even though *P*'s truth-value is nevertheless still knowable a posteriori. What is highly questionable, however, is the possibility that a proposition *P* can be known a priori to have either the general modal status of its being necessary or the specific modal status of its being necessarily true, while *P*'s truth-value is also known *only* a posteriori. But we will see shortly that this is precisely what the scientific essentialist tries to argue.

(2) But necessarily and analytically:

(KB) If Kant is a bachelor, then Kant is not married.

It is part of the meaning of "bachelor" that it excludes the meaning of the predicate "married," hence there is no logically possible world in which Kant is a bachelor and is also married. So (KB) is not only true but necessarily true.

(3) Therefore (KB) is analytically necessary a posteriori.
(4) Therefore some analytic propositions are known only a posteriori.

Obviously, both (1) and (2) are true, but (3) and (4) are false: all parties to this debate will agree that (KB) is analytic a priori and that analytic propositions, if knowable at all, are knowable a priori. So the argument is invalid. *Neither* the necessity *nor* the apriority of (KB) is affected by the indisputable facts that:

(i) (KB) contains ineliminable empirical constituents (because "Kant" is directly referential);
(ii) (KB) is made true by an empirical fact (as it happens, Kant is indeed unmarried);
(iii) (KB) is knowable a posteriori (its consequent is empirically verifiable); and
(iv) it is "epistemically possible" that Kant is married (the actual empirical evidence is at least logically consistent with that belief).

Despite (KB)'s empirical elements, its necessity is grounded in the fact that the semantic connection between the predicate-concept of the antecedent and the predicate-concept of the consequent is one of *conceptual containment*. And, despite (KB)'s being knowable a posteriori, its apriority is grounded in the fact that any thinker who can understand the whole proposition can also know it to be necessarily true merely by grasping the meanings of its constituent terms. As Kant points out, "all analytic judgments are *a priori* even when the concepts are empirical" (P 4: 267). And Kripke even implicitly agrees:

Let's just make it a matter of stipulation that an analytic statement is, in some sense, true by virtue of its meaning and true in all possible worlds. Then something which is analytically true will be both necessary and *a priori*.[53]

---

[53] Kripke, *Naming and Necessity*, p. 39.

Now what is the relevant difference between the form of the obviously invalid argument that some analytic propositions are a posteriori, and the form of the scientific essentialist argument for the claim that true essentialist identity statements about natural kinds, such as (GE), are metaphysically necessary and yet a posteriori? Answer: *there is no relevant difference!* As we shall see, (GE) is relevantly analogous to (KB). Hence neither (GE)'s having ineliminable empirical constituents nor the fact that its truth is knowable a posteriori entails its being knowable *as necessary* only through experience. Like (KB), (GE) is knowable a priori and its specific modal status is known only a priori.

To show this, we need only look closely at what actually justifies the claim that (GE) is not only true but necessarily true. The case of (KB) already shows us that the fact that a necessary proposition is discovered to be true a posteriori does not entail that its specific modal status is known only a posteriori. For (KB)'s being a necessary truth is known only a priori. Now the justification for the belief in (GE)'s being a necessary truth depends, as we have seen, ultimately on the Metaphysical Doctrine, according to which the nature and identity of natural kinds is determined by their essential microphysical constitution. We continue to grant the truth of that doctrine just for the moment. Is that doctrine an *empirical* theory? Are the essential microphysical properties that it ascribes to natural kinds *empirical* properties? Is any necessary semantic identification of the referents of a given natural kind name and a given scientific natural kind term—an identification based entirely on the Metaphysical Doctrine—an *empirical* identification? Is the knowledge of any such necessary identity *empirical* knowledge? Obviously not: the Metaphysical Doctrine is an a priori philosophical theory; essential microphysical properties are non-empirical properties; the modal connections ascribed to particular identity propositions about natural kinds are non-empirical connections; and the knowledge of those modal connections is correspondingly a priori. In a precisely analogous way, the doctrine of analyticity, the analytic connection between the concepts in (KB), and the knowledge of that connection, are all a priori. The presence of ineliminable empirical factors, and the fact of empirical knowability, are both totally irrelevant to the essential apriority of both (GE) and (KB).

In this regard, Kripke's remarks about another class of metaphysically necessary a posteriori propositions, identity propositions about individuals, are highly pertinent. He says:

We have concluded that an identity statement between names, when true at all, is necessarily true, even though one may not know it *a priori*.⁵⁴

Some of the problems which bother people in these situations, as I have said, come from an identification, or as I would put it, a confusion, between what we can know *a priori* in advance and what is necessary. Certain statements—and the identity statement is a paradigm of such a statement on my view—if true at all must be necessarily true. One does know *a priori*, by philosophical analysis, that *if* such an identity statement is true it is necessarily true.⁵⁵

In other words, then, take a true individual identity proposition such as the familiar "Hesperus is Phosphorus" (HP). Kripke points out that (HP) is about contingent empirical objects, that (HP)'s truth can be discovered through empirical means, and that the general connection between an ordinary identity proposition's being true and its being necessarily true is known a priori "by philosophical analysis." From a Kantian point of view, Kripke should then have concluded that (HP) is a priori precisely because its being about this or that contingent empirical object, and its being knowable as true a posteriori, are quite irrelevant to its specific modal status—its being a necessarily true identity proposition. (HP)'s specific modal status is grounded in the internal semantic connections ascribed to it by the theory of identity. Modal features deriving specifically from the concept of identity (say, those reflected in the Indiscernibility of Identicals or the Identity of Indiscernibles) are non-empirical, and known only a priori "by philosophical analysis." Hence since (HP) is known to be necessarily true only by appeal to the Logical Doctrine, then despite its empirical constituents and the empirical acquisition of its truth-value, (HP) is known to be necessarily true only a priori.

Just as in the case of identity propositions involving ordinary proper names, so too in the case of (GE) and other scientific essentialist identity propositions about natural kinds, the modal features of those propositions, which are grounded in the Metaphysical Doctrine, are known in a strictly a priori way. At one point Kripke even implicitly admits this:

All the cases of the necessary *a posteriori* advocated in [*Naming and Necessity*] have the special character attributed to mathematical statements: Philosophical analysis

---

⁵⁴ Kripke, *Naming and Necessity*, p. 108.   ⁵⁵ Ibid. p. 109.

tells us that they cannot be contingently true ... This characterization applies, in particular, to the cases of identity statements and of essence.[56]

But, significantly, in the phrase occurring in my ellipsis he states, "so any empirical knowledge of their truth is automatically empirical knowledge that they are necessary." This, again, is the crucial non sequitur. What the scientific essentialist's argument in fact shows is that knowledge of the specific modal status of any essentialist identity proposition about natural kinds is grounded in modal factors within the semantic content of the proposition—factors necessarily identifying the reference of the natural kind name with the reference of the scientific natural kind term. The explanation for the necessary identification is a shared microphysical essence. The thesis that microphysical essences play this semantical role in general is known only a priori by "philosophical analysis." Therefore knowledge of the necessity of any particular proposition that involves a specific microphysical essential identification, despite its ineliminable empirical components and the empirical knowability of its truth-value, is itself a priori knowledge.[57]

Once we are able to see this point, a basic error of essentialism can then be exposed by employing an extremely useful Kantian distinction between "pure" and "impure" a priori judgements (*CPR* B3). Pure a priori judgements, such as "$(P) \sim (P \,\&\, \sim P)$" or "$2 + 2 = 4$," are such that their semantic contents contain only strictly formal or non-empirical constituents. But there are also "empirically infected" or impure a priori judgments such as

(KB) If Kant is a bachelor, then Kant is unmarried (= a non-logically analytic truth),

---

[56] Kripke, *Naming and Necessity*, p. 159

[57] That a priori "philosophical analysis" is presupposed in Kripke's account of the necessary a posteriori has been pointed up by Bealer in "The Philosophical Limits of Scientific Essentialism," pp. 292, 310–17; by Casullo in "Kripke on the A Priori and the Necessary," p. 164; and by Sidelle in *Necessity, Essence, and Individuation*, pp. 86–104. Neither Bealer nor Casullo nor Sidelle actually goes so far as to argue what I have argued on Kant's behalf here: that knowledge of the specific modal status of a true scientific essentialist identity proposition is itself a priori. But, interestingly enough, a very similar claim to mine has been made by Australian critics of "a posteriori physicalism" in the philosophy of mind: see, e.g., Chalmers, *The Conscious Mind*, ch. 2; Jackson, "Finding the Mind in the Natural World"; Jackson, *From Metaphysics to Ethics: A Defense of Conceptual Analysis*; and Stoljar, "Physicalism and the Necessary a Posteriori," esp. pp. 49–53.

(SM) It is not the case that Socrates is both mortal and not mortal (= a logically analytic truth),

(TA) Two apples added to two apples make a total of four apples (= an applied arithmetical truth),

and more controversially:

(EC) Every event has a cause (= a metaphysical truth).

Each of these judgments is such that although it is necessary and experience-independent, it nevertheless contains ineliminable empirical content in some of its conceptual constituents. And each can be learned or acquired empirically. But, since the empirical content does not enter into the specifically *modal* character of the judgments, it does not in any way determine their necessary truth or falsity. Consequently, the knowledge of the necessity-determining features of their propositional contents is not itself empirical knowledge but rather a priori knowledge. In effect, what scientific essentialism provides are examples of necessary truths—namely, (GE) and other essentialist identity propositions about natural kinds—that like (KB), (SM), (TA), and (EC), are at one empirical remove from wholly pure necessary a priori truths. But just as (KB), (SM), (TA), and (EC) are simply impure a priori, not necessary a posteriori, so too are (GE) and other propositions like it. That is, *there is strictly speaking no such thing as the scientific essentialist's* "necessary a posteriori." Essentialists have mistakenly applied this oxymoronic label to what is in fact only the "empirically infected" or impure a priori. It follows, then, that contrary to its official claims, scientific essentialism does not provide any plausible counterexamples to the Kantian doctrine that all necessary truths are a priori. Kant's doctrine stands.

3.4. Objection III: The Antinomy of Essentialism

Bracket for a moment the important epistemological worries expressed in the first and second objections. Suppose just for the purposes of argument that the scientific essentialist is correct about gold's being, as a matter of actual fact, the element with atomic number 79, and even suppose further that this is (somehow) empirically knowable. Still, it will not follow that the essentialist is correct that gold is necessarily the element with atomic number

79. That is, scientific essentialism will be the source of a "fifth cosmological antinomy" in the Kantian sense, because by logically acceptable processes of reasoning it leads to contradictory conclusions about the natural world from the same set of metaphysical assumptions (*CPR* A420–5/B448–53).

I have already unpacked the argument for the essentialist "thesis": that necessarily gold is the element with atomic number 79, on the assumption that it is an actual scientific fact that gold is the element with that atomic number. Now let us consider the argument for its "antithesis," the negation of the essentialist thesis: that it is not necessary that gold is the element with atomic number 79, on the assumption that it is an actual scientific fact that gold is the element with that atomic number.

Suppose that gold is indeed the element with atomic number 79 in the actual world. Given the scientific essentialist picture of the physical world as metaphysically constituted by a vast system of sub-apparent intrinsic non-relational dispositional physical microproperties and micro-entities, however, it is perfectly possible that the material world has levels of micro-constitution even deeper than that of molecular, atomic, or even sub-atomic micro-constitution.[58] Let us call these deeper levels the levels of "physical micro-micro-constitution." Consider now the deepest possible level of physical micro-micro-constitution; call it the level of "XXX-constitution." Such a fundamental micro-micro-constitution of physical matter need not be in any sense atomic or particulate in character; in principle, it could be of some other wholly unheard-of but non-atomic character—constituted by a primordial micro-microcosmic "cheese," perhaps.[59] This is of course not to suggest the absurdity that the micro-micro-constitutional level of matter might be constituted by actual macroscopic cheese; instead it is only to indicate by an inadequate "macro-analogy" the sort of underlying composition that the XXX-level *might* realize. The point is that for all the essentialist knows, the "cheeselike," viscous, continuous XXX-level is in fact both the metaphysically deepest level of essentialist physical micro-constitution and also essentially distinct from the "nutlike," granular, discontinuous, atomic/quantum micro-constitution. Moreover,

---

[58] This has also been pointed out by Zemach in "Putnam's Theory on the Reference of Substance Terms," p. 121, and by Mellor in "Natural Kinds," p. 311.

[59] For a fascinating account of an ordinary man—a miller—who dared to advocate just this sort of highly unconventional microphysical cosmology in sixteenth-century Italy, see Ginzburg, *The Cheese and the Worms*. Sadly, the speculative miller was put to death by the Inquisition for heresy.

this hypothesis is not a violation of the widely held thesis of "mereological supervenience," which says that the physical entities occurring at any level of natural reality are wholly composed of proper parts consisting of physical entities occurring at the next lowest level of nature, and are also strictly determined by the total set of properties and relationships of those proper parts.[60] This is because the atomic/quantum-level entities, for all the essentialists know, might be nothing but mereological sums of *strings* in the (as it were) Primordial Cheese.

Now it will probably have already been noticed by Kant-afficionados that my "micro-microcosmic cheese" possibility is importantly analogous to the possibility of a fundamental physical fluid aether that Kant explicitly offers against the atomic theory of matter in *Metaphysical Foundations of Natural Science*:

[O]ne would not find it impossible to think of a matter (as one perhaps represents the aether) that entirely filled its space without any void and yet with incomparably less quantity of matter, at an equal volume, than any bodies we can subject to our experiments... And the only reason why we merely assume such an aether, because it can be thought of, is as a foil to a hypothesis (that of empty spaces) which depends only on the assertion [made by the atomic theory] that such rarified matter cannot be thought of without empty spaces. (*MFNS* 4: 534)

Both the cheese-hypothesis and the aether-hypothesis offer us metaphysical pictures of a cosmologically fundamental "atomless gunk." But for the purposes of my argument in *this* section, I prefer the cheese. This is because, in order to keep things relatively simple here, I want to treat the aether theory as a logically separate issue to which I will return in sections 4.2 and 8.3.

In any case, what is crucial about any micro-micro-constitutional hypothesis is that by the general theory of scientific essentialism, physical properties deriving from the *Ur*-level of XXX-composition will, by causal strong supervenience, ontologically and causally determine the properties of all higher levels of physical composition, without being identified with those higher levels. That is, the level of atomic composition will be ontologically and causally related to the XXX-level just as the level of sense-perceivable macrophysical facts is supposed by the essentialist of the thesis to be ontologically and causally related to the level of atomic composition.

---

[60] See Kim, "Epiphenomenal and Supervenient Causation," pp. 96–7, 101–2.

The only difference between the XXX-world of the antithesis and the scientific-essentialist-world of the thesis is that whereas the thesis-world is ontologically bi-leveled (atomic composition vs. perceivable macrophysical facts) the antithesis-world is ontologically tri-leveled (XXX-composition vs. atomic composition vs. perceivable macrophysical facts).

And here's the rub. On the hypothesis of the antithesis, while gold is as a matter of fact the element with atomic number 79, it is *really* the element with some totally different categorical micro-micro-constitutional property. So let us suppose further, just to fill in the antithesis story a little, that XXX-composition really *is* somehow essentially cheeselike in nature. Let us also suppose that different types or sub-types of the Primordial Cheese correspond to different "consistencies"—on the very rough macro-analogy of the different consistencies of, say, Brie, Cheddar, Swiss cheese, and Stilton—each of which is assigned some numerical value on the basis of its fundamental micro-constitutional properties (whatever they are). And finally let us suppose that gold is in fact the consistency with Primordial Cheese number 1000 ("PC-number 1000" for short). But while gold's being the consistency with PC-number 1000 determines in the actual world that gold is also the element with atomic number 79 (and in turn that actual-world gold is yellow and metallic), it does not follow that this is *necessarily* so. Think of the causal strong supervenience relation between Primordial Cheese composition and atomic composition on the strict analogy of the essentialist's causal strong supervenience relation between atomic composition and perceivable macrophysical facts. While gold's being the element with atomic number 79 causally strongly superveniently determines in the actual world that gold is a yellow metal, there are possible worlds in which the element with atomic number 79 is neither yellow nor metallic. For in such worlds the contingent causal conditions required for the emergence or instantiation of color-properties, or of the surface-properties of metal in the actual world, could be lacking. By the same reasoning, then, it follows that there are also possible worlds in which the element with PC-number 1000 exists, but it does not happen to be the element with atomic number 79: perhaps the special causal conditions required for the production of higher-level atomic micro-constitution are lacking in those worlds, and only the Primordial Cheese exists there. Perhaps—to give this story a Twin Earth-like twist—this was even the cosmological state of our own actual world (causally) prior to the Big

Bang![61] What could the scientific essentialist of the thesis say to rule this weird possibility out of court, without begging the question?

The bottom line is that on the hypothesis of the antithesis, since gold at the micro-micro-constitutional level is the consistency with PC-number 1000, then gold is not necessarily the element with atomic number 79. For, by scientific essentialist reasoning from that hypothesis, gold necessarily is the consistency with PC-number 1000, which causally strongly superveniently determines all its higher-level properties including the micro-constitutional property of having atomic number 79. Therefore, since by virtue of the basic assumptions of scientific essentialism it can be argued both that it is necessary (according to the thesis) and that it is not necessary (according to the antithesis) that gold is the element with atomic number 79, essentialism must be an inherently incoherent and paradoxical philosophical theory.

3.5. Objection IV: The Logical Contingency of the Laws of Nature

The argument-strategy of Objection III—to look for possible worlds in which gold is not the element with atomic number 79 while accepting (GE) as an actual fact—suggests an even more direct Kantian objection to the metaphysics of scientific essentialism.

Suppose, as before, that it is somehow known to be a truth in the actual world that gold is the element with atomic number 79. Suppose further that the atomic sort of composition is the ontologically deepest level of micro-constitution, so that it is also true that gold is fundamentally the element with atomic number 79. Given those two suppositions, we can even allow, finally, that *in at least some sense* (and I will offer a Kantian interpretation of that sense in the next few paragraphs) it is a "necessary law of nature" that gold is the element with atomic number 79.

---

[61] For a good non-technical account of Big Bang cosmology, see Hawking, *A Brief History of Time*. On this view, both matter and space-time are supposed to be causally produced by the primordial categorical properties and primordial interactions of some original particles. But how the original particles arose, how they acquired those primordial categorical properties, and how the primordial interactions were caused, is totally mysterious. So conceptually speaking, and for all *we* know, the causal predecessor to the Big Bang could have been Primordial Cheese.

Even granting all that, the problem is that the metaphysical character of the microphysical constitution of gold is not nearly as straightforward as the scientific essentialist likes to make out. For, on the one hand, there is *this actual stuff or substratum that is gold*, bits of which are lying around here, over there, and way over there too, as it were. But, on the other hand, this total mass of stuff which is originally picked out in a purely indexical way by indicating some samples of it, takes on the properties of being an element with atomic number 79 *only* by virtue of the fact that the actual physical world contains, or realizes, a certain actual set of necessary natural laws governing atomic-level physical relations and causal interactions: call it set L^*. In other words, the intrinsic non-relational dispositional properties of microphysical matter, on the scientific essentialist metaphysics, are necessarily and indeed even constitutively connected with the laws of nature that obtain in our actual world, or in any other possible world considered as actual. And this leads directly to a deep modal problem for essentialism: the contingency of the *actual set or package* of natural laws.[62]

As Kant points out in the Analogies of Experience, without a fixed system of laws there simply is no natural world as such. But not all laws that govern nature are precisely of the same type:

By nature (in the empirical sense) we understand the combination of appearances, as regards their existence, in accordance with necessary rules, that is, in accordance with laws. There are therefore certain laws, and indeed *a priori*, which first make a nature possible; [by contrast] the empirical laws can only obtain and be found by means of experience, and indeed in accord with its original laws, in accordance with which experience itself first becomes possible. (*CPR* A216/B263)

In other words, physical nature is a law-governed totality or unity. But only some of the laws are both transcendental and a priori (for example, the Analogies of Experience). By contrast, the natural laws governing causal processes and dynamic interactions in actual physical nature are only *empirically* given to us, in the sense that they presuppose the contingent existence of inert or mechanistically causal matter in this actual world (*CPR* A218/B266, A226/B279, A723/B751) (*MFNS* 4: 470), and are discovered by empirical science: hence Kant calls them "empirical laws." Nevertheless

---

[62] The Kantian modal point to be developed in the next few paragraphs has also been made by Putnam in "Is Water Necessarily H$_2$O?," pp. 68–70. And in the same paper, Putnam also explicitly links his objection with Kant's theory of causality (pp. 74–6).

this is not to say that the empirical laws of nature, as laws, do not *necessarily* govern causal-dynamic activity in the actual world. They *do* necessarily govern it: indeed that is precisely what it is to *be* a causal-dynamic natural *law*, as opposed to a merely regular *rule* of appearances: all rules are either necessary (hence they are laws, or true generalizations across all possible worlds) or else contingent (hence they are merely true generalizations about the actual world alone) (*JL* 9: 12).

How then can causal-dynamic laws of nature be in any sense be "necessary" and "governing" for the actual empirical world, and yet not obtain in every logically possible *empirical* world, because those laws presuppose a contingent fact (the actual existence of inert matter), much less in every logically possible world *überhaupt*, including the non-empirical ones? The Kantian answer, in a nutshell, is that the "dynamic necessity" or "material necessity" of natural laws and scientific propositions about natural laws is simply a stronger *synthetic* modality (where modal strength is inversely proportional to the size of the class of possible worlds in which a necessary truth holds—so stronger necessities hold in *smaller* classes of worlds) than the synthetic necessity of mathematical truths and metaphysical or transcendental propositions (which are true in every world in which human experience is possible), which in turn is itself a stronger modality than analytical necessity (which is truth in every logically or conceptually possible world).[63] Kant's account of dynamic or material necessity is given in the third "Postulate of Empirical Thought":

That whose connection with the actual (*Wirklichen*) is determined in accordance with general conditions of experience, is (exists [existiert]) **necessarily**. (*CPR* A218/B266)

As I understand it, Kant's claim here is that a proposition *P* is dynamically or materially necessary if and only if *P* is true in every experienceable world that contains the same physical substratum or totality of matter under the same causal-dynamic laws, and also within the same unique spatiotemporal framework, as the actual world. In this way, Kantian dynamic or material necessity is in certain ways similar to what modal logicians sometimes call

---

[63] For a general account of Kant's theory of propositional modality, see Hanna, *Kant and the Foundations of Analytic Philosophy*, ch. 5. There I used the terminology of modal strength in a different way; in that context, a stronger modality was one that holds in a *larger* class of logically possible worlds.

"physical necessity,"[64] with the important difference that Kantian dynamic or material necessity is not only constrained by and relativized to "physical-law postulates" (i.e., a set of propositions expressing the several natural laws) but also constrained by and relativized to an indexical "existence postulate" (i.e., a proposition to the effect that this actual empirical world's underlying stuff, or matter, actually exists), *and also* falls under propositions expressing formal-transcendental conditions for the possibility of experience (i.e., transcendental principles). Because both dynamic or material necessity itself and our knowledge of it ultimately fall under formal-transcendental conditions for the possibility of experience, and because these conditions are known a priori either by means of pure intuition or by means of transcendental arguments involving pure concepts of the understanding, Kant is able to hold that necessary laws of nature are synthetic a priori *despite* their being contingently conditioned and empirically learned:

Natural science (*Physica*) contains within itself synthetic *a priori* judgments as principles. (CPR B17)

Even laws of nature, if they are considered as principles of the empirical use of understanding, at the same time carry with them an expression of necessity (*Ausdruck der Notwendigkeit*), thus at least the supposition (*Vermutung*) of a determination by grounds that are *a priori* and valid prior to all experience. (CPR A159/B198)[65]

That is, for Kant, the laws of nature provide yet another domain of examples of the impure or "empirically infected" a priori.

In the *Critique of the Power of Judgment*, Kant draws an illuminating distinction between (i) *universal* natural laws (for example, Newton's Laws), which are a priori because they are directly derivable from transcendental laws despite their containing an empirical concept, namely, "the empirical concept of a body (as a movable thing in space)" (*CPJ* 5: 181); and (ii) *particular* natural laws (for example, "Falling gold accelerates towards the earth at the rate of 10 meters per second/per second"), which are only indirectly derivable from transcendental laws because they contain an irreducibly empirical, indexical component, namely, direct reference to particular "things (of nature)" (*CPJ* 5: 185), which needs to be established prior to the recognition of the specific modal status of that law

---

[64] See, e.g., Montague, "Logical Necessity, Physical Necessity, Ethics, and Quantifiers."
[65] See also *MFNS* 4: 469–78.

(*CPJ* 5: 181–7). Particular natural laws are thus only *indirectly* a priori. For the rest of the chapter, however, in order to keep things fairly simple, by the phrases "natural laws," "empirical laws," and "causal-dynamic laws", I will mean only *universal* natural laws.

Now it must be frankly admitted that Kant's theory of the modal and epistemic status of causal-dynamical laws of nature is a hotly debated topic in the Kant literature.[66] Hence my brief interpretation of it in the last three paragraphs is bound to be controversial. But for my purposes here I need only insist (a) that Kant *does* have a theory of causal-dynamic laws of nature, and (b) that it is at least internally self-consistent for Kant to hold that necessary natural laws can be at once existentially conditioned, empirically acquired, and yet transcendentally conditioned or a priori. Thus it is possible to hold that natural laws are necessary in their own unique way—so "nomologically" or "naturally" or "physically" necessary—*without* being automatically committed to the thesis that they are logically or strictly metaphysically necessary. But the crucial point is this. From a Kantian point of view, for the purposes of argument it is possible to grant to the essentialist the truth of (GE), together with the thesis that without the actual set of necessary natural laws, L^*, there could be no atomic micro-constitution in the actual world, and hence no elements with particular atomic micro-constitutions in this actual world. Nevertheless L^*, as the "actual set or package" of laws, is not even required for *every* possible natural empirical world, far less for every logically possible world. And here is why.

Assuming the general validity, for Kant, of inferences from conceivability to possibility,[67] then the plain philosophical fact is that it seems

---

[66] See, e.g., Allison, "Causality and Causal Laws in Kant: A Critique of Michael Friedman"; Buchdahl, *Metaphysics and the Philosophy of Science*, pp. 651–65; Buchdahl, "The Conception of Lawlikeness in Kant's Philosophy of Science"; Guyer, "Kant's Conception of Empirical Law"; Friedman, *Kant and the Exact Sciences*, chs. 3–4, pp. 136–210; Friedman, "Causal Laws and the Foundations of Natural Science"; Harper, "Kant on the A Priori and Material Necessity"; Walker, "Kant's Conception of Empirical Law"; and Watkins, "Kant's Justification of the Laws of Mechanics."

[67] In fact, it is controversial whether inferences from conceivability to possibility *are* generally valid: see Chalmers, *The Conscious Mind*, pp. 131–40; and Gendler and Hawthorne (eds.), *Conceivability and Possibility*. But there are two replies I can make to this worry. The first reply is that there is no problem about the step from conceivability to possibility *for Kant*, because he is a conceptualist about the nature of possible worlds. More precisely, for Kant, possible worlds are nothing but distinct maximally consistent set of concepts logically constructed from our total conceptual repertoire: hence cognitive access to concepts just *is* cognitive access to possible worlds (see Hanna, *Kant and the Foundations of Analytic Philosophy*, pp. 85–7, 239–42). The second reply is that basic objections to the inference from conceivability to possibility all assume the truth of scientific essentialism in particular and the existence

quite easy to conceive of logically possible, strictly metaphysically possible, experienceable worlds—that is, natural worlds meeting the general transcendental conditions of the possibility of experience—where some or all of the laws of nature are different. Suppose, for instance, that in some such world FORCE = MASS *divided by* ACCELERATION, or FORCE = MASS times ACCELERATION *raised to the power of three*. This general observation about logically and metaphysically possible worlds with very different sets of natural laws could perhaps be accepted by the scientific essentialist with a certain equanimity, if it also remained the case that gold showed up *only* in the restricted class of possible worlds realizing L^*. But given the contingency of L^*, and given the further Kantian assumption that the intrinsic non-relational properties of gold (namely, the actual stuff in this world picked out by ordinary uses of the directly referential term "gold") are necessarily and even constitutively connected with the laws of nature that obtain in our actual world, or in any given possible world considered as actual,[68] it follows that gold is logically distinct from L^* and thereby also from the restricted class of worlds realizing L^*. That is, a possible world with the natural kind substratum of gold—the actual stuff lying around here and there in our world—in that world *need not* also contain or realize L^*: it is easy to conceive of worlds in which a very different set of laws (say, L^{**}) obtains but the actual gold-stuff is nevertheless also there. In such L^{**}-governed worlds, however, gold would no longer *be* the

---

of the necessary a posteriori more generally. So, since I am arguing on Kant's behalf *against* scientific essentialism, and also hold *that there is no such thing as* the necessary a posteriori, if I were to concede that the objections have a rational force sufficient to switch the burden of proof over to me to provide a defense of the inference from conceivability to possibility, then that would be tantamount to begging the question in favor of scientific essentialism. Therefore, I think that in order to keep the playing field level, I must be permitted provisionally on Kant's behalf to assume that inferences from conceivability to possibility are generally valid, until some independent reasons are offered for doubting this.

[68] It should be remembered that the modal semantics of scientific essentialism is Kripkean: see n.25. So propositions are always truth-evaluated at worlds accessible from our actual world, or from a given possible world considered as actual—i.e., from a possible world taken *as if* it were our actual world. Or, in other words, Kripkean modal evaluation always minimally includes *the human point of view*. So, oddly enough, even though Kripke's modal metaphysics is scientific essentialist, nevertheless his modal semantics is basically Kantian: possible worlds are nothing but different total ways our actual world could have been, for creatures minded like us. If I were forced to speculate as to why, my hypothesis would be that Kripke's modal semantics has been significantly influenced by the later Wittgenstein's anthropocentrism, which in turn is a development of the Tractarian Wittgenstein's transcendental solipsism, which in turn is a development of Schopenhauer's monistic idealism of the will, which in turn is a development of Kant's transcendental idealism. See Hanna, "Kant, Wittgenstein, and the Fate of Analysis." But cf. Soames, *Philosophical Analysis in the Twentieth Century*, esp. vol. 1, Introduction to the Two Volumes and part 3, and vol. 2, part 1 and Epilogue.

element with atomic number 79, because in those worlds there would be a foreign causal-dynamic nomological system of nature in place. Because the laws of nature were very different in those worlds, then the microphysical properties of gold would be correspondingly very different. Indeed, the set of laws could be so radically different that gold might even be *alive* in some L^{**}-governed worlds, and not inert.[69] Here, to pump our conceivability-intuitions about the radical possibility of living gold, we need only think of the living, thinking ocean in Andrei Tarkovsky's brilliant sci-fiction film study of animism, nostalgia, and human commitment, *Solaris* (1972). Therefore our gold-stuff would have a very different or even a radically different *microphysical constitution* in the L^{**}-governed worlds.

And therefore, in the strict or Leibnizian metaphysical sense of necessity, gold is *not* necessarily the element with atomic number 79. So, again, by Kantian lights, scientific essentialism is just plain wrong.

3.6. Concluding Anti-Scientific-Essentialist Postscript

It should be emphasized that while each of the anti-scientific-essentialist arguments in the preceding four sections can indeed be treated as an independent line of objection, the truth is that the four of them are ultimately integral parts of a single comprehensive Kantian argument against scientific essentialism. That is, supposing that the scientific essentialist is allowed to reply to these Kantian criticisms, the justification for the cogency of each Kantian objection will eventually make an appeal to one or more of the other objections. This is not to concede vicious circularity, but only that at the end of the day the four arguments are going to turn out to be

---

[69] For a similar point, see Westphal, *Kant's Transcendental Proof of Realism*, ch. 6. The world of living gold is a limit case of natural laws, however, since the causal-dynamic laws of matter in that world would all be *one-time laws*, that is, laws governing natural causal singularities of organismic activity. So living gold could not be correctly described by *general* mechanistic causal natural laws, or support true mechanistic-law-expressing counterfactuals (*MFNS* 4: 544). This in turn is intimately connected with Kant's thesis that there exists an irreducible explanatory gap between inert nature and living nature: 'It is quite certain that we can never adequately come to know the organized beings and their internal possibility in accordance with merely mechanical principles of nature, let alone explain them; and indeed this is so certain that we can boldly say say that it would be absurd for humans even to make such an attempt or to hope that there may yet arise a Newton who could make comprehensible even the generation of a blade of grass according to natural laws that no intention has ordered' (*CPJ* 5: 400). See also sections 8.3–8.4.

rationally corporate or holistic, not strictly autonomous from one another. So I think that the overall Kantian case against scientific essentialism is best regarded as cumulative rather than as driven by one or another of the objections on its own.

For example: suppose that the scientific essentialist objects to the Antinomy of Essentialism (= the third objection) by suggesting that the essentialist thesis is intended to include or imply the ancillary thesis that the level of physical micro-constitution discovered by actual empirical scientific inquiry (say, the atomic/sub-atomic level) is by hypothesis the *deepest* or *fundamental* level. Then the Kantian can reply that his contrary proposal to the effect that it is possible that there is an even deeper, physical micro-micro-constitutional level beneath (say) the atomic level is only meant as something conceivable, not as an *empirical* hypothesis.

If the scientific essentialist then counter-replies by stating that he meant that the level discovered by empirical science is *necessarily* deepest, then in turn the Kantian can counter-counter-reply as follows.

First, as the first objection shows, the essentialist has only perceptual empirical evidence for his discovery and cannot get perceptual cognitive access to any level of non-apparent micro-constitution, much less the necessarily deepest.

Second, as the second objection shows, any appeal to necessity is based on a priori insight, not empirical evidence, even when there is significant empirical content in the relevant propositions.

And, finally, as the fourth objection shows, even if the scientific essentialist did somehow manage to show "empirically" that (say) the atomic/sub-atomic level is necessarily deepest or fundamental, and in this way—modally as it were "from below"—universal across all the possible worlds in which gold exists, nevertheless that compositional level of nature would *still* be necessarily dependent on a contingent actual set or package of natural laws, and in that way—modally "from above"—*not* universal across the possible worlds in which gold exists.

This leads me to one last observation. Perhaps the most telling Kantian point against the scientific essentialist is that even if there *is* some constitutive physical microstructure for a given natural kind in the actual world, the substratum of the kind will not be strictly identical to that microstructure *except* under some further enabling assumptions about the actual set or package of natural laws. And this is because the intrinsic non-relational dispositional properties of microphysical matter, according to the scientific essentialist metaphysics, are necessarily and indeed even constitutively connected with the laws of nature that obtain in our actual world, or in any other possible world considered as actual. This would not of course be true on a Humean conception of causal laws of nature, according to which such laws are nothing but empirical generalizations describing extrinsic relational dispositional properties of material objects. And it would also not be true on the well-known "Armstrong-Dretske-Tooley" (ADT) conception of laws of nature, according to which laws of nature are logically contingent relations of necessary entailment between universals, so that actual causal relations are extrinsic non-dispositional entailment-relations holding between instances of those universals.[70] But scientific essentialists, like Kant, are *neither* Humeans *nor* ADT-ists about the causal laws of nature: both Kantians and scientific essentialists agree that natural causal laws are *intrinsic* to individual material things and natural kinds. For Kant, natural causal laws capture intrinsic structural dynamic causal powers of individual macrophysical material things and natural kinds, which in turn essentially express specific relations of the primitive attractive and repulsive forces in nature.[71] By contrast, for the scientific essentialist, natural causal laws express intrinsic non-relational microphysical causal powers of those things and kinds. But, despite the differences between Kant and the scientific essentialists, one crucial consequence of their sharing the notion of the intrinsicness of causal laws is that the essentialists' microphysical metaphysics of natural kinds, insofar as it is even *intelligible*, is in this way deeply subject to certain existential, essentially indexical, logically contingent, and anthropocentric—in a Kantian word, "synthetic"—conditions. Therefore whatever sort of meaning-content or modality a scientific microstructural identity proposition about natural kinds such as (GE) really has, it *cannot* involve logical or strict metaphysical Leibnizian necessity: that is, (GE) must be non-analytic, and also not logically or strictly metaphysically necessary, in the plain-and-simple sense that there are some logically possible worlds

---

[70] See, e.g., Tooley, "Causation: Reductionism vs. Realism."
[71] See Watkins, *Kant and the Metaphysics of Causality*, ch. 4. Watkins's account of Kant's theory of causal laws differs from mine however. For Watkins, Kant situates the metaphysical ground of the laws in the individual things and kinds themselves. By contrast on my view Kant holds that the metaphysical ground of the laws is the total structural complex of attractive and repulsive forces, which is the same same as the One Big Material Substance of the First Analogy of Experience, which in turn is the same as the fundamental fluid aether of the *Opus postumum*. See sections 8.1 and 8.3.

in which gold exists but (GE) is not true. Now we have also seen that knowledge of the specific modal status of any necessary proposition is a priori. This means that if the scientific essentialist wants to maintain that propositions like (GE) have any sort of strong modal force at all, then he had better dust off his copy of the first *Critique* and start thinking about synthetic a priori propositions all over again.

4

Manifest Realism II: Why Gold is Necessarily a Yellow Metal

> At least Kant thinks it's a *part* of the concept that gold is to be a yellow metal. He thinks that we know this *a priori*, and that we could not discover it to be empirically false ... Is Kant right about this?
>
> Saul Kripke [1]

> Gold [is] ... a yellow malleable ductile high density metallic element resistant to chemical reaction.
>
> Oxford English Dictionary [2]

> Nature considered materially is the totality of all objects of experience.
>
> P 4: 295

4.0. Introduction

According to Kant in the *Prolegomena*, the proposition (GYM) "Gold is a yellow metal," is analytic, a priori, and necessary:

> All analytic judgments are a priori even if the concepts are empirical, e.g., "Gold is a yellow metal"; for to know this scientifically (*wissen*) I require no experience beyond my concept of gold, which had as its content that this body is yellow and metal. For just this constituted (*machte...aus*) my concept; and I need only decompose (*zergliedern*) it, without looking beyond it elsewhere. (P 4: 267)

What is declared to be cognized a priori is thereby announced as necessary. (P 4: 278)

---

[1] Kripke, *Naming and Necessity*, p. 117.
[2] Hawkins and Allen (eds.), *Oxford Encyclopedic English Dictionary*, p. 606.

For mainstream analytic philosophers in the latter half of the twentieth century and the early parts of the twenty-first, however, this Kantian doctrine has seemed obviously—even outrageously—wrong. As we saw in chapter 3, Kripke puts it this way in *Naming and Necessity*:

Kant... gives as an example [of an analytic statement] 'gold is a yellow metal,' which seems to me an extraordinary one, because it's something I think that can turn out to be false.[3]

Note that Kant's example, 'gold is a yellow metal,' is not even *a priori*, and whatever necessity it has it established by scientific investigation; it is thus far from analytic in any sense.[4]

In turn, again as we saw in chapter 3, Kripke grounds his criticism of Kant on the doctrine of scientific essentialism:

Scientific investigation generally discovers characteristics of gold which are far better than the original set. For example, it turns out that a material object is (pure) gold if and only if the only element contained therein is that with atomic number 79. Here the 'if and only if' can be taken to be *strict* (necessary). In general, science attempts, by investigating basic structural traits, to find the nature, and thus the essence (in the philosophical sense) of the kind.[5]

According to scientific essentialism, we will remember, natural kinds are essentially constituted by intrinsic non-relational properties at the microphysical and more specifically atomic or quantum level of physical reality, which according to scientific essentialism is also the fundamental level of physical reality; and these microphysical essences are already known (or at least are knowable in principle) by means of contemporary fundamental physics or fundamental chemistry. Further, expressions standing for natural kinds—whether names of natural kinds such as "gold" or "water," or essence-disclosing scientific natural kind terms, that is, definite descriptions such as "the element with atomic number 79" or "H_2O"—are rigid designators, or referring terms that pick out a certain object in every possible world in which that object exists, and never refer to anything else in any other possible world.[6] From this it follows that propositions such as (GE) "Gold is the element with atomic number 79" and (WH) "Water

---

[3] Kripke, *Naming and Necessity*, p. 39.   [4] Ibid. p.123, n.63.
[5] Ibid. p. 138.
[6] See Kripke, *Naming and Necessity*, p. 48; and Stanley, "Names and Rigid Designation," p. 556.

is H₂O" are true essence-disclosing identity statements about the natural world acquired through empirical scientific investigation alone. And from the plausible assumption that all true identity statements are necessary,[7] it follows in turn that (GE) and (WH) are logically or "strictly" metaphysically necessary in the Leibnizian sense that they are true in all possible worlds[8] but also a posteriori in the sense that their being justifiably believed

---

[7] See Kripke, "Identity and Necessity," p. 162–3.

[8] This glosses over some fairly subtle points; see ch. 3, nn.11–15. But to be more specific, suppose that it is metaphysically necessary that water is H₂O: even granting that hypothesis, it still seems conceivable and logically or strictly metaphysically possible that there are worlds in which our actual world water is something other than H₂O. See Putnam, "Is Water Necessarily H₂O?" But if this is so, then either it is not metaphysically necessary that water is H₂O (i.e., scientific essentialism is false or paradoxical), or else not all metaphysical necessity is the same as logical or strict metaphysical necessity. See Farrell, "Metaphysical Necessity is not Logical Necessity." Orthodox Kripkeans can handle this difficulty in one or both of two ways. First, they can skeptically reject the step from conceivability to logical or metaphysical possibility; see Chalmers, *The Conscious Mind*, pp. 131–40. Second, they can introduce a second kind of metaphysical necessity over and above logical or strict metaphysical or Leibnizian necessity, call it "restricted metaphysical necessity," or "non-logical necessity," or "synthetic necessity." See Chalmers, *The Conscious Mind*, ch. 4. Chalmers's term for restricted, non-logical, or synthetic necessity is "strong metaphysical necessity," in contradistinction to the "weak metaphysical necessity" which is the same as logical or strict metaphysical (Leibnizian) necessity. This second option also includes one further branching choice point. Having isolated restricted, non-logical, synthetic, or strong metaphysical necessity, the orthodox Kripkean can then either (i) go *modal dualist* and accept the existence of two irreducibly different types of necessity, or (ii) go *modal monist* and claim that all necessity is nothing but restricted, non-logical, synthetic, or strong metaphysical necessity. As Chalmers points out, all actual orthodox Kripkeans who have so far gone for the second option, have also opted for (ii), and also reject the step from conceivability to possibility. This move is then usually construed as expressing some sort of posteriori physicalism or scientific naturalism. But it is by no means clear to me that it would not be more in the spirit of Kripke himself, who is *neither* a physicalist *nor* a scientific naturalist, and indeed is explicitly a non-reductivist in the philosophy of mind and also a modal rationalist in epistemology, to opt for (i), and also *accept* the step from conceivability to possibility. That would amount to a position very like Kant's. Neo-Kripkeans like Chalmers, in any case, insist that necessity is univocal (modal monism) but then distinguish between two different types of intensions or concepts of the word "water." The "primary" intension or concept (which expresses narrow or individualist cognitive content) is purely descriptive and takes us from possible worlds "considered as actual" (i.e., worlds that are egocentrically indexically centered on the here and now of conscious minds) onto whatever conforms to the relevant description in those worlds. By contrast, the "secondary" intension (which expresses wide or externalist cognitive content) is directly referential or anyhow rigidly designating and takes us from possible worlds "considered as counterfactual" variants on the microphysically defined actual world, to whatever is essentially microphysically identical to the relevant actual world referent in those worlds. See Davies and Humberstone, "Two Notions of Necessity"; and Chalmers, *The Conscious Mind*, pp. 52–69. On this two-dimensional modal semantics. "Water is the watery stuff" is a priori and necessary according to the primary intension, while "Water is H₂O" is a posteriori and contingent; and "Water is H₂O" is a posteriori and necessary according to the secondary intension, while "Water is the watery stuff" is a priori but contingent. Kant's modal semantics is significantly different from both orthodox Kripkean and neo-Kripkean modal semantics: it ties apriority tightly to necessity; it accepts the validity of the step from conceivability to possibility (for a Kantian argument supporting that validity, see ch. 3, n.67); and it is modally *dualistic* rather than modally *two-dimensional*. For details, see Hanna, *Kant and the Foundations of Analytic Philosophy*, esp.

(and thereby known to be true) depends on specific factual conditions and sensory evidence. Obviously, if scientific essentialism is right about natural kind propositions and our knowledge of them, then Kant's general theory of meaning necessity and apriority is wrong; but contrapositively, if Kant is right, then scientific essentialism is wrong.

In chapter 3, I argued that the scientific essentialist challenge to Kant's theory of meaning, necessity, and apriority ultimately fails because of some fundamental flaws in the essentialists' epistemology and metaphysics. The present chapter, however, is constructive rather than critical. My aim is to work out the positive theory behind Kant's surprising claim that gold is necessarily a yellow metal, using scientific essentialism only as a critical foil.

Scientific essentialism is a paradigmatic version of *maximal scientific realism*, which in turn is committed to *microphysical noumenal realism*. As Kant shows in the Refutation of Idealism, and as we also saw in section 1.0, all metaphysical realists of any stripe hold minimally that some knowable things exist outside the mind and not merely in one's phenomenal consciousness or inner sense. But noumenal realism, we will recall, is minimal metaphysical realism plus the two-part thesis that all knowable things are:

(a) metaphysically constituted by a set of intrinsic non-relational properties; and

(b) *transcendent*, in the triple sense that: (i) possibly the knowable things exist even if human minds do not or cannot exist; (ii) possibly the knowable things exist even if all human cognizers do not know or cannot know them; and (iii) necessarily the knowable spatial things are neither directly humanly perceivable nor observable but at best only indirectly humanly perceivable or observable.

In turn, for Kant the entities that would, *if* they existed, uniquely satisfy the conditions of noumenal realism are things-in-themselves or noumena in the positive sense: namely, entities whose essences consist in a set of intrinsic non-relational properties such that, if those entities existed, they would be mind-independent, non-sensory, unperceivable and unobservable ("colorless"), and knowable in the very special sense that only a being equipped with "intellectual intuition"—that is, only a being having an

---

chs. 2, 3, and 5. On the other hand, for the purposes of this chapter it is important to note a close logico-semantic similarity between Kant's notion of analyticity and the neo-Kripkeans' deep or a priori conceptual necessity according to the primary intension.

intuitional faculty which falls altogether outside the spatiotemporal forms of all human or non-human sensible intuition[9]—could ever cognize them (*CPR* B71–2, B305–8).

In order to understand Kant's notion of the thing-in-itself or positive noumenon properly, however, we must also draw the following set of Kantian background distinctions between:

(1) Noumenal *subjects* (= pure rational knowers or pure rational persons).
(2) Noumenal *objects* (= whatever is directly known by a pure rational knower or directly acted upon by a pure rational person, but is not itself a rational knower or rational person).
(3) Noumenal *substances* (= independently existing noumenal individuals or kinds possessing non-relational essences).
(4) Noumenal *properties* (= noumenal attributes or determinations, picked out by noumenal concepts).
(5) *Positive* noumena (= things-in-themselves).
(6) *Negative* noumena (= things possessing intrinsic non-sensible properties):

> If by a noumenon we mean a thing **insofar as it is not an object of our sensible intuition,** because we abstract from the manner of our [sensible] intuition of it, then this is a noumenon in the **negative** sense. (*CPR* 307)

One important point to note is that the distinction between (3) and (4) is a specific version of the more general metaphysical distinction, familiar

---

[9] This special Kantian notion of non-spatiotemporality raises a tricky question about the nature of space and time. Kant's view is that nothing can ever count as real space or real time unless it is properly representable by the pure or formal intuition of rational human animals. This is an analytic consequence of the weak transcendental ideality of space and time; and it also has some interesting affinities with the controversial Anthropic Principle in contemporary physics (see ch. 2, n. 85). But, in any case, for Kant the words "space" and "time" (I mean of course *Raum* and *Zeit*) express forms of intuition and pure or formal intuitions, and thereby function as *directly referential proper names* of space and time, not *descriptions* of them, or mere concept-terms. See Hanna, *Kant and the Foundations of Analytic Philosophy*, pp. 223–4. Hence, as long as we are using "space" and "time" correctly, then there cannot be a noumenal space or a noumenal time, although it is logically possible that there are two positive noumenal analogues of space and time, as it were *schpace* and *schtime*, which share the *non-designated* properties (see section 2.4 above) of space and time: extension, continuity, relationality, etc. More generally, these are precisely the spatial and temporal properties that can be represented in the second-order classical polyadic predicate logic of *Principia Mathematica*, and which abstract away from uniquely identifying differences between space and time. So in fact *nothing* could ever adequately discriminate between so-called noumenal space (i.e., schpace) and so-called noumenal time (i.e., schtime). Or in other words, to twist an apt phrase from the Introduction to Hegel's *Phenomenology of Spirit*, the world of positive noumena is the night in which all spatial and temporal cows are black.

to students of the mind–body problem, between *substance dualism* (= the thesis that two distinct classes of substances exist) and *property dualism* (= the thesis that two distinct classes of properties exist). Substance dualism entails property dualism, but property dualism does not entail substance dualism: it is possible for there to be one and only one class of substances that have two irreducibly distinct classes of properties. The distinction between substance dualism and property dualism in turn implies that there are two sharply distinct ways to construe the phenomena vs. noumena distinction: as a form of substance dualism (= the classical Two World or Two Object Theory), or as a form of property dualism (= what I call *the Two Concept or Two Property Theory*). Going in for the latter and rejecting the former, however, one can accept a sharp distinction between phenomenal properties and noumenal properties, while also (i) affirming the existence of phenomenal substances, and (ii) refusing to affirm the existence of noumenal substances. As I will argue in section 8.2, it seems to me that the Two Concept or Two Property Theory is the philosophically most acceptable interpretation of Kant's noumena vs. phenomena distinction, and therefore the one that I will charitably ascribe to him.

Another and closely related important point to note is that the distinction between (1) and (2) cross-cuts the distinction between (5) and (6). Thus things-in-themselves or positive noumena can be either noumenal subjects or noumenal objects. Similarly, negative noumena can be either noumenal subjects or noumenal objects. Combining this recognition with the distinction between (1) and (2), we can immediately see that while things-in-themselves or positive noumena are necessarily *not* empirical phenomenal entities, nevertheless negative noumena can in principle *also* be empirical phenomenal entities, in the sense that an empirical subject or empirical object can consistently *also* possess some non-sensible properties. Or, in other words, *non*-sensible properties are not necessarily *super*-sensible properties. For example, the empirical phenomenal subject and individual human animal ETH, in addition to possessing many phenomenal properties, also possesses the negatively noumenal properties of *satisfying the law of non-contradiction, being numerically one, being spatiotemporal, being a material substance*, and so on. Correspondingly, the empirical bodily event of ETH's lifting her hands in the air (to wave them like she just don't care), again in addition to possessing many phenomenal properties, also possesses the negatively noumenal properties of *satisfying the law of cause and effect, satisfying the law of the simultaneous reciprocal dynamic interaction of material substances, being permissible according to the Categorical Imperative, being psychologically free,* and *being transcendentally free.* The last three of these properties cannot be scientifically cognized, but they can be known to apply to phenomenal beings via *practical reason.* The applicability of at least some negatively noumenal properties to phenomenal subjects or objects will play a seminal role in chapter 8.[10]

In any case, assuming those three pairs of distinctions, we can then see how examples of things-in-themselves or positive noumena will include:

(i) Platonic Forms and other Platonic abstracta (for example, Pythagorean numbers-as-objects, Leibnizian possible worlds, Fregean numbers-as-objects, Cantorian transfinite numbers, Meinongian possibilia and impossibilia, early Russellian universals, early Moorean concepts, etc.);

(ii) the Cartesian God and other divine beings of all sorts;

(iii) Cartesian minds, Leibnizian monads, and other non-relational *mental* essences of all sorts;

and last but not least, but indeed most relevantly in the present context:

(iv) Lockean real essences and other non-relational *physical* essences of all sorts.

Now things-in-themselves or positive noumena are consistently thinkable and therefore logically possible (*CPR* Bxxvi). Indeed, the very concept of an appearance analytically entails the concept of something X that appears. But, as Kant's critique of the ontological argument in the Ideal of Pure Reason (*CPR* A592–603/B620–31) clearly shows, analytic entailments of the concept of existence guarantee at most the logical possibility of instantiations of the concept of the thing to which the concept of existence necessarily applies—for example, the analytic entailment of the concept of the existence of a perfect being, given its possession of all perfections and the assumption that existence is a perfection—but not that thing's actual existence or reality, its *Realität* or *Wirklichkeit*, which also requires a sensory intuition of that thing, and

---

[10] In section 8.2, I will argue that according to (what I will call) Kant's *embodied agency theory of freedom*, all human noumenal subjects, as moral persons, are negative noumena but *not* positive noumena or things-in-themselves.

which of course is notoriously lacking in the case of a perfect being. So things-in-themselves are, at most, logically possible values of this "generic representational object," or "transcendental object $= X$," or "object in general" (*Gegenstand überhaupt*) (*CPR* A108–9, A253), which, in Kant's transcendental framework, is the precise cognitive-semantic equivalent of Quine's logico-linguistic *bound variable* of ontic commitment: "to be is to be the value of a bound variable."[11] Furthermore, as Kant's critique of noumenal thinking in the chapter "On the Ground of the Distinction of All Objects in General into *Phenomena* and *Noumena*" also clearly shows, both the existence and the non-existence of things-in-themselves are equally wholly uncognizable and thereby equally wholly unassertible. And for this reason, as Critical philosophers, we must remain aggressively and consistently agnostic about them (*CPR* A255/B310, A286–7/B343). This aggressive and consistent agnosticism about things-in-themselves, in turn, is equivalent to what I will call Kant's *methodological eliminativism* about things-in-themselves:

[O]bjects in themselves are not known to us at all, and ... what we call outer objects are nothing other than mere representations of our sensibility, whose form is space, but whose true correlate, i.e., the thing in itself, is not and cannot be cognized through them, but is also never asked after in experience. (*CPR* A30/B45)

In other words, things-in-themselves are conceptually intelligible to us, and correspondingly we possess concepts of them, but at the same time the things-in-themselves themselves are practically, explanatorily, epistemologically, and metaphysically superfluous. *We just don't need them.*

By sharp contrast to noumenal realism, as we have seen in chapters 1–3, Kant's empirical realism is a minimal metaphysical realism plus two further theses:

(1) Every self-conscious human cognizer has direct veridical perceptual or observational access to some actual dynamic material things in objectively real space and time (direct perceptual realism).

---

[11] See Hanna, *Kant and the Foundations of Analytic Philosophy*, pp. 106–8; Quine, *Ontological Relativity*; and Quine, "On What There Is." Bound variables of course imply quantifiers, and the crucial difference between open (unquantified) and closed (quantified) sentences: only closed sentences can have truth-values. Similarly, for Kant, only judgments can have truth-values, and thus the generic object of representation, the transcendental object $= X$, must always occur within the context of complete judgments if it is to contribute to the truth or falsity of a judgment.

(2) All the essential properties of knowable dynamic material things are nothing but their directly humanly perceivable or observable intrinsic structural macrophysical properties (manifest realism).

On Kant's view, as I am construing it, things are directly perceivable or observable if and only if they can at least in principle be made intuitionally present via perception or memory to the conscious human senses without a necessary dependence on concepts, beliefs, inferences, or theories. This in turn of course yields the *non-conceptuality* of direct perception or observation. It is crucial to note that the non-conceptuality of direct perception or observation is perfectly consistent with the evolutionary modification or refinement of the human senses (such that at least in principle, e.g., we might develop a sensitivity to magnetic fields, or develop a capacity for batlike sonar), and also with the "consumerist" or conceptually and theoretically unsophistcated use of various sorts of detection technology (e.g., binoculars, telescopes, electronic sonar, cloud chambers, electron microscopes, Geiger counters, or cyclotrons). In modifying, refining, and supplementing our perceptual capacities in these ways, we never leave the realm of appearances but instead only discover more and more about the fine-grained law-governed details of natural appearances. *Cyclotrons are not noumenotrons.* Things that are neither directly perceivable nor observable in any of these ways are nothing but "entities of the understanding" (*Verstandeswesen*) (*CPR* B306), that is, purely theoretical entities, including of course all things-in-themselves.

It is again Kant's manifest realism that will particularly concern me in this chapter. In a nutshell, manifest realism says that, cognitively and ontologically speaking, *nothing is hidden* in the material world that is the object of true empirical scientific cognition. Part of what I mean by this is that fundamental physics for Kant is wholly and exclusively about *what lies right out there in front of us*, on the dappled surfaces of physical nature. This is not folk physics. Folk physics, or our common-sense physical know-how, is neither mathematized nor nomological. It is just the set of pragmatic rules of thumb with actual contextualized conditions of success explicitly added, or *ceteris paribus* laws,[12] that express in a post hoc way how we actually get around in the material world, navigate by the stars, build pyramids and

---

[12] See Fodor, "Making Mind Matter More."

Gothic cathedrals, or build and operate telescopes, microscopes, air pumps, steam engines, gasoline engines, radios, light bulbs, vacuum cleaners, air conditioners, television sets, fridges with built-in ice-cube-makers, Velcro fasteners, laptop computers, cellphones, CDs, DVDs, iPods, and "smart" missile systems. On the contrary, however, fundamental physics, according to Kant, is *a mathematized nomological science of matter and motion, which presupposes a background metaphysics of causal-dynamic structuralism together with a strictly macrophysical ontology.*

I will come back to Kant's causal-dynamic structuralism in section 4.2. But for the moment the crucial point to recognize is that, according to his strictly macrophysical ontology, the objectively real material world is nothing but *our* material world, the natural-law-governed causal-dynamic apparent spatiotemporal world in which we live, move, and have our embodied being. As far as philosophical method and natural science are concerned, we can simply ignore the logical possibility of there being any deeper ontological levels. So for Kant the being or *Sein* (being) of physical nature is nothing but its appearance or *Erscheinung* (*CPR* A49/B66, B69).[13] Indeed, Kant consistently and systematically—but unfortunately, not always fully explicitly—employs two sharply different concepts of appearance:

(1) *Mere Appearance or* **Schein**. Sometimes X appears to be F even though X is in fact not F. For example, in the dusk and to perceivers with mild myopia and a vivid imagination that bush over there (alarmingly) appears to be a crouching person. This sort of appearance is mere appearance or subjective illusion.

(2) *Authentic Appearance or* **Erscheinung**. Sometimes X appears to be F because X really is F. For example, in broad daylight and to perceivers with properly functioning visual systems that bush over there appears to be just what it really is—a bush! This sort of appearance is a true appearance or natural phenomenon.

---

[13] If I am correct about Kant's manifest realism, then Wilson is wrong when she claims that "for Kant, what is empirically real is primarily the material world of the science of his time—a world that does not possess colors, tastes, and the like in any irreducible sense" ("The 'Phenomenalisms' of Kant and Berkeley," p. 161). But to reject Wilson's influential interpretation is *not* to say that Kant is a Berkeleyan phenomenalist who thinks that the spatiotemporal world is a private construct from mental qualia: see Aquila, "Kant's Phenomenalism." On the contrary, manifest realism is a form of metaphysical realism that gets relevantly between both contemporary scientific realism and also anti-realism (whether empirical idealism, phenomenalism, or the weaker variety that reduces truth-conditions to assertibility-conditions).

So for Kant to say that the *Sein* of physical nature is its *Erscheinung* is to say that physical nature truly appears to be what it really is. In this way, if Kant is right, then the very idea of a necessarily hidden colorless world of intrinsic non-relational microphysical properties, to which everything else explanatorily or ontologically reduces, or anyhow upon which everything else strongly supervenes, is nothing but a metaphysical myth.

As we have seen, scientific essentialism is a version of noumenal scientific realism that is specifically committed to the knowability, transcendence, and explanatory primacy of microphysical entities, properties, facts, events, processes, and forces. Among other things, this entails that the essences of natural kinds are *necessarily not* perceivable or observable. Correspondingly and oppositely, however, Kant's theory of natural kind terms, taken together with his manifest realism, implies his commitment to the direct human perceivability and observability of the intrinsic structural properties of natural kinds. In other words, for Kant all natural kinds are *manifest kinds*,[14] not microphysical kinds. At least on the face of it, then, Kant's manifest realism is a strong vindication of common sense against the anti-commonsensical scientific essentialist. But faces can be misleading; and as the first epigraph in the Introduction I quoted a characteristically brilliant Nietzschean remark which may make us think twice about this all-too-easy conclusion:

Kant's Joke. Kant wanted to prove in a way that would dumbfound the common man that the common man was right: that was the secret joke of this soul.[15]

Playing a riff now on Nietzsche's aphorism, from a contemporary standpoint we can regard Kant's manifest realism as an especially *ironical* kind of secret joke, in that his common-sense-friendly theory *itself* would undoubtedly "dumbfound" today's college-educated common man or common woman. For today's college-educated commoner shares with the scientific essentialist the firm belief that the natural world is ultimately microphysical and necessarily hidden from our perceptual capacities. Otherwise put, today's college-educated commoner happily takes in Eddington's metaphysical myth of the Two Tables (see section 3.2) along with his morning coffee and the *Science* section of his daily newspaper. So very ironically the *vox populi*, in concert with contemporary noumenal scientific realists, would together insist that Kant's empirical scientific realism is outrageously wrong.

---

[14] I borrow this apt term from Johnston, "Manifest Kinds."
[15] Nietzsche, *The Portable Nietzsche*, p. 96.

Given the heavy-duty cultural and philosophical artillery aimed directly at his empirical scientific realism, can Kant then offer a plausible account of the meaningfulness and truth of propositions about natural kinds? In my opinion, yes. And to support that claim, I will unpack and argue for two Kantian theses: (1) that natural kind terms are *decomposable phenomenological*[16] *indexical predicates*, terms whose cross-possible-worlds extensions are partially determined by subject-centered actual spatiotemporal contexts but also partially determined by a priori analyzable conceptual modes of identification; and (2) that Kant's manifest realism is both internally consistent and—despite its incompatibility with noumenal scientific realist accounts of the foundations of the exact sciences—in certain basic respects independently philosophically compelling. If I am right, then Kant shows us how we can consistently be both scientific realists and conceptual analysts. Indeed, if I am right, then Kant shows us that his is the *only* way we can consistently be both scientific realists and conceptual analysts.

Q: And please tell me very quickly, just before you get started, why conceptual analysis is so philosophically important?

A: Because in a natural world in which nothing is hidden (1) true empirical cognitions in the natural sciences tell us directly about the essential nature of that manifest world; (2) analytic propositions are true by virtue of relations between and within the intrinsic conceptual microstructures ("logical essences") of the objectively valid concepts that are the basic semantic constituents of those propositions;[17] (3) the act of conceptual analysis imaginatively recapitulates and makes explicit the implicit microstructures of the concepts we use to do this (see section 7.2 below), thereby telling us how the manifest world is metaphysically put together—

But given this answer: "But you know how [true empirical cognitions in the natural sciences] do it, for nothing is concealed" one would like to retort "Yes, but it all goes by so quick, and I should like to see it as it were laid open to view.";[18]

---

[16] Here I am using "phenomenological" in *Newton's* sense, not Husserl's (*MFNS* 4: 554–65). In Newton's *Principia*, "phenomenology" is the mathematical theory of apparent physical motion. And a vestige of this usage remains in the contemporary scientific "phenomenology of metals": i.e., the experimental surface-level study of the physico-chemical properties of metals. So, in my usage here, "phenomenological" means *macrophysical, mathematically describable, and causal-dynamical.*
[17] See Hanna, *Kant and the Foundations of Analytic Philosophy*, ch. 3.
[18] Wittgenstein, *Philosophical Investigations*, §435, p. 128e, text modified slightly.

and (4) a cognitive capacity for conceptual analysis is partially constitutive of our human theoretical rationality itself (see section 4.3). I hope that helps.

4.1. Kant's Theory of Natural Kind Terms

In sections 3.0 and 3.1, I sketched in broad outline the semantics of natural kind terms according to scientific essentialism. From a Kantian point of view, however, a natural kind term is any word, phrase, propositional constituent or judgment-constituent that expresses an empirical concept and refers to a natural kind. A natural kind, in turn, is a mass or totality of not-necessarily-contiguous-or-unmixed[19] inorganic or organic physical stuff (for example, gold, water, cats, and elms) sharing a unifying physical substrate of some sort. For simplicity's sake (and because of some important Kantian subtleties about the content and applicability of the concepts LIFE and ORGANISM, which I want to set aside for separate treatment later in section 8.3), I will focus exclusively on terms standing for inorganic, inert natural kinds such as gold or water.

Since on Kant's account natural kind terms are empirical concepts, I need first of all to say something about his theory of empirical concepts. For my purposes here, these three texts convey the most basic and relevant points:

> [The logical essence] includes nothing further than the cognition of all the predicates in regard to which an object is determined *through its concept*... If we wish to determine, e.g., the logical essence of body, then we do not necessarily have to seek for the data for this in nature; we may direct our reflection to the characteristics which, as essential points (*constitutiva rationes*) originally constitute the basic concept of the thing. For the logical essence is nothing but *the first basic concept of all the necessary characteristics of a thing (esse conceptus)*. (JL 9: 61)

> Every concept, *as partial concept*, is contained in the representation of things; as *ground of cognition*, i.e., *as characteristic*, these things are contained *under* it. In the former respect, every concept has an *intension* (*Inhalt*), in the other a *comprehension* (*Umfang*) ... As one says of a ground in general that it contains the *consequence* under

---

[19] I mean that the stuff can occur in separated bits or pieces (gold rings), and also impurely or in alloy (white gold).

itself, so one can say of the concept that as *ground of cognition* it contains all those things under itself from which it has been abstracted, e.g., the concept of metal contains under itself gold, silver, copper, etc. (*JL* 9: 95–6)

If I say, for instance, "All bodies are extended," this is an analytic judgment. For I do not require to go beyond the concept which I connect with the word 'bodies,'[20] in order to find extension as bound up with that concept. To meet with this predicate, I have merely to decompose the concept, that is, to become conscious to myself of the manifold (*Mannigfaltigen*) which I always think in it. (*CPR* A7/B11)

Here now are those points redescribed. Empirical concepts are made up of "characteristics" or *Merkmale*. Characteristics, in turn, are partial concepts that express complex or simple attributes of sensory things. Every empirical concept has a "logical essence" or *esse conceptus*. This is a logically ordered set of necessary characteristics, involving genus-to-species or "determinable-to-determinate"[21] hierarchical relations (*JL* 9: 96–8), making up the semantic core of that representation. In short the logical essence or *esse conceptus* of an empirical concept is a *conceptual microstructure* that constitutes and individuates that concept, even if by itself it does not exhaust the total conceptual content (more on this below). Further, the logical essence or *esse conceptus* of an empirical concept is the same as the "intension" or *Inhalt* of that concept.[22] And the primary function of the intension of an empirical concept is uniquely to determine its "comprehension" or *Umfang*, its cross-possible-worlds extension: the

---

[20] Here I am using the A edition's version of this phrase instead of the B's less explicit version.

[21] See Sanford, "Determinates vs. Determinables."

[22] See section 2.2 for the distinction between (1) the semantic content or intension (*Inhalt*) of a conscious cognition, and (2) the sensory qualitative content (*Materie*) of a conscious cognition. The crucial point is that the former is conscious and representational, hence can be either semantically opaque (concepts) or semantically transparent (intuitions); while the latter is conscious and non-representational, hence non-semantic, and transparent in a metaphysical sense that is even stronger than semantic transparency. Since the Kantian conceptual *Inhalt* has both a decompositional microstructure and an immanence in human consciousness, it is both a fine-grained and a hyper-fine-grained intensional entity: two *Inhalte* sharing the same cross-possible-worlds extension, or *Umfang*, can still differ in internal structure (*CPR* B4); and two *Inhalte* sharing the same internal structure and *Umfang* can still differ in "viewpoints" (*Gesichtspunkte*) (*JL* 9: 147) or modes of presentation. See also section 4.2 for the distinction between (2) above, and (3) the inert physical matter (*Materie*) of an individual empirical object or natural kind. Sensory qualitative content or matter is inside the conscious mind of a cognizing animal, while the inert physical matter of an empirical individual object or natural kind is outside the conscious cognizing animal's body in objectively real space. The organismic body of a conscious animal is, as we would now say, a living-and-lived body (*Leib*), whereas the bodies presupposed by fundamental physics are inert mechanical bodies (*Körper*) (*CPR* A7–8/B11–12).

open-ended set of actual or possible items identified by[23] that concept and to which the concept thereby correctly applies. It is a direct consequence of this doctrine that any cognitive "decomposition" (*Zergliederung*) of the essence or microstructure of a given empirical concept also entails the necessary truth of a corresponding analytic proposition. For if the predicate-concept of the judgment is contained in the subject-concept, then the corresponding judgment is necessarily true by virtue of conceptual content alone: it is analytic. Moreover, assuming intensional containment, then every actual or possible item included in the comprehension of the subject-concept is also automatically included in the comprehension of the predicate concept.[24]

We are now in a position to look more directly at Kant's theory of natural kind terms. In addition to the crucial text from the *Prolegomena*, which explicitly asserted the analyticity and apriority of (GYM) despite its containing empirical concepts (*P* 4: 267), the other seminal texts can be found in the first *Critique's* chapter on the "Discipline of Pure Reason," in the Transcendental Doctrine of Method:

If we are to judge synthetically in regard to a concept, we must go beyond the concept and appeal to the intuition in which it is given. For should we confine ourselves to what is contained in the concept, the judgment would be merely analytic, serving only as an explanation of the thought, in terms of what is actually contained in it. But I can pass from the concept to the corresponding pure or empirical intuition, in order to consider it in that intuition *in concreto*, and so to cognize, either *a priori* or *a posteriori*, what the properties of the object of the concept are. The *a priori* method gives us our rational and mathematical cognition through the construction of the concept, the *a posteriori* method our merely empirical (mechanical) cognition, which is incapable of yielding necessary and apodeictic

---

[23] Identification by means of concepts is not complete determination or individuation: identification picks out things as actual or possible tokens of a type, not as fully determinate or individuated items. Otherwise put, identification does not satisfy Leibniz's Laws (of the indiscernibility of identicals, and of the identity of indiscernibles) because it picks out only necessary features of things, not sufficient features. Indeed, Kant holds that complete cognitive determination of a thing can occur only by way of intuition (*CPR* A581–2/B609–10; see also *JL* 9: 99).

[24] In other words, analyticity for Kant is determined not solely by concept-decomposition, but also by comprehensional inclusion. In most cases, these are equivalent. But in a few crucial cases, propositions are analytic through comprehensional inclusion, but not through concept-decomposition: e.g., "Triangulars are trilaterals" (*CPR* A716/B744) (*BL* 24: 115) (*JL* 9: 60–1). In turn, both concept-decomposition and comprehensional inclusion fall under Kant's general criterion of analyticity, according to which a proposition is analytic if and only if its denial leads to a contradiction (*CPR* A151/B190–1). See Hanna, *Kant and the Foundations of Analytic Philosophy*, section 3.1.

propositions. Thus I might analyse my empirical concept of gold without gaining anything more than merely an enumeration of everything that I actually think in using the word, thus improving the logical character of my cognition, but not in any way adding to it. But I take the matter, which comes forward under this name (*die Materie, welche unter diesem Namen vorkommt*), and obtain perceptions by means of it; and these perceptions yield various propositions which are synthetic but empirical. (*CPR* A721–2/B749–50)

To *define* (*Definieren*), as the word itself suggests, really only means to present the complete (*ausführlichen*), original concept of a thing within the limits of its concept. If this be our standard, an *empirical* concept cannot be defined at all, but only *made explicit*. For since we find in the concept only a few characteristics of a certain sort (*Art*) of sensible object, it is never certain whether, by using the word which indicates the very same object (*denselben Gegenstand bezeichnet*), we do not think sometimes more and sometimes less characteristics of the same object. Thus one man can think in the concept of gold, in addition to its weight, colour, malleability, also its objective quality (*Eigenschaft*) of resisting rust, while another man can perhaps scientifically know (*wissen*) nothing of this objective quality. We make use of certain characteristics only insofar as they are adequate for making distinctions; new observations remove some characteristics and add others; therefore the limits of the concept are never fixed. And indeed what useful purpose could be served by defining an empirical concept, such, for instance, as that of water? When we speak of water and its properties, we do not stop short at what is thought in the word 'water,' but proceed to experiments. The word, with the few characteristics that we attach to it, should be regarded as only an *indication* (*Bezeichnung*) and not a concept of the thing; the so-called definition is no more than a determination of the word. (*CPR* A728/B756)

When we place these texts against the background of Kant's theory of empirical concepts, what immediately comes forward is that his theory of concepts appears to be directly contradicted by his own claim that "an empirical concept cannot be defined at all" and the closely connected rhetorical question, "what useful purpose could be served by defining an empirical concept, such, for instance, as that of water?" What Kant seems to be saying here is that no decompositional analysis of any empirical concept is possible, and in particular that no analytic propositions whatsoever can be framed about natural kind concepts such as GOLD and WATER. But then in other contexts, as we have seen, he explicitly asserts the existence of analytic propositions based on natural kind concepts, for example (GYM).

This difficulty has led Fred Kroon and Robert Nola to suggest that Kant implicitly, although inconsistently, operates with two distinct empirical concepts of gold—and correspondingly, that he implicitly and inconsistently allows for two radically distinct types of natural kind concept: a rigidly designating concept, and a stereotype-concept.[25] Rigid designators, as I have already mentioned, are terms that have semantic hooks into their actual world referents in every possible world in which they exist, and never have their hooks into anything else. A stereotype (also called a "prototype" in the recent psychological literature on concepts[26]) is an identifying description containing only phenomenological predicates, none of whose descriptive components is strictly necessary or strictly sufficient for determining the reference of that description.[27] Kroon and Nola propose that the rigidly designating concept, call it GOLD$_1$, directly and invariantly picks out the natural kind gold, but does not contain any phenomenological descriptive content, and hence cannot support analytic propositions about manifest properties of gold. By contrast, the merely stereotypical or prototypical concept, GOLD$_2$, reflects only the concept-user's "individual recognition criterion" for empirical objects falling under the concept, and supports quasi-analytic propositions about those manifest properties, yet does not directly pick out the underlying microphysical gold-stuff.[28] According to Kroon and Nola, the surface incoherence in Kant's doctrine is thus explained by appeal to a deeper confusion about the nature of empirical concepts: a confusion caused, in effect, by

---

[25] See Kroon and Nola, "Kant, Kripke, and Gold," p. 449–51.

[26] See, e.g., Smith, "Concepts and Thought."

[27] See Putnam, "The Meaning of 'Meaning'," pp. 249–52 and 269. If by the word "concept" one intends Putnam's notion of a *property* (that is, a strictly rough-grained intension, defined by its cross-possible-worlds extension), then a stereotype or prototype is not a concept. On the other hand, if by "concept" one means a Kantian empirical concept (or its close relative, Frege's notion of the "sense" or *Sinn* of an ordinary predicate expression, i.e., the mode-of-presentation of the members of the set of values of the Fregean unsaturated *Begriff*, which in turn is the *Bedeutung* of that predicate-expression), then a stereotype or prototype is at least concept-*like*. The nearest analogue to a stereotype or prototype in Kant's theory is the "empirical schema" or "representation of a universal procedure of imagination in providing an image for a concept" that must be added to the intension of an empirical concept in order to apply it to actual cases (*CPR* A140–2/B179–81).

[28] Kroon and Nola, "Kant, Kripke, and Gold," p. 451–6. Propositions reflecting the decompositional structure of a stereotype or prototype cannot be genuinely analytic. For, if they were, then they would be necessarily true and genuinely a priori. On the Kroon-Nola account, however, and also on Putnam's account, propositions based on stereotype- or prototype-content are at best contingent and contextually a priori. So, again, stereotypes and prototypes are not concepts in the Kantian sense. These features also sharply distinguish the stereotype or prototype from the "primary intension" of two-dimensional modal semantics: see n. 8.

his regrettable failure to anticipate and recognize the truth of scientific essentialism.

Well *maybe*. But if the Kantian arguments we studied in chapter 3 are sound, then scientific essentialism is false. And even if those Kantian arguments are unsound, at the very least scientific essentialism cannot be used automatically as a sufficient reason for rejecting Kant's theory, without independent support. In any case, it seems clear enough that the appearance of a contradiction—or at least of a sharp dichotomy—within Kant's theory of concepts is based mainly on taking Kant's remarks about the indefinability of empirical concepts out of context. What we need to concentrate on is Kant's framing observation: "To define, as the word itself suggests, really only means to present the complete, original concept of a thing within the limits of its concept. Given such a standard (*Forderung*), an empirical concept cannot be defined at all, but only *explicated*" (*CPR* A727/B755).

What I want to suggest is that, in this particular context, Kant's notion of a definition is that of a strict analytic definition, a definition that fully determines the content of a concept by revealing all its necessary and sufficient characteristics. As he points out in an accompanying footnote, "*completeness* signifies the clarity and sufficiency of characteristics"; *limits* signifies that the set of characteristics is maximal; and *originality* signifies that the determination of the limits is self-evident and needs no further proof (*CPR* A728/B756 n.). Moreover, he tells us in the *Jäsche Logic* that a representational content, be it concept or intuition, has clarity if and only if it is entirely, immediately, and discriminably available to the cognizer (*JL* 9: 33–4). "Given such a standard" (*CPR* A727/B755), no empirical concept can be analytically defined. But we need not adopt such an impossibly high standard. As we will see shortly, Kant explicitly holds elsewhere that an analytic definition can be legitimate even if it is not a strict one. Hence there is a sense in which empirical concepts *cannot* be analytically defined, and another sense in which empirical concepts *can* be analytically defined. More precisely, then, an empirical concept cannot be analytically defined in the strict sense (that is, it is not possible to reveal all its necessary and sufficient characteristics), nevertheless, an empirical concept can, quite legitimately, be non-strictly analytically defined (that is, it is possible to reveal at least some of its necessary characteristics, hence at least a part of its logical essence). Indeed, it turns out that to provide a non-strict analytic definition is precisely the same as to "explicate" (*CPR* A727/B755) an empirical concept.

The point I am making can be elaborated by an appeal to Kant's distinction between the "analytic" and "synthetic" characteristics of a concept:

Analytic or *synthetic* characteristics. The former are partial concepts of my *actual* concept (characteristics that I already think there), while the latter are partial concepts of the *merely possible* complete concept (which is supposed to *come to be* through a synthesis of several parts). (*JL* 9: 59)

Analytic characteristics are intrinsic to a concept and constitute its conceptual microstructure or *core*, while synthetic characteristics are added to the conceptual core by means of a non-discursive or non-intellectual synthesis. All non-discursive synthesis requires human sensory intuition. This synthesis can be either empirical, via empirical intuition, or non-empirical, via pure intuition. Hence synthetic characteristics can be added to the concept either a posteriori (for example, in empirical scientific investigation) or a priori (for example, in metaphysics or mathematics).

Consider now the content of any empirical concept. This content includes both analytic characteristics in its conceptual core, and synthetic empirical characteristics in its conceptual periphery.[29] Analytic characteristics of empirical concepts are necessarily contained in those concepts and "lie in them" (*in ihnen liegt*), while synthetic empirical characteristics are contingently appended to them through sensory experience and merely "belong to them" (*zu ihnen gehört*) (*JL* 9: 141). For example, UNMARRIED is an analytic characteristic of BACHELOR, while UNTIDY and RARELY AT HOME are synthetic empirical characteristics.[30] No empirical concept considered in relation to all and both its analytic or necessary and synthetic empirical or

---

[29] As I mentioned in n.27, Kant holds that the intension of an empirical concept must be supplemented by an empirical schema in order to be applicable to actual objects. So for Kant every objectively real empirical concept contains: (i) a logical essence or conceptual core made up of analytic characteristics; (ii) a conceptual periphery made up of synthetic empirical characteristics; and (iii) a supplementary schema, stereotype, or prototype.

[30] Kant is not, however, committed to the view that it is going to be easy—even relatively easy—to tell whether a given characteristic is analytic or synthetic. That is an epistemic issue which must be worked out case-by-case, and there will be tricky or borderline cases. Is ADULT an analytic or synthetic characteristic of the concept BACHELOR? I'm strongly inclined to say that it is an analytic characteristic, but obviously there is room for discussion. What Kant is committed to is that every ordinary empirical concept has some absolutely necessary analytic characteristics (and presumably no one would ever challenge the analyticity of "Bachelors are unmarried males," even if how BACHELOR applies to, say, Catholic priests is unclear) and also some contingent synthetic empirical characteristics.

contingent characteristics will ever be strictly analytically definable: "One cannot become certain through any test whether one has exhausted all the characteristics of a given concept through a complete analysis" (*JL* 9: 142). This is obviously because, even in principle, only the analytic or necessary characteristics of an empirical concept such as GOLD (namely, YELLOW + METAL + HEAVY/HIGH DENSITY + MALLEABLE/DUCTILE + RUST-RESISTANT) could be made available to a priori decompositional analysis, whereas its empirical synthetic or contingent characteristics (SHINY + FOUND IN LARGE DEPOSITS IN ALASKA AND CALIFORNIA DURING THE NINETEENTH CENTURY + EXPENSIVE IN THE 1970S + METAL PREDOMINANTLY USED IN THE CONSTRUCTION OF A CLASSIC ROLEX WATCH RECEIVED BY MY GRANDFATHER AS TOKEN OF A QUARTER CENTURY OF HONEST TOIL AT EATON'S OF CANADA, etc.) are open to free variation over time and space.

So the complete, original, limited set of necessary and sufficient, or analytic and synthetic, characteristics of GOLD, WATER, or any other empirical concept is never available to strict definitional analysis. Still—and this is the crucial point—it does not follow that there are not perfectly legitimate, epistemically accessible, non-strict analytic definitions of GOLD, WATER, and so on. Kant calls these non-strict analytic definitions of concepts "descriptions" (*Beschreibungen*) of them (*JL* 9: 143). Concept-descriptions result from "expositions" (*Erörterungen*) or partial decompositions of a concept's logical essence. In turn, a concept-description resulting from an exposition is essentially the same as a "nominal definition," "which contain[s] the meaning that one wanted voluntarily to give to a certain name, and which therefore signif[ies] merely the logical essence of [its] object, or which serve[s] merely for distinguishing it from other objects" (*JL* 9: 143).

Nominal definitions are carefully distinguished by Kant from "real definitions," which "suffice for cognition of the object according to its inner determinations" and thereby pick out its "real essence" (*JL* 9: 143). It follows that nominal definitions will not provide cognitive access to things-in-themselves or positive noumena. Nevertheless, it remains perfectly possible for nominal definitions to provide cognitive access to phenomenal objects or appearances. Moreover, according to Kant, nominal definitions are neither stipulative nor constructive. By means of a stipulative definition, one generates "arbitrarily thought" concepts, according to which "I can always define my concept... since I deliberately made it up," but about which "I cannot say that I have thereby defined a true object" (*CPR* A729/B757). So making a concept by stipulation is the same as *deliberately making it up:* it generates a mere conceptual fabrication or pseudo-concept. By contrast, according to the Kantian taxonomy of definitions, all genuinely "made" concepts are the result of what he calls "synthetic constructive" definitions (*JL* 9: 141). Synthetic constructive definitions, in turn, are all a priori and mathematical, because the construction occurs within the special semantic framework of the representational content of pure intuition (see sections 2.3, 6.4, and 7.4). The crucial point for my purposes here is avoiding the natural mistake of confusing a Kantian nominal definition with either an arbitrarily thought concept or a synthetic constructive definition. For, as Kant is at pains to state explicitly, a nominal definition expresses a "logical essence," and every definition that expresses a logical essence is automatically an analytic definition. But an arbitrarily thought concept can never support a genuine analytic judgment—which is absolutely or logically necessary—but at best supports only a "miserable tautology" (*CPR* A597/B625). And synthetic constructive definitions are, obviously, *synthetic* a priori, not analytic (*CPR* A718/B746, A731–2/B759–60).

What does Kant mean, then, by saying that all nominal definitions involve an element of voluntariness or choice? All he intends, I think, is the truism that the word-sign which is given a nominal definition—say, "water" or *Wasser*—is conventionally adopted by us: in principle, we could have used "schwater" or *Schwasser* or any other phonologically acceptable sequence in English or German. Nevertheless, the fact that the concept WATER describes just that sort of wet, clear, odourless, tasteless, drinkable, etc., stuff, and not some other sort of thing, is not at all chosen by us:

The concept that water is a fluid element, without odour or taste... etc., is the logical essence of water[;] for if I have mastered physical cognitions about something, then I think of all of this as soon as I mention the word 'water'. (*BL* 24: 118)

Water by any other name would be just as odorless. Thus a nominal definition invokes "the word, with the few characteristics that we attach to it," and yet also provides "an enumeration of everything that I actually think in using the word." A nominal definition of GOLD or WATER is a concept-description, which in turn results from an exposition, or partial and non-strict analytic decomposition, of the logical essence or intensional microstructure of the relevant concept. But every such concept-description, hence every nominal definition of an empirical concept, is an "approximation" to an ideally complete and strict analytic definition, and therefore a "true and useful exhibition of a concept" (*JL* 9: 143).

We can now see clearly that, on Kant's view, GOLD is semantically and logically legitimately decomposable or analyzable, despite the fact that empirical concepts are not strictly analytically definable and despite the further fact that they support nominal definitions rather than real definitions.

Q: But aren't Kantian natural kind concepts just like any other old empirical concept?

A: No, sorry, not by a long shot. Their content is semantically "thick" and uniquely structured. Let me show you what I mean.

Just as Locke carefully distinguished between (i) general ideas (and general names) of "substances" and (ii) the more common-or-garden-variety general ideas (and general names) of "mixed modes,"[31] so too Kant regards natural kind concepts as a special case of his theory of empirical concepts. He explicitly describes "my concept of gold, which had as its content that this body is yellow and metal" (*P* 4: 267); hence GOLD partially decomposes to this conceptual microstructure:

<THIS BODY + YELLOW + METAL>[32]

---

[31] See Locke, *An Essay Concerning Human Understanding*, bk. II, chs. 22–3, pp. 288–317, and bk. III, chs. 5–6, pp. 428–71.

[32] There is a small interpretive puzzle concerning Kant's views as to precisely which characteristics count as analytic characteristics of GOLD. He says that "one man can think in the concept of gold, in addition to its weight [that is, HEAVY or HIGH DENSITY], colour [that is, YELLOW], malleability, also its property of resisting rust, while another man can perhaps scientifically know nothing of this quality" (*CPR* A728/B756). What he seems to be saying is that the concept GOLD analytically contains the characteristics HEAVY/HIGH DENSITY, YELLOW, and MALLEABLE/DUCTILE, while RUST-RESISTANT remains a merely synthetic and contingent characteristic of GOLD. Does this make any sense? Why would "low density gold," "colorless gold" and "unmalleable gold" be analytic contradictions, while "rusty gold" is analytically consistent? I am strongly inclined to think that Kant believes that RUST-RESISTANT is an analytic characteristic of GOLD. Here we must remember that Kant's theory of concepts is not solipsistic: the content of a given empirical concept is determined for an indefinitely large intersubjective community of human cognizers spread out over time and space, each of whom shares the same formal unity of apperceptive consciousness and the same set of cognitive capacities—not merely for an actual individual cognizer at a time. This is at least part of what Kant means when he says that every concept is a *conceptus communis* (*CPR* B133–4). So my reading of this passage is that Kant is saying that heaviness/high density, yellow, malleability/ductility, and rust-resistance all equally belong as analytic

It can be immediately seen that for Kant GOLD contains two distinct components: (a) a referential component, THIS BODY; and (b) an attributive or descriptive component that reflects some (but obviously, not all) of the manifest identifying properties of gold, namely, YELLOW and METAL. What I want to argue is that for Kant this fusion of distinct components yields the very feature of natural kind concepts that sets them radically apart from all other sorts of empirical concepts, and furthermore that this feature remains perfectly consistent with the thesis that (GYM) is analytic.

In his various characterizations of natural kind concepts, Kant gives us four tantalizing clues that all appear to centre on the same notion:

(1) He says that his concept GOLD "had as its content that this body (*dieser Körper*) is yellow and metal" (*P* 4: 267).

(2) He says that once I have by means of analytic decomposition provided an "enumeration of everything that I actually think in using the word" standing for gold, then "I take the matter which comes forward under this name (*die Materie, welche unter diesem Namen vorkommt*) and obtain perceptions by means of it" (*CPR* A721/B749).

(3) He says that, in using the word "gold," although I have in my possession a concept which contains "only a few characteristics of a certain sort (*Art*) of sensible object," nevertheless I am thereby "using the word which indicates the very same object (*denselben Gegenstand bezeichnet*)" (*CPR* A728/B756).

(4) He says that "the word" which expresses a natural kind concept, along "with the few characteristics that we attach to it, should be regarded as only an indication (*Bezeichnung*) and not a concept of the thing" (*CPR* A728/B756).

So what is Kant driving at in these texts? My proposal, in a nutshell, is that the referential component of the concept GOLD—namely, THIS BODY—is an essentially indexical component shared by all natural kind

---

characteristics to GOLD: but just which of those characteristics is or are self-consciously taken to be important or notable—and therefore part of the stereotype or prototype of the concept—In fact varies across individual concept-users and situations. The second concept-user in Kant's example is ignorant or forgetful of some basic definitional knowledge about gold (i.e., that it is rust-resistant), which shows that his own personal grasp, and therefore the stereotype or prototype, of the concept GOLD is somewhat loose.

concepts,[33] which restricts them to the totality of matter found in any given possible world considered as the actual world; but in any such world, the extension of GOLD is all and only those bodies that possess the analyzable or decomposable phenomenological identifying features of GOLD, namely, YELLOW, METAL, HEAVY/HIGH DENSITY, MALLEABLE/DUCTILE, RUST-RESISTANT, and so on.

Let us focus for a moment on the idea that THIS BODY is an essentially indexical component showing up in all natural kind concepts. A semantic term is *indexical* just insofar as the determination of its semantic value (= its truth-value, extension, or referent) is systematically context-sensitive, egocentric, and actual-world dependent. And it is *essentially* indexical[34] just in case its indexicality is irreducible to any corresponding description that attempts to capture the full modal or semantic force of the relevant indexical term: for example, a rigidifying definite description of the form "the actual F." More specifically, an essential indexical is a term which holds a certain formal meaning—an a priori rule of use for that term—fixed, yet regularly varies in objective reference across actual spatiotemporal, subject-centered contexts of use.[35] All essentially indexical terms are directly referential in that:

(a) they determine their objective reference in a non-attributive or non-descriptive manner;
(b) the way the actual world—the actual subject-centred context of the use of the term—just happens to be, plays an ineliminable role in the determination of reference; and
(c) the cognitive meaning or semantic value of the term is exhausted by its objective reference.

In turn, the distinction between directly referential essentially indexical terms (for example, "this") and rigidifying definite descriptions (for example, "the actual table at which RAH is sitting") is precisely the distinction between terms that are *necessarily* indexical, and those that are only *accidentally* indexical. For, once we have fixed a context of utterance,

---

[33] My attention was drawn to the presence of the indexical component in Kant's analysis of natural kind concepts by Erik Anderson. See also his "Kant, Natural Kind Terms, and Scientific Essentialism," in which he gives an account of its semantic role that is somewhat different from the one I work out here.

[34] See Perry, "The Problem of the Essential Indexical."

[35] See Kaplan, "Demonstratives" and "Afterthoughts"; and Perry, "Indexicals and Demonstratives."

this is logically necessarily the same as *this* by the semantics of indexicals. Nevertheless, even once we have fixed a context of utterance, *this* is not necessarily the same as the actual table at which RAH is sitting. For the actual table at which RAH is sitting is, as it happens, inert and not alive: but *this* actual table at which RAH is sitting *might have been a living organism in a logically possible world in which the causal-dynamic laws of nature were radically different*. As we saw in section 3.5, empirical natural worlds with very different or even radically different causal-dynamic laws are conceivable and therefore possible. I will come back to this subtle but crucial modal point about essentially indexical terms again at the beginning of section 4.2.

In any case, it is arguable that Kant has a theory of essential indexicality. This is because it is arguable that in the Transcendental Aesthetic Kant offers a theory of irreducibly context-dependent, subject-centered, and non-conceptual (hence non-descriptive) direct singular intuitional representation of actual individual objects, and that he assigns the representational function for demonstration (that is, the indexical pointing function expressed by the demonstrative terms *this* or *that*) to empirical intuitions, and that he also assigns the corresponding formal meaning or a priori rule of use (sometimes also called the "character") of demonstratives to the forms of inner and outer empirical intuition, our representations of time and space.[36] Kant says:

> In whatever mode and by whatever means a cognition may refer to object, intuition is that through which it immediately refers to them, and to which all thought is mediately directed. (*CPR* A19/B33)

> [The representation of] space is not an empirical concept which has been derived from outer experiences. For in order that a certain sensation be referred to something outside me (that is, in another region of space from that in which I find myself), and similarly in order that I be able to represent them as outside and alongside one another, and accordingly as not only different but in different places, the representation of space must be presupposed. (*CPR* A23/B38)

> [The representation of] space is a necessary representation, *a priori*, which is the ground of all outer intuitions. (*CPR* A24/B38)

> [The representation of] time is not an empirical concept that is somehow drawn from an experience. For simultaneity and succession would not themselves come into perception if the representation of time did not ground them *a priori*. Only

---

[36] See Hanna, *Kant and the Foundations of Analytic Philosophy*, ch. 4.

under its presupposition can one represent that several things exist at one and the same time (simultaneously) or in different times (successively). (*CPR* A30/B46)

[The representation of] time is a necessary representation that grounds all intuitions. (*CPR* A31/B46)

Commenting on these texts, Peter Strawson remarks:

Now what of the doctrine that space and time are forms of intuition?... The duality of intuition and concept is merely the epistemological aspect of the duality of particular instance and general type ... Clearly the thought, at its most general, is of some peculiarly intimate connection between space and time, on the one hand, and the idea of the particular item, the particular instance of the general concept, on the other... Spatio-temporal position provides the fundamental ground of distinction between one particular item and another of the same general type, hence the fundamental ground of identity of particular items... J. L. Austin says that empirical statement requires the existence of two kinds of *conventions*, which he describes as follows:

"*Descriptive* conventions correlating the words with the *types* of situation, thing, event, etc., to be found in the world; *Demonstrative conventions* correlating the words with the *historic situations*, etc., to be found in the world."

Austin's duality of semantic conventions corresponds to Kant's duality of cognitive faculties. By the demonstrative (i.e., particularizing) conventions, correlation is said to be achieved with historic situations. Since 'historic' is evidently a time-word and 'situations' fundamentally at any rate, a space-word, this balances very happily [with] Kant's doctrine that the forms of intuition are space and time.[37]

It seems to me that Strawson's interpretation of Kant's theory of forms of intuition is basically correct.[38] It follows that the essentially indexical component THIS in THIS BODY expresses our representations of time and space, the subjective forms of inner and outer sense, the necessary and a priori conditions of demonstrative empirical intuition.

It might appear puzzling that Kant's theory of essential indexicals, which is focused on our demonstrative capacity for inner and outer sensory intuition, can also apply to the characteristics of natural kind concepts. For did I not say just above that essential indexicals are directly referential and

---

[37] Strawson, *The Bounds of Sense: An Essay on Kant's Critique of Pure Reason*, pp. 48–9 and p. 49 n.1.
[38] I do not mean to imply that I completely accept Strawson's interpretation of the Transcendental Aesthetic.

hence non-conceptual (non-descriptive or non-attributive)? The solution to the puzzle is contained in Strawson's remark that "the duality of intuition and concept is merely the epistemological aspect of the duality of particular instance and general type." The upshot is that natural kinds are picked out *essentially* indexically, but not *purely* indexically; instead, they are picked out essentially indexically under the further constraint of a certain associated description. Otherwise put, the referential component of a natural kind term is an essentially indexical *predicate*.[39]

In this way, according to Kant every natural kind concept contains a semantically mixed or hybrid component he variously describes as THIS BODY and THE MATTER WHICH IS BROUGHT FORWARD BY THIS NAME. This constituent is both demonstratively essentially indexical (THIS, WHICH IS BROUGHT FORWARD BY THIS NAME) and also descriptive or predicative (BODY, THE MATTER). By virtue of the essentially indexical component, a natural kind concept systematically shifts in its objective reference as a function of the subject-centered spatial and temporal context of concept-application. So, to this extent, any word expressing a natural kind concept functions merely as an "indication (*Bezeichnung*) and not a concept of the thing" (*CPR* A728/B756). But since THIS BODY or THIS MATTER is also partially predicative or descriptive, it narrows the scope of the demonstrative THIS by restricting it to items falling within the comprehension of the concept BODY or MATTER. Thus the application of the concept GOLD varies systematically by actual subject-centred context, but under the special qualification that it "indicates the very same object (*denselben Gegenstand bezeichnet*)" (*CPR* A728/B756), namely, body or matter, insofar as it stands in a certain intuitional relation to the concept-user.

This yields two further points. First, the reference of the essentially indexical predicate THIS BODY or THIS MATTER (meaning the same as, roughly, THIS EXTENDED, DIVISIBLE,[40] etc., STUFF) is not ordinary singular reference.

---

[39] See Heal, "Indexical Predicates and their Uses."

[40] Here is another small interpretive puzzle. In *Metaphysical Foundations of Natural Science*, Kant argues that infinite divisibility is a synthetic feature of actual matter (*MFNS* 4: 503–4); elsewhere, of course, he holds that divisibility is an analytic characteristic of the concept BODY. Assuming charitably that Kant is not simply contradicting himself, then either BODY and MATTER are different concepts, or else he is implicitly drawing a distinction between infinite divisibility and divisibility per se. Since the former seems unlikely, I opt for the latter. Hence body's or matter's being divisible is analytic; but its being infinitely divisible is synthetic. That makes sense, because quantitative properties for Kant always depend upon pure forms of intuition, and there are logically possible worlds in which body or matter is only finitely divisible.

I mean that the concept GOLD is not an individual concept—a concept of a uniquely particular item, a given individual body with limited extension in space or time, for example, some particular piece of gold—but rather a mass term. Thus THIS BODY or THIS MATTER must have the same content as the concepts THIS CORPOREAL TOTALITY or THIS TOTALITY OF MATTER (or THIS TOTALITY OF EXTENDED, etc., STUFF). Second, since THIS BODY or THIS MATTER picks out the same totality of bodily or material stuff across the actual world of the perceiving subject, no matter how scattered its bits may be, its objective reference appears to be precisely what Kant in the First Analogy of Experience calls "that which persists," that is, the underlying material substratum of all determinations and temporal change in the natural world (*CPR* A182–9/B224–32; see also *MFNS* 4: 503). In the next section we shall see that for Kant, ontologically speaking, that which persists in empirical nature as a whole is nothing but *an objectively real nomological spatiotemporal causal-dynamic structure, consisting entirely of attractive and repulsive forces related in an indefinitely large number of determinate ways, ultimately to be identified with a universal fluid aether*, and not an individual material thing; but that crucial point can be held in reserve for the time being.

In any case, this brings us to the crux of Kant's theory of natural kind terms. Since the objective reference of THIS BODY or THIS MATTER, taken on its own, is the entire actual material world in egocentrically structured space and time, the essentially indexical predicate is obviously not sufficiently precise to pick out gold, as opposed to water or air or any other specific actual-world inorganic natural kind. Such a "precisification" is therefore the function of the rest of the content of GOLD. Otherwise put, Kant's view is that the objective reference of GOLD in any possible world is the material substratum of that given world considered as actual—the actual-world spatiotemporal totality of physical matter, the underlying law-governed causal-dynamic substratum of all determinations and change in the empirical natural world—*as determined thus-and-so* by the various phenomenological defining features of gold (YELLOW + METAL + HEAVY/HIGH DENSITY + MALLEABLE/DUCTILE + RUST-RESISTANT, etc.). So whatever the actual-world totality of matter turns out to be (and it can in principle be different in different possible worlds, even when the subject-centred context of use is held fixed, depending on that world's set or package of natural laws) the identifying component of the natural kind concept inevitably picks out the yellow, metallic, heavy/high density, malleable/ductile, rust-resistant, etc., nomological "position" or "role" within that actual-world law-governed causal-dynamic totality. The concept GOLD identifies the particular causal-dynamic career of gold in the actual natural history of matter in that possible world.

But, perhaps most importantly, it follows that on Kant's view the natural kind gold is *compositionally plastic over many different possible sets or packages of natural laws* (see section 3.5). It is conceivable and therefore logically and strictly metaphysically possible that the actual world might have been other than it is, and thus that the actual-world totality of matter might have been other than it is, just because the actual set or package of natural laws could have been other than it is. But in every logically and strictly metaphysically possible world accessible from the actual world, gold is yellow, metallic, and so on. Thus for Kant the decomposable phenomenological identifying component of GOLD fully supports analytic truths about any actual-world totality of matter that is picked out by the essentially indexical predicative component of GOLD.

By contrast, on the scientific essentialist view, the essence-disclosing term "the element with atomic number 79" reflects a microphysical fact about gold that cannot be analytically derived from the concept GOLD, but instead is only to be learned by means of empirical natural science. In every logically possible world in which gold exists, gold is the element with atomic number 79, no matter what its manifest properties turn out to be in those worlds. Now let me ask you, which strikes you as more plausible:

(1) Kant's manifest realist view that necessarily gold is any actual-world material stuff that is yellow, metallic and so on, whether or not it is the element with atomic number 79?

Or:

(2) The Kripke-Putnam scientific essentialist version of scientific realism, to the effect that necessarily gold is the element with atomic number 79, whether or not it is yellow, metallic and so on?

As I suggested in section 4.0, our contemporary conventional wisdom certainly favours the latter: but this is based on the highly questionable assumption that microphysical noumenal realism is true. What I want to do in the next section is to motivate a case for Kant's alternative view by sympathetically unpacking his conception of empirical scientific realism and his corresponding thesis of manifest realism.

4.2. Scientific Realism in the Manifest Image

As we have seen, the natural kind concept GOLD for Kant partially decomposes to this conceptual microstructure:

<THIS BODY + YELLOW + METAL>

And, as we have also seen, the first component of GOLD—THIS BODY or THIS MATTER—plays a descriptively qualified essentially indexical role, by pegging the *Umfang* or comprehension (cross-possible-worlds extension) of the natural kind concept at that world to the totality of physical matter in the actual subject-centered spatiotemporal context. Then, in turn, a decomposable set of phenomenological identifying conceptual characteristics picks out the specific natural kind gold as a definite position or role within the total causal-dynamic structure of attractive and repulsive forces that constitutes actual-world matter. So for Kant gold is necessarily the yellow (and also heavy/high density, malleable/ductile, rust-resistant, and so on) metal having precisely this actual-world totality of physical stuff as a substratum.

It is a crucially important feature of this account, that the actual referent of the essentially indexical predicate in a given context of utterance can in principle vary across possible worlds. If the totality of physical matter in the actual world had been very different or even radically different, precisely because the actual set or package of causal and dynamical laws of nature had been very or radically different—say, if metals had been all somehow *alive* and not *inert*—then gold would also have been correspondingly very or even radically nomologically different.[41] Gold is logically and strictly metaphysically necessarily a yellow metal, but it might have been alive.

[41] See ch. 3, n.69.

That may seem odd and excessively science-fictional. But first-rate sci-fi always vividly illuminates and exposes sublimated features of our actual lives and actual world. And in any case the possibility of living gold is far less odd than what the scientific essentialist is committed to, the downright paradoxical view that gold—namely, that which we correctly identify in the actual world as being a yellow metal—might have been *neither yellow nor a metal*. That seems absurd. I fully grant that it is not hard to see how I and everyone else might have used the word "gold" for something that is not a metal: "gold" might have been the word we used to pick out water—or real ale, or Kentucky bourbon whiskey, or Scottish shortbread cookies for that matter. But how could our natural kind *gold* have been non-metallic?[42]

In any case, Kant's view is only superficially odd. As it happens, gold is actually inert or dynamically mechanistic, and all our actual natural scientific laws are modally dependent on the contingent inertness of matter (*MFNS* 4: 544).[43] Nevertheless, conceptually and logically it could have been otherwise. There is no conceptual contradiction in the very idea of living matter (hylozoism), hence no contradiction in the very idea of living metal. But even if gold is alive in some logically possible worlds accessible from the actual world, it remains a yellow metal in every logically possible world accessible from the actual world.

Now at this point our critical, burden-of-proof-shifting scientific essentialist would want to insist on something like the following:

The type of property identity used in science seems to be associated with *necessity*, not with aprioricity, or analyticity.[44]

---

[42] Kripke fudges on this crucial point; see *Naming and Necessity*, pp. 117–18, and 123. Putnam on the other hand assigns METAL to what he calls the "semantic marker" of the word "gold" ("The Meaning of 'Meaning'," pp. 266–8). More specifically, he says that the semantic marker is a central feature of the speaker's linguistic competence for that word; but even so, according to him, in principle something could be gold but fail to be a metal.

[43] For Kant, causal laws are synthetically necessary and a priori, but their necessity is constrained by something actual: "That which *in its connection with the actual* is determined in accordance with universal conditions of experience is ... *necessary*" (*CPR* A218/B266, first emphasis added). My reading of this text is that it implies, e.g., that for Kant our general mechanistic causal-dynamic laws are true in all and only those possible worlds of human experience that also contain the inert matter found in our actual world. So the strong modality of causal laws is constrained by the empirical concept BODY or MATTER (*MFNS* 4: 472). For *living* matter, the causal-dynamic laws would have to be radically different. In fact they would have to be what in ch. 8 I will call *one-time-only or "one-off" laws of natural causal singularities*. It must be admitted, however, that Kant's theory of causal laws is very controversial; see also ch. 3, n.66, section 3.5, and sections 8.2–8.3.

[44] Kripke, *Naming and Necessity*, p. 138.

The interesting fact is that the way the reference [of a natural kind term] is fixed seems sometimes overwhelmingly important to us in the case of sensed phenomena ... The fact that we identify [a given natural kind] in a certain way seems to us to be *crucial*, even though it is not necessary: the intimate connection may create an *illusion* of necessity. I think that this observation, together with the remarks on property-identity above, may well be essential to an understanding of the traditional disputes over primary and secondary qualities.[45]

In other words, a crucial part of scientific essentialism's objection to the Kantian approach to natural kinds is that any theory about natural kinds that is based on mere secondary qualities—for example, the phenomenological identificational properties expressed in analytic propositions such as (GYM)—is hopelessly inadequate for the sort of sophisticated "theoretical identification" solely in terms of microphysical primary qualities, that is characteristic of the fundamental natural sciences.[46] An appeal to manifest features of kinds is acceptable for recognitional[47] and practical purposes, and accurately reflects broadly human and more narrowly individual interests, but it is semantically, epistemically, and metaphysically misguided. It confuses the familiar reference-fixing features, belonging to the use of natural kind terms, with the essential underlying microphysical properties of kinds, and so generates the apriorist "illusion of necessity" that its claims have strong modal force. Kant's positive reply to this important line of criticism lies in his empirical scientific realism, and more precisely in his manifest realist theory of manifest kinds. And, to motivate that reply, I now want to situate this theory in relation to the classical distinction between primary and secondary qualities.

A very great deal can be said—and has already been said—about the primary versus secondary quality distinction, and I cannot possibly undertake an adequate discussion of it here.[48] So I will restrict myself to three basic points that bear directly on Kant's empirical scientific and manifest realism.

---

[45] Kripke, *Naming and Necessity*, pp. 139–40. [46] See, e.g., Quine, "Natural Kinds."
[47] See Brown, "Natural Kind Terms and Recognitional Capacities." Brown associates human recognitional capacities for natural kinds with an ability to pick out microphysical kinds. But in fact her account would work just as well for manifest kinds.
[48] For a start, however, see: Bennett, *Locke, Berkeley, Hume: Central Themes*; Hacker, *Appearance and Reality*; Hirst, "Primary and Secondary Qualities"; and McGinn, *The Subjective View*.

(1) The first point has to do with the precise formulation of the primary versus secondary quality distinction. From the contemporary perspective of scientific essentialism, primary qualities are taken to be microphysical, neither directly perceivable nor observable, mind-independent, mathematizable, causally efficacious, intrinsic non-relational properties of things (also known as fundamental physical properties); and secondary qualities are taken to be macroscopic, directly perceivable, mind-dependent, non-mathematizable, causally inert, intrinsic non-relational properties of conscious states (also known as phenomenal qualia).[49] These phenomenal qualia are, in turn, instantiated in cognitive states, and those cognitive instances are what Locke would have called "ideas of sensation," what Hume would have called "sensory impressions," and what early Moore and Russell would have called "sense data." But this all-too-familiar philosophical picture, recycled by scientific essentialism with a nice shiny new frame, fails to respect or reflect the primary versus secondary quality distinction originally created by Boyle and Locke, in the two following respects:

(i) The Boyle-Locke distinction construes both primary and secondary qualities alike as purely dispositional causal powers of material things to produce sensory ideas of those qualities in human minds by means of physical effects on our sense organs,[50] hence as nothing but

---

[49] This needs a further comment. Kripkean scientific essentialists are non-reductionist about secondary qualities in general and phenomenal consciousness in particular. See Kripke, *Naming and Necessity*, pp. 144–55; and Chalmers, *The Conscious Mind*, chs. 3–4. Now it may seem possible to defend both scientific essentialism and either physicalism (the thesis that mental properties or facts are identical to fundamental physical properties or facts) or eliminativism (the thesis that mental properties and mentalistic concepts are nothing but mythic constructs of folk psychology that will eventually wither away, like phlogiston and the aether). See Smart, *Philosophy and Scientific Realism*, chs. 4–5; and Churchland, *Matter and Consciousness*, pp. 43–9. But on the one hand physicalism and scientific essentialism are inconsistent, since the former holds that true psychophysical identity statements are either contingent or else at best non-logically necessary (a.k.a. "strong metaphysical necessity"), while the latter holds that all true identity statements are logically or strictly metaphysically necessary. And on the other hand, eliminativism reduces to a speculative wager on the epistemic and cultural efficacy of future neuroscience. I should add that I do not mean to assert that scientific essentialism cannot be consistently combined with other forms of reductive materialism based on logical strong supervenience instead of identity (e.g., reductive functionalism). But the philosophical fate of that combination is very much a moot point; see Chalmers, *The Conscious Mind*.

[50] Even more precisely, on the Boyle-Locke view, primary qualities of material things produce sensory ideas exactly resembling their causes, while secondary qualities of material things produce non-resembling and privately individuated sensory ideas.

extrinsic relational causal properties obtaining between material things and human minds, and *not* as intrinsic non-relational properties either of material things or of human minds.

(ii) While both Boyle and Locke do indeed also ground the primary qualities in a substratum consisting of intrinsic non-relational properties, in the hidden or unperceivable and unobservable *micro*physical constitution of material things ("corpuscles" and their compositional complexes or "textures"), nevertheless their actual working list of primary qualities—extension (size), figure (shape), motion or rest, and solidity (impenetrability)—picks out exclusively a set of *macro*physical, manifest, directly perceivable and observable properties, and *not* microphysical properties.

Berkeley and Hume of course radically exploit both (i) and (ii). Their first move is to assimilate the purely dispositional extrinsic relational causal powers of material things to produce sensory ideas of those qualities in human minds by means of physical effects on our sense organs, to the sensory ideas themselves that are produced by those powers, via (i). Their second move is to assimilate primary qualities to secondary qualities, via (ii). And their third move, via (i) and (ii) together, is to reduce causation to extrinsic relational constant-conjunction properties of the sensory ideas. The result is classical Berkeley-Hume phenomenalism, with its characteristic causal anti-realism.

But one could reject the classical Boyle-Locke primary versus secondary quality distinction and *also* resist the Berkeley-Hume phenomenalist reduction and its causal anti-realism, by: (a) refusing to accept Locke's theory of ideas of sensation; (b) refusing to accept the idea that there are any intrinsic non-relational unperceivable unobservable properties of material things; and, most fundamentally, (c) refusing to accept the assumption that the causal powers of material things are nothing but extrinsic relational dispositional properties of those things. Then, in turn, one could positively propose that in the material world nothing is ever cognitively or ontologically hidden, that the direct objects of sense perception are real material objects outside the human body in space, and that the causal powers of material things are intrinsic structural properties of those things. For one could hold along with Kant instead that:

everything in our cognition that belongs to intuition ... contains nothing but mere relations, of places in one intuition (extension), alteration of places (motion), and laws in accordance with which this alteration is determined (moving forces). But what is present in the place, or what it produces in the things themselves besides the alteration of place, is not given through these relations. Now through mere relations no thing-in-itself is cognized; it is therefore right to judge that since nothing is given to us through outer sense except mere representations of relation, outer sense can contain in its representation only the relation of an object to the subject, and not that intrinsic nature (*das Innere*) which attaches to the object in itself (*CPR* B67);

and that "to our outer intuitions there corresponds something real in space" (*CPR* A375).

The result would then be Kant's manifest realism, according to which so-called primary and so-called secondary qualities alike belong to a single unified class of intrinsic relational structure-dependent spatiotemporal causal-dynamic nomological macrophysical essentially unhidden properties that are ascribed directly to material things in space, whether material individuals or natural kinds. Kant's *non-reductive manifest realism* would then sharply contrast with Berkeley's or Hume's *reductive phenomenalism*, according to which primary and secondary qualities alike are nothing but causally inert intrinsic non-relational phenomenal qualia instanced in mental items—be they ideas, mental representations, intentional states, or whatever their label—falling contingently under a set of causally irreal extrinsic relational constant-conjunction generalizations.[51]

(2) The second point is that there is an obvious problem with the Berkeley-Hume phenomenalist reduction of the classical Boyle-Locke primary vs. secondary quality distinction. For this reduction, of course, problematically and dichotomously splits the natural world into a "Really Real" objective and non-human domain on the other hand (the world of unreduced primary qualities, construed now as Lockean microphysical real essences), and an "illusory or merely apparent" subjective and all-too-human domain on the other (the phenomenalized world of mere appearance or *Schein*). As Kant points out, all such views are explicitly

---

[51] For phenomenalist readings of Kant's empirical realism, see Langton, *Kantian Humility*; Van Cleve, *Problems from Kant*; and Wilson, "The 'Phenomenalisms' of Kant and Berkeley."

or implicitly committed to a skeptical and ultimately self-undermining noumenal realism, according to which the material world investigated by natural science has a knowable intrinsic non-relational essence which nevertheless exists beyond all possible human sensory experience:

> The absolutely intrinsic nature (*Innerliche*) of matter, as it would have to be conceived by the pure understanding, is nothing but a phantom (*Grille*); for matter is not among the objects of pure understanding, and the transcendental object which may be the ground of this appearance that we call matter is a mere something of which we should not understand what it is, even if someone were in a position to tell us. (*CPR* A277/B333)

> The [noumenal] realist therefore represents outer appearances (if their reality is conceded) as things-in-themselves, which would exist independently of us and our sensibility and thus would also be outside us according to pure concepts of the understanding. It is the [noumenal] realist who afterwards plays the empirical idealist; and after he has falsely presupposed about objects of the senses that if they are to exist they must have their existence in themselves even apart from sense, he finds that from this point of view all our representations of sense are insufficient to make their reality certain. (*CPR* A369)

(3) The third and final point is another consequence of the classical primary versus secondary quality distinction, a consequence that is arguably even philosophically worse than epistemic skepticism:

Q: But what could *ever* be philosophically worse than epistemic skepticism?

A: Well, as Kant and Wittgenstein both pointed out, we could be at risk of losing our souls. I'm utterly serious about this. Kant's eighteenth-century near-contemporary, the dramatic poet Heinrich Kleist, fully grasped this point and it nearly drove the poor man mad.[52] I mean that if we unconsciously and uncritically belong to a world culture driven by a false philosophy of the exact sciences, we are at serious risk of losing the embodied, vital, affective-desiderative, conscious, and finally moral dimension of human nature. We could become, in effect, mere zombie-counterparts of ourselves: or, as Kleist put it, nothing but *mechanical puppets* lacking any phenomenal consciousness or moral capacities. And that's why Kant had to deny scientific knowing (*Wissen*) in order to make room for **belief** (*Glauben*).

---

[52] See Kleist, "Über das Marionettentheater." Taking the *Critique of Pure Reason* utterly seriously, and convinced that Kant's notion of appearance or *Erscheinung* should be interpreted in a strictly mechanistic and phenomenalist way, Kleist suffered a nervous breakdown generally known as his "Kant crisis."

This point cannot be overemphasized. To use McDowell's formulation, the classical Boyle-Locke primary versus secondary quality distinction, as phenomenalistically reduced by Berkeley and Hume, utterly "disenchants nature" by placing everything that might be of direct value or concern to human lives, commitments, and moral practices inside the psychological subject, narrowly construed.[53] Then, since the natural world is really nothing but microphysical matter, it follows that *nothing really matters* in the natural world. Of course, whatever affects physical nature, also affects human nature, since rational human beings are also rational *animals*: incarnate rational agents and specifically corporeal conscious animals—the prisoners of gravity. So the disenchantment of nature bounces right back at the embodied psychological subject. He thereby becomes Musil's *Mann ohne Eigenschaften*, the man without any phenomenal consciousness or moral capacities, a human zombie, and a direct twentieth-century descendant of Kleist's morbid marionettes. And, from a more narrowly Kantian perspective, this tragically unhappy metaphysical situation also directly implies a *determinist* conception of human nature that makes a human morality based on the Categorical Imperative impossible by making human freedom impossible (Bxxvii–xxx; see also *GMM* 4: 446–63).

In any case, according to Kant, the empirical realist conception of scientific objectivity does not imply that any such metaphysically, epistemologically, axiologically, or morally invidious primary versus secondary quality distinction ever needs to be drawn between "Really Real" intrinsic non-relational properties of microphysical matter on the one hand and "Phenomenalistically Unreal" intrinsic non-relational properties of the conscious mind on the other. Instead, all the properties of material objects are fully manifest and belong to precisely the same single ontological type:

Long before Locke's time, but assuredly since him, it has been generally assumed and granted without detriment to the actual existence of external things that many of their predicates (*Prädikate*) may be said to belong, not to things themselves, but to their appearances (*Erscheinungen*), and to have no proper existence outside our representation. Heat, colour, and taste, for instance, are of this kind. Now if I go further and, for weighty reasons, rank as mere appearances also the remaining qualities (*Eigenshaften*) of bodies, which are called primary—such as extension, place, and in general, space, with all that belongs to it (impenetrability

---

[53] See McDowell, *Mind and World*, lect. 4.

or materiality, shape, etc.)—no one in the least can adduce the reason of its being inadmissible ... I find that more, nay, *all the qualities which constitute the intuition of a body* belong to its appearance. (P 4: 289)

To hold that all properties of physical objects are fully apparent or manifest, however, is *not* to hold that they are all illusory:

> If I say: in space and time intuition represent both outer objects as well as the self-intuition of the mind as each affects our senses, i.e., as it **appears**, that is not to say that these objects would be a mere **illusion** (*Schein*). For in the appearance (*Erscheinung*) the objects, indeed even the qualities (*Beschaffenheiten*) that we attribute to them, are always regarded as something really given (*wirklich Gegebenes*). (CPR B69)

In other words, Kant holds that all natural qualities of things are nothing but authentically apparent or manifest properties *and* that they are objectively real. But how can he consistently hold both theses? There are three parts to the answer.

First, as we have seen already, Kant is a minimal metaphysical realist who thereby accepts the idea that knowable things exist in space and not merely in one's phenomenal consciousness or inner sense. But, as the Refutation of Idealism clearly shows, minimal metaphysical realism is also consistent with transcendental idealism (CPR xxxix–xli, B274–9). And at a deeper level, if my charitable reconstruction of the Refutation is sound, then it also implies that an effective response to Cartesian external world skepticism is possible only if noumenal realism is replaced by direct perceptual realism (see chapters 1–2) and by manifest realism (see also chapter 3)—hence also by empirical realism (CPR A366–80).

Second, Kant secures objectivity for manifest properties by way of human intersubjectivity:

> We commonly distinguish in appearances that which is <u>essentially inherent in their intuition and holds for sense in all human beings</u>, from that which belongs to their intuition accidentally only, and is not valid in relation to sensibility in general but only in relation to a particular standpoint or to a particular organization of this or that sense. The former kind of cognition is then declared to represent the object in itself (*Gegenstand an sich selbst*), the latter its appearance only. But this distinction is merely empirical. (CPR A45/B62; underlining added)

Otherwise put, for a property in empirical nature to be objectively real—that is, for it to belong to an "empirical thing-in-itself"—is for it to obtain invariantly in relation to an idealized human sensibility that is found more or less realized or implemented in every actual human cognizer. A real property is invariant for all actual and possible finite sensory cognizers like us, under the assumption that our individual sensibilities are treated at a sufficiently high level of generality, prescinding from all idiosyncrasies of cognitive situation and all idiosyncrasies of the organization of our sensory equipment. In this connection, Kant carefully distinguishes "idealization" from the formation of ideas of pure reason (*CPR* A567–71/B595–9): an ideal is a concrete individual viewed under the aspect of some idea. So the ideal sensibility is my actual sensibility *ideally regarded*, hence not regarded as a "normal perceiver" under "normal conditions."[54] In this way, gold for example is objectively metallic because it meets all the experimental criteria for being a metal for any actual or possible human sensibility considered archetypically, or as an ideal under that type. Analogously, the Principle of Non-Contradiction is a strictly universal law of logic because it theoretically and categorically normatively constrains the thoughts and inferential processes of any actual or possible human reasoner considered ideally, or without regard to his or her individuating consciousness and without regard to any special topic of his or her thought (*JL* 9: 11–14). Hence the Kantian conception of objectivity is a conception of that which is at once directly observable and yet also humanly intersubjectively invariant at an appropriate level of idealization.

Third, and perhaps most importantly, Kant firmly rejects Locke's theory of ideas of sensation, hence he also rejects Hume's sensory impressions, and would equally reject Moorean-Russellian sense data. According to the Lockean-Humean-Russellian-Moorean conception, a sense perception is the direct mental grasp of a sensum or phenomenal qualia-token. For Kant, however, as we saw in section 2.2, an objective or exogenously caused sensation is nothing but "the effect of an object on the capacity for representation, insofar as we are affected by it" (*CPR* A19–20/B34), and "refers to the subject as the modification of its state" (*CPR* A320/B376), which is to say that it is a purely subjective response to the causal impact of something external on the sense organs of a cognizing animal. An objective sensation therefore does not represent an object: it is instead nothing but

---

[54] For a plausible critique of the thesis that perceptual objective qualities are how things merely appear to normal observers under normal conditions, see Hacker, *Appearance and Reality*, ch. 3.

the subject's immediate phenomenally conscious awareness in inner sense of her own empirical intuitional representation of an outer object, given a causal or anyhow dynamical triggering of her intuitional capacities. An objective sensation is simply "what it is like" for the intuiting subject to intuit something in the empirical object: it is phenomenally conscious, but cognitively transparent. Or, to put it more precisely, for Kant objective sensations are non-representational subjective experiences or *Erlebnisse* inherent in perceptual activity, whose sole cognitive function is to alert the representing subject to the outer (and therefore real) presence of distinct and therefore discriminable manifest properties of objects. He writes:

Sensation in itself is not an objective representation, and in it neither the intuition of space nor that of time is to be encountered. (*CPR* A165/B208)

That in the intuition which corresponds to the sensation is reality (*realitas phaenomenon*). (*CPR* A168/B209)

Intuition is referred to the object, sensation merely to the subject. (R XII E 15–A20; 23: 21)

Sensations ... constitute the peculiar subjective quality (*eigentliche Qualität*) of empirical representations (appearances). (*P* 4: 307)

Sensation [is] the subjective quality (*Qualität*) of an empirical intuition in respect of its specific difference from other sensations. (*P* 4: 309)

In this way, Kant is implicitly but consistently working with a crucial distinction between (a) subjective qualities (*Qualitäten*) or qualia on the one hand, and (b) objective qualities (*Beschaffenheiten, Eigenschaften*), determinations (*Bestimmungen*) or predicates (*Prädikate*)—in a word, *manifest properties of empirical things-in-themselves*—on the other. Sensations are nothing but immanently reflexive phenomenal qualia or phenomenal characters of empirical intuitions. But the representational target of an empirical intuition is an objective quality, determination, predicate, or manifest property. Thus the sensation of taste or the sensation of colour falls strictly under (a), while the *intuited taste or colour of the object itself falls strictly under* (b):

The [sensation of the] taste of a wine does not belong to the objective determinations (*Bestimmungen*) of the wine, not even if by the wine as an object we mean the wine as an appearance, but to the special constitution of sense in the subject that tastes it. [Sensations of] colors are not objective qualities (*Beschaffenheiten*) of the bodies to whose intuition they are attached, but only modifications of the sense of sight, which is affected in a certain manner by light. (*CPR* A28)

The sensations of colors, sounds, and warmth...are merely sensations and not intuitions, [and] do not in themselves allow any object to be cognized. (*CPR* B44)

[Sensations of] colors, tastes, etc., cannot rightly be regarded as objective qualities (*Beschaffenheiten*) of things, but only as changes in the subject, changes which may, indeed, be different for different human beings. For in this case that which is originally itself only appearance, e.g., a rose, counts in an empirical sense as a thing-in-itself. (*CPR* A29–30/B45)

The predicates of the appearance can be ascribed to the object itself, in relation to our sense, e.g., the red colour or fragrance to the rose. (*CPR* B69–70 n.)

I freely admit that my interpolations into the first and third texts (both of which occur in the A edition) are controversial. But they also have the following exegetical virtues: they conform to the second and fourth texts (both of which are new in the B edition); they conform to Kant's theory of sensation as I have construed it; they conform to his explicit distinction between subjective qualities or phenomenal qualia, and objective qualities or manifest properties; and—what is perhaps most important—they make all the texts consistent with one another.

I will also concede that the entire text at *CPR* B69–70 n., from which the fourth text is excerpted, is fiendishly difficult to interpret. For in that text Kant seems to say that redness is both a real *and* an illusory property: he seems to say that to ascribe redness to a rose is both unlike and like the mere fallacy of ascribing "handles" to Saturn. And of course Kant may simply be confused here. On the other hand, interpreting him charitably as usual, I think that if we sharply distinguish between (1) the "object itself" (*Objekte selbst*) or real empirical object (the "empirical thing in itself"), and (2) the "object in itself" (*Objekte an sich selbst*) or positively noumenal object (the "transcendental thing-in-itself"), then it is quite correct and intelligible to say that colour properties are real properties of the object itself (for an idealized human sensory constitution) yet only merely illusory properties of the object-in-itself or positive noumenon. The same would of course be true of spatial and temporal properties, as Kant explicitly points out:

On the contrary, if I ascribe redness to the rose **in itself**, the handles to Saturn or extension to all outer objects **in themselves**, without looking to a determinate relation of these objects to the subject and limiting my judgment to this, then illusion first arises. (B70 n.)

And this tendency to regard all manifest properties as illusory belongs solely to noumenal realism, which in turn is the flip side of problematic idealism—the Cartesian external world skepticism that is the special target of the Refutation of Idealism (see chapter 1).

In any case, it should be clear by now just what sort of thing Kant is taking to be objectively and yet also phenomenally real, namely, material substances, whether material individuals or natural kinds: "all outer appearances are *phenomena substantiata* because we treat them as substances" (*R* 17: 572); and "matter is *substantia phaenomenon*" (*CPR* A277/B333). But what precisely is the "matter" here? As I proposed in sections 1.0, 2.0, and 3.0, my view is that Kant rejects both traditional "corpuscularian" or "rigid atom" atomism (*MFNS* 4: 532–5), and also his own pre-Critical "force-shell atom" or "virtual atom" atomism in the *Physical Mondaology* (*MFNS* 4: 521), as he progressively develops his radical anti-microphysicalist critique of noumenal scientific realism. And in its place, in the Critical period, radically extending both Newton's theory of law-governed dynamic forces acting at a distance and also Boscovich's structuralist kinematic atomic theory, Kant substitutes his *causal-dynamic structuralist* theory of physical matter, according to which:

(I) matter as a whole, namely the single "that which persists" of the First Analogy of Experience, is essentially relationally constituted by a real spatiotemporally organized total structure of attractive and repulsive forces determined by synthetically necessary causal and interactive dynamic laws;

and

(II) material substances in particular, be they individual things or natural kinds, are nothing but positions in that total structure, or determinate dynamic causal roles without any hidden causal role-players.

It follows from (I) and (II), together with points (1)–(3), that the essential properties of dynamic material things in objectively real space and time are nothing but their directly humanly perceivable or observable intrinsic structural macrophysical properties. In other words, these points together yield manifest realism.

But those are just *my* words. My distinction between (i) intrinsic non-relational properties of things-in-themselves and (ii) intrinsic structural properties of dynamic empirical things is mirrored by Kant's distinction between (i*) "absolutely intrinsic determinations" and (ii*) "comparatively intrinsic relational determinations." So here is how Kant puts it himself:

Everything in our cognition that belongs to intuition ... contains nothing but relations, of places in one intuition (extension), alterations of places (motion), and laws in accordance with which this alteration is determined (moving forces). (*CPR* B66–7)

The intrinsic determinations of a *substantia phaenomenon* in space are nothing but relations, and it is entirely made up of mere relations. We are acquainted with substance in space only through forces which are active in this and that space, either bringing other objects to it (attraction), or preventing them penetrating into it (repulsion and impenetrability). We are not acquainted with any other qualities constituting the concept of the substance which appears in space and which we call matter. (*CPR* A265/B321)

Matter is *substantia phaenomenon*. That which intrinsically belongs to it I seek in all parts of space which it occupies, and in all effects which it exercises, though admittedly these can only be appearances of outer sense. I have therefore nothing absolutely but only comparatively intrinsic, which itself in turn consists of outer relations. (*CPR* A277/B333)

A permanent appearance in space (impenetrable extension) contains mere relations and nothing absolutely intrinsic, and nevertheless can be the primary substratum of all outer perception. (*CPR* A284/B340)

Things insofar as they are given in intuition with determinations that express mere relations without having anything [absolutely] intrinsic at their ground ... are not things in themselves but simply appearances. (*CPR* A284–5/B341)

Whatever we cognize only in matter is pure relations (that which we call their intrinsic determinations is only comparatively intrinsic); but there are among these some self-sufficient and permanent ones, through which a determinate object is given to us. (*CPR* A285/B341)

The concept of matter is reduced to nothing but moving forces; this could not be expected to be otherwise, because in space activity and no change can be thought of but mere motion. (*MFNS* 4: 524)

And it gets even better than this. In the *Metaphysical Foundations of Natural Science,* Kant points out that, contrary to any version of the atomic theory, his causal-dynamic structuralist theory of matter is conceptually consistent with the existence of a fundamental fluid aether (*MFNS* 4: 534). A Kantian fundamental fluid aether, if it existed, would be a spatiotemporally continuous plenum of attractive and repulsive forces, and rule out the possibility of discontinuous indivisible fundamental particles. Later, in the sketches for his "Transition" project in the *Opus postumum,* moreover, Kant insists that not only is his causal-dynamic metaphysical structuralist theory of matter conceptually consistent with a fundamental fluid aether, but also in fact it *transcendentally requires* the actual existence of such an aether, in order to complete the three Analogies of Experience by precisely determining the nature of objectively real material substances (*OP* 21: 215–33, 22: 239–42).[55] Otherwise there would be an explanatory and ontological gap in Kant's transcendental philosophy of nature between (a) the a priori *formal* conditions for the possibility of experience (the categories, the forms of intuition, and transcendental apperception), and (b) the a priori *material* condition of possible experience, namely, the given material substance that affects the cognitive subject via sensory intuition. According to Kant in the *Opus postumum,* there is one and only one such material substance (the fundamental fluid aether), and yet there is also a plurality of such substances (individual material objects and natural kinds). This seeming paradox withers away as soon as we understand Kant's causal-dynamic structuralism about matter. *Material substances, whether individuals or kinds, are nothing but particular causal-dynamic positions or roles in the total structural fundamental fluid aether of moving forces.*

This may seem to be at best a minor footnote in the history of Newtonian theories, and at worst an indication of Kant's senility. But in the context of a causal-dynamic structuralist theory of matter, the notion of a dynamic fundamental fluid aether seems a priori to be an utterly brilliant way of overcoming the seeming unintelligibility of the notion of *action-at-a-distance*. In this way, physical contact over space and time, which is a

---

[55] See: Edwards, *Substance, Force, and the Possibility of Knowledge;* Förster, *Kant's Final Synthesis;* Förster (ed.), *Kant's Transcendental Deductions,* part IV (with papers by Tuschling, Förster, and Vuillemin); Friedman, *Kant and the Exact Sciences,* ch. 5, esp. pp. 290–341; Guyer, *Kant's System of Nature and Freedom,* ch. 4; Hall, "A Reconstruction of Kant's Ether Deduction in *Übergang* 11"; and Hall, "Understanding *Convolut* 10 of Kant's *Opus postumum.*"

metaphysical mystery in a world of atoms, becomes essentially a function of the intrinsic structural characteristics of a spatiotemporally continuous dynamic plenum of attractive and repulsive forces. There is no need for action-at-a-distance, nor is there any need for a passive medium of action (that is, an *inert* fluid aether, which as we know was disconfirmed by the Michelson-Morley experiments), because the action is *already primitively everywhere*. Now fast-forward 120 years ahead in the history of physics. As Einstein and Eddington both point out, the relativity theory, which gives up both absolute simultaneity and action-at-a-distance, is in fact *perfectly consistent* with the existence of a dynamic fluid aether.[56] In turn, the notion of a dynamic fluid aether ultimately drops out of Einstein's doctrine only in view of the Bohr-Rutherford theory of matter, which requires *absolute gaps* in physical space and *quantum leaps* in physical time, in opposition to the notion of a dynamic "atomless gunk." But both the strange phenomenon of *quantum entanglement* and the correspondingly strange doctrine of *quantum field theory* seem to work much better on Kant's metaphysical model of a dynamic atomless gunk.[57] Hence it may well be the case that if we independently question the intelligibility of the microphysicalist and scientific essentialist metaphysical interpretation of the Bohr-Rutherford theory of matter and want to give it a Kantian regulative-hypothetical fictionalist interpretation instead, but *also* want to hold onto Special Relativity *and* General Relativity *and* quantum mechanics, then a reconsideration of Kant's dynamic fluid aether theory of matter is in order.

Back to Kant's theory of matter? That may seem absurd. But in my opinion there is almost always a significant philosophical pay-off to be gained when we take Kant charitably and seriously. So, following this strategy, it seems to me that perhaps a significant conceptual advance in the metaphysics of matter could be made if some theoretical reconciliation can be managed between (1) Kant's thesis that our a priori representation of space as 3-D, Euclidean or rectilinear, homogeneous, and oriented is objectively valid, and (2) Einstein's thesis that actual space-time is 4-D, non-Euclidean or non-rectilinear, of variable curvature, and continuously filled with matter and energy: or, in short, that actual space-time is a dynamic "field." One way of doing this is just to identify Kant's *dynamic fluid*

[56] See Eddington, *The Nature of the Physical World*, pp. 30–2; and Rynasiewicz, "Absolute versus Relational Space-Time: An Outmoded Debate?," esp. pp. 293–9.
[57] See, e.g., Stapp, *Mind, Matter, and Quantum Mechanics*.

aether with Einstein's *dynamic field*. Then we could treat Einstein's thesis as *an empirical theory about the intrinsic structural characteristics of the dynamic fluid aether, but not, properly speaking, a theory about space and time themselves.* In turn, according to this hybrid Kantian-Einsteinian picture, space and time themselves would be weakly transcendentally ideal unique formal-structural constraints on the possible existence of causal-dynamic material objects in the manifest world, or objectively real a priori conditions of the possibility of multiple egocentric causal-dynamic frames-of-reference. So the fundamental dynamic fluid aether would be Einsteinian-relativistic, while space and time remained fully Kantian-anthropocentric. Whatever the ultimate theoretical merits of this suggestion, it shows at the very least that the all-too-familiar but hugely influential early Russellian and logical positivist objections to Kant's Critical philosophy, to the effect that it is "refuted" by the combined theoretical advances of non-Euclidean geometry, Relativity, the Bohr-Rutherford atomic theory of matter, and Heisenberg-Schrödinger quantum mechanics, is fairly over-rated.[58]

As I have noted several times already, Kant's radically original conception of matter is explicitly and intrinsically combined by him with his manifest realism: all the essential properties of individual dynamic material substances or events and natural kinds are nothing but their directly humanly perceivable intrinsic structural macrophysical properties. So once again, for Kant, cognitively and ontologically speaking, *nothing is hidden* in the material or natural world. He thereby effectively detaches what he regards as the perfectly acceptable inherent cognitive movement, within the ongoing practice of natural science, towards an increasingly complete and unified causal explanation of nature (*CPR* A642–68/B671–96), from what he regards as the epistemically, metaphysically, cognitively, and morally disastrous appeal to microphysical noumenal scientific realism. For he believes that fundamental physics can focus consistently and solely on the world of appearances without in any way compromising the epistemic, metaphysical, or methodological integrity of natural science:

Through observation and decomposition (*Zergliederung*) of appearances we penetrate to nature's intrinsic nature (*das Innere*), and no one can scientifically know

---

[58] The Russellian and logical positivist critique of Kant is also seriously compromised by its failure to understand Kant's distinction between analytic and synthetic necessary truths, and correspondingly by its strategy of merely assuming the truth of *modal monism* in the face of Kant's explicit *modal dualism*. See Hanna, *Kant and the Foundations of Analytic Philosophy*, esp. ch. 5.

how far this may in time extend. But even with all this, and even if the whole of nature were revealed to us, we should never be able to answer those transcendental questions which go beyond nature. (*CPR* A277–8/B334–5)

Natural science will never be able to reveal to us the absolutely intrinsic nature of things (*das Innere der Dinge*), which, though not appearance, yet can serve as the ultimate ground for explaining appearances. Nor does science need this for its physical explanations... For these explanations must only be grounded upon that which as an object of sense can be brought into connection with our actual perceptions according to empirical laws. (*P* 4: 353)

In this way, fundamental physics can learn increasingly more, for example, about *how* manifest physical bodies gravitationally or magnetically attract one another: but it cannot have a microphysical insight into *why* they attract one another (*P* 4: 349–50). At no point does the endless natural scientific advance towards the discovery of the complete truth about the manifest world by means of empirical methods of investigation pass over into a knowledge of microphysical essences, since these are positively noumenal and therefore unknowable. If it were claimed that in principle two natural kinds could share all the manifest properties of gold, relative to all actually available and really possible empirical tests, while still differing essentially in some microphysical, and categorically unperceivable or unobservable and non-manifest way, then Kant would insist that this is a purely speculative metaphysician's difference that cannot make an authentic difference to creatures minded like us:

> If by the complaints—*that we have no insight into the absolutely intrinsic nature of things* (*das Innere der Dinge*)—it be meant that we cannot conceive by pure understanding what the things that appear to us may be in themselves, they are entirely illegitimate and unreasonable. For what is demanded is that we should be able to cognize things, and therefore to intuit them, without senses, and therefore that we should have a faculty of cognition altogether different from the human, and this not only in degree but as regards intuition likewise in kind—in other words, that we should not be human beings but beings of whom we are unable to say whether they are even possible, much less how they are constituted. (*CPR* A277–8/B333–4)

This raises a final, but rather delicate, set of issues. Something is non-apparent or non-manifest in the Kantian sense if and only if it is neither directly humanly perceivable nor observable. In turn, something is directly humanly perceivable or observable in the Kantian sense if and only if it can in principle be made either intuitionally present to the conscious human senses or quasi-intuitionally represented via conscious memory (that is, the reproductive synthesis of the imagination), without a necessary dependence on concepts, judgments, inferences, or theories (*CPR* A90–1/B122–3, A320/B376–7; see also chapter 2). These two Kantian definitions have three consequences.

First, while the noumenal scientific realist is committed to the view that what is noumenally real about nature cannot ever be directly perceived or observed, he might also insist that it is indirectly perceivable or indirectly observable by means of concepts, judgments, inferences, theories, or high-powered detection technology. What Kant would deny is that such cognitive activity is in any sense genuine perception or observation of noumenal things: on the contrary, it is an act of theoretical postulation that imposes a speculative or transcendent metaphysical interpretation on empirical inputs. The entities supposedly indirectly perceived by such means are in fact *unperceived unperceivables*: nothing but "entities of the understanding" (*Verstandeswesen*) (*CPR* B306), or purely theoretical entities. Only God with her intellectual intuition could ever "perceive" them. All such entities are positively noumenal, or things-in-themselves. And, as humanly unperceivable and uncognizable, things-in-themselves are prime candidates for methodological *elimination*. In any case, only cognition that is grounded in direct or immediate empirical intuition is objectively real perception. Kant makes this quite explicit in his gloss on the second Postulate of Empirical Thought:

In the *mere concept* of a thing no characteristic of its existence can be encountered at all. For even if this concept is so complete that it lacks nothing required for thinking of a thing with all of its inner determinations, still existence has nothing in the least to do with this, but only with the question of whether such a thing is given to us in such a way that the perception of it could in any case precede the concept. For that the concept precedes the perception signifies its mere possibility; but perception, which yields the material for the concept, is the sole characteristic of actuality. (*CPR* A225/B272–3)

Second, however, all directly perceivable things are such that we can also have concepts, judgments and theories about them, make inferences concerning them, and pick them out with theory-driven interpretations of the data yielded by direct perception. So all directly perceivable things are also indirectly perceivable. But their direct perceivability implies that they can be made intuitively present to the conscious human senses even if there are no occurrent concepts, beliefs, and so on held by the subject, or even if all occurrent concepts, beliefs, and so on held by the subject are incorrect. In short, as we saw in chapter 2, direct perceivability implies the very strong and fairly strong non-conceptuality of perceptual cognition and observation. So indirect perception piggybacks on non-conceptual direct perception. By sharp contrast, microphysical entities and *Verstandewesen* more generally are not even in principle directly perceivable and therefore are not indirectly perceivable either.

Third, just because something is not currently directly perceived or observed, it does not follow that it is *not* manifest or directly perceivable. Kant is committed to the view that even things that are now too far away or too small for us to observe directly are directly observable if they can in principle be made intuitively present to human perception. Such very distant and very little things are unperceived direct perceivables. This is true of all things that are detectable only through cloud chambers, Geiger counters, or cyclotrons, or are visible only through telescopes or microscopes. Such devices all extend the limits of the directly perceivable, and they detect previously undetected fine-grained details of the causal-dynamic natural world, but they do not go beyond the domain of the manifest. Thus entities accessible to human perception only via detection technology are still essentially manifest, unhidden entities. They are *natural phenomena*. This is in sharp contrast with microphysical noumenal entities, which as *Verstandewesen* are consistently thinkable but never apparent. For example, electron microscopes—whose technology requires the *theory* of electrons—do not *detect* electrons. Locke's fantasized "microscopical eyes" could never, even in principle, have given him a glimpse of the "real internal constitutions" of things, if we construe that phrase in terms of microphysics. *Microscopes are not noumenoscopes.*[59]

In this way, micro*physical* noumenal material individuals, kinds, events, processes, and forces should be carefully distinguished from micro*scopic* phenomenal material individuals, kinds, events, processes, and forces—although, to be sure, both are small and invisible. Microscopic phenomenal

---

[59] See Wilson, *The Invisible World: Early Modern Philosophy and the Invention of the Microscope*, esp. chs. 1 and 6–8.

items are small in an *anthropocentric* sense and *contingently* invisible. That is, they are small in relation to us, to the extent that they are not currently visible to the naked human eye. But in principle, via favourable human evolution and improved detection technology, microscopic items are directly perceivable by us and thereby count as members of the manifest world. As vivid illustration of this, we need only think here of that classic Cold War sci-fi film study of sublimated science-anxiety, Jack Arnold's *The Incredible Shrinking Man* (1957). By contrast, microphysical items are small in a strictly *metaphysical* sense and *necessarily* invisible. That is, by virtue of some physical theory together with some background noumenal metaphysical theory, they are assigned quantities that bear no determinate lesser-than relations to the quantitative boundaries assigned by the really possible extended boundaries of our empirically applicable systems of measurement. They therefore fall beyond the outer limits of the causal and dynamical scale and scope of our bodies. An example would be something so far away that it falls outside the light cone, on the Einsteinian assumption that the speed of light is a logically contingent but metaphysically necessary limit on the causal-dynamic scale and scope of the human body. But we cannot give any sense to the notion of making scientific measurements except in some determinate relation to the causal-dynamic scale and scope of our own living bodies. Similarly, microphysical noumenal items cannot be seen by us even through the most powerful detection devices and are not even in principle directly observable, since they are nothing but entities of the understanding or *Verstandeswesen*.

This last point makes it possible to give a plausible interpretation of a very tricky passage that has sometimes been taken to support the (in my view quite wrong-headed) idea that Kant is committed to a doctrine of indirect observability equivalent to that of the noumenal scientific realist:

One can also cognize the existence of something prior to the perception of it, and therefore cognize it comparatively *a priori*, if only it is connected with some perceptions in accordance with the principles of their empirical connections (the analogies). For in that case the existence of the thing is still connected with our perceptions in a possible experience, and with the guidance of the analogies we can get from our actual perceptions to the thing in the series of possible perceptions. Thus we cognize the existence of a magnetic matter penetrating all bodies from the perception of iron filings, although an immediate perception of this matter is impossible for us given the constitution of our organs. For in accordance with the laws of sensibility and the context of our perceptions we could also happen upon the empirical intuition of it if our senses, the crudeness of which does not affect the form of possible experience in general, were finer. Thus whenever perception and whatever is appended to it in accordance with empirical laws reaches, there too reaches our cognition of the existence of things. (*CPR* A225–6/B273; underlining added)

As I read it, the Kantian text does not in any way support the noumenal scientific realist claim that we can somehow indirectly perceive or observe noumenally microphysical things that are inherently directly unperceivable and unobservable. On the contrary, I take it to be saying that, even though we currently do not directly perceive or observe "magnetic matter" (that is, magnetic fields) but rather only indirectly perceive it, it is nevertheless in principle directly perceivable, directly observable, or manifest. The iron filings function like a simple "magnetoscope," broadly analogous to a telescope or microscope. And, owing to various evolutionary pressures or cosmic accidents or cognitive technology, perhaps future humans will develop a visual (or a tactile, or a proprioceptive, or whatever) sensitivity for magnetic fields. Perhaps some actual humans already have it: there are stranger things out there! Certainly no law of nature or a priori rule of human sensibility rules this out. Here we need only consider the easy step of physical and cognitive conceivability presupposed by another classic Cold War sci-fi film, Roger Corman's *The Man with the X-Ray Eyes* (1963). But if human vision, or some other mode of perception, of magnetic fields is really possible, then "magnetic matter" is directly perceivable or manifest. More generally, the scope of the human senses is not strictly fixed by the current physical constitution of our bodies but rather only by natural laws together with the non-empirical functional architecture of our cognitive capacities. Assuming that we humans manage to avoid annihilating ourselves,[60] then the human body will no doubt either gradually evolve or be surgically or bio-technologically modified, or else we will develop the appropriate

---

[60] Here the obvious irony is that the exact sciences have provided the theoretical basis for doing precisely this: there is no more effective way to "disenchant nature" than the Manhattan Project.

mind-extending cognitive technology, in such a way as to make future humans capable of directly perceiving many things that are not directly perceived by current humans.[61] But, more directly to the point, all those things that actual or possible humans *could in principle* directly perceive still fall within the limits of the world of authentic appearances or manifest nature, whether future humans ever actually directly perceive them or not.

4.3. Kant's Other Joke

As we have seen in this chapter, according to Kant the conceptual analysis of natural kind concepts is possible, although the sort of knowledge required by the microphysical noumenal scientific realist is *im*possible:

> We must not think at all of the *real* or *natural essence* of things, into which we are never able to have insight... In this science [of conceptual analysis][62] the talk can only be of the *logical* essence of things. And into this we can easily have insight. For it includes nothing further than the cognition of all the predicates in regard to which an object is determined *through its concept*; whereas for the real essence of the thing (*esse rei*) we require cognition of all the predicates on which, as grounds of cognition, everything that belongs to the existence of the thing depends. (*JL* 9: 61)

If Kant is right about the nature of fundamental physics, then truths about natural kinds such as gold or water gained a posteriori through natural science do not in any way contradict analytic truths about natural kinds known a priori by decompositional analysis of their concepts. On his strongly non-reductive, anthropocentric, and realistic conception of fundamental physics, scientific investigation as applied to the study of natural kinds is directed to all (even if not only[63]) the objects that

---

[61] I am not talking about magic or miracles, for example, clairvoyance and mental telepathy (*CPR* A222/B270). The sensory developments I am talking about must at least be consistent with causal-dynamic laws of nature. See, e.g., Clark, *Mindware*, ch. 8.

[62] The term Kant actually uses is "logic." But it is important to recognize that pure general logic for him includes not only bivalent first-order monadic logic, but also a second-order intensional logic of fine-grained decomposable concepts with cross-possible-worlds extensions as well as irreducible modal operators, and this in effect amounts to conceptual analysis. See Hanna, *Kant and the Foundations of Analytic Philosophy*, ch. 3; and Hanna, "Kant's Theory of Judgment."

[63] This is to allow for mixtures and alloys. For example, water is a clear liquid, but we also learn by empirical investigation that there is a lot of water in tea and human blood. More precisely, what we

are identified by the phenomenological intensions of natural kind concepts and fall within their cross-possible-worlds extensions. To be sure, this does not imply that experimental physical investigations into natural kinds can or should be somehow carried out a priori by means of conceptual analysis, from the philosopher's armchair: "when we speak of water and its properties, we do not stop short at what is thought in the word 'water', but proceed to experiments" (*CPR* A728/B756). So, by means of methods of inquiry involving a significantly a posteriori dimension, according to Kant's causal-dynamic structuralist conception of physical matter together with his manifest realism, and thus according to his empirical scientific realism, fundamental physics gives causal-dynamical explanations of the very properties of natural kinds that show up a priori as analytic characteristics in the microstructures of their physical concepts. It is knowable a priori, for example, by means of conceptual decomposition of GOLD, that gold has the analytically necessary property of being some sort of metal; we learn a posteriori through experiments that it is a heavy or high-density metal; and the causal explanation for this feature of gold is knowable through fundamental physics by means of the theory of specific gravities. In this way, causal-dynamic explanations of natural kinds, presupposing some irreducibly a posteriori truths and knowledge, are subsumable under a priori conceptual analyses, although they are not of course logically derivable from those analyses.[64]

What is above all cognitively and epistemically important about conceptual analysis for Kant is that it reveals to our reflective consciousness

---

learn is that tea and blood are mostly composed of what would otherwise be pure water. Nevertheless we call lake water "water" because it is a (relatively) clear liquid, but do not call tea or blood "water" because it is opaque—that is, dark brown or dark red—even though there is much more water in tea or blood than any other element. (Weak tea is not a counterexample, only a borderline case; indeed, we generally speak of "watery tea" in such cases.) These points provide a rather clear and crisp Kantian solution to a recent debate as to whether the everyday application of a natural kind term tracks microphysical science or is relativized to current human interests. See Chomsky, "Language and Nature," at 22–3; Abbott, "A Note on the Nature of 'Water'"; and LaPorte, "Living Water." The Kantian answer is "neither": the application of the concept follows the phenomenological cross-world identification conditions built decompositionally into the concept expressed by the natural kind term.

[64] It would be a mistake to think that conceptual analysis historically precedes scientific knowledge of natural kinds: in fact, they arise more less simultaneously. That is, natural kind concept-formation and the scientific investigation of natural kinds are two sides of the same cognitive process. See LaPorte, "Chemical Kind Term Reference and the Discovery of Essence."

the otherwise merely implicit and contingently cognitively hidden internal structures of the basic constituents of our own thoughts:

> The greatest part of the business of our reason consists in decompositions of the concepts we already have of objects. This supplies us with a multitude of cognitions that, while they are nothing more than clarifications or elucidations of what is already thought in our concepts (although in a confused way), are, at least as far as their form is concerned, valued as new insights. (*CPR* A5–6/B9)

Armed with this non-trivial self-knowledge, the physicist then proceeds to the world in order to discover facts about what falls under her natural kind concepts.

An extremely unpalatable cognitive and epistemic consequence of scientific essentialism, by contrast, is that conceptual analysis is dichotomously detached from scientific explanation. But this undermines a central element of the first-person epistemic authority, or "conviction" (*Überzeugung*), that is necessary for human rationality (*CPR* A820–3/B848–51) (*JL* 9: 65–73).[65] For, if there is no basic coherence between the rational thinker's clear and distinct conscious grasp of her own concepts and the scientific facts (both necessary and contingent) about the world, then there can be no rational certainty (see chapter 7 below). But without rational certainty, there can be no such thing as natural science in the proper sense (*MFNS* 4: 468).

The basic coherence of conceptual analysis and natural science, and, in particular, fundamental physics, implies that for Kant the natural sciences (*Naturwissenschaften*) are necessarily *also* human sciences or moral sciences (*Geisteswissenschaften*). Still, to hold this Kantian doctrine is not to make physics into (as a contemporary Edinburgh-trained philosopher of science might say) nothing but the most culturally privileged, well-funded, politically powerful, and methodologically straitjacketed language-game we have yet devised for the social construction of knowledge.[66] Rather, it is to insist that the very idea of natural scientific objectivity cannot be divorced from transcendental conditions for the possibility of human experience. Kant is

---

[65] See Burge, "Reason and the First Person." Oddly enough, although Burge's account of first-person epistemic authority is clearly indebted to Kant's theory of apperception, he takes a rather dim view of conceptual analysis in Kant's sense (p. 261 n.11). But I think Burge overestimates both the cogency and the scope of Quine's critique of analyticity. For a defence of Kant's theory and a critique of Quine, see Hanna, *Kant and the Foundations of Analytic Philosophy*, sections 3.1 and 3.5.

[66] See, e.g., Bloor, *Knowledge and Social Imagery*.

an empirical scientific realist, not an anti-realist. Kantian natural science is strongly non-reductive and anthropocentric, yet what it investigates is objectively real despite being based on "appearances": the appearances are *authentic* (*Erscheinungen*), not *illusory* (*Schein*), and this is *as real as it gets* for creatures minded like us.

One last remark, to run down the curtain on this chapter and also Part I of the book. What Kant's account excludes is the idea that the natural sciences can legitimately constitutively invoke what Bernard Williams aptly calls an "absolute"[67] conception of the natural world—as provided, for example, in the microphysical noumenal scientific realist picture that is built into scientific essentialism. This picture ultimately and unintentionally entails that the objects of physics are unperceived unperceivables or *Verstandeswesen*, and that inevitably makes the knowledge-claims of the natural sciences into fat, slow-moving targets for skeptics and relativists.[68] So the absolute conception of the natural world cannot be taken absolutely.

But even if we do not take microphysical things-in-themselves absolutely, and adopt a methodological eliminativism about positive noumena, they can still be regarded *ironically*. What I mean is that Kant's account does not in any way rule out a purely regulative-hypothetical, or fictionalist, invocation of scientific theories motivated by the absolute or noumenal conception (*CPR* A642–68/B670–96)—for example, Bohr-Rutherford atomic theory. So Kant explicitly proposes that noumenal scientific theories, construed as *ideas of reason*, can quite legitimately function as heuristic models, or investigation-guiding schemata, for promoting the systematicity (unity, lawfulness, and coherence) of the practice of scientific inquiry:

If we survey the cognitions of our understanding in their entire range, then we find that what reason quite uniquely prescribes and seeks to bring about concerning it is the *systematic* in cognition, i.e., its interconnection based on one principle. This unity of reason always presupposes an idea... One cannot properly say that this idea is the concept of an object, but only that of the thoroughgoing unity of these concepts, insofar as the idea serves the understanding as a rule. Such concepts of reason are not created by nature, rather we question nature according

---

[67] See Williams, *Descartes: The Project of Pure Inquiry*, pp. 65–8.
[68] The skeptic says: if the objects of physics are merely theoretically posited or inferred entities, then nothing will conclusively show that they exist in the external world (*CPR* A368). And the relativist says: if the sensory given is a myth, that is, if all theories are underdetermined and ultimately unconstrained by empirical evidence, then anything goes (*CPR* A236–8/B295–8).

to our ideas... Admittedly, it is hard to find *pure earth, pure water, pure air,* etc. Nevertheless, concepts of them are required ... in order appropriately to determine the share that each of these natural causes has in appearance; thus one reduces all materials to earths (mere weight, as it were), to salts and combustibles (as force), and finally to water and air as vehicles (machines, as it were, by means of which the aforementioned operate), in order to explain the chemical effects of materials in accordance with the idea of a mechanism. (*CPR* A645–6/B673–4)

In this way Kant's empirical scientific realist pragmatically adopts a purely methodological and deflationary analogue of the absolute conception of nature.[69] Microphysical things-in-themselves are methodologically eliminable from a philosophical point of view, but in doing natural science we usefully *pretend* that we believe they exist. So, by a surprising inversion of our conventional wisdom, the manifest image finally absorbs the scientific image. As Nietzsche tells it, Kant's private joke is that he vindicates the standpoint of common sense in a way that completely flummoxes the ordinary person. That is correct. But Kant's Other Joke—the one that really has them rolling in the aisles—is that he also vindicates both natural science in general and fundamental physics in particular in a way that would similarly flummox the noumenal scientific realist.

---

[69] Karsten Harries first drew my attention to this important point. He also argued that, according to the Kantian conception of science, natural science must adopt a methodological absolute conception of nature in order to have an adequate theory of truth. But it seems to me that, according to Kant, an adequate theory of truth requires only the much weaker notion of an externally actual world—a "truth-maker"—that provides a necessary condition of cognitive orientation. And the minimal notion of a truth-maker is consistent with noumenal realism and empirical realism alike. It also seems to me that the concept of truth must ultimately answer not to the demands of pure theoretical reason but instead to the demands of practical reason. Only sincerity or truthfulness can explain what the *point* of truth-as-correspondence is. See ch. 5.

PART II
The Practical Foundations of the Exact Sciences

5

Truth and Human Nature

Reality is compared (*wird ... verglichen*) with the proposition. Propositions can be true or false only by being pictures of reality.

Ludwig Wittgenstein[1]

[My] methodology can be described on the negative side by saying it offers no definition of the concept of truth, nor any quasi-definitional clause, axiom schema, or other brief substitute for a definition. The positive proposal is to attempt to trace the connections between the concept of truth and the human attitudes and acts that give it body.

Donald Davidson[2]

5.0. Introduction

Here is the story so far. The goal of this book is to work out a Kantian solution to the Two Images Problem: How is it possible to reconcile *the manifest image* of human beings and their world, with *the scientific image* of human beings and their world? The manifest image is the subjective, phenomenal, perspectival, first-personal, value-laden, purposive, and moral metaphysical picture of the world yielded by the conscious experience of rational human beings. And the scientific image is the objective, non-phenomenal, perspectiveless, impersonal, value-neutral, mechanistic, and amoral metaphysical picture of the world delivered by pure mathematics and fundamental physics. The head-on collision between these sharply opposed metaphysical frameworks, first noted by Husserl in the 1930s in his *Crisis of European Sciences* and canonically formulated by Sellars in the 1960s in his "Philosophy and the Scientific Image of Man," constitutes the basic

---

[1] Wittgenstein, *Tractatus Logico-Philosophicus*, props. 4.05 to 4.06, p. 71.
[2] Davidson, "The Folly of Trying to Define Truth," p. 276.

problem of analytical philosophy after 1950, and is perhaps *the* fundamental problem of modern philosophy.

Now, according to *scientific or reductive naturalism*, the manifest image explanatorily and ontologically reduces to the scientific image. But on the contrary, according to Kant, the scientific image is fully explicable in terms of the manifest image. The manifest image in turn both explanatorily and ontologically presupposes rational human nature, which is both explanatorily and ontologically *irreducible*, in that it is underdetermined by all possible empirical (that is, sensory-experiential and contingent) facts. And this in turn is because rational human nature is constituted by a unified network of capacities for cognition and volitional action according to absolutely necessary and categorically normative a priori principles. So, in a nutshell, the Kantian solution to the Two Images Problem is that the exact sciences have a non-reductive explanatory and ontological grounding in rational human nature.

In the first part of this book I developed and defended an account of Kant's empirical realism, and then critically contrasted it with contemporary scientific realism. Empirical realism consists of the conjunction of the epistemological thesis of direct perceptual realism and the metaphysical thesis of manifest realism:

Direct Perceptual Realism: Every self-conscious human cognizer has direct veridical perceptual or observational access to some macrophysical dynamic material objects in objectively real space and time.

Manifest Realism: All the essential properties of individual dynamic material substances, natural kinds, events, processes, and forces are nothing but their directly humanly perceivable or observable intrinsic structural macrophysical properties.

The key to direct perceptual realism is Kant's theory of non-conceptual content, and the slogan here is *that it's all about location*: by which I mean that non-conceptual content is proto-rational representational content whose essential structure is spatiotemporal, and whose cognitive role is to situate and track individual material objects in immediate relation to an oriented and (at least in principle) motile, embodied, conscious perceiving subject. Correspondingly, the key to manifest realism is Kant's causal-dynamic metaphysical structuralist and radical macrophysical ontology of matter, and the corresponding Kantian slogans here are *that matter is causal-dynamic structure* and *that nothing is hidden*: by which I mean:

(i) that material objects are nothing but positions or roles in a nomologically determined total systematic complex of attractive and repulsive forces in objectively real space and time—which constitutes matter as a whole, which is the One Big Substance or "that which persists" of the first Analogy of Experience, and which Kant ultimately identifies with the universal fluid aether in the *Opus postumum*; and
(ii) that all such material objects are directly humanly perceivable or observable.

The existence of a set of putatively "Really Real" atomic or sub-atomic microphysical entities, as defined by their intrinsic non-relational microphysical properties, can be neither asserted nor denied and therefore must be methodologically eliminated, precisely because they are nothing but unknowable things-in-themselves or positive noumena. The upshots of all this are, first, that Kant's epistemology-and-metaphysics of the exact sciences is fully non-reductive, anthropocentric, and also realistic; and, second, that Kant's empirical scientific realism opens up a new intermediate place in philosophical space that is importantly distinct from those occupied by noumenal scientific realism and scientific anti-realism alike, and thereby has significant philosophical advantages over each of those all-too-familiar doctrines. So much for part I.

And now for something *not* completely different: I mean of course part II. In this part of the book I move from epistemological and metaphysical issues about scientific and empirical realism, to the philosophy of scientific rationality. In particular, I will focus on Kant's theory of the fundamental connection between scientific rationality and practical rationality.

By "scientific rationality" or "scientific reason," I mean the human capacity for judging and inferring according to a priori principles, as it is manifest in the exact sciences; and by "practical rationality" or "practical reason," I mean the human capacity for willing and acting according to a priori principles, as it is manifest in either productive (i.e., skill-based, technical-practical) or prudential (i.e., interest-based, pragmatic) human action. Scientific rationality in this sense should be sharply contrasted with the capacity for judging and inferring according to *merely empirical rules*; and correspondingly practical rationality in this sense should be sharply contrasted with the capacity for willing and acting according to *merely instrumental rules*. In other words, both scientific rationality and practical rationality in the senses I am discussing, unless otherwise noted, are *non-Humean*.[3] Given that backdrop, I want to develop and defend the following Kantian thesis:

The Practical Foundations of the Exact Sciences: Practical reason has both explanatory and ontological priority over theoretical reason.

The relevant Kantian slogan here is *that practical rationality rules*. This means, at bottom, two things. First, theoretical reason in the exact sciences presupposes practical reason (this yields the explanatory priority of the practical). Second, there cannot be a scientifically knowable world in which human value, human action, and human morality are really impossible (and this yields the ontological priority of the practical). In order to work out Kant's Practical Foundations thesis, in the next four chapters I will look successively at Kant's theories of truth, mathematics, a priori knowledge, causation, and freedom of the will.

So what is the truth about truth? In the third chapter of the Transcendental Logic in the *Critique of Pure Reason*, Kant poses for himself the very hard question cynically and skeptically asked by Pontius Pilate:

The old and famous question with which logicians were to be driven into a corner and brought to such a pass that they must either fall into a miserable circle or else confess their ignorance, hence the vanity of their entire art, is this: **What is truth** (*Wahrheit*)? (*CPR* A57–8/B82)

In reply to Pilate's question, Kant says this:

The nominal definition (*Namenerklärung*) of truth, namely that is the accordance (*Übereinstimmung*)[4] of cognition with its object, is here granted and presupposed; but one demands to know what is the universal and certain criterion (*Kriterium*) of the truth of any cognition? (*CPR* A58/B82)

Truth, it is said, consists in the accordance of cognition with its object. In consequence of this mere verbal definition (*Worterklärung*), my cognition, to count as true, is supposed to accord with its object… The question here is, namely, whether and to what extent there is a criterion of truth that is certain, universal,

---

[3] Cf. Hume, *Treatise of Human Nature*; and Blackburn, *Ruling Passions: A Theory of Practical Reasoning*.
[4] An alternative and more familiar translation of *Übereinstimmung* is "agreement"; but here for consistency's sake I adopt John Wisdom's term. See n. 5.

and useful in application. For this is what the question, *What is truth?*, ought to mean. (*JL* 9: 50)

But this Kantian reply to Pilate, on the face of it, seems stale, flat, and unprofitable.

And here is why. According to the merely "nominal" or "verbal" definition of truth, truth is the accordance of cognition with its object. This leads directly to the further and logically distinct question, whether there is a certain, universal, and effective test or criterion of truth. But Kant's way of formulating the issue about truth—in terms of its nominal definition and its criterion—produces three unhappy interpretive results:

(1) In saying that cognition-to-object accordance is only the nominal or verbal definition of truth, Kant seems to be saying that this definition is minimally acceptable but ultimately vacuous. As John Wisdom puts it:

> The accordance theory of truth is true. Indeed it is only *too* obviously true. This is a paradoxical way of saying that the definition offered, though correct, does nothing to reveal the fundamental structure of what we express about a statement when we say that it is true.[5]

This suggests that Kant mentions the nominal definition of truth only in order to set it aside, so that he can move on to the thicker-and-richer issue of the criterion of truth.

(2) It seems that we would naturally want a *real* definition of truth, and not merely a nominal one. But Kant never explicitly tells us here or anywhere else just what the real definition of truth is. This suggests that Kant thinks that in the last analysis nothing substantive can be said about the concept of truth.

(3) Although in the first *Critique* Kant explicitly raises the question of "the universal and certain criterion of the truth of any cognition," two paragraphs later he explicitly rejects the existence of any such criterion or test: "a sufficient and yet at the same time universal criterion of truth cannot possibly be provided" (*CPR* A59/B83). This suggests that Kant thinks that the question of the criterion of truth cannot be satisfactorily answered.

---

[5] Wisdom, *Problems of Mind and Matter*, pp. 194–5.

These three interpretive results, when taken together, imply that Kant has no philosophically significant theory of truth to oppose to Pilate's cynical skepticism. But I think that each of the results expresses at best a superficial reading, and at worst a serious misunderstanding, of Kant's truth-doctrine. When Kant's truth-doctrine is properly reinterpreted, the inference to its insignificance can be effectively blocked. Beyond that, and even more importantly, I also think that Kant's theory of truth has both striking connections with and also genuine relevance for recent and contemporary work on truth in particular, and on the nature of human rationality more generally.

In order to show all this, I develop and defend four claims in this chapter. First, in section 5.1, I argue that Kant holds that "the accordance of my cognition with its object" is the strictly and analytically correct, *yet metaphysically and epistemologically neutral*, explication of the concept of truth. This Kantian account is *almost* "deflationist" but not quite. The truth-deflationist thinks that there is nothing of any philosophical significance whatsoever to be said about truth beyond what is self-evidently expressed in the classical schema:

"P" is true if and only if P.

Now Kant is committed to a plausibly weak version of the correspondence theory, a version which has strong similarities with both Wittgenstein's Tractarian "picture" theory of truth and also Alfred Tarski's "semantic" conception of truth. So what distinguishes Kant from the deflationist is that Kant holds that the *only* things of any philosophical significance whatsoever to be said about truth are those *implicit* in and *presupposed* by the classical schema. So Kant is a *minimalist* about truth, and therefore most certainly not an *inflationist* about truth, even if he is also not quite a deflationist about truth.

Second, in section 5.2, I argue that the idea of a "real definition" of truth is equivalent to Kant's notion of a criterion of truth. Kant holds that there is no real definition of truth, in that there can be no single or univocal universal effective test of truth. So he would be in fundamental agreement with Davidson's important claim that it is "folly" to attempt to give a formally adequate, materially adequate, and recursive definition of truth for an entire natural language.

But, third, in section 5.3, I argue that, in denying that there is a real definition of truth, Kant is not saying that there are no criteria of truth whatsoever. On the contrary, what he is saying is that there is *no single or univocal universal* criterion of truth, because in fact there are *several irreducibly different types of truth*, each of which reflects a distinct human cognitive capacity, and that accordingly truth has several different, equally legitimate, and systematically specifiable criteria. So Kant is what I will call a *constrained pluralist* about truth.

Finally, in section 5.4, I briefly explore some broader and deeper implications of Kant's theory of truth. Here I adumbrate the Kantian idea that truth has an essential connection with rational human nature, which is to say that no theory of truth is adequate unless it explicitly and intrinsically relates the concept of truth to (in Davidson's nice phrase) "the human attitudes and acts that give it body." Bernard Williams has made the same point by stressing the need for a philosophical investigation into "the value of truth."[6] In this connection, the distinction between *truth* and *truthfulness* becomes fundamental, and I argue that for Kant practical truthfulness is the enabling presupposition of all propositional truth.

5.1. The Definition of Truth

Truth, for Kant, is properly predicated of judgments, and not of singular cognitions or intuitions: "truth … [is] not in the object, insofar as it is intuited, but in the judgment about it insofar as it is thought" (*CPR* A293/B350). And, although a judgment always includes concepts, a judgment is neither a single concept nor a mere list or bundle of concepts:

> I have never been able to satisfy myself with the explanation the logicians give of a judgment in general: it is, they say, the representation of a relation (*Verhältnisses*) between two concepts … I remark only that it is not here determined what this *relation* consists in. (*CPR* B141)

But if a judgment is neither an intuition, nor a single concept, nor even an unstructured aggregate of concepts, then what is it? In section 19 of the B Deduction, Kant says:

> I find that a judgment is nothing other than the way to bring given cognitions to the **objective** unity of apperception. This is the aim of the copula **is** in them: to

---

[6] See Williams, *Truth and Truthfulness*, esp. chs. 1–6.

distinguish the objective unity of given representations from the subjective. For this word designates the relation of given representations to the original apperception, and its **necessary unity**.... Only in this way does there arise from this relation a **judgment**, i.e., a relation which is **objectively valid** (*objektiv gültig*), and that is sufficiently distinguished from the relation of these same representations in which there would be only subjective validity.... According to the latter I could only say "If I carry a body, I feel a pressure of weight", but not "It, the body, is heavy," which would be to say that the two representations are combined in the object (*Objekt*), i.e., regardless of any difference in the condition of the subject, and are not merely found together in perception (however often as that might be repeated). (*CPR* B142)

In this way, a judgment is essentially an objectively valid logical (i.e., a predicative, truth-functional, and monadic-quantificational[7]) synthesis of sensory intuitions and concepts, carried out by a conscious and also self-conscious or apperceptive subject. This overall structure in turn divides into two complementary sides: the subjective side, or the judgment-act, -state, or -process (the *judging*); and the objective side, or the judgment-content (the *judged*).

On the judging side, we find first the subject's conscious activity of synthesizing the judgment. "Synthesis" in general[8] is the generative (i.e., formal, explicit, and structure-conferring), creative (i.e., recursive, and input-underdetermined or a priori) mental processing of original worldly perceptual inputs according to innate rules. The output of the synthesis is the *semantic content* of a cognition, that is, a cognition's determinate representational specification of some object or another (*CPR* A108–9). In judging, the conscious synthesizing subject undergoes or "lives" a series of different conscious mental acts, sensory states, and processes in time, in inner sense (see section 1.1). Kant also calls this series "the determinations (*Bestimmungen*) of our state in inner perception" (*CPR* A107) and "my existence (*Daseins*) as determined in time" (*CPR* B275). As apperceptive or self-conscious, the conscious subject is able to ascribe these

---

[7] Monadic quantification is quantification into one-place predicates (or concepts). The fact that Kant's logic is essentially monadic has several important philosophical implications. See section 6.1 below; and Hanna, "Kant's Theory of Judgment," section 2.1.2.

[8] Kant contrasts "productive" synthesis and "reproductive" synthesis (*CPR* A100–3, A118, B151–2). The former—which is basically a capacity for generating and creating spatiotemporal representations—is a condition of the possibility of all empirically meaningful cognition, while the latter—which is roughly equivalent to memory—is more specialized in its operations, and can be merely empirical or associative. But even the empirical synthesis of reproduction presupposes productive synthesis.

inner determinations to herself. The self-conscious subject is also able to ascribe to herself the semantic or representational contents of those inner determinations. This latter function is expressed by what Kant calls the "I think," which can be appended to any objective representational content of the mind (*CPR* B131–40). So, for him, all objective representations are self-consciously synthesized. Judgmental synthesis, however, is more specifically further bound up with a self-conscious logical intention or attitude whereby a subject asserts something; and this is what Kant calls "holding-to-be-true" or *Fürwahrhalten* (*CPR* A820/B848). This intentional truth-interest or truth-conation is what makes the act of judgment not merely an act of synthesis but also a *rational action*: "judgments are actions of the understanding and of reason" (*BL* 24: 844).

By contrast, on the side of the judged we find first the *logical syntax*, or underlying logical form, of the judgment-content, as described in the Table of Judgments (*CPR* A70–6/B95–101).[9] In every simple or atomic judgment—to take Kant's example, "It, the body, is heavy"—a general concept (HEAVY) functions as a predicate applied to an object, which in turn is designated by an indexical or intuitional singular term, often juxtaposed with a supplementary description (IT, THE BODY). These elements are respectively what Peter Strawson calls the "attributive" and "referential" parts of basic judgments.[10] In atomic judging, then, the subject asserts something (attribution) about something (reference). The function of predication, as expressed by the copula, is to take conceptual or attributive terms together with intuitional or referential terms, over into complete well-formed logico-syntactical units, or judgments in the purley syntactical sense. Molecular or non-basic judgments result from truth-functional or quantificational operations recursively applied to one or more atomic or basic judgments. Contrary to the frequent complaints by critics of Kant's theory of judgment, Kant does not "privilege" the categorical judgment except in the *constructive* sense that all molecular judgments are built up by repeated logical operations over a base of categorical atomic judgments.

Nevertheless, not every judgment in the purely syntactical sense is a *truth-bearer*—that is, an item taking one of the two classical truth-values, true and false. For in the first place a judgment may be merely entertained,

---

[9] See also Hanna, "Kant's Theory of Judgment."
[10] See Strawson, "On Referring," section IV; and Strawson, *Introduction to Logical Theory*, p. 145.

reserved, or suspended, and thereby not actually asserted or put forward as a truth-candidate (*JL* 9: 73–5). If a judgment is indeed asserted or anyhow is assertible, however, then it is a "proposition" or *Satz* (*VL* 24: 934). And, in the second place, even if a judgment is also a proposition, it may fail to be "objectively valid" or "objectively real":[11]

> If cognition is to have objective reality (*objective Realität*), that is, to refer to an object, and is to have meaning (*Bedeutung*) and sense (*Sinn*) in [that reference], the object must be able to be given in some way. Without that the concepts are empty, and through them one has, to be sure, thought, but not in fact cognized anything through this thinking, but rather merely played with representations. To give an object, if this is not again meant only mediately, but it is rather to be exhibited immediately in intuition, is nothing other than to refer its resentation to experience (whether this is actual or still possible). (*CPR* A155–6/B194–5; see also A239/B298)

Objective validity or reality includes four basic features: first, an objectively valid or real semantic content remains invariant across the changing inner or phenomenally conscious mental determinations of the conscious subject; second, it is communicable across the inner or phenomenally conscious mental determinations of different conscious subjects; third, it determinately specifies an object and thereby refers to that object; and, fourth, that object must be empirically accessible via human sensory intuition. In a word, then, objective validity or reality is the necessary and sufficient condition of a judgment's empirical meaningfulness, hence of a judgment's being able to take (if asserted or assertible) a classical truth-value. Whenever a judgment fails to be objectively valid or real, then it is truth-valueless or "empty" (*leer*) (*CPR* A51/B75), and thereby opens up a "truth-value gap." More positively put, however, the objective validity or reality of a judgment constitutes the judgment-content's applicability to the empirical world: it determines *precisely which* actual or possible empirical object is being talked about by the judgment.

Even so, an assertoric judgment is not true *just because* it is objectively valid:

> If truth consists in the accordance of cognition with its object, that object must thereby be distinguished from others: for cognition is false if it does not agree with

---

[11] Objective validity and objective reality, strictly speaking, are non-equivalent concepts: everything that is objectively real is also objectively valid, but not conversely. Nevertheless, the difference between them will not become important until further on in this section.

the object to which it is referred, even though it contains something which could well be valid of other objects. (*CPR* A58/B83)

The empirical object picked out by the judgment may be merely possible, but not actual; and in that case, the judgment is false.[12] Since an objectively valid judgment can be false, objective validity is a necessary but not sufficient condition for truth.

In addition to logico-syntactic well-formedness and objective validity, there is yet another necessary but not sufficient condition of truth, namely, formal logical correctness. As Kant puts it:

We will be able to advance three principles here as universal, merely formal or logical criteria of truth:

1. *the principle of contradiction and of identity* (*principium contradictionis and indentitatis*), through which the internal possibility of a cognition is determined:
2. *the principle of sufficient reason* (*principium rationis sufficientis*), on which rests the (logical) *actuality* of a cognition, the fact that it is grounded;
3. *the principle of the excluded middle* (*principium exclusi medii inter duo contradictoria*), on which the (logical) necessity of a cognition is grounded—what we must necessarily judge thus and not otherwise, i.e., that the opposite is false. (*JL* 9: 52–3; see also *CPR* A59–60/B84, A150/B190)

According to Kant's "principle of contradiction and of identity," every true judgment must be such as neither to include nor entail both a proposition and its negation. According to his "principle of sufficient reason," a judgment must be "grounded" in two senses: "that it (a) have grounds and (b) not have false consequences" (*JL* 9: 52). So (a) a true proposition must follow logically from some other true propositions; and (b) a proposition is true if and only if all its logical consequences are also true. Finally, according to his "principle of excluded middle," if a judgment is true then its negation must be false (and conversely). Thus the three formal or logical necessary conditions of truth are self-consistency, logical consequence or validity, and excluded middle or bivalence.[13]

---

[12] Kant is not committed to the existence of "negative facts" as the falsity-makers of false judgments (or the truth-makers of negative judgments), but rather only to the distinction between actual objects and merely possible objects. For the doctrine of negative facts, see Russell, "The Philosophy of Logical Atomism," p. 211; and Wittgenstein, *Tractatus Logico-Philosophicus*, p. 37, prop. 2.06.

[13] Although Kant does not in fact do so, one should distinguish between the law of excluded middle (LEM), which says that every truth-bearer of the form "P v ∼P" is true, and the law of bivalence

Do syntactical well-formedness, objective validity, and logical correctness exhaust the content of the concept of truth? They do not; and here is where the nominal definition of truth (i.e., the accordance of cognition with its object) comes into play. Nevertheless, before we can fully understand what "accordance" is, we must get clearer on what Kant means by saying that this is the "nominal definition" of truth.

The notion of a nominal definition can be understood only by way of a direct comparison and contrast with the notion of a "real definition" (*Realdefinition*):

> By mere *definitions of names*, or *nominal definitions*, are to be understood those that contain the meaning that one wanted voluntarily (*willkürlich*) to give to a certain name, and which therefore signify merely the logical essence of the object, or which serve merely for distinguishing it from other objects. *Definitions of things*, or *real definitions*, on the other hand, are ones that suffice for cognition of the object according to its intrinsic determinations (*innern Bestimmungen*), since they present the possibility of the object from intrinsic characteristics ... Logical nominal definitions of given concepts of the understanding are derived from an attribute, real definitions, on the other hand, from the essence of the thing, the first ground of possibility. Thus the latter contain what always belongs to the thing—its real essence. (*JL* 9: 143–4)

> All *definitiones* are either *nominales* or *reales definitiones*. *Nominales definitiones* are ones that contain everything that is equal to the whole concept that we make for ourselves of the thing; *reales definitiones*, however, are ones that contain everything that belongs to the thing in itself. (*BL* 24: 268)

> I mean here the real definition, which does not merely supply other and more intelligible words for the name of a thing, but rather contains in itself a clear **characteristic** by means of which the **object** (*definitum*) can always be securely cognized, and that makes the concept that is to be explained usable in application. A real definition would be that which does not merely make the concept, but at the same time its **objective reality**, distinct. (*CPR* A241–2 n.)

(LB) which says (i) that the two classical truth-values, true and false, jointly exhaust the class of possible truth-values (this avoids non-classical truth-values), and (ii) that every truth-bearer is assigned one and only one of the two classical truth-values (this avoids "truth-value gluts" or "true contradictions"). LEM and LB are logically distinct, since one might —like the Intuitionists—deny LEM but still affirm LB. Truth-value gaps are consistent with LEM and LB alike, since gaps are wffs or non-wffs that are not truth-bearers.

Kant says here that a nominal definition supplies the "logical essence" of an object, which serves merely for "distinguishing it from other objects"; and he also says that to give a nominal definition is to give a definition that "contain[s] the meaning that one wanted voluntarily to give a certain name" in the sense that one merely "suppl[ies] other and more intelligible words for the name of a thing." Elsewhere he also tells us explicitly that the logical essence of a thing is nothing other than the *concept* of a thing, as articulated into its necessary characteristics or *Merkmale*:

[The logical essence of a thing] includes nothing further than the cognition of all the predicates in regard to which an object is determined *through its concepts*; whereas for the real essence of the thing (*esse rei*) we require cognition of those predicates on which, as grounds of cognition, everything that belongs to the existence of the thing depends. If we wish to determine, e.g., the logical essence of body, then we do not necessarily have to seek for the data for this in nature; we may direct our reflection to the characteristics which, as essential points (*constitutiva rationes*) originally constitute the basic concept of the thing. For the logical essence is nothing but *the first basic concept of all the necessary characteristics of a thing (esse conceptus)*. (*JL* 9: 61)

In this way a Kantian nominal definition exposes a conceptual microstructure: a concept's intensional essence or intrinsic connotative architecture. And that is why a nominal definition can "contain everything that is equal to the whole concept that we make for ourselves of the thing." By sharp contrast, a real definition does not expose the intensional essence of the concept of a thing; rather it supplies a set of "intrinsic determinations" which give the "essence of the thing, the first ground of possibility … [or in other words] what always belongs to the thing—its real essence." The real definition is thus a metaphysically and epistemically *loaded* definition. It is metaphysically loaded because in order to cognize "the real essence of the thing (*esse rei*) we require cognition of those predicates on which, as grounds of cognition, everything that belongs to the existence of the thing depends." And it is epistemically loaded because it "suffice[s] for cognition of the object according to its intrinsic determinations (*innern Bestimmungen*), since [it] present[s] the possibility of the object from intrinsic characteristics" and "contains in itself a clear **characteristic** by means of which the **object** (*definitum*) can always be securely cognized, and ... makes the concept that is to be explained usable in application."

In short, then, it is wrong to regard a Kantian nominal definition as a *vacuous* definition. Indeed, a nominal definition in Kant's sense is what we would now call the "explicit definition" of a concept: it unfolds the concept, and makes its internal structure clearly and distinctly available to thinkers, by means of a "decomposition" (*Zergliederung*) of the intensional content of the concept into its several necessary characteristics. It thereby provides a way of effectively discriminating between objects that are subsumable under that concept and objects that are not subsumable. But such a definition does not guarantee insight into the object's intrinsic non-relational nature, nor does it guarantee the actual application of the concept (that is, the objective reality of its intensional content). A real definition, by contrast, advances from concept to intuition, exposes the underlying constitution or essence of an object, and provides an effective test or criterion for cognizing an actual object. The real definition clearly does important metaphysical and epistemological work for the cognizer; but just because it is in this way a substantive definition, it does not follow that a Kantian nominal definition is empty or inconsequential.

How does Kant's nominal vs. real definition distinction apply to the concept of truth? When he says that the nominal definition of truth is the accordance of cognition with its object, he means that characterizing truth in terms of accordance gives a *strictly correct analysis of the concept of truth*. But conceptual analysis only takes you just so far. It exposes and clarifies the intensional architecture of a thinking subject's conceptual repertoire, and it gives a way of identifying the sorts of things that can fall under the concept; but it does not in and of itself determine the meaningful application of concepts to objects (*CPR* A5–6/B9). Hence the nominal definition of truth, although it is analytically correct and perspicuous, does not at the same time yield actual instances of truth or provide a decision-procedure for determining in particular cases whether a given judgment is true. In short, the nominal definition does not supply a real definition, effective test, or criterion of truth. That is why, even having first explicitly defined the concept of truth, one must nevertheless advance beyond it to the further question of a truth-criterion.

Before I do that on Kant's behalf in the next section, however, we must delve further into the intensional microstructure of the concept of truth. We have seen that the strictly correct analysis of the concept of truth is that it is the accordance of a judgmental cognition with its object. And I think that we can safely say, by virtue of mere paraphrase of what is meant by the word "accord", that if a judgment *J* "accords" with its object O, then *J* "corresponds" or "conforms" to O. Kant's theory of truth is therefore accurately described as a version of *the correspondence theory*. But which version? Kant clearly realizes that he must deal directly with a fundamental problem built into the traditional version of the correspondence theory. On the traditional correspondence theory, it is held that a judgment *J* corresponds to an object O if and only if *J* accurately depicts O. Accurate depiction in turn requires that there be a resemblance-relation between *J* and O. In turn, again, a resemblance between these two entities is established by *comparing* them to one another. But this leads to an explanatory loop:

> Now I can compare the object with my cognition, however, only *by cognizing it*. Hence my cognition is supposed to confirm itself, which is far short of being sufficient for truth. For since the object is outside of me, the cognition in me, all I can ever pass judgement on is whether my cognition of the object accords with my cognition of the object. The ancient called such a circle in explanation a *diallelon*. And actually the logicians were always reproached with this mistake by the skeptics, who observed that with this explanation of truth it is just as when someone makes a statement before a court and in doing so appeals to a witness with whom no one is acquainted, but who wants to establish his credibility by maintaining that the one who called him as witness is an honest man. (*JL* 9: 50)

If the correspondence-relation involves a comparison between judgment $J1$ and its object O, then another judgment $J2$ is needed to establish the comparison between $J1$ and O. But the establishment of such a comparison consists in the *truth* of $J2$; and truth is of course precisely what we are trying to explain by appealing to correspondence! Hence any attempt to explain correspondence by appealing to a comparison between a judgment and its object implicitly uses and presupposes the notion of truth.[14] For this

---

[14] Interestingly, and not entirely coincidentally, variants of this regress argument were later used by both Frege and G. E. Moore: see Frege, "Logic [1897]," pp. 128–9, and "Thoughts," pp. 352–4; and Moore, "The Nature of Judgement," pp. 6–7. Both Frege and Moore want to show that truth is a simple or unanalyzable, non-natural (i.e., irreducible) property. But for Kant truth is a logically complex or *analyzable* non-natural property whose non-naturalness consists in its being a non-empirical or transcendental concept.

reason, as Gerold Prauss points out, Kant's theory of truth does not attempt to explain correspondence in terms of comparison.[15]

If truth is the correspondence of a judgment to its object, but an appeal to comparison only circles back to the concept of truth, then in order to explain truth in terms of correspondence what we must do is find a way to talk about correspondence without invoking the notion of a comparison, or more generally without presupposing truth. In my opinion, Kant satisfies this formal adequacy condition by construing accordance or correspondence as *an isomorphic semantic projection by a judger from the propositional content of a judgment onto an actual or real object.*[16] On this Kantian view, then, a judgment accords with its object, or bears a correspondence-relation to an object, if and only if each syntactically distinct semantic constituent of the propositional content of the judgment is systematically paired *by a judger* one-to-one with a feature of an actual or real object, in the very same order in which it occurs in that propositional content.[17],[18]

---

[15] Prauss, "Zum Wahreitsproblem bei Kant," pp. 167–8.

[16] Kant often uses the notions of "actuality" (*Wirklichkeit*) and "reality" (*Realität*) interchangeably. In this sense, something is actual or real if and only if it is necessarily bound up with the empirical, spatiotemporal, dynamical or causal world (CPR A144–5/B184, A218/B265–6). Somewhat confusingly, both notions are semantically distinct from the notion of "being" (*Sein*) or "existence" (*Dasein*), which expresses a second-order concept predicated of all and only first-order concepts that are *instantiated*, i.e., have at least one individual object falling under that first-order concept, i.e., have at least one individual object subsumed under that first-order concept (CPR A599/B627). See also n. 36.

[17] Roughly speaking, all isomorphism involves one-to-one correlation and a shared ordering of correlated elements. A semantic isomorphism is discontinuous and not "smooth," however, since it is not the case that every single element of the judgment is paired one-to-one with a part of the object, and conversely. In this regard, semantic isomorphism contrasts with geometric or figural isomorphism, which is indeed continuous or point-for-point. So, assuming that the notion of "picturing" means roughly the same as "isomorphic projection," then there are two sharply different types of picturing: (i) semantic; and (ii) geometric or figural. Tractarian propositions semantically picture their corresponding states of affairs but do *not* geometrically or figurally picture their corresponding states of affairs. And I think that the failure to recognize this point has often vitiated interpretations of Wittgenstein's theory of meaning and truth, and in the process has covered over the important extent to which Wittgenstein's theory recapitulates Kant's theory of meaning and truth. It should nevertheless be noted that this is *not* to say that there is no associated mental imagery in the mind of the Tractarian or Kantian judger that geometrically or figurally pictures the proposition: indeed there most certainly *is* such imagery, but that is a sharply different sort of fact belonging to the epistemology of logic and also to the phenomenology of meaning-experience, not to the semantics of logic—see *Tractatus Logico-Philosophicus*, prop. 4.1121, p. 77; and ch. 7.

[18] The semantic isomorphism projection approach to correspondence is sometimes contrasted with a purely linguistic approach which conventionally correlates parts of sentences with parts of the world: see, paradigmatically, Austin's famous essay "Truth," pp. 117–33. Austin's essay, in turn, led to an important general debate about the concept of truth; see Pitcher (ed.), *Truth*. Pitcher says in his Introduction (pp. 9–11) that Austin's version of correspondence involves mere one-to-one correlation without any structural congruity; but that seems to me wrong. The successive stages and distinct types

So, for example, if the judgment is "This rose is red," then it bears an accordance or correspondence relation to the relevant object (i.e., this red rose) if and only if:

(1) The subject-term THIS ROSE is paired by a judger with an actual or real object O, and O indeed instantiates the property of being a rose, which in turn is type-identical with the concept of being a rose (so $O = O_{rose}$).
(2) The predicate-term RED is paired by a judger with another real attribute or property of O_{rose} which in turn is type-identical with the concept of being red.
(3) The copula IS, which binds together the subject-term and predicate-term of the judgment, is paired by a judger with the object (a state of affairs, fact, or situation) in the actual or real world which consists of O_{rose} and its being red.[19]

There are several important affinities and parallels between Kant's theory of meaning and truth, and Wittgenstein's so-called "picture" theory of meaning and truth in the *Tractatus Logico-Philosophicus*. According to Wittgenstein, something is a meaningful proposition if and only if it is a syntactically well-formed sentence or propositional sign that is isomorphically projected by the user of that propositional sign onto a possible elementary or complex "state of affairs," "fact," or "situation" (*Sachverhalt, Tatsache, Sachlage*).[20] And for a proposition to be true is simply for this possible state-of-affairs, fact, or situation to be something that also "is the case" (*der Fall ist*) or "existing" (*Bestehende*).[21] Wittgenstein's use of the term "picture" or "*Bild*" is unhappy however. For it all too easily suggests the circular accurate-depiction or "comparison" theory of correspondence; and the Tractarian proposition quoted as the first epigraph of this chapter only reinforces this very unfortunate choice of terminology. At bottom, the

---

of conventional correlation express the element of structural conformity. So Austin's theory of truth is actually a semantic isomorphism projection theory too; he has merely moved the structural component to the level of linguistic acts, thereby avoiding talk of propositions.

[19] The correlates of judgments for Kant are "objects" in a very broad sense which covers entities of many different sorts. Indeed, there are as many different sorts of objects as there are distinct categories in the Table of Categories: singular, universal, or plural (quantity); positive, negative, or privative (quality); substances, events, or disjunctive complexes (relation); and actual, possible, or necessary (modality) (CPR A80/B106).

[20] See Wittgenstein, *Tractatus*, pp. 39–49; props. 2.1 to 3.221.

[21] Ibid. pp. 31–9; props. 1 to 2.063.

comparison theory of truth is what results when we treat the proposition or propositional sign as *a phenomenal mental image, in linguistic format, of the actual fact*. But in fact this is directly contrary to what I take to be Wittgenstein's real intentions: his theory of the judging and language-using subject in the *Tractatus* is explicitly *transcendental* and not *psychologistic*.[22] What is relevant to truth and meaning for Wittgenstein is not how the propositional sign looks visually in relation to how the actual or possible fact looks or would look visually, but instead simply how the user of the propositional sign semantically projects *from* the syntactically well-formed sentence *onto* the actual fact; and this is what Wittgenstein calls the "form of representation" (*Form der Abbildung, Form der Darstellung*).

Similarly, for Kant the possibility of isomorphic first-person projections by the judger from the propositional content of the judgment onto an actual or real object is guaranteed by the conjunction of his theory of synthesis, his theory of judgment, his theory of logical or grammatical form, and his theory of objective validity or empirical meaningfulness. For, according to the conjunction of those theories, nothing will *count* as a genuine object of judgmental representation except that which systematically lines up with the intuitional and conceptual constituents of the judgment, with its underlying logico-syntactical form, and with its overall unified semantic content under the synthetic unity of apperception. This is what Kant calls the "transcendental truth" of a judgment (*CPR* A62–3/B87, A146/B185, A221–2/B269). In turn, Wittgenstein's "form of representation" is essentially a recapitulation of Kant's notion of the transcendental truth of a judgment, in the post-Kantian context of Frege's and Russell-Whitehead's "new" logic — that is, mathematical logic in the mode of the *Begriffsschrift, Basic Laws of Arithmetic*, and *Principia Mathematica*. In any case, for Kant all that needs to be added to the transcendental truth of a judgment, in order to yield truth per se, is the judging act of a particular judger, together with actuality or reality: "when I cognize the thing as it actually is, then my cognition is true" (*BL* 24: 56).

---

[22] See Wittgenstein, *Tractatus Logico-Philosophicus*, pp. 77 and 149–53; props. 4.1121 and 5.6 to 5.641. One way of formulating this contrast is to say that what is transcendental essentially has to do with a priori cognitive faculties, structures, and functions, and the necessary possibility of minds like ours; whereas what is psychologistic essentially has to do with empirical or natural cognitive contents, and actually existing human minds. That having been said, it must also be admitted that there is a genuine sense in which a transcendental approach to logic and cognition is "the higher psychologism." See Hanna, *Rationality and Logic*, chs. 1–4.

The unique contribution of the judger to truth-as-correspondence is to present the object of the judgment under one or another of the many possible specific "points of view" (*Gesichtspunkte*) that could be associated with the constituent concepts of the judgment by some rational human animal in that context (*OT* 8: 134–7) (*JL* 9: 57, 147) (*DWL* 24: 779). A Kantian point of view includes what Frege later called a "mode of presentation" or *Art des Gegebenseins*[23] — that is, a package of conceptual descriptive informational content about the object — but it is more than that, since it also indexically relates the object *to* the judger, and thereby individuates that object in relation to that very judger. A true judgment thus *fulfills a rational expectation of the judger*, who presents the world in a certain way from her own cognitive point of view.[24] So, by means of the addition of the cognitive point of view of the judger, together with the actuality or reality of the object represented, the transcendental truth of the judgment thereby passes over into its "material truth": "material truth must consist in the accordance of a cognition with just that determinate object (*demjenigen bestimmten Objekte*) to which it refers" (*JL* 9: 51).

For convenience, I will call the doctrine that truth is an isomorphic semantic projection by a judger from the propositional content of the judgment onto an actual or real object, from a particular cognitive point of view, *the Projection Theory*. We are now in a position to see how the Projection Theory, invented by Kant and rediscovered by Wittgenstein in the *Tractatus* in the wake of the invention of mathematical logic by Frege, Russell, and Whitehead, overcomes the difficulties of the traditional or naive version of the correspondence theory of truth, and satisfies Kant's formal adequacy condition. The crucial point is this: on the Projection Theory, *nothing whatsoever* intervenes between the propositional content of the true judgment, and the actual or real object. So in this sense the true judgment is not at all aptly described as a "picture" of reality whose accuracy is determined only by its being "compared" to reality. Instead it is *a user-friendly semantic map of reality*:[25] it is how an individual

---

[23] See Frege, "On Sense and Meaning."
[24] Husserl later rediscovers this Kantian idea in *Logical Investigations* VI when he claims that all truth includes "fulfilled intentions." See Husserl, *Logical Investigations*, vol. 2, Investigation VI, chs. 1–5.
[25] The distinction between pictures and maps is a sub-species of the important Kantian distinction between "images" (*Bilder*) and "schemata" (*CPR* A140–1/B179–80). Images are mental or externalized visual copies of objects; but schemata are formal spatiotemporal models of objects or manifolds of content, and need not visually resemble those objects or manifolds in any way.

rational human being cognitively and propositionally orients herself in the actual world by directly applying her judging activity to the actual things themselves.[26] Indeed, according to the Projection Theory, whenever a judgment is true, then the semantic content of the judgment is literally identical with its "truth-maker." Or, in other words, the possible object that is isomorphically mapped by the judger's projection from the propositional content of the judgment, is literally identical with an actual or real object.[27]

This notion of a "user-friendly semantic map" is, I think, a particularly apt metaphor for explicating Kant's truth-doctrine. As a partial expression of his deep interest in dynamic natural phenomena of all kinds, Kant lectured on physical geography throughout his teaching life and frequently used geographical metaphors and analogies in both his theoretical and practical philosophies. Moreover, the philosophical deployment of geographical imagery is explicitly described by Kant in the important 1786 essay "What is Orientation in Thinking?," which I have already mentioned in section 1.2. Here Kant compares and contrasts (a) perceptual orientation or orientation in space, and (b) speculative orientation or orientation in thinking (*OT* 8: 134–137). Perceptual orientation or orientation in space is finding the compass points, and then other material objects, relatively

---

[26] Just as a map can have a greater or lesser degree of resolution or fine-grainedness, so too for Kant a true judgment can be more or less exact: "Cognition is *exact* when it is adequate to its object, or when there is not the slightest error in regard to its object, and it is *rough* when there can be errors in it yet without being a hindrance to its purpose.... Whether a cognition is to be determined roughly or exactly always depends on its purpose". (*JL* 9: 54). What this means is that the phenomenon of vagueness is perfectly consistent with Kant's thesis that truth is governed by the laws of excluded middle and bivalence. In context, the judgment is always decidable, whether the degree of resolution under which the predicate's actual application/non-application is decided is itself exact or rough.

[27] My interpretation of Kant's theory of truth is quite close to one promoted by Sellars: "Now from the Kantian point of view, [these] concepts pair up in an interesting way: *judging* with *state of affairs*, and *truth* with *actuality*. Indeed, to say that they pair up is to understate the closeness of their relationships. For Kant argues, in effect, that the pairs turn out, on close examination, to be identities." See Sellars, "Some Remarks on Kant's Theory of Experience," p. 636. Essentially the same doctrine of truth was developed almost 70 years earlier by Moore, although in a Platonic-realistic context, in "The Nature of Judgement" and "Truth and Falsity," pp. 6–8, 18, and 21. See also Baldwin, "The Identity Theory of Truth." There is a subtle but important difference between the Moore/Sellars idea and Kant's, however. Moore and Sellars hold that the whole true judgment is identical with an actual state-of-affairs. Kant's idea, by contrast, is that a judgment is true if and only if its propositional content (i.e. the possible fact), as projected by the judger under a certain mode of presetation, is literally identical with an actual fact. But since a judgment is irreducibly two-sided—consisting of a "judging" (the mental act, state, or process) plus a "judged" (the propositional content)—it follows that it is not the *whole* judgment which is identical to the actual fact, but rather only its propositional content-side. I am not sure about Sellars, but certainly Moore's tendency to fuse the whole judgment to the actual fact was greatly promoted by his peculiar conception of consciousness as utterly "transparent" in relation to its intentional objects. See Moore, "The Refutation of Idealism," p. 37.

to the three dimensional axes of one's own egocentrically centered body (*DS* 2: 378–80). But in both "What is Orientation in Thinking?" and in the *Jäsche Logic* Kant says that speculative orientation or orientation in thinking is how one situates one's pure theoretical reason in relation to the empirically meaningful cognition of the understanding:

> This is what it means *to orient* oneself in *thought* or in the speculative use of reason by means of the common understanding, when one uses the *common* understanding as a test for passing judgment on the correctness of the *speculative* use. (*JL* 9: 57)

In other words, orientation in thinking is how pure speculative reason is constrained by our anthropocentric common sense. Orientation in perception and thinking both fall squarely under a more general notion of cognitive orientation, which Kant describes in the *Dohna-Wundlacken Logic* as finding a certain effectively humane and commonsensical standpoint on the actual world of real things: "to *orient oneself*, means to put oneself in a certain standpoint where one can easily consider the things *in concreto*" (*DWL* 24: 779).

This notion of cognitive orientation can be smoothly extended to the particular case of judgment. If we start with Kant's Projection Theory of truth, whereby the propositional content of a judgment is isomorphically mapped by the individual judger onto some object in the actual or real empirical world, and furthermore explicitly take the notion of truth-in-a-projection to necessarily include a judger's egocentric point of view or *Gesichtspunkt*, then we can clearly see that for Kant truth always provides an *orientation* in judging. Through repeated acts of true judgment, I gradually construct a local, regional, and then finally global map of my surrounding actual world. And these progressively larger-scale propositional maps are then systematically compiled by me, over the course of my rational life, into a *cognitive world atlas*: "The World as I Found It."[28]

To orient oneself in judging thus means effectively to locate one's own true beliefs in the actual world by constructing an egocentrically-organized systematic true theoretical representation of the world, and thereby a fixed first-personal discursive standpoint on reality as a whole. Is this an Archimedean point, or the standpoint of the absolute Cartesian ego? No. It must be emphasized that for Kant truth provides only a *relatively* fixed first-personal discursive standpoint. That is, the first-personal discursive

---

[28] See Wittgenstein, *Tractatus Logico-Philosophicus*, prop. 5.631, p. 151.

standpoint of Kantian truth is fixed relatively to the nature of our human cognitive capacities and to the conditions of the possibility of human experience. So it necessarily is a perspectival "view from somewhere," and thus it necessarily is *not* a perspective-less "view from nowhere."[29] In other words, for Kant, the concept of truth is an essentially anthropocentric concept: *it's truth with a human face.*

I have just argued that the Projection Theory is Kant's fully unpacked nominal definition of the concept of truth, which in turn is an analytically correct account of truth. But there are four crucial features of this doctrine that are somewhat hidden by this characterization.

The first feature is that despite Kant's official doctrine of strong transcendental idealism—which states that all the proper objects of our cognition are nothing but mind-dependent sensory appearances, not things-in-themselves, and that those objects necessarily conform to the a priori generative rules of our productive cognitive capacities (*CPR* Bxvi–xvii, A369)—his nominal definition of truth also contains an important dimension of realism. Dummett has argued that the proper characterization of realism, as against anti-realism, is not determined solely by whether the external world is taken to be either mind-independent or mind-dependent.[30] Anti-realism consists merely in the reduction of truth-conditions to assertibility-conditions. So realism denies that. But in addition to this, a crucial element in realism is whether, relative to certain class of judgments, what makes the judgment true is a *determinate* object. This determinateness is essentially bound up with the law of excluded middle, because if excluded middle holds, it follows that the world can be carved up into identifiable objects and their well-defined properties or relations. By contrast, a theory of truth is anti-realistic if truth is partially or wholly grounded on some non-objective feature of the act of judging (say, an intuitionist, verificationist, or pragmatic feature), for in this case both the law of excluded middle and the world of determinate objects are undermined. According to Intuitionism, for example, if a finite constructive proof of a given proposition is not possible, then neither the proposition nor its negation is true: the proposition does not tell us about a determinate object. We might think of a realism or anti-realism that

---

[29] See Nagel, *The View from Nowhere*.

[30] See, e.g., Dummett, "Realism" (1963), in *Truth and other Enigmas*; and also Dummett, "Realism" (1982) and "Realism and Anti-Realism," both in *The Seas of Language*.

is based primarily on one's views about the law of excluded middle, as a *minimal* realism or *minimal* anti-realism. Kant's nominal definition of truth then yields a form of minimal realism in Dummett's sense, since (as we have seen) Kant is also fully committed to the law of excluded middle.

The second feature is that although Kant's nominal definition of truth is realistic in the minimal or Dummettian sense, nevertheless at the same time it is otherwise *metaphysically neutral*. For no special constraints, apart from the broad requirements of logico-syntactic form, are laid down by the truth-definition itself as to what will *count* as an object. For the purposes of Kant's truth-doctrine, strong transcendental idealism can be regarded as a set of mere side-constraints. Leaving those side-constraints aside, however, the actual or real object of the truth-definition could in principle be wholly mind-dependent or wholly mind-independent. *Truth* is essentially anthropocentric, but the *truth-maker* can in principle be either accidentally or essentially non-anthropocentric. The object could be simple or complex. The object could be concrete (spatiotemporal) or abstract (non-spatiotemporal). And so on. Of course, according to Kant's own doctrine of manifest realism, an empirical judgment is true precisely because it accords with or conforms to the way the unhidden actual or real material world determinately is, as focused at or (recalling now Kant's causal-dynamic metaphysical structuralism: see section 4.2 above) *as positioned at* one of its constituent structure-dependent objects, just because its content is a user-friendly semantic map of the actual or real world. But that is not built into the nominal truth-definition. The Kantian doctrine of a true judgment and its truth-maker is in fact metaphysically anodyne, or *boringly correct*, saying only that a judgment is true if and only if things actually are the way we judge them to be. Who could seriously disagree? In this sense, Kant's nominal definition of truth can be adopted with a suppressed yawn by metaphysical idealists or anti-realists and also by metaphysical realists, by the friends of substantial individuals and of "trope-bundles," by materialists and platonists, and so on, alike.

This metaphysical neutrality is closely connected with the third feature, which is that Kant's nominal truth-definition is also *epistemologically neutral*: it explicitly refuses to pick out any particular method for justifying beliefs. This is directly entailed by Kant's distinguishing sharply between the nominal definition of truth and the criterion or test of truth. Only if you have, ready-to-hand, an explicit criterion of truth, can you attempt to justify your truth-claims.

Taken together, these three features point up an important elective (or anyhow prescient) affinity of Kant's nominal definition of truth. When Kant says that the nominal definition of truth is the accordance of a human cognition with its object, he is in effect saying in the "material mode" just what Tarski says in the "formal mode" with his famous "convention-T" or "T-schema" (in its simplest version [31]):

'S' is true if and only if S.

This point becomes obvious when we juxtapose Kant's terse or unexpanded description of truth with Tarski's informal characterization of his semantic conception of truth. Kant says: "when I cognize the thing as it actually is, then my cognition is true" (BL 24: 56). And Tarski says: "a true sentence is one which says that the state-of-affairs is so and so, and the state-of-affairs indeed is so and so." [32] Given that a sentence cannot say something unless it is used by a particular cognizing speaker in a particular context, Kant and Tarski are in effect secret sharers about the concept of truth. Moreover, there are two other significant parallels between Kant's theory of truth and Tarski's theory of truth.

First, by way of what I called his "formal adequacy condition," Kant designs his truth-theory so as to avoid the vicious truth-regress of the comparison-theory of correspondence. Similarly, Tarski lays down a formal adequacy condition to the effect that a truth-theory must rule out all semantic paradoxes, and in particular the Liar.[33] Semantic paradoxes may or may not necessarily involve explicit self-reference, but they do all involve a logically vicious circle in Russell's sense:

Thus all our contradictions [including the Liar paradox] have in common the assumption of a totality such that, if it were legitimate, it would at once be enlarged by new members defined in terms of itself. This leads us to the rule: "Whatever involves *all* of a collection must not be one of the collection"; or

---

[31] The more complicated version allows for the quoted sentence to be syntactically different (e.g. in a different language) from the disquoted one.
[32] Tarski, "The Concept of Truth in Formalized Languages," p. 155.
[33] See Tarski, "The Semantic Conception of Truth and the Foundations of Semantics."

conversely: "If, provided a certain collection had a total, it would have members definable in terms of that total, then the said collection has no total."[34]

The comparison theory of truth defines truth in terms of truth. And the Liar sentence implies that sentences about truth or falsity can independently determine their own truth-conditions. So, in respectively ruling out a truth-regress and the Liar, Kant and Tarski both implicitly hold versions of the Vicious Circle Principle.

Second, Kant's expanded or fully analyzed definition of truth, as the isomorphic semantic projection by an individual judger from the propositional content of a judgment onto an object that is causal-dynamically and structuralistically embedded in the total actual or real world, from a certain point of view—that is, under an egocentrically oriented mode-of-presentation—is surprisingly similar to Tarski's technical notion of "satisfaction." Technical details aside, truth-as-satisfaction is simply the idea that a true sentence is one which, given its particular logical form, is semantically guaranteed by a certain correlative ordering of *all* the objects in the domain. That is, the sentence not only has a model (an interpretation which makes it true) but also a corresponding open-ended class of distinct models, each of which contains the original truth-maker, that ultimately exhausts all the objects in the world.[35] This fits Kant's causal-dynamic metaphysical structuralism about the material world to a (convention-) T.

Now Tarski has three special motivations, not shared by Kant, for developing a minimally realistic, but also metaphysically and epistemically neutral, conception of truth:

- Because Tarski is an associate member of the Vienna Circle and thus a fellow traveler of logical positivism, he wants a conception of truth which is broad enough to handle all the central uses of the word "true" in everyday and scientific discourse, yet imports no heavy-duty metaphysical theorizing, and also is in principle scientifically or reductively naturalizable.

---

[34] Russell, "Mathematical Logic as Based on the Theory of Types," p. 63.

[35] The Loewenheim-Skolem theorem says that any sentence satisfied by a denumerably infinite model is satisfied by every finite model which is a subset of it, and also by every non-denumerably infinite model which contains the denumerably infinite model as a subset. What this means is that given the fact of satisfaction alone, it will not be possible to know precisely *which* model is the sentence's *unique* model. But in a very real sense this is only to remake, in the formal mode, Kant's material-mode point that the *criterion* of truth cannot be read directly off the *definition* of truth.

- Because Tarski is a post-Gödelian logicist—that is, a logicist who must accommodate Gödel's famous proof that some true (valid) sentences of classical-logic-plus-the-Peano-axioms-for-arithmetic are unprovable (and undecidable)—he also wants a conception of truth which sharply detaches it from provability (and from decidability).
- Because Tarski is a formal semanticist, he wants to develop a materially adequate recursive truth-definition for an entire language, that is, a truth-definition which both fits the informal gloss of satisfaction mentioned in the just-previous full paragraph and also recursively constructs truth-conditions for every (indicative) sentence in the language, considered as instances of the T-schema.

Yet, despite these differences in philosophical motivation, both Kant and Tarski isolate a root conception of truth which is rationally unexceptionable; both isolate a conception of truth which carries at least the general outline of realism; and both isolate a conception of truth which sharply disjoins it from the quite different issues of the metaphysics of truth and the epistemology of truth.

The fourth feature of Kant's nominal definition of truth goes significantly beyond Tarski's semantic conception, however. According to Kant, the concept of truth is by no means an ordinary or first-order concept. So far, of course, Tarski would agree because he thinks that the truth-predicate for a logically consistent or "closed" language always belongs one level higher up in the hierarchy of languages and meta-languages. But Kant also holds that:

(1) The concept of truth is a "logical" (second-order, syncategorematic) predicate and *not* a "determining" or "real" (first-order, categorematic) predicate.[36]

(2) The concept of truth is a "pure concept of the understanding" or a "category," not an empirical concept.

---

[36] Determining or real predicates are categorematic or independently meaningful in that they have both an "intensional content" or *Inhalt* and a "comprehension or *Umfang* (JL 9: 95–6); but logical predicates are syncategorematic or non-independently meaningful, and second order, in that they are essentially operators *on* determining predicates and thereby essentially apply *to* determining predicates. To take the best-known example, according to Kant the unschematized logical predicate EXISTS is satisfied if and only if some determining or first-order predicate is instantiated in a domain of individual objects, and the schematized logical predicate EXISTS is satisfied if and only if some determining or first-order predicate is instantiated in the domain of real or actual individual objects (CPR A597–602/625–30).

The second-order, syncategorematic character of the concept of truth is captured in Kant's Table of Judgments by the "assertoric" modality of judgments: "the assertoric proposition deals with logical actuality or truth (*logischer Wirklichkeit oder Wahrheit*)" (CPR A75/B101). And the pure conceptual character of the concept of truth is captured in the Table of Categories by the corresponding category of "reality" (*Realität*) (CPR A80/B106). Since, according to Kant, pure concepts are dedicated procedural rules of our innate cognitive capacity for thinking and judging, it follows that the Kantian nominal definition of truth is built innately into our overall rational human capacity for cognition.

Taken together, these features provide a neat Kantian explanation for why the predicate IS TRUE is such a philosophically *peculiar* predicate. For the truth-predicate is simply *not* like IS RED or IS EXPENSIVE: it is *second-order, syncategorematic, and transcendental*. It does not apply directly to things, it is not independently significant, and it does not just accidentally apply to judgments: it expresses a type-theoretic, non-independent, a priori, innate condition for the possibility of judgment itself.

5.2. The Criterion of Truth

Let's now turn to the question of the criterion of truth. Here, however, we must immediately face up to the interpretive problem mentioned earlier. According to Kant's own words there is no such thing as a "general criterion of truth":

Now a general criterion (*Kriterium*) of truth would be that which was valid of all cognitions without any distinction among their objects. But it is clear that since with such a criterion one abstracts from all intensional content (*Inhalt*) of cognition (reference to its object), yet truth concerns precisely this content, it would be completely impossible and absurd to ask for a mark of the truth of this content of cognition, and thus it is clear that a sufficient and yet at the same time universal criterion (*Kennzeichen*) of truth cannot possibly be provided. (CPR A59/B83)

A universal material criterion of truth is not possible; it is even self-contradictory. For as a *universal* criterion, valid for all objects in general, it would have to abstract fully from all differences among objects, and yet at the same time, as a material criterion, it would have to deal with just this difference, in order to be able to determine whether a cognition agrees with just that object to which it refers and not with any object in general, in which case nothing would really be said. ... Hence it would be absurd to demand a universal material criterion of truth, which should abstract and at the same time not abstract from differences among objects. (*JL* 9: 50–1)

Kant's point is easily misinterpreted. He is not saying that the concept of truth has no universal or necessary features. Nor, more importantly, is he saying that there is no such thing as an effective test or criterion for truth, that is, no such thing as a real definition of truth. What he *is* saying, however, is that there is no such thing as *the* criterion of truth: that there is *no single or univocal universal criterion of truth*. And this is because no single or univocal criterion could be at once sufficiently general and yet sufficiently specific to take account a priori of all the different sorts of objects of cognition. Kant is thereby implicitly specifically rejecting the monolithic reductive truth-theory of the Cartesians to the effect that any and every proposition is true if and only if it is grasped by means of a clear and distinct intuition; and he is also implicitly specifically rejecting the monolithic reductive truth-theory of the Leibnizians to the effect that any and every proposition is true if and only if it is ultimately reducible to an instance of the principle of identity. But by rejecting the possibility of a monolithic reductive truth-theory, Kant is of course leaving open the possibility of a *non*-monolithic or pluralist and *non*-reductive truth-theory.

In this regard, and perhaps not surprisingly in view of the elective affinities between Kant and Tarski, there are some important similarities between the negative side of Kant's theory of truth, and the later Davidson's reflections on Tarski's theory of truth.[37] Kant as we have seen argues that truth has an unexceptionable non-vacuous nominal definition, and an explicit minimalistic analysis in terms of the Projection Theory. Similarly, Davidson agrees with Tarski that truth has both an unexceptionable non-vacuous characterization, namely convention-T, and also a formal analysis in terms of satisfaction. But the later Davidson also thinks that it is pointless to try to define truth in the strong sense such that for each and every possible (indicative) sentence S in a given language L there is some way of systematically specifying just *which* objects will satisfy S, or in just *which* ways S will be satisfied by those objects. For there is just no way of anticipating

---

[37] See Davidson, "The Structure and Content of Truth"; and Davidson, "The Folly of Trying to Define Truth."

a priori what will count as the relevant objects or the relevant ways. This, I think, almost perfectly matches Kant's worries about the possibility of a single or univocal universal real definition or criterion of truth.

But at this point, just as Kant goes beyond Tarski, so too he goes beyond Davidson. For it is perfectly consistent with Kant's negative thesis about the criterion of truth that there still be *several* distinct effective tests, criteria, or real definitions for truth, correlating to the several distinct basic types of judgment there can be. For each criterion of truth, and for each basic type of judgment, there is a correspondingly different basic way of according with the object—a different basic way for a judger to project isomorphically onto the actual world from a point of view, that is, under an egocentrically oriented mode-of-presentation of that world.[38] That Kant in fact holds this view is clear from this important remark in the Dohna-Wundlacken Logic:

> A material criterion of truth cannot possibly be universal and hold for all objects. <u>One can of course have material criteria of truth but they cannot be universal.</u> (*DWL* 24: 719, underlining added)

5.3. The Criteria of Truth

Here is a brief sketch of Kant's doctrine of the several criteria of truth. We can instructively begin with his account of the principle of non-contradiction (PNC). The PNC not only functions as a negative condition for all truth, but also as a positive or determining principle for all analytic propositions:

> We must allow the **principle of contradiction** to count as the universal and completely sufficient **principle of all analytic cognition;** but its authority and usefulness does not extend beyond this, as a sufficient criterion of truth. (*CPR* A151/B190-1)

---

[38] In "The Trouble with Truth in Kant's Theory of Meaning," p. 18, n.3 and p. 19, n.19, I mistakenly claimed that Kant's theory as accordance (agreement, correspondence) could not be generalized from empirical judgments to a priori judgments. That mistake was, I think, based on my false assumption that to every true judgment there corresponds a merely empirical object. But that is not in fact Kant's view: his view is instead that every true judgment agrees with its corresponding *actual* or *real* object—an object which may or not itself be merely concrete or contingent. For the object might be abstract (e.g. numbers, concepts, pure forms of intuition, etc.); and also it might be necessary in the sense that it exists not only in the actual or real world but in every possible world as well.

Therefore, a judgment is analytically true if and only if its denial entails a contradiction; and this criterion of truth is to be restricted to all and only the members of this special class of judgments. A fuller discussion of the ins-and-outs of Kant's doctrine of analyticity can be found in chapters 3 and 5 of *Kant and the Foundations of Analytic Philosophy*. The basic idea, however, is that a judgment is analytically true if and only if it is isomorphically projected by a judger from a point of view, not only onto the actual world but also onto every logically possible world, *by virtue of intrinsic intensional connections between the concepts constituting the propositional content of the judgment*. Projection onto possible worlds from a point of view, in turn, is the same as defining necessary truth in terms of *cognitive accessibility from the actual world*. A proposition P is analytically necessarily true if and only if P is true in every logically possible world—which in turn for Kant is nothing but a maximally consistent set of different conceivable ways the actual world of the judger might have been. By contrast, a proposition P is synthetically necessarily true if and only if P is true in all and only every world of possible human experience—which in turn for Kant is nothing but a maximally consistent set of different humanly experienceable ways the actual world of the judger might have been.[39]

What is most salient for us here, in any case, is simply that Kant unambiguously states that the PNC is a secure, universal, effective, and restricted test of analytical truth. So Kant holds that there *are* some genuine criteria of truth; and the fact that the PNC is indeed a genuine criterion of truth depends entirely on restricting the scope of its application to analytic propositions *only*. From this we can derive the general lesson that the possibility of a genuine criterion of truth consists in tying it to a specially delimited class or domain of judgments.

This lesson can then be directly applied to the other basic types of judgment: synthetic a priori, and synthetic a posteriori. And here we find that Kant again unambiguously states that each class has its own effective test or criterion of truth, and that each criterion is possible only because the scope of its application is restricted to all and only judgments of the relevant type.

In the first *Critique*'s System of the Principles of Pure Understanding, in the section entitled "The Highest Principle of all Synthetic Judgments,"

---

[39] See Hanna, *Kant and the Foundations of Analytic Philosophy*, sections 5.1–5.3.

Kant describes the criterion for the truth of synthetic a priori judgments as follows:

[S]ince experience, as empirical synthesis, is in its possiblity the only kind of cognition that gives all other synthesis reality, as *a priori* cognition it also possess truth (accordance with the object) only insofar as it contains nothing more than what is necessary for the synthetic unity of experience in general. The supreme principle of all synthetic judgments is, therefore: Every object stands under the necessary conditions of the synthetic unity of the manifold of intuition in a possible experience. (*CPR* A157–8/B196–7)

We will remember that the criterion of truth for analytic propositions invokes conceptual features of the judgment only. Here, by contrast, the criterion also explicitly invokes (pure or formal) intuition and the transcendental imagination, as well as the pure concepts of the understanding or categories, and transcendental apperception.[40] Kant is saying that a proposition *P* is synthetic a priori true if and only if *P* is isomorphically projected by a judger from a point of view not only onto the actual world, but also onto every possible world of human experience, *because* the transcendental conditions of human experience—including pure or formal intuition, the transcendental imagination, the categories, and the original synthetic unity of apperception—collectively carve out precisely that class of worlds (*CPR* B159–61, A155/B194).[41]

Kant also holds that there is a real definition or criterion of synthetic a posteriori truth, or empirical truth. This is evident from his referring to what he calls "the formal conditions of empirical truth" (*CPR*: A191/B236),

---

[40] Indeed, Kant elsewhere explicitly defines a synthetic a priori proposition as one that is true by virtue of its conceptual content together with pure intuition (*CPR* B73) (*PC* 11: 38). Here, as always, it is important to distinguish between (i) pure or formal intuition, which includes an a priori apperceptive component and is only moderately non-conceptual, and therefore intrinsically involves concepts and the categories even though it is not fully determined by them; and (ii) the forms of intuition, which are very strongly non-conceptual, and do not necessarily imply the categories. See sections 2.3 and 6.2–6.3.

[41] For simplicity's sake, I am here overlooking a controversial issue about Kant's account of synthetic a priori propositions: i.e., whether there is only one type of synthetic a priori proposition or more than one. It is certainly arguable that Kant distinguishes between the synthetic a priori truths of mathematics and of the metaphysics of experience on the one hand, and the synthetic a priori truths expressing causal laws of nature on the other. See Friedman, "Causal Laws and the Foundations of Natural Science," pp. 161–9; and section 3.5. Precisely how this distinction should be characterized, however, is a difficult task. But for some further discussion, see Hanna, *Kant and the Foundations of Analytic Philosophy*, pp. 260–1, where I formulate a distinction between "strongly" and "weakly" synthetic a priori truths, and also section 3.5, where I characterize the special "dynamical" or "material" necessity of natural causal laws, and correspondingly of propositions about natural causal laws.

"the criterion of empirical truth" (*das Merkmal empirischer Wahrheit*) (*CPR* A451/B479), and the "sufficient criterion of empirical truth" (*zureichendes Merkmal empirischer Wahrheit*) (*CPR* A651/B679). He spells out his doctrine of the criterion of empirical truth in these texts:

Now which given intuitions actually correspond (*korrespondieren*), to outer objects, which therefore belong to outer **sense**, to which they are to be ascribed rather than to the imagination—that must be decided in each particular case according to the rules through which experience in general (even inner experience) is to be distinguished from imagination. (*CPR* Bxli n.)

Since the accordance of cognition with the object is truth, only the formal conditions of empirical truth can be inquired after here, and appearance, in contradistinction to the representations of apprehension, can thereby be represented only as the the object that is distinct from them if it stands under a rule that distinguishes it from every other apprehension, and makes one way of combining the manifold necessary. That in the appearance which contains the condition of this necessary rule of apprehension is the object. (*CPR* A191/B236)

The coherence (*Zusammenhang*) of appearances necessarily determining one another according to universal laws, which one calls nature, and with it the criterion of empirical truth ... [is what] distinguishes experience from dreaming. (*CPR* A451/B479)

The difference between truth and dreaming is not ascertained by the nature of the representations which are referred to objects (for they are the same in both cases), but by their connection according to those rules which determine the coherence (*Zusammenhang*) of the representations in the concept of an object, and by ascertaining whether they can subsist together in an experience or not (*P* 4: 290)

Criterion of empirical truth. Ordering (*Ordnung*) of nature or the ordering in itself, i.e. combination (*Verbindung*) according to rules, proves the reference to an object and not merely something arbitrary. (*R* 5563; 18: 234)

Thus Kant's criterion of empirical truth is based on a direct appeal to the synthesizing operations of mind underlying a synthetic a posteriori judgment. If the mind applies a necessary conceptual rule to a manifold of actual or real perceptions via judgmental synthesis, in such a way that this particular rule tightly coheres with the universal synthetic a priori rules for the organization of appearances, then the judgment is true; otherwise it is false.[42] Again, a proposition *P* is synthetic a posteriori true if and only if *P* is isomorphically projected by a judger from a point of view onto the actual or real world, but not onto every possible experienceable world, *because* it expresses a necessary rule effectively applied to some actual or real perceptions in time, which in turn is directly subsumable under the principles of pure understanding.[43]

The upshot is that Kant holds that there are three distinct criteria of truth, correlating with the three distinct types of proposition. The criterion of analytic truth is that the proposition's denial either is or entails a conceptual contradiction; the criterion of synthetic a priori truth is the proposition's strict dependence on the pure intuitional conditions for the possibility of experience; and the criterion of empirical truth is coherent rule-application to the law-governed manifold of actual perceptions.

What is most important for our present purposes, however, is not so much the several criteria themselves, but rather the fact that each truth-criterion correlates directly with a distinct primitive cognitive faculty or *Vermögen*. As we saw in section 0.2, a cognitive faculty is an innate, spontaneous human cognitive capacity or *Fähigkeit*; and each such faculty operates by means of synthesis or mental processing as applied to raw cognitive data, the capacity-triggering informational inputs from the external world (*CPR* A50–1/B74–5). And, according to Kant there are at least five primitive or underived cognitive faculties: (1) sensibility; (2) understanding; (3) imagination; (4) reason; and (5) apperception.[44] Now analytic truth is conceptual

---

[42] This particular criterion of truth shares some but not all features with the standard or traditional coherence theory of truth. See Hanna, "The Trouble with Truth in Kant's Theory of Meaning," pp. 10–15. Similarly, Kant's nominal definition of truth shares some but not all features with the standard or traditional (i.e., comparison-theoretic) correspondence theory of truth. Therefore Kant's theory of truth falls somewhere *between* the all-too-familiar traditional coherence and correspondence doctrines. Kemp Smith seems to think that this indicates a deep confusion on Kant's part: see Kemp Smith, *Commentary to Kant's "Critique of Pure Reason,"* p. 36. On the contrary however, I think that it is a definite philosophical advantage of Kant's theory that it cannot be neatly slotted into one of the standard pigeon-holes.

[43] See n. 39. As the title of my paper indicates, I think that there are some serious problems with this particular criterion of truth. But this does not affect the point I am making right now, which is simply that Kant does indeed explicitly formulate a criterion of empirical truth.

[44] These five faculties operate in "bottom-up" coordination in order to constitute a non-primitive or derived cognitive faculty, the power of judgment (*Urteilskraft*), which is the *central* cognitive faculty of the human mind. Conversely, the operations of all the basic and non-basic cognitive faculties are executively coordinated "top-down" by apperception. See Hanna, "Kant's Theory of Judgment," section 1.1.

truth and therefore connects most directly with our primitive spontaneous innate capacity for conceptualization and having thoughts—the faculty of understanding and its *synthesis intellectualis* (*CPR* B151). Synthetic a priori truth is based on the possibility of human sensory experience and therefore connects most directly with our primitive spontaneous innate capacity for pure intuition—the faculty of sensibility and its "synthesis of apprehension" (*CPR* A99–100). And synthetic a posteriori truth is based on the application of conceptual and judgmental rules to particular actual objects of perception, and therefore connects most directly with our primitive spontaneous innate capacity for interpreting general conceptual rules in terms of more specific figural spatiotemporal forms and sensory images—the faculty of imagination and its *synthesis speciosa* (*CPR* B151). So Kant's doctrine of the criteria of truth entails a pluralist or multi-grade doctrine of truth that is grounded directly on his faculty-based cognitive psychology (*CPR* B150–2).

But at the same time it is a *constrained pluralism*. There are precisely three types of truth, correlating one-to-one with the three basic semantic types of propositions, and then in turn with three of the primitive cognitive faculties making up our total capacity for cognition. Each of the three types of truth is a type of (non-traditional) correspondence or accordance. Moreover, since Kant is committed to the law of excluded middle or bivalence, each of the three types of truth is also a type of classical truth. In this way, Kant shows us that one can be a pluralist about truth without becoming a truth-relativist or unconstrained pluralist, without giving up the rational core of the correspondence theory, and without adopting non-classical truth-values.

5.4. The Sense of Truth: Truth and Truthfulness

By way of concluding this chapter, I want to point up some broader and deeper implications of Kant's doctrine of truth. I will do this by extracting five "principles of truth" from Kant's theory. Each of these principles expresses an important Kantian thesis about the concept of truth. Furthermore the conjunction of the five theses, it seems to me, offers us *a generative analysis of truth,* that is, a simple program for constructing particular theories of truth that could be adopted by many philosophers who are otherwise non-Kantian or even anti-Kantian in their philosophical orientations. In any case, here are the five Kantian truth-principles:

- (T1) Truth is predicated first and foremost of rational human judgments, and every judgment has the dual aspect of (i) the "judging," or assertoric rational performance, and of (ii) the "judged," or objective propositional semantic content of that performance.
- (T2) The truth of a judgment implies a corresponding truth-maker.
- (T3) Something is a truth-maker for a judgment if and only if it is (a) isomorphic with the semantic structure of the propositional content of the judgment from the judger's point of view, and (b) also actual or real.
- (T4) There are several distinct criteria of truth, each of which specifies a distinct way in which judgments can be determinately applied to actual or real objects.
- (T5) The several criteria of truth are each criteria of classical truth that correlate systematically with the several different basic semantic types of judgment and in turn with the several different basic human faculties for cognition.

(T1) yields an anthropocentrism about truth, but also ties it directly to human rational action and to objective semantic content. (T2) and (T3) together yield a broadly Tractarian, Tarskian, and Dummettian, or metaphysically and epistemically neutral and yet minimally realistic, theory of truth. (T4) yields a pluralism about truth. (T5), however, sharply constrains this pluralism by tying it to classical truth, to a systematic semantics of judgment, and to a systematic theory of human cognition.

This Kantian program for theories of truth, it seems to me, teaches us some crucial philosophical lessons. First, the concept of truth should be initially detached both from metaphysics and from epistemology: truth is first and foremost a *semantic* conception. Second, this cognitive-semantic conception of truth can incorporate *the pure form of realism* as a transcendental feature while avoiding the non-minimalist forms of metaphysical realism, and also metaphysical idealism, alike. Third, the semantic conception of truth can, if desired, be reunited with both metaphysics and epistemology through a supplementary doctrine of criteria for truth; but this supplementary doctrine must be *independently grounded and argued-for*. And, fourth and finally, both the semantic conception of truth and the theory of truth-criteria should be anchored in an overarching theory of *human cognition*. Truth is, intrinsically, a *cognitive*-semantic conception.

Nevertheless, Kant's conception of truth does not terminate in an appeal to human cognition. According to him, the capacity for cognition is only the first of two first-order constitutive psychological capacities of the nature of a human person, the other being the capacity for conative volition or effective desire, sometimes also called the "power of choice" or *Willkür*. Both the faculties of cognition and effective desire are subsumed under an overarching and unified capacity for theoretical and practical reason (*GMM* 4: 391).[45] Now an essential feature of Kant's unified conception of reason is what in the Introduction, following Onora O'Neill, Susan Neiman, and Nicholas Rescher,[46] I called "the priority of practical reason over theoretical reason." As we also saw in the Introduction and as every reader of the first *Critique* knows, Kant finds it necessary to constrain the scope of the theoretical rather strictly:

I have therefore found it necessary to deny scientific knowing (*Wissen*), in order to make room for **belief** (*Glauben*). (*CPR* Bxxx; see also A828/B856)

[E]ven after [theoretical] reason has failed in all its ambitious attempts to pass beyond the limits of experience, there is still enough left to satisfy us, so far as our practical standpoint is concerned. (*CPR* A828/B856)

Kant limits the theoretical not just because unrestricted scientific knowledge-claims can lead to paradox and metaphysical nonsense, but also because unless the claims of scientific knowledge, and especially those of noumenal scientific realism and Newtonian mechanistic fundamental physics, are strictly critically bounded, no place in logical space would remain for practical human agency and morality. Without that strict critical boundary in place, then the world of valueless natural facts and mechanistic deterministic causation would smother the world of human values, rational agency, and freedom. So we would have committed not only cognitive but also ethical suicide. We must therefore admit, says Kant, that practical concepts have intrinsically broader scope than theoretical concepts.

But there is another way in which the practical has priority over the theoretical. The very *sense* or *point* of the concept of truth ultimately derives

---

[45] See Neiman, *The Unity of Reason*; and O'Neill, "Vindicating Reason."
[46] See O'Neill, *Constructions of Reason*, chs. 1–4; Neiman, *The Unity of Reason*, ch. 3; and Rescher, *Kant and the Reach of Reason*, chs. 5–9.

from the special role it plays in human practical agency and morality.⁴⁷ This idea is implicit in what Kant has to say about truth in his *Lectures on Ethics*:

Value of the love of truth: It is the basis of all virtue; the first law of nature, Be truthful!, is a ground (1) of virtue towards others, for if all are truthful, a man's untruth would be exposed as a *disgrace*. (2) of virtue to oneself, for a man cannot hide from himself. (*LE* 27: 60)

In human social life, the principal object is to communicate our attitudes, and hence it is of the first importance that everyone be truthful in respect of his thoughts, since without that, social intercourse ceases to be of any value. (*LE* 27: 444)

The all-important moral concepts of sincerity and insincerity, and veracity and mendacity, include the concept of truth. Sincerity implies a respect for the truth, and insincerity implies a failure to respect the truth. Moreover, and most importantly, however, truthfulness is in fact the enabling presupposition of all truth. It is not the case that, strictly speaking, I must always be sincere in order to state a truth: for I could intend to tell a lie or to dissemble, and yet accidentally say what is true. Nor is it the case that my judgments strictly speaking have to be true in order to treat myself or others non-exploitatively or respectfully, and to satisfy the other stringent demands of the Categorical Imperative: for the demands of morality can in principle be met even if I judge falsely.⁴⁸ It is rather that we must be *truthful*, that is, *non-cognitively aimed* at truth (so this is truth-conation, or "the love of truth"), because this is a necessary condition of all the other intellectual and moral virtues that together constitute an ideally successful rational human life. *Rational human animals are truthful animals*. No human animal could at once be rational and yet also have an innate cognitive and volitional disposition to lie or dissemble,⁴⁹ not even *a race of liars*, who

---

⁴⁷ The same general thesis has also been developed in a non-Kantian—more specifically, Humean and Nietzschean—framework by Williams in *Truth and Truthfulness*.

⁴⁸ To modify a famous fictional example slightly, a southern sheriff might falsely believe that by his refusing to turn over a black prisoner to a lynch mob, a violent and terrible riot will result in the deaths of many innocent black people—for, unbeknownst to the sheriff, the National Guard is just around the corner. Still, his refusal is (arguably, from a Kantian point of view) the right thing for him to do.

⁴⁹ This does not of course mean that rational human animals never lie or dissemble, or endorse deceitfulness—we do!—but instead only that rational human animals *do not always lie*, and *always care deeply about telling the truth*. This in turn raises tricky questions about the status of the moral duty to truthfulness in Kantian ethics: are we morally required by the Categorical Imperative *never to lie or dissemble period*, or are we instead morally required by it only *always to take truthfulness into deeply*

would have to assume others' minimal commitment to truthfulness in order to make their lies successful. So Kant is telling us that truthfulness, not the truth, shall make you free.

Correspondingly, truth is not itself a practical concept, but rather (as we have seen) a cognitive-semantic concept. True judgments are user-friendly cognitive-semantic maps of the actual or real world: maps that provide self-conscious subjects with discursive orientation in that world by virtue of their egocentric isomorphic projections from the propositional meanings of the judgments onto the actual or real facts. And the very idea of a true judgment is of course built into the very idea of a science in general and of an exact science in particular, as a body of true judgments. Yet only practical concepts such as truthfulness can give the concept of truth a real grip or purchase on us. This is because only practical concepts such as truthfulness can guarantee that truth is after all a "thick," normative, or value-laden concept, and not merely a "thin," descriptive, or value-neutral concept. Truth is often useful, and it also sometimes matters for its own sake. But truthfulness is a central constitutent of our practical and theoretical rationality alike. The sense or point of being true to the facts is that it cognitively orients us in a world of values, a world in which we must ultimately act freely and accept causal and moral responsibility for our actions, and this in turn is impossible without sincerity or truthfulness. For this reason alone, even if we could find no others, the concept of truth cannot be eliminable.[50] To eliminate the concept of truth would be to eliminate all the rational human animals: it would be to eliminate *us*.

---

serious consideration, even when it unluckily turns out in some particular context (as, e.g., in the famous Nazi-at-the-door example) to be my duty to choose the lesser of several evils and lie or dissemble in that context? See Hill, *Human Welfare and Moral Worth*, ch. 12; and Williams, *Truth and Truthfulness*, p. 122.

[50] Eliminativists about truth are radical deflationists who hold: (1) that to say that *P* is true is merely to assert *P*; (2) that the concept of truth is therefore vacuous; and (3) that apparently real facts about truth are fully reducible to facts about something else, and therefore are eliminable. See Ramsey, "Facts and Propositions"; Field, "The Deflationary Conception of Truth"; and Kalderon, "The Transparency of Truth." If the concept of truth is required by all Kantian and more generally deontological conceptions of morality, and if deontology is the essence of morality, then contrapositively (and oddly enough) eliminativist theories of truth constitute a serious threat to morality. Pilate was no fool.

6

Mathematics for Humans

I teach at Harvard that the world and the heavens, and the stars are all *real*, but not so *damned* real.

Josiah Royce [1]

In Kant we find an old form of intuitionism, now almost completely abandoned, in which space and time are taken to be forms of conception inherent in human reason.

L. E. J. Brouwer [2]

The concept *natural number* cannot itself be categorically characterized in pure logic. We can only say that the natural numbers are those which come in the sequence 1, 2, 3, …. We do have an intuition of this sequence. Perhaps, as Kant supposed, it is connected to the intuition of succession in time.

Ian Hacking [3]

6.0. Introduction

According to Kant, mathematics is the pure formal science of quantity or magnitude, and the very paradigm of an exact science (*CPR* Bx, A712–3/B740–1). In turn, quantities or magnitudes are of two fundamentally different kinds: numerical and spatial. Arithmetic is the pure science of numbers, and geometry is the pure science of space. Whether arithmetic or geometry, however, mathematics for Kant is synthetic a priori, not analytic a priori, which is to say that it is a substantive or world-dependent science, not a

---

[1] Royce, *The Letters of Josiah Royce*, p. 217.
[2] Brouwer, "Intuitionism and Formalism," p. 67.
[3] Hacking, "What is Logic?," p. 316.

purely logical science. But how can mathematics be at once pure a priori (that is, devoid of all empirical content, experience-independent, and necessary) and also substantive or world-dependent? As Brouwer correctly observes, for Kant mathematics is possible because it presupposes the innate human cognitive capacity for pure temporal and spatial representation, the innate human cognitive capacity for *pure or formal intuition* (CPR A38–9/B55–6, B160–1n.) (P4: 280–3). In turn, as the Transcendental Aesthetic shows, our pure or formal intuitions of time and space are at once (a) the non-empirical necessary subjective forms of inner and outer human sensibility, and also (b) the representations of space and time themselves as unique abstract relational totalities or singular infinite given wholes. In this chapter, I revisit Kant's much-criticized views on mathematics in general and arithmetic in particular. In so doing, I make a case for the claim that Kant's theory of arithmetic is *not* subject to the most familiar and forceful objection against it, namely, that his doctrine of the dependence of arithmetic on time is plainly false, or even worse, simply unintelligible. On the contrary, Kant's doctrine about time and arithmetic is highly original, fully intelligible, and with qualifications due to the inherent limitations of his conceptions of arithmetic and logic, to an important extent defensible. But what, from the standpoint of the Two Images Problem, is most philosophically striking about Kant's doctrine is the fact that arithmetic turns out to be a paradigm of *the exact sciences* (*exacten Naturwissenschaften*) only by virtue of its ultimately being one of *the human or moral sciences* (*Geisteswissenschaften*).[4]

This striking fact does *not* however mean that for Kant arithmetic can be "psychologized," or that arithmetic is somehow "socially constructed." Nevertheless, according to him, at the same time arithmetic is also *non*-Platonic: the natural numbers are not abstract things-in-themselves. So what *are* the numbers? For Kant, as we shall see, the natural numbers are nothing but players of roles in an empirically applicable and humanly graspable a priori time-structure under special logical-conceptual constraints, which

---

[4] The term "human sciences" derives from the work of Wilhelm Dilthey, and the term "moral sciences" derives from the work of J. S. Mill. Both of these nineteenth-century thinkers sharply contrast the human or moral sciences with the exact sciences. Social constructivist philosophers of science in the twentieth century—e.g., post-positivists like David Bloor, Bruno Latour, and Paul Feyerabend—go one step further by reducing fundamental physics to the human or moral sciences in Dilthey's and Mill's senses. And the later Wittgenstein's philosophy of mathematics comes very close to reducing pure mathematics to the human or moral sciences in the same way. By contrast, Kant wants to say that the exact sciences, *whose propositions are irreducibly necessary, a priori, and objective*, are such precisely because they are *also* fundamentally realistic, anthropocentric, and categorically normative.

is delivered to us solely by means of pure or formal intuition. Or, in other words, for Kant arithmetic is *an a priori cognitive science with categorically normative foundations*, just like pure general logic (*CPR* A52–5/B76–9) (*JL* 9: 11–16).[5] This thesis, in turn, is essentially bound up with the nature of Kant's transcendental idealism. To twist Royce's nice line slightly, the natural numbers are all *real*, but not so *damned* real.

My account has four stages. In the first stage, I reconstruct Kant's general argument for the possibility of mathematics and reconsider the traditional "neglected alternative" objection against the Three Alternatives Argument for the transcendental ideality of space and time (section 6.1). Here I draw a fundamental distinction between (1) strong transcendental idealism and (2) weak transcendental idealism, and offer some reasons why, when considering Kant's writings critically and historico-philosophically and not merely historically, we should charitably ascribe to him only *weak* transcendental idealism and not strong. In the second stage, I reconstruct Kant's argument for the synthetic apriority of arithmetic (section 6.2). In the third stage, I develop a new account of his notorious doctrine of the dependence of arithmetic on time (section 6.3). And finally, in the fourth stage, I develop a correspondingly new account of the Kantian notion of arithmetical construction (section 6.4).

6.1. Mathematics and Transcendental Idealism

Brevity may be the soul of wit, but frequent repetition is the piper's price of clarity. I have said (see sections 0.2 and 1.0) that Kant's transcendental idealism is the conjunction of the following two theses:

(1) that all the representational contents of cognition are strictly determined in their underlying forms or structures by the "synthesizing," or generative-and-productive, activities of the a priori cognitive faculties of the human mind, insofar as those faculties are applied to original perceptual inputs from the world (the transcendentalism thesis) (*CPR* Bxvi, A11/B25) (*P* 4: 373 n.),

and

(2) that all the proper objects of human cognition are nothing but sensory appearances or phenomena, and not things-in-themselves or noumena (the idealism thesis) (*CPR* A369) (*P* 4: 293).

---

[5] See also Hanna, *Rationality and Logic*, esp. chs. 1 and 7.

Correspondingly, I have also said that these two theses in turn jointly imply that for Kant empirical objects are token-identical with the contents of sensory representations (*PC* 11: 314) and type-identical with the a priori forms or structures that are innately specified in our cognitive faculties (*CPR* Bxvii). In addition, Kant holds that both theses (1) and (2) are directly entailed by this thesis:

(3) that space and time are neither things-in-themselves nor ontologically dependent on things-in-themselves, but instead are nothing but a priori necessary subjective forms of all empirical intuitions of appearances (*CPR* A26–8/B42–4, A32–6/B49–53) (*P* 4: 286–88).

The explicit conjunction of (1), (2), and (3), together with an implicit assumption I called *the intrinsicness of space and time*, which says:

(4) that spatiotemporal properties are intrinsic structural properties of all appearances (*CPR* B66–7),

then yields the general Kantian thesis of *strong transcendental idealism*:

Human beings can cognize and know only either sensory appearances or the forms or structures of those appearances—such that sensory appearances are token-identical with the contents of our objective sensory cognitions, and such that the essential forms and structures of the appearances are type-identical with the representational forms or structures generated by our own cognitive faculties, especially the intuitional representations of space and time—and therefore we can neither cognize, nor scientifically know,[6] nor even meaningfully assert or deny, anything about things-in-themselves (*CPR* A369, B310–11).

Kant's theory of mathematics is necessarily bound up with his transcendental idealism, via its intimate entanglement with his theory of space and time. As I pointed out in section 2.3, Kant says in the first *Critique* that "we have already traced the concepts of space and time to their sources by means of a transcendental deduction, and explained and determined their a priori objective validity" (*CPR* A87/B119–20), and then later in the *Prolegomena* he says that there is a "transcendental deduction of the concepts of space and time" which "explains also at the same time the possibility of pure mathematics" (*P* 4: 285). I also pointed out in section 2.3 that, according to my

---

[6] Kant distinguishes quite sharply between "cognition" (*Erkenntnis*) and "scientific knowing" (*Wissen*). See Hanna, *Kant and the Foundations of Analytic Philosophy*, pp. 18 and 30; and ch. 7 below.

cognitive-semantic approach to Kant's transcendental idealism, I take a transcendental deduction to be a demonstration of the objective validity—the empirical meaningfulness or cognitive significance—of an a priori representation R (whether R is an a priori concept, an a priori intuition, an a priori necessary proposition, or a systematic corpus of a priori necessary propositions), by means of demonstrating that R is the presupposition of some other representation R^*, which is assumed for the purposes of the argument to be objectively valid (*CPR* A84–94/B116–27, A156/B195). It follows from all these points, that Kant believes that a single line of transcendental argumentation displays, in one fell swoop, both the objective validity of the a priori representations of space and time and also the objective validity of mathematics.

Sadly for Kant's readers, however, his argument is a loose, baggy monster spread out over two books and four sections, including the Introduction and Transcendental Aesthetic of the first *Critique*, and the Introduction and first part of the *Prolegomena*. Even so, leaving aside Kant's admittedly inelegant way of presenting the argument, I think that it can be reconstructed fairly straightforwardly as a four-step proof that takes us from our representations of space and time to the possibility of synthetic a priori truth in mathematics. As in chapter 2, I will abbreviate "the representation of space" as "r-space" and "the representation of time" as "r-time." Here then is an outline of the four-step proof and corresponding conclusion.

First, Kant argues that r-space and r-time are the a priori necessary subjective forms of all empirical intuitions of appearances (see the "Metaphysical Exposition of the Concept of Space" and the "Metaphysical Exposition of the Concept of Time").

Second, he argues that space and time are strongly "transcendentally ideal," i.e., that space and time are "nothing but" r-space and r-time, the a priori necessary subjective forms of all empirical intuitions of appearances (see the "Conclusions from the above Concepts [of Space]," "Conclusions from these Concepts [of Time]," and "Elucidation"). NB: for the purposes of rational reconstruction, I will construe phrases of the form "X is nothing but Y" to mean the same as "X is identical to Y."[7]

---

[7] There is a fine point that could be raised here about interpreting the phrases "nothing but" or "nothing other than" or "nothing over and above" as parts of reductive theses in analytical metaphysics. The question is whether the use of such phrases always implies a strict identity thesis, or instead implies only some form of strong supervenience, which is consistent with non-identity. See the Introduction,

Third, he argues that geometric and arithmetic truths alike are synthetic a priori (see section V of the Introduction, and section 13 and "Remark I" of the first part of the *Prolegomena*).

And, *fourth*, he argues that true mathematical propositions are possible if and only if r-space and r-time are the a priori necessary subjective forms of all empirical intuitions of appearances (see the "Transcendental Exposition of the Concept of Space," section 1, chapter I of the "Transcendental Doctrine of Method," and sections 10–12 of the first part of the *Prolegomena*).

From these four premises Kant concludes:

R-space and r-time, which are both the a priori necessary subjective forms of all empirical intuitions of appearances and also identical to space and time respectively, are the conditions of the possibility of the fact that mathematical truths are synthetic a priori.

But that is only the macrostructure of Kant's argument. Each of the four premises I just spelled out follows as an intermediate conclusion from several sub-premises. So in the step-by-step reconstruction to follow shortly, I will also unpack the argument's microstructure by spelling out the main sub-premises and intermediate conclusions, along with the proof-goal and general conclusion. And for each step in the argument I will offer some textual support. It needs to be stressed that the point of this detailed reconstruction is only to demonstrate that Kant's argument for the possibility of mathematics is internally coherent and formally valid. This provides the basic framework I will need for discussing Kant's philosophy of arithmetic in sections 6.2 to 6.4.

The further question of soundness, however, is subtle. Indeed I want to argue that the second step of the argument—the strong transcendental ideality of space and time—is false as it stands. This in turn entails the falsity of Kant's strong transcendental idealism. But on the other hand, a weaker version of transcendental idealism seems to me to be true. So at the end of this section I will say what I think is wrong with the original or strong

---

n.30. In any case, to simplify matters, I will assume that when Kant says that "space is nothing other than merely (*nichts anders, als nur*) the form of appearances of outer sense" (*CPR* A26/B42) and that "time is nothing other than (*nichts anders, als*) the form of inner sense"(*CPR* A33/B49), he is stating strict identity theses and not merely strong supervenience theses.

transcendental ideality thesis and propose replacing it with a weaker and commensurately more plausible transcendental ideality thesis. This patch-up job will permit Kant's argument to carry us safely to our final destination, a defensible version of his theory of arithmetic. But, most importantly of all, the charitable ascription to Kant of weak transcendental idealism has direct and fundamental implications for my earlier discussion of Kant's empirical realism in part I. In my opinion the ascription to Kant of *strong* transcendental idealism would render both his theoretical and practical philosophies arguably *false*; but I also believe that with the substitution of *weak* transcendental idealism for strong transcendental idealism, both his theoretical and practical philosophies are arguably *true*. I am utterly serious about this.

In any case we must now get down to the nitty-gritty details of Kant's argument for the possibility of mathematics. The proof of *Step I* of the argument has already been spelled out with a running commentary in section 2.3, so I will simply repeat that proof here without the commentary.

A Step-by-Step Reconstruction of Kant's Argument for the Possibility of Mathematics

Prove: R-space and r-time, which are the a priori necessary subjective forms of all empirical intuitions of appearances and also identical to space and time respectively, are the conditions of the possibility of the fact that mathematical truths are synthetic a priori.

Step I: Prove that r-space and r-time, as the forms of intuition, are the a priori necessary subjective forms of all empirical intuitions of appearances.

(1) Empirical intuitions are singular representations of undetermined apparent or sensible objects, and those representations in turn possess both matter and form.

The undetermined object of an empirical intuition is called **appearance**. (*CPR* A20/B34)

I call that in the appearance which corresponds to sensation its **matter**, but that which allows the manifold of appearance to be intuited as ordered in certain relations I call the **form** of appearance (*Form der Erscheinung*). (*CPR* A29/B34)

(2) Appearances or objects of the senses are represented in empirical intuition by means of either outer (or spatial) sense or inner (or temporal) sense. R-space and r-time are the mutually distinct and jointly exhaustive (although not mutually exclusive) forms of intuition, and also the subjective forms of outer and inner sense respectively.

> By means of outer sense (a property of our mind) we represent to ourselves objects as outside us, and all as in space. In space their shape, magnitude, and relation to one another is determined, or determinable. Inner sense, by means of which the mind intuits itself, or its inner state, gives, to be sure, no intuition of the soul itself, as an object; yet it is still a determinate form, under which the intuition of its inner state is alone possible, so that everything that belongs to the inner determinations is is represented in relations of time. (*CPR* A22–3/B37)

> Time can no more be intuited externally than space can be intuited as something in us. (*CPR* A23/37)

> [R-]space is nothing other than merely the form of all appearances of outer sense, i.e., the subjective condition of sensibility, under which alone outer intuition is possible for us. (*CPR* A26/B42)

> [R-]time is nothing other than the form of inner sense, i.e., of the intuition of our self and our inner state. (*CPR* A33/B49)

(3) R-space and r-time are necessary conditions for the empirical intuition of appearances in outer and inner sense.

> [R-]space is a necessary representation, a priori, which is the ground of all outer intuitions. One can never represent that there is no space, although one can very well think that there are no objects to be encountered in it. (*CPR* A24/B38)

> [R-]time is a necessary representation that grounds all intuitions. In regard to appearances in general one cannot remove time, though one can very well take the appearances away from time. (*CPR* A31/B46)

(4) R-space and r-time, the forms of intuition, by means of an act of self-consciousness, can also be treated as "pure intuitions" or "formal intuitions," that is, singular non-conceptual representations of themselves as unique abstract relational totalities or formal-structural frameworks, thereby in turn representing space and time as singular infinite given wholes.

[R-]space is not a discursive or ... general concept of relations of things in general, but a pure intuition. (*CPR* A24–25/B39)

Space is represented as a given infinite magnitude. (*CPR* A25/B39)

[R-]time is no discursive or ... general concept, but a pure form of sensible intuition. (*CPR* A31/B47)

The infinitude of time signifies nothing more than that every determinate magnitude of time is only possible through limitations of a single time grounding it. The original representation, [r-]time, must therefore be given as unlimited. (*CPR* A32/B48)

[R]-space and [r]-time and all their parts are **intuitions**, thus individual representations along with the manifold that they contain in themselves (see the Transcendental Aesthetic), thus they are not mere concepts by means of which the same consciousness is contained in many representations, but rather are many representations that are contained in one and in the consciousness of it; they are thus found to be composite, and consequently the unity of consciousness, as **synthetic** and yet as original, is to be found in them. This singularity of theirs is important in its application. (*CPR* B136 n.)

[R]-space, represented as **object** (as is really required in geometry), contains more than the mere form of intuition, namely the **putting-together** (*Zusammenfassung*) of the manifold given in accordance with the form of sensibility in an **intuitive** representation, so that the **form of intuition** (*Form der Anschauung*) merely gives the manifold, but the **formal intuition** (*formale Anschauung*) gives unity of the representation. (*CPR* B160 n.)

(5) R-space and r-time are a priori. (From (3), (4), and the definition of "a priori" as absolute experience-independence, or essential under-determination by all sets and sorts of sensory experiences. That is: to say that X is a priori is to say that X is not strongly supervenient[8] on sensory experiences.)

[W]e will understand by a priori cognition not those that occur independently of this or that experience, but rather those that occur *absolutely* independently of all experience. (*CPR* B3)

---

[8] See the Introduction, n. 30.

(6) Since r-space and r-time are: (a) mutually distinct and jointly exhaustive (although complementary) necessary forms of the empirical intuition of appearances; (b) subjective forms of outer and inner sense; and (c) able to to be treated, via self-consciousness, as pure a priori non-conceptual intuitions of themselves as unique relational totalities or formal-structural frameworks, they are therefore the a priori necessary subjective forms of all empirical intuition of appearances. (From (1)-(2) and (5).) **QED**

Step II: Prove that space and time are strongly transcendentally ideal, i.e., that space and time are nothing but the a priori necessary subjective forms of all empirical intuition of appearances, which is also to say that space and time are identical to r-space and r-time respectively.

(7) Space and time are either: (a) things-in-themselves, (b) ontologically dependent on things-in-themselves (either as intrinsic non-relational properties of things-in-themselves or as extrinsic relations between things-in-themselves), or else (c) strongly transcendentally ideal, i.e., nothing but the a priori necessary subjective forms of all empirical intuitions of appearances, which is also to say that space and time are identical to r-space and r-time respectively. And there are no other alternatives.

Now what are space and time? Are they real essences (*wirkliche Wesen*)? Are they only determinations or relations of things, yet ones that would pertain to them even if they were not intuited, or are they relations that attach only to the form of intuition alone, and thus to the subjective constitution of our mind, without which these predicates could not be ascribed to anything at all? (*CPR* A23/B37–8)

(8) But space and time are neither things-in-themselves nor ontologically dependent on things-in-themselves (either as intrinsic non-relational properties of things-in-themselves or as extrinsic relations between things-in-themselves).

Those ... who assert the absolute reality of space and time, whether they assume it to be subsisting or only inhering, must themselves come into conflict with the principles of experience. For if they decide in favor of the first ... then they must assume two eternal and infinite self-subsisting nonentities (space and time), which exist (yet without there being anything real) only in order to comprehend everything real within themselves. If they adopt the second position...and hold space and time to be relations of appearances...that are abstracted from experience...then they must dispute the validity or at least the apodictic certainty of a priori mathematical doctrines in regard to real things (e.g., in space), since this certainty does not occur a posteriori. (*CPR* A39–40/B56–7)

(9) Therefore space and time are strongly transcendentally ideal, i.e., space and time are nothing but the a priori necessary subjective forms of all empirical intuition of appearances, which is to say that space and time are identical to r-space and r-time respectively. (From (7) and (8).) **QED**

Step III: Prove that mathematical truths are synthetic a priori.

(10) Mathematical truths are a priori and necessary, not a posteriori and contingent.

[M]athematical propositions are always a priori judgments and are never empirical, because they carry necessity with them, which cannot be derived from experience. But if one does not want to concede this...I will restrict my proposition to **pure mathematics**, the concept of which already implies that it does not contain empirical but merely pure a priori cognition. (*CPR* B14–15)

Here [in pure mathematics] is a great and established branch of knowledge ...carrying with it thoroughly apodeictical certainty, i.e.,...necessity, which therefore rests on no empirical grounds. (*P* Ak. iv. 280)

(11) Mathematical truths are synthetic, not analytic.

The concept of twelve is by no means already thought merely by my thinking of that unification of seven and five, and no matter how long I analyze my concept of such a possible sum I will not find twelve in it.... That 7 **should** be added to 5 I have, to be sure, thought in the concept of a sum = 7+5, but not that this sum is equal to the number 12. The arithmetic proposition is therefore always synthetic. (*CPR* B15–16)

Just as little is any principle of pure geometry analytic. That the straight line between two points is the shortest is a synthetic proposition. For my concept of **the straight** contains nothing of quantity, but only a quality. The concept of the shortest is therefore entirely additional to it,

298   THE PRACTICAL FOUNDATIONS OF THE EXACT SCIENCES

and cannot be extracted out of the concept of the straight line by any decomposition (*Zergliederung*). (*CPR* B16)

(12) Therefore mathematical truths are synthetic a priori. (From (10) and (11).) **QED**

Step IV: Prove that r-space and r-time are the conditions of the possibility of mathematical truths.

(13) R-space and r-time are necessary conditions of the objective validity of the truths of geometry and arithmetic.

Geometry is a science that determines the properties of space synthetically and yet a priori. (*CPR* A25/B41).

Now the intuitions which pure mathematics lays at the foundation of all its cognitions and judgments which appear at once apodeictic and necessary are space and time.... Geometry is based on the pure intuition of space. Arithmetic attains its concepts by the successive addition of units in time.... Both representations, however, are merely intuitions. (*P* 4 : 283)

(14) R-space and r-time are sufficient conditions of the objective validity of the truths of geometry and arithmetic.

To determine an intuition a priori in space (shape), to divide time (duration), or merely to cognize the universal in the synthesis of one and the same thing in time and space and the magnitude of an intuition in general (number) which arises from that: that is a **rational concern** through construction of the concepts, and is called **mathematical**. (*CPR* A724/B752)

[T]he intuitions which pure mathematics lays at the foundation of all its cognitions and judgments which appear at once apodeictic and necessary are [r-]space and [r-]time. For mathematics must first exhibit all its concepts in intuition, and pure mathematics in pure intuition, i.e., it must construct them. (*P* 4: 283)

The ground of mathematics actually is pure intuitions, which make its synthetic and apodeictically valid propositions possible. (*P* 4: 285)

(15) Therefore r-space and r-time are the conditions of the possibility of mathematical truths. (From (13) and (14).) **QED**

Conclusion

(16) R-space and r-time, which are the a priori necessary subjective forms of all empirical intuitions of appearances and also identical to space and time respectively, are the conditions of the possibility of the fact that mathematical truths are synthetic a priori. (From (6), (9), (12), and (15).) **QED**

> Time and space are accordingly two sources of cognition, from which different synthetic cognitions can be drawn a priori, of which especially pure mathematics in regard to the cognitions of space and its relations provides a splendid example. Both taken together are, namely, the pure forms of all sensible intuition, and thereby make possible synthetic a priori propositions. But these a priori sources of cognition determine their own boundaries by that very fact (that they are merely conditions of sensibility), namely that they apply to objects only so far as they are considered as appearances, but do not present things-in-themselves. (*CPR* A38–9/B55–56)

In my opinion, Kant's argument as I have just reconstructed it is logically valid. But is it sound? *Step II* is clearly the weak link in the chain of argumentation. It tells us that space and time are *nothing but* or *identical to* the a priori necessary subjective forms of our sensibility. But this seems to me plainly false, for three reasons.

First, as I mentioned in section 3.1, Kripke has persuasively argued that propositions or statements about identity, if true, are necessarily true. Yet it is perfectly conceivable and therefore logically and metaphysically possible[9] for space and time to have existed, even if no creatures minded like us also existed. This conceivability is based on two highly plausible premises. (i) The existence of minded creatures like us is clearly logically and metaphysically *contingent*, and therefore neither logically nor metaphysically necessary. Metaphysicians have sometimes argued that God is a necessary being, or that mathematical entities are necessary beings, but no one, not even the most radical solipsist, has ever seriously argued that *finite human cognizers* are necessary beings. (ii) There is clearly no *logical* entailment either from the existence of creatures minded like us to the existence of space and time, or conversely. There is no law of logic that connects our existence

---

[9] I am assuming that inferences from conceivability to possibility are generally valid. For a Kantian defense of this, see ch. 3, n. 67.

to the existence of space and time. So our existence and the existence of space and time are mutually logically and metaphysically independent.

Second, if space and time are strongly mind-dependent in the way required by strong transcendental idealism, then since all natural empirical things, as appearances, are in space and time, and since spatial and temporal properties are immanent in all appearances as intrinsic structural properties of those appearances, then it follows that all natural empirical things are strongly mind-dependent too. But that also seems to me plainly false. It is perfectly conceivable and therefore logically and metaphysically possible for natural empirical things to have existed even if no creatures minded like us also existed. What if the Big Bang had occurred in a different way, so as to generate a causal-dynamic world that operated under a different actual package of laws of nature? If human and other animal beings on this planet are an evolutionary product of nature, and if as it so happens we are the only minded creatures in the physical universe, and if cosmological and biological evolution could have gone differently from the Big Bang forward, then space, time, and natural empirical things could *all* have existed even though no creatures minded like us also existed.

Third, and finally, as one of my all-time cleverest freshman students pointed out to me during my first year of lecturing on Kant,[10] if Kant's thesis of the strong transcendental ideality of space and time were true, then supposing (as seems perfectly conceivable and logically possible) that all minded creatures in the actual world accidentally perished due to some terrible cataclysm or disease, it would follow that space and time themselves would wink out of existence as the last conscious animal died. But that is absurd. Space and time themselves do not wax and wane merely as a function of the mental operations of some conscious animal's brain processes. Space and time can exist even if we have all gone out of existence.

Three strikes and you're out. Surely these conceivability arguments are strong evidence against the identity thesis expressed by the strong transcendental ideality of space and time.

But of course it is still possible to doubt conceivability arguments. One man's inference from conceivability to possibility is another man's fallacious confusion between conceivability and mere imaginability. So, in addition to this threefold worry about the strong transcendental ideality

---

[10] Many thanks to Michael Handelman.

thesis, is another worry about the soundness of Kant's argument *for* the thesis. Because this argument, or the Three Alternatives Argument, is a constructive trilemma of the form "Either *P* or *Q* or *S*; not-*P* and not-*Q*; therefore *S*," it is obviously open to the objection that the trilemma itself is unjustified because Kant has not canvassed all the relevant alternatives. Not surprisingly, therefore, most of the criticism of this step has traditionally focused on the possibility of a fourth or "neglected" alternative which is less idealistic than the strong transcendental ideality thesis, but also avoids the difficulties of the first two prongs of the trilemma.

On the other hand, however, it has proved to be surprisingly difficult to formulate a defensible neglected alternative. The best-known candidate goes back to Kant's eighteenth-century contemporaries.[11] It surfaced again in the famous nineteenth-century neo-Kantian controversy between F. A. Trendelenburg and Kuno Fischer, and was exhaustively treated by Hans Vaihinger in his *Commentar*.[12] This is of course the proposal that Kant has ignored the possibility that space and time are at once a priori necessary subjective forms of sensibility and *also* things-in-themselves. Let us call this *the classical neglected alternative*. But, as H. J. Paton and Henry Allison have pointed out, the classical neglected alternative is not logically coherent.[13] The problem lies in the simple fact that things-in-themselves are positive noumena, and positive noumena are defined, in part, precisely as entities *not* subject to the sensible conditions of human experience (*CPR* B306–307). Therefore, whatever has the property of being an a priori necessary subjective form of human sensibility is also necessarily *not* a thing-in-itself. So to the extent that, according to the classical neglected alternative, space and time are a priori necessary subjective forms of human sensibility, they also *cannot* be things-in-themselves. The classical neglected alternative is logically and metaphysically impossible.

Here is another argument against the classical neglected alternative, which I will mention not because we really need it but rather because it brings out an interesting feature of the contrast between our representations of things-in-themselves and our representations of space and time. In the positively noumenal world, for all we know, and indeed for all we do not know, there

---

[11] See, e.g., J. G. C. Kiesewetter's letter to Kant on 20 April 1790 (*PC* 11: 157).
[12] Vaihinger, *Commentar zu Kants Kritik der reinen Vernunft*, vol. 2, pp. 134–51 and 290–313. See also Allison, *Kant's Transcendental Idealism*, pp. 111–14.
[13] Paton, *Kant's Metaphysic of Experience*, vol. 1, p. 167; and Allison, *Kant's Transcendental Idealism*, p. 114.

might conceivably be a "schpace" and a "schtime," that is, supersensible entities instantiating some very high-level properties of space and time—for example, extension, continuity, relationality, etc., and more generally all those properties of space and time that can be represented in the second-order classical polyadic logic of Whitehead's and Russell's *Principia Mathematica*. But such entities could not really *be* space and time, for those shared properties would not uniquely identify either real space or real time, by failing to capture either the 3-D Euclidean egocentric orientability of real space, or the past <—→> future directional asymmetry of real time. Indeed those shared properties would literally fail *to distinguish space from time*. But space and time are primitively different in their natures, even if they are necessarily complementary with one another:[14] hence schpace and schtime cannot be the referents of *our* words "space" and "time." So once again the classical neglected alternative is not logically coherent.

But recognizing the impossibility of the classical neglected alternative provides us with an important conceptual clue. Go back now to the thesis of the strong transcendental ideality of space and time, and ask ourselves about some of its merely *necessary* conditions. Consider for example this one. Whatever else space and time may be, assuming that they actually exist, they must at the very least be *the very objects that semantically satisfy our pure or formal intuitions of space and time*, which in turn are merely self-conscious or apperceptive representations of r-space and r-time, the a priori necessary subjective forms of our sensibility. Again, this necessary condition of the transcendental ideality thesis says merely that if space and time are to exist at all, then it must be possible for us to represent them adequately through our pure or formal intuition of them. And that seems entirely correct, if we are inclined to accept *anything* that Kant has argued in the Transcendental Aesthetic. If so, then it follows that actual space and actual time must be at the very least, in their own natures, formally and structurally identical (that is, isomorphic) with our pure or formal intuitions of space and time.

This crucial point in turn yields us a new claimant for the office of neglected alternative. And this is the possibility that space and time are *neither* things-in-themselves, *nor* ontologically dependent on things-in-themselves, *nor* "nothing but" or identical to r-space and r-time, but instead, whatever else they may be, *intrinsically the objectively real objects that actually satisfy our*

---

[14] See sections 2.3 and 2.4.

pure or formal intuitions of space and time. I will call this the "cognitive-semantic alternative," since according to it space and time are intrinsically those very things that, as objectively real entities, do actually meet the satisfaction-conditions of our pure or formal intuitions of them.

I specifically highlight the notion of semantic *satisfaction* here, because it is a broader notion than truth. As we saw in section 5.1, satisfaction according to Tarski is a relation according to which the structure of given representation is precisely modelled by an ordered sequence of all the objects in the domain. So truth entails satisfaction, and truth is necessarily equivalent with satisfaction as restricted to sentences, propositions, or judgments. But satisfaction, as such, covers more types of representation than that by means of sentences, propositions, or judgments alone. In particular, any sort of objective representation, including empirical intuitions or sense perceptions, memories, mental images, and pure or formal intuitions, can have satisfaction conditions. This means that even though pure or formal intuition is not a sentential, propositional, or judgmental type of representation, the notion of semantic satisfaction still applies to it. On this general cognitive-semantic picture then, space and time are *not* identical to our representations of space and time, but they nevertheless do correspond to (in the sense of semantically satisfying) our pure or formal intuitional representations of space and time. By the same token, as we also saw in section 5.1, for Kant facts in the world correspond to true sentences, true propositions, or true judgments about those facts, but those facts are not identical with either those sentences, propositions, or judgments. The comparison-theory of correspondence, which requires the cognitive establishment of various similarity-relations between sentences, propositions, or judgments and the facts they represent, is paradoxical. The relation of correspondence however, properly understood as *isomorphic semantic projection from a judger's point of view*, is a relation of actual isomorphism. So to say that space and time correspond to our pure or formal intuitions of them, in the sense of semantically satisfying them, is just to say that objectively real space and time are *actually precisely isomorphic* with our pure or formal intuitions of space and time, even though they are not identical to them and even though no relation of similarity needs to be cognitively established between them.

What this cognitive-semantic point about the intrinsic linkage between semantic satisfaction, correspondence, projection, actual isomorphism, and the Transcendental Aesthetic ultimately implies is that objectively real space and time are neither wholly ontologically *dependent on* minds like ours, nor wholly ontologically *independent of* minds like ours. Since, according to the cognitive-semantic alternative, space and time do indeed correspond to our pure or formal intuitions of them, and since we build this fact into their very nature as an intrinsic feature, *they also necessarily possibly correspond to our pure or formal intuitions of them*. Hence, according to this new neglected alternative, space and time, whatever else they might be, *are necessarily possibly isomorphic with our pure or formal intuitions of space and time*.

I should also explicitly mention in this connection that I am assuming on Kant's behalf, in addition to his theory of truth-as-correspondence, two theses about the logic and metaphysics of modality:

(i) whatever is actual is thereby also possible; and
(ii) whatever is possible is thereby also necessarily possible.

In other words, I am assuming that the best modal logic for discussing metaphysical issues in a Kantian context is either C. I. Lewis's system S4 alone, or else some conservative extension of S4. For this is one very plausible way of interpreting the familiar Kantian idea that some proposition Q is a "necessary condition of the possibility of" another proposition P: given the truth of P, together with a sound transcendental argument that shows how Q is a necessary condition of the possibility of P, then it follows that necessarily possibly Q.[15] Interestingly, and relatedly, Kant also explicitly accepts a weaker axiom of S4, namely, "if P then possibly P," but rightly resists the non sequitur that therefore the ontological scope of possibility exceeds the ontological scope of actuality (*CPR* A231–2/B283–4). Hence Kant, as a modal conceptualist (about the nature of possible worlds), as a modal dualist (about the analytic-synthetic distinction), and as a close personal friend of S4, is also at bottom a modal *actualist*.

In any case, here is the crucial point. Just because space and time necessarily possibly correspond to our pure or formal intuitions of space and time, and just because space and time are necessarily possibly isomorphic with our pure or formal intuitions of space and time, it does *not* follow that if r-space and r-time failed to exist—that is, if *our representations of space and time* failed to

---

[15] See Hughes and Cresswell, *Introduction to Modal Logic*, chs. 3–7. Thanks to David Bell for helping me see this point.

exist—then necessarily space and time would fail to exist. As I pointed out before, it is perfectly conceivable and therefore logically and metaphysically possible that space and time could exist even if minded creatures like us did not exist. What the cognitive-semantic alternative thesis implies is only that space and time could not have existed if minded creatures like us had somehow been metaphysically impossible. But assuming that we *do* actually exist as minded creatures who can adequately represent objectively real space and time, then the nature of objectively real space and time must not have metaphysically ruled us *out* as such. On the contrary, the nature of objectively real space and time must have metaphysically ruled us *in* as such. To this weak but still metaphysically non-trivial extent, objectively real space and time are *minimally anthropocentric*.

Otherwise put, according to the cognitive-semantic alternative, the metaphysical foundational architecture of objectively real space and time intrinsically includes some intrinsic structural mental properties which, as it happens (for better or worse!), were eventually instantiated by actual rational human animals. But those properties were for a very long time uninstantiated, and they might well have gone forever uninstantiated. Rational human animals did not actually exist for a very long time, and might never have existed at all. Nevertheless, *here we are, just as we are*—finite, embodied, alive, conscious, rational, and desperately crooked timbers all—in this natural world and of this natural world, until our inevitable natural death stops the Big Parade. And once every such Parade is over, and there are no more creatures minded like us, then objectively real space, time, and material nature will no doubt continue to exist indefinitely just as they are, always containing our necessary possibility but—sadly enough—not automatically guaranteeing our actual existence.

So more explicitly then, according to the cognitive-semantic alternative, space and time are:

(i) the actual semantic satisfiers of our pure or formal intuitions of space and time;
(ii) the necessarily possible semantic satisfiers of our pure or formal intuitions of space and time, hence necessarily possibly structurally identical or isomorphic with our a priori necessary subjective forms of sensibility, r-space and r-time; and yet also

306   THE PRACTICAL FOUNDATIONS OF THE EXACT SCIENCES

(iii) ontologically distinct from r-space and r-time, hence *not* "nothing but" or identical to r-space and r-time.

To mark the important difference between the cognitive-semantic alternative and the third alternative of the original Kantian trilemma, according to which space and time are nothing but or identical to r-space and r-time, I will say that if space and time meet conditions (i) to (iii), then they are *weakly transcendentally ideal*. By sharp contrast, if space and time are either nothing but or identical to r-space and r-time, then they are *strongly transcendentally ideal*.

Assuming now the intelligibility of the cognitive-semantic alternative to the strong transcendental ideality of space and time—that is, assuming now the intelligibility of the thesis of the weak transcendental ideality of space and time—we can then make the appropriate soundness-reinstating changes to *Step II* in the reconstruction of Kant's argument for the possibility of mathematics.

(7*) Space and time are either: (a) things-in-themselves; (b) ontologically dependent on things-in-themselves (either as intrinsic monadic properties of things-in-themselves or as extrinsic relations between things-in-themselves); (c) strongly transcendentally ideal; or else (d) weakly transcendentally ideal. There are no other alternatives.

(8*) But space and time are neither things-in-themselves, nor ontologically dependent on things-in-themselves (either as intrinsic monadic properties of things-in-themselves or as extrinsic relations between between things-in-themselves), nor strongly transcendentally ideal.

(9*) Therefore space and time are weakly transcendentally ideal. (From (*7) and (8*).)

Obviously, the entire reconstructed argument for the possibility of mathematics now goes through just as before, with only one trivial change in the proof-goal and overall conclusion:

R-space and r-time, which are the a priori necessary subjective forms of all empirical intuitions of appearances and correspond to weakly transcendentally ideal space and time respectively, are the conditions of the possibility of the fact that mathematical truths are synthetic a priori.

What are the implications of the weak transcendental ideality of space and time for my discussion of Kant's empirical realism in part I? The basic issue is *whether Kant's empirical realism requires strong transcendental idealism, or not*. Kant's original official argument for strong transcendental idealism, as I have construed it, requires the strong transcendental ideality of space and time, together with the intrinsicness of space and time. I accept the intrinsicness of space and time. But in my opinion the strong transcendental ideality of space and time is false. So strong transcendental idealism is false. By fundamental contrast, the weak transcendental ideality of space and time seems to me to be true. So my proposal is that we substitute the weak transcendental ideality of space and time for the strong transcendental ideality of space and time in Kant's original official argument for strong transcendental idealism. What we get is Kant's *weak transcendental idealism*,[16] the argument for which then runs as follows:

> (1) Space and time are (i) the actual semantic satisfiers of our pure or formal intuitions of space and time, (ii) the necessarily possible semantic satisfiers of our pure or formal intuitions of space and time, hence necessarily possibly structurally identical or isomorphic with our a priori necessary subjective forms of sensibility, r-space and r-time, and yet also (iii) ontologically distinct from r-space and r-time, hence *not* "nothing but" or identical to r-space and r-time (= the weak transcendental ideality of space and time).

And:

> (2) Spatiotemporal properties are intrinsic structural properties of all appearances (= the intrinsicness of space and time).

Therefore:

> (3) The forms or structures of all those objective appearances are necessarily possibly isomorphic with mental representations whose forms or structures are strictly imposed upon those representations by the

---

[16] Several other weakened versions of transcendental idealism can be found in, e.g., Allison, *Kant's Transcendental Idealism*, chs. 1–6; Collins, *Possible Experience*; Dicker, *Kant's Theory of Knowledge*, pp. 46–8; and Guyer, *Kant and the Claims of Knowledge*, ch. 2. But all of these versions are explicitly epistemological, not metaphysical. I count it as a theoretical advantage of my more metaphysically oriented cognitive-semantic approach that if it is true, it is both consistent with and also a necessary condition of all the epistemological versions: the knowability of empirical objects via judgments *also* requires the necessary possibility of minds like ours that are capable of adequately representing space and time through pure or formal intuition.

"synthesizing" or generative and productive activities of the a priori cognitive faculties of the human mind, as applied to original perceptual inputs from the world (= weak transcendentalism).

And also:

(4) All the proper objects of human cognition are necessarily directly humanly perceivable or observable, and necessarily not things-in-themselves or positive noumena (= weak idealism).

And when conjoined these also entail:

(5) Human beings can cognize and know all and only those objects which are such that they are necessarily directly humanly perceivable or observable, and also such that the essential forms or structures of those objects are necessarily possibly structurally identical or isomorphic with the representational forms or structures generated by our own cognitive faculties, especially the intuitional representations of space and time. Therefore we can neither cognize, nor scientifically know, nor even meaningfully assert or deny, anything about things-in-themselves or positive noumena (= weak transcendental idealism). **QED**.

Now strong transcendental idealism obviously entails weak transcendental idealism. So, since Kant explicitly holds the former, he must also implicitly hold the latter too, even if weak transcendental idealism also just as obviously does not entail strong transcendental idealism. Kant's strong transcendental idealism says that space, time, and matter are not objectively real, precisely because they are necessarily mind-dependent, and also because their essential forms and structures are literally imposed upon them by us. So, according to strong transcendental idealism, *if creatures minded like us do not exist, then space, time, and matter don't exist either, and we have also constructed them formally—even if we have not materially constructed them.* By sharp contrast, Kant's weak transcendental idealism says that although the essential forms and structures of our mental representations are literally imposed on those *representations* by the innate spontaneous cognitive faculties of our minds, nevertheless space, time, and matter are all objectively real and mind-independent, indeed as real and mind-independent as things can ever get for creatures like us, even if they are not so *damned* mind-independent. So, according to weak transcendental idealism, *even if creatures minded like us do not exist, nevertheless space, time, and matter can still exist, and we have not constructed them in any sense—instead we directly encounter them in intuition*. Nevertheless, at the same time, the very existence of space, time, and matter also implies *that representational rational minds like ours cannot be impossible*: that is, space, time, and matter have to be, at the very least, in principle cognizable and knowable by minded creatures like us.

It should be particularly noted that the weak transcendental idealism thesis is *not* a version of what Kant calls "the preformation-system of pure reason":

If someone still wanted to propose a middle way between the only two, already-named ways [of accounting for the agreement of experience with the concepts of its objects], namely that the categories were neither **self-thought** *a priori* first principles of our cognition nor drawn from experience, but were rather subjective predispositions of our thinking, implanted in us by our author in such a way that their use would agree exactly with the laws of nature along which experience runs (a kind of **preformation-system** of pure reason) then ... this would be decisive against the supposed middle way: that in such a case the categories would lack the **necessity** that is essential to their concept. (*CPR* B167–8)

The pre-formation-system of pure reason is obviously a cognitive version of the Leibnizian notion of a *pre-established harmony*, which thereby requires the existence of a rational and benevolent God. And weak transcendental idealism does not require the existence of any sort of God. But the *fundamental* difference between the preformation-system of pure reason and weak transcendental idealism is that on the preformation-system of pure reason, the connection between the mind's transcendental capacities and the world is only *extrinsic and relational*, whereas according to weak transcendental idealism, the forms of intuition and the categories are *intrinsic structural properties of the world*. Hence, on the preformation-system of pure reason, the applicability of the forms of intuition and the categories to the world is at best merely *contingent*, whereas on weak transcendental idealism the *necessity* of their applicability to the world is fully preserved.

At the end of the day, Kant's weak transcendental idealism seems to me to be a true philosophical thesis. To be sure, reality did not have to be like this; and there is no reason whatsoever to think that reality was in any sense designed to be like this by some non-deceiving all-powerful, all-knowing, all-good God. But as a matter of sheer brute fact, objective natural reality simply *is* like this. What I mean is that it seems to me true that, as a matter of sheer brute fact the necessary possibility of minds like ours is built into objective natural reality primitively and intrinsically, by virtue of the inclusion of uninstantiated but instantiable mental properties at the fundamental level of the natural world, in the constitution of space and time themselves. Again, I am utterly serious about this. Why should we be *more* surprised to discover that as a matter of fact objective natural reality is essentially both mental-and-physical, than to discover that as a matter of fact objective natural reality is made up of unperceivable unobservable fundamentally non-mental microphysical entities all the way down? If that latter were in fact true, then where in heaven's name did *the conscious, causally efficacious, and free rational minds of persons* come from? Weak transcendental idealism says, very plausibly, that they all came from the natural material world. But if weak transcendental idealism is true, then the natural material world is deeper, richer, and thicker than scientific and other reductive naturalists have allowed. Weak transcendental idealism therefore offers a thoroughly *liberal* naturalism as a reasonable alternative to the explanatory and ontological illiberalism of scientific and other reductive naturalists.

Given then the two charitable methodological principles I have adopted in my interpretation of Kant (see section 0.0), this implies that we should ascribe to Kant *only* weak transcendental idealism (and therefore also liberal naturalism) and *not* strong transcendental idealism, while of course also fully admitting that the historical Kant himself was, at least in some texts, explicitly committed to strong transcendental idealism. But even Kant gets it wrong occasionally. And, just because we think that Kant is a very deep thinker who can teach us much, it does not follow that as contemporary Kantians we should make all his mistakes too. Indeed, for contemporary Kantians to treat Kant merely historically and not critically and historico-philosophically, would seem to be in direct contravention of Kant's own meta-philosophy:

He who has properly **learned** a system of philosophy, e.g., the Wolffian system, although he has in his head all of the principles of, explanations, and proofs together with the division of the entire theoretical edifice, and can count everything off his fingers, still has nothing other than a complete **historical** cognition of the Wolffian system; he knows and judges only as much as has been given to him. If you dispute one of his definitions, he has no idea where to get another one. He has formed himself according to an alien reason, but the faculty of imitation is not that of generation, i.e., the cognition did not arise **from** reason for him, and although objectively it was certainly a rational cognition, subjectively it is still merely historical. He has grasped well and preserved well, i.e., he has learned, and is a plaster cast of a living human being. Rational cognitions that are objectively so (i.e., could have arisen originally out of the reason of human beings themselves), may also bear this name subjectively only if they have been drawn out of the universal sources of reason, from which critique, indeed even the rejection of what has been learned, can also arise, i.e., from principles. (*CPR* A836–7/B864–5, underlining added)

Those are pretty strong words—especially that bit about being a mere plaster cast of a living human being. Kant is *not*, however, telling us that it is philosophically wrong to learn from other philosophers, but rather only that in order to learn philosophy from others we must also be fully prepared to justify their claims rationally for ourselves, accepting only what is arguably true and rejecting what is arguably false. So I think that we can charitably ascribe to Kant the liberal naturalistic thesis of weak transcendental idealism (because it is arguably *true*), but not the thesis of strong transcendental idealism (because it is arguably *false*). My proposal is therefore this: that Kant's empirical realism requires *only* his weak transcendental idealism and *not* his strong transcendental idealism. And the same goes, *mutatis mutandis*, for Kant's theory of theoretical and practical rationality. Since strong transcendental idealism entails weak transcendental idealism, and since in part I of this book I also explicitly adopted the practice of never substantively appealing to the truth of any Kantian premise that presupposes more than weak transcendental idealism, it follows that weak transcendental idealism is not only consistent with, but also a necessary condition of, Kant's direct perceptual realism and manifest realism. So now and for the rest of this book I shall explicitly reject Kant's strong transcendental idealism, but also explicitly accept and defend Kant's weak transcendental idealism. In any case, my guiding thought right from the start has been that weak transcendental idealism, direct perceptual realism, and manifest realism are *all the basic metaphysics and epistemology that Kant needs in order to solve the Two Images Problem*. How this guiding thought also specifically allows for an adequate theory of our a priori cognition of necessary truths in the exact sciences is the topic of the rest of this chapter and the next.

6.2. Why Arithmetic is Synthetic A Priori

By the notion of elementary arithmetic I mean elementary logic (that is, bivalent first-order quantified polyadic predicate calculus including identity) plus the five Peano axioms:

(1) 0 is a number.
(2) The successor of any number is a number.
(3) No two numbers have the same successor.
(4) 0 is not the successor of any number.
(5) Any property which belongs to 0, and also to the successor of every number which has the property, belongs to all numbers,

taken together with the primitive recursive functions over the natural numbers—the successor function, addition, multiplication, exponentiation, etc. According to Gödel's first incompleteness theorem, elementary arithmetic is incomplete, which is to say that there are sentences of elementary arithmetic that are true and unprovable. One way of making sense of this from a Kantian point of view is to say that elementary arithmetic is incomplete because it is *synthetic* and not analytic. Frege argued that elementary arithmetic is analytic because its truths are derivable from general logical laws together with "logical definitions."[17] But Fregean logicism foundered on Russell's set-theoretic paradox, the deep unclarity of Frege's notion of a logical definition,[18] and Gödel-incompleteness. Neo-logicists[19] argue that if we drop the Fregean identification of numbers with sets of equinumerous sets and adopt second-order logic (that is, elementary logic plus quantification over properties and functions) plus Hume's principle,[20] then elementary arithmetic is after all analytic. But neo-logicists must appeal to a logic stronger than elementary logic in order to show this. And even in view of the provability of elementary arithmetic in second-order logic

---

[17] See Frege, *Foundations of Arithmetic*.
[18] See Benacerraf, "Frege: The Last Logicist."
[19] See, e.g., Hale, *Abstract Objects*; Tennant, "On the Necessary Existence of Numbers"; and Wright, *Frege's Conception of Numbers as Objects*.
[20] Hume's principle says that the number of Fs = the number of Gs if and only if there are just as many Fs as Gs.

plus Hume's principle, still no one can deny that Gödel-incompleteness entails that elementary arithmetic is not analytic, on the assumption that the criterion for analyticity is provability in elementary logic.

Now, to be sure, Kant's logic is very different from the logics used by logicists and neo-logicists, for his logic is significantly weaker than elementary logic. Kant's logic includes only truth-functional logic, Aristotelian syllogistic, and a theory of (fine-grained, decomposable) monadic concepts, which is to say in more modern terms that it includes only monadic logic[21] and a partial anticipation of higher-order intensional logic. And this may lead us to think, as Alan Hazen has put it, that "Kant had a terrifyingly narrow-minded and mathematically trivial, conception of the province of logic."[22] Well, yes: Kant's conception of the province of logic does not include polyadic predicate logic. But, on the other hand, Kant's logic certainly captures a fundamental fragment of elementary logic.[23] Furthermore, since we already know from Gödel-incompleteness that elementary arithmetic is not analytic on the assumption that the criterion for analyticity is provability in elementary logic plus the Peano axioms, and since Kant's logic is weaker than elementary logic, there seems to be little or no reason to believe that Kant's argument for the syntheticity of arithmetic will be vitiated by the limited character of his logic alone.[24] For, if a *stronger* logic shows that elementary arithmetic is not analytic, then Kant's thesis that elementary arithmetic is not analytic surely cannot depend merely on the relative weakness of his logic.

That having been said by way of finessing familiar worries about the limited scope of Kant's logic, in this section I will develop an argument for the synthetic apriority of arithmetic by unpacking Kant's argument for a slightly narrower thesis: namely, that a fundamental fragment of elementary arithmetic is synthetic a priori. There are two reasons for this.

---

[21] Monadic logic is a restricted form of elementary logic that permits quantification into one-place predicates only. Interestingly, first-order monadic logic is not only consistent and complete but also effectively decidable. See Boolos and Jeffrey, *Computability and Logic,* chs. 22, and 25. Similarly interestingly, second-order monadic logic is also consistent, complete, and decidable; see Denyer, "Pure Second-Order Logic."

[22] Hazen, "Logic and Analyticity," p. 92.

[23] And we certainly should not undervalue the fact that Kant's logic partially anticipates higher-order intensional logic; indeed, this is essential to understanding his theory of analyticity. See Hanna, *Kant and the Foundations of Analytic Philosophy,* pp. 80–3.

[24] This is a standard complaint against Kant's argument for the synthetic apriority of mathematics, going at least as far back as Russell's *Principles of Mathematics.* See Friedman, *Kant and the Exact Sciences,* pp. 55–135.

First, since Kant's logic is only a monadic logic, and therefore contains no theory of either multiple first-order quantification or multiple second-order quantification, he would not be able to formulate Peano's axioms (2) through (5). So he would not be able to formulate classical first-order Peano arithmetic or PA, much less a second-order reading of the principle of mathematical induction, axiom (5).[25] On the other hand, however, even though lacking a fully general theory of quantification, presumably Kant would still be able to formulate primitive recursive arithmetic or PRA, the quantifier-free theory of the natural numbers and the primitive recursive functions.[26] But, second, if a fundamental fragment of elementary arithmetic is synthetic a priori, then obviously elementary arithmetic as a whole is also synthetic a priori. For convenience, I will call the fundamental fragment of elementary arithmetic that was studied by Kant "arithmetic*." And to give my argument some theoretical bite, I assume that arithmetic* includes PRA and monadic logic but falls short of PA.

I turn now to Kant's argument for the synthetic apriority of arithmetic*. The argument has four crucial background assumptions that we need to make explicit before surveying it step-by-step.

First, according to Kant, a true proposition is analytic if and only if its denial leads to a logical contradiction:

The contrary of that which as a concept is contained and is thought in the cognition of the object, is always correctly denied, while the concept itself must necessarily be affirmed of it, since its opposite would contradict the object. Hence we must allow the **principle of contradiction** to count as the universal and completely sufficient **principle of all analytic cognition**. (*CPR* A151/B191)

Therefore the negative criterion of a synthetic proposition is that its denial is logically or analytically consistent.

Second, for Kant, the positive mark of the syntheticity of a proposition is its semantic dependence on intuition. In turn, as we have already seen in section 2.2, an intuition for Kant is an immediate or non-descriptive, sense-related, singular, object-dependent, non-discursive or non-conceptual and non-propositional representation of some actual spatial or temporal object, or of the underlying spatial or temporal form of all such objects (*CPR*

---

[25] See Parsons, "Mathematics of Foundations," p. 194.
[26] See Skolem, "The foundations of elementary arithmetic established by means of the recursive mode of thought, without the use of apparent variables ranging over infinite domains"; and also Troelstra and Dalen, *Constructivism in Mathematics: An Introduction*, vol. 1, pp. 120–6.

A19–22/B33–6, A68/B93, B132, A320/B377) (*P* 4: 280–1). More precisely, then, a proposition is synthetic if and only if its objective validity and truth require an intuition in this special sense:

[A]ll concepts, and with them all principles, however a priori they may be, refer nevertheless to empirical intuitions, i.e., to *data* for possible experience. Without this reference they have no objective validity ... One need only take as an example the concepts of mathematics, and first, indeed, in their pure intuitions .... Although all these principles, and the representation of the object with which this science occupies itself, are generated in the mind completely a priori, they would still not mean anything at all, if we could not always exhibit their meaning in appearances (empirical objects). (*CPR* A239–40/B299)

If one is to judge synthetically about a concept, then one must go beyond this concept, and indeed go to the intuition in which it is given. (*CPR* A721/B749)

Third, on this Kantian picture of syntheticity as semantic intuition-dependence, synthetic a posteriori propositions are dependent on empirical intuitions, and correspondingly synthetic a priori propositions are dependent on pure or formal intuitions:

This principle [of syntheticity] is completely unambiguously presented in the whole *Critique*, from the chapter on the schematism on, though not in a specific formula. It is: *All synthetic judgments of theoretical cognition are possible only through the reference of a given concept to an intuition.* If the synthetic judgment is an experiential judgment, the intuition must be empirical; if the judgment is a priori synthetic, there must be a pure intuition to ground it. (*PC* 11: 38)

Fourth, and finally, according to Kant synthetic a priori propositions are necessary, but unlike analytic propositions they are not absolutely necessary or true in every logically possibly world. Rather they are *restrictedly necessary*. This is to say that they are true in all and only the humanly experienceable worlds; in worlds that are not experienceable, they are objectively invalid, "empty" (*leer*), or truth-valueless:

Here we now have one of the required pieces for the solution of the general problem of transcendental philosophy: **how are synthetic a priori propositions possible?**—namely, pure a priori intuitions, space and time, in which, if we want to go beyond the given concept in an a priori judgment, we encounter that which is to be discovered a priori and synthetically connected with it, not in the concept but in the intuition that corresponds to it: but on this ground such a judgment never extends beyond the objects of the senses and can hold only for objects of

316   THE PRACTICAL FOUNDATIONS OF THE EXACT SCIENCES

possible experience. (*CPR* B73) Our theoretical cognition never transcends the field of experience .... [I]f there is synthetic cognition a priori there is no alternative but that it must contain the a priori conditions of the possibility of experience. (*RP* 20: 274)

To be sure, this barely scratches the surface of an adequate discussion of Kant's analytic/synthetic distinction.[27] But, for my purposes here, the bottom line is this: from these four points, it follows that a truth of arithmetic is synthetic a priori just in case it is (1) consistently deniable; (2) semantically dependent on pure intuition; and (3) necessarily true in the restricted sense that it is true in every experienceable world and never false otherwise, because it lacks a truth-value in every unexperienceable world.

So much for the cognitive-semantic framework. My reconstruction of Kant's argument for the synthetic apriority of arithmetic* is based on the notorious "finger-counting" passage at B14–16, the all-too-brief definition of the concept NUMBER at A142–3/B182, and §10 of the *Prolegomena*. As usual, for clarity's sake I will spell out the argument step-by-step; and again where it is relevant, I will also quote the texts upon which my reconstruction is based.

A Reconstruction of Kant's Argument for the Syntheticity of Arithmetic

(1) Assume the existence of arithmetic*. (Implicit premise.)
(2) It is a priori and thereby necessary that $7+5=12$. (From (1) and the definition of apriority.)

It must ... be noted that properly mathematical propositions are always a priori judgments and are never empirical, because they carry necessity with them, which cannot be derived from experience. (*CPR* B14)

(3) There is at least one logically possible world in which $7+5 \neq 12$. (Premise.)

The concept of twelve is by no means already thought merely by my thinking of that unification of seven and five. (*CPR* B15)

That 5 **should** be added to 7, I have, to be sure, thought in the concept of a sum $= 7+5$, but not that this sum is equivalent to the number 12. (*CPR* B15–16)

---

[27] Hanna, *Kant and the Foundations of Analytic Philosophy*, chs. 3–5.

MATHEMATICS FOR HUMANS   317

(4) So it is not necessary that 7+5=12. (From (3).)

To be sure, one might initially think that the proposition 7+5=12 is a merely analytic proposition that follows from the concept of a sum of seven and five in accordance with the principle of contradiction. Yet if one considers it more closely, one finds that the concept of the sum of 7 and 5 contains nothing than the unification of both numbers in a single one, through which it is not at all thought what this single number is which combines the two of them. (*CPR* B15)

(5) If and only if the pure or formal intuition of time is invoked, can (2) and (4) be made consistent with one another. For, in any possible world representable by our pure or formal intuition of time, there is a sufficiently large and appropriately structured supply of "stuff" (*Stoff*)—the total set of homogeneous temporal moments generated in the sempiternal successive synthesis of the sensory manifold—to constitute a truth-maker of the arithmetic proposition in question. That is, (2) is made true by all the experienceable worlds. And in some worlds that are not representable by our pure or formal intuition of time, nothing suffices to be a truth-maker of the arithmetic proposition in question. That is, (4) is made true by some unexperienceable worlds. So (2) is consistent with (4), assuming pure or formal intuition; otherwise they are inconsistent. (Premise.)

[N]umber [is] a representation that summarizes the successive addition of one homogeneous unit to another. Number is therefore nothing other than the unity of the synthesis of the manifold of a homogenous intuition in general, because I generate time itself in the apprehension of the intuition. (*CPR* A142–143/B182)

Now, the intuitions which pure mathematics lays at the foundation of all its cognitions and judgments which appear at once apodeictic and necessary are space and time. For mathematics must first exhibit all its concepts in pure intuition… If it proceeded in any other way, it would be impossible to make a single step; for mathematics proceeds, not analytically by dissection of concepts, but synthetically, and if pure intuition is wanting there is nothing in which the stuff (*Stoff*) for synthetic *a priori* judgments can be given.(*P* 4: 283)

(6) Since the proposition that 7+5=12 is both necessary *and* also consistently deniable, and its necessity is grounded on our pure or formal

intuitional representation of time, it follows that it is synthetic a priori. (From (2), (4), (5), and the definitions of syntheticity, apriority, and synthetic apriority.)

(7) The argument applied to the proposition that 7+5=12 can be applied, *mutatis mutandis*, to any other truth of arithmetic* involving larger numbers. (Generalization of (6).)

> One becomes all the more distinctly aware of that if one takes somewhat larger numbers, for it is then clear that, twist and turn our concepts as we will, without getting help from intuition we could never find the sum by means of the mere analysis of our concepts. (CPR B16)

(8) Therefore all truths of arithmetic* are synthetic a priori. (From (7).)

Pretty obviously, the most controversial move in this argument occurs at step (3), in which the consistent deniability of truths of arithmetic* is asserted. But what sort of a logically possible world could fail to be a truth-maker for something as apparently unexceptionable as "7+5 = 12"? We must find a possible world that lacks at least one of the underlying structural features of arithmetic.* One sort of world that will do the job is a *radically finite world*, and, in particular, a world containing no structure rich enough to include 12 or more elements—that is, a world containing less than 12 objects—which thereby lacks the requisite supply of "stuff" for satisfying "7+5 = 12."[28] Another different sort of world that would do the same job is a *radically*

---

[28] Similar points about finite countermodels for arithmetic are made by Parsons, "Kant's Philosophy of Arithmetic," p. 131), and Shapiro, "Induction and Indefinite Extensibility: The Gödel Sentence is True, But Did Someone Change the Subject?," p. 604. This is not to say that appeals to such countermodels are uncontroversial, however. Indeed, someone might claim that there are at least two big worries about radically finite worlds: (1) it is well known that there are inferential gaps between imaginability and conceivability, and it is also often claimed that there is a basic gap between conceivability and possibility; and (2) radically finite worlds fail to verify what seem to be obvious truths based on the extensional law of identity, e.g., $12 \neq 13$. Obviously I cannot adequately rebut these objections in a footnote: but here are very brief indications of possible replies. First, the imaginability-conceivability and conceivability-possibility gaps are alike double-edged swords, in the sense that *both* the critics and the defenders of the radically finite worlds thesis must appeal to conceivability arguments. Indeed it seems to me that the resistance to the possibility of radically finite worlds depends mostly on the challengeable thesis—challengeable because, presumably, justified by the step from the inconceivability of its denial to its necessity—that the natural numbers exist necessarily. Second, as I mentioned in ch. 3, n.67, on my interpretation of Kant's modal theory, possible worlds are formal constructions on our concepts—see Hanna, *Kant and the Foundations of Analytic Philosophy*, pp. 85, 241–2—so for Kant the step from conceivability to possibility is automatically guaranteed. And, third, for Kant extensional identity is not a purely logical notion (see Hanna, *Kant and the Foundations of Analytic Philosophy*, p. 142, n.57), and if he is right then it is not surprising that the extensional law of identity fails in some logically possible worlds.

unrecursive world, or a world in which even if there is the requisite supply of "stuff" for assigning reference to all the number words, nevertheless there is no way of iteratively operating on that stuff. This would be a world with enough structure and enough objects to satisfy arithmetic*al propositions but without any primitive recursive functions over those objects, including of course the successor function and addition.[29]

Now because, according to Kant, our pure or formal intuitional representation of time yields an infinite given whole (*CPR* A32/B47–8) in which the members of the unidirectional series of homogeneous moments are successively summed up to the magnitude of any later moment, it follows that both the radically finite and radically unrecursive worlds just described will not conform to our pure or formal intuition of time. Indeed, it seems plausible to believe that any countermodel to arithmetic* will also fail to conform to our pure or formal intuition of time. And since our pure or formal intuitional representation of the time-series is a necessary condition of the possibility of all sensory representation of objects (*CPR* A31/B46), it then follows that all the countermodels to arithmetic* will also be unexperienceable worlds. In turn, the recognition that all the countermodels to arithmetic* also violate the conditions of the possibility of human experience yields the further recognition that we must ground the necessary truth of arithmetic* directly on the primitive innate spontaneous human capacity for pure or formal temporal intuition. For arithmetic* is true only in worlds that include either the time-structure itself or else something isomorphic to the time-structure. And in every temporally-structured world not only is arithmetic* true, but also its truth-maker is cognizable a priori, and furthermore it has a direct application to objects of human experience—neither of which is guaranteed in timeless worlds. So while there are some conceivable and therefore possible timeless or noumenal worlds that make arithmetic* true, none of them will count in favor of arithmetic*'s synthetic necessity, a priori cognizability, or applicability. Only temporally structured worlds can count in favor of these; and only pure or formal temporal intuition gives us direct cognitive and semantic access to all those special worlds.

---

[29] A third sort of countermodel world would be a *quus*-world, i.e., a world in which some sort of non-Peano addition-function holds. See Shapiro, *Thinking about Mathematics*, pp. 89–90; and Kripke, *Wittgenstein on Rules and Private Language*, ch. 2.

6.3. The Meaning of the Concept NUMBER

At this point, you might well wonder what precisely is going on. In arguing that arithmetic* is synthetic a priori, is Kant arguing that arithmetic* is the science of time as we represent it in pure intuition, just as he argues that geometry is the science of space as we represent it in pure intuition? No. In the "transcendental exposition" of the representation of space in §3 of the B edition version of the Transcendental Aesthetic, he argues explicitly that our pure or formal intuition of space is necessary and sufficient for the objective validity of geometry. But in the corresponding transcendental exposition of the representation of time Kant very pointedly does *not* focus on arithmetic* but instead on the "general doctrine of motion" (*CPR* B49) or universal classical Newtonian mechanics. I think that we may take this to be an indication of an important asymmetry between Kant's theories of geometry and arithmetic, as Philip Kitcher points out: "Kant did not believe, as is often supposed, that arithmetic stands to time as geometry does to space."[30]

In his classic *Commentary* on the first *Critique*, Norman Kemp Smith also makes a pertinent remark in this connection:

> Though Kant in the first edition of the *Critique* had spoken of the mathematical sciences as based on the intuition of space and time, he had not, despite his constant tendency to conceive space and time as parallel forms of experience, based any separate mathematical discipline upon time.[31]

In one sense this is quite correct, but in another sense it is misleading. The problem lies in a certain ambiguity in Kemp Smith's phrase "based on." That phrase has both a logico-metaphysical sense and a semantic sense. According to the logico-metaphysical sense, X is based on Y if and only if Y is a necessary and sufficient condition of X (where this covers everything from strict identity through necessary equivalence to logical strong supervenience). But the semantic sense of "based on" is different. According to the semantic sense, X is based on Y if and only if X is *about* Y. In turn, X is *about* Y if and only if Y is the semantic value of X, which is to say that Y determines the extension of X, which is also to say that Y determines what semantically satisfies X. So if X is a concept-term, and X is

---

[30] Kitcher, "Kant and the Foundations of Mathematics," pp. 33–4.
[31] Kemp Smith, *A Commentary on Kant's "Critique of Pure Reason,"* p. 133.

about Y, then Y determines what objects X applies to; if X is a propositional term, and X is *about* Y, then Y determines the truth-maker(s) of X; if X is a theory, and X is *about* Y, then Y determines the model(s) of X. It should be noted that Y certainly can (although it does not necessarily have to) determine the extension of X by being identical with the extension of X. But the crucial point is that once we have isolated the semantic sense of "based on" as *semantic aboutness*, it is then quite correct to say that Kant does not conceive of arithmetic* as a science that is *about* time and its formal-structural features in the way that geometry is *about* space and its formal-structural features. Instead, arithmetic* is *about* the natural numbers and their formal-structural features, but not *about* time. Still, as Michael Friedman aptly puts it, "there is no doubt that [for Kant] arithmetic *involves* time."[32] So how can it be true that for Kant arithmetic is not *about* time, yet arithmetic still presupposes time as a necessary and sufficient condition of its objective validity?

The interpretation I favor is that our pure or formal intuition of the infinite unidirectional successive time-series supplies a fundamental semantic condition for arithmetic*, but does not fully determine the semantics of arithmetic* until it is combined with a second representational factor—namely, a purely logical factor. I will discuss this purely logical factor at the end of this section. But right now we need to see how pure or formal intuition manages to supply a fundamental semantic condition for arithmetic*. The answer is revealed in these texts, the first of which we have seen already:

[N]umber [is] a representation that summarizes the successive addition of one homogeneous unit to another. Number is therefore nothing other than the unity of the synthesis of the manifold of a homogenous intuition in general, because I generate time itself in the apprehension of the intuition. (*CPR* A142–3/B182)

Time is in itself a series (and *the formal condition* of all series). (*CPR* A411/B438, emphasis added)

Arithmetic attains its concepts of numbers by the successive addition of units in time. (*P* 4: 283)

Time [is] the successive progression as form of all counting and of all counting and of all numerical quantities; for time is the basic condition of all this producing of quantities. (*PC* 11: 208, underlining added)

---

[32] Friedman, *Kant and the Exact Sciences*, p. 105, n.16.

322   THE PRACTICAL FOUNDATIONS OF THE EXACT SCIENCES

Here is what I think Kant is driving at: the pure or formal intuition of time sharply constrains what can count as a model for arithmetic*, but does not itself determine the extension of number terms or arithmetic propositions. Arithmetic*, by means of number concepts, represents the natural numbers, their intrinsic relational properties, and the recursive functions over them. But all models of arithmetic* are non-conceptually structurally restricted or limited by means of our pure or formal intuitional representation of the infinite unidirectional successive time-series. So nothing will count as a model of arithmetic* unless it is at least isomorphic with the infinite unidirectional successive time series delivered by pure or formal intuition. Pure or formal intuition does not therefore tell us just what the intended or standard model of arithmetic* *is*—pure intuition does not tell us what the numbers *are*—but it does tell us what the numbers *cannot be*, and it lays down a basic condition for something's being a referent of numerical terms or a truth-maker for arithmetic*al propositions.

So, to repeat, I am saying that Kant's thesis about the role of our pure or formal intuition of time in arithmetic* is that our pure intuition of the infinite unidirectional successive time-series supplies a fundamental semantic condition for the objective validity or meaningfulness of the concept NUMBER by partially determining what will count as a referent for numerical terms or a truth-maker for arithmetic*al propositions: such terms and propositions cannot be *about* the natural numbers unless their extensions are isomorphic with time. A similar point is made by Charles Parsons:

> Time provides a universal source of models for the numbers... What would give time a special role in our concept of *number* which it does not have in general is not its necessity, since time is in some way necessary for all concepts, nor an explicit reference to time in numerical statements, which does not exist, but its sufficiency, because the temporal order provides a representative of the number which is present to our consciousness if any is present at all.[33]

Of course the pure or formal intuition of time does more than merely constraining the class of models for arithmetic *. By virtue of the threefold fact that our pure or formal intuition of the infinite unidirectional time-series (i) is

---

[33] Parsons, "Kant's Philosophy of Arithmetic," p. 140. Nevertheless Parsons also thinks that "Kant did not reach a stable position on the place of the concept of number in relation to the categories and the forms of intuition" ("Arithmetic and the Categories," p. 152). But if I am right about the interplay between intuitional and logical factors in Kant's analysis of the meaning of concept NUMBER, then Kant's account is in fact far more stable (and cogent) than Parsons supposes.

built innately and therefore also dispositionally into human representational capacities, (ii) is a necessary condition of all sensory experience of objects, and also (iii) picks out time, which is partially constitutive of the empirical world, it follows then that Kant can neatly explain not only (i*) why arithmetic* is synthetically necessary, or true in all experienceable worlds and never false otherwise (i.e., because time is included in every experienceable world and every model of arithmetic* is isomorphic with time), but also (ii*) why arithmetic* is cognizable a priori for creatures like us (i.e., because our capacity for pure or formal temporal intuition is innate), and (iii*) why arithmetic* is guaranteed to have empirical application (i.e., because the representation of time is guaranteed to have empirical application). The dimension of applicability, moreover, is a particularly crucial factor, as Frege points out in *Basic Laws of Arithmetic*:

It is applicability that raises arithmetic from the rank of a game to that of a science. Applicability therefore belongs to it of necessity.[34]

All of this adds up to an important point. As Michael Potter has observed, two fundamental and intimately related problems in the philosophy of arithmetic are: (1) how to explain arithmetic's necessity?; and (2) how to explain arithmetic's empirical applicability?[35] But there is also a second pair of similarly fundamental and intimately-related problems: (3) a uniform "standard" or Tarskian semantics of natural language implies that numbers are platonically abstract truth-makers of arithmetic truths, but on the one hand the plausible assumption that numbers are causally inert abstracta implies that they are *unknowable*, when combined on the other hand with the plausible assumption that a "reasonable" epistemology requires causal contact between the knower and the known, as for example in ordinary sense perception;[36] and (4) what *are* the numbers?[37] Paul Benacerraf has influentially and plausibly argued for the salience and interrelatedness of the

---

[34] Frege, *Basic Laws of Arithmetic*, vol. II, section 91.
[35] See Potter, *Reason's Nearest Kin*.
[36] See Benacerraf, "Mathematical Truth."
[37] This problem arises in several ways. In its most general form it is Quine's problem of "what there is" ("On What There Is," pp. 14–15); in the context of first-order logic it is Hacking's problem about categoricity ("What is Logic?"); and in the context of second-order logic it is both Frege's Caesar problem (*Foundations of Arithmetic*, p. 68) about identifying the numbers with objects, and also Benacerraf's problem of the indeterminacy of the reference of number terms ("What Numbers Could Not Be").

third and fourth problems.³⁸ The deep significance of Kant's philosophy of arithmetic then lies in the fact that he adumbrates *a unified solution to all four problems*.

Now I want to wrap up this section by taking a look at a very puzzling letter that Kant wrote to his friend and disciple Johann Schultz in 1788, one year after the publication of the B edition of the first *Critique*. Schultz was then working on the manuscript of a book entitled *Prüfung der kantischen Kritik der reinen Vernunft* ("Examination of the Kantian Critique of Pure Reason"), which he had shown to Kant. In that manuscript, Schultz had anticipated Frege by defending the idea that all the truths of arithmetic are purely logical or analytic.³⁹ Here is the key part of Kant's response to the manuscript:

> Time, you correctly notice, has no influence on the properties of numbers (considered as pure determinations of quantity), as it may have on the character of those alterations (of quantity) that are possible only relative to a specific state of inner sense and its form (time). The science of numbers, notwithstanding the succession that every construction of quantity requires, is a pure intellectual synthesis, which we represent to ourselves in thought. But insofar as specific quantities (quanta) are to be determined according [to this science of numbers], they must be given to us in such a way that we can grasp their intuition successively; and thus this grasping is subjected to the time condition. (*PC* 10: 556–7)

Part of what Kant is doing here is simply reiterating his view that while arithmetic* presupposes our pure intuition of the time-series, arithmetic* is not itself the science of "alterations" (*Veränderungen*) or events—that is, arithmetic* is not *about* time. But for our purposes the crucial question is, what does Kant mean by his remark that "the science of numbers...is a pure intellectual synthesis, which we represent to ourselves in thought"? What he seems to be saying is that arithmetic* is fundamentally based on pure conceptualization, which of course in his terms would make it purely logical or analytic in nature. So by 1788 has Kant quietly switched over to some version of logicism?

---

³⁸ See Benacerraf, "What Numbers Could Not Be"; Benacerraf, "Mathematical Truth"; and Benacerraf, "What Mathematical Truth Could Not Be—I."

³⁹ To be sure, Leibniz had already anticipated this idea. But Schultz was apparently the first philosopher to float it after the publication of the *Critique of Pure Reason*. Significantly, the published version of the *Prüfung* does not contain this thesis. This could just be a matter of Schultz's deferring to his teacher and master. But it could also be a more rational matter of Schultz's believing that Kant's reply had adequately handled his objection.

No. One way of seeing this is to return to a subtle point he makes in the B edition of the *Critique of Pure Reason*. There Kant explicitly commits himself to the thesis that all mathematics is strictly constrained by pure logic, in that "the inferences of the mathematician all proceed in accordance with the principle of contradiction" (*CPR* B14). But this constraint on inference and proof "is required by the nature of any apodictic certainty" (*CPR* B14), so it is not special to mathematics. More generally, it does not follow that mathematics is essentially logic just because its proofs must meet some minimal pure logical requirements:

> Since one found that the inferences of the mathematician all proceed in accordance with the principle of contradiction ..., one was persuaded that the principles could also be cognized from the principle of contradiction, in which, however, they erred; for a synthetic proposition can of course be comprehended in accordance with the principle of contradiction, but only insofar as another synthetic proposition is presupposed from which it can be deduced, never in itself. (*CPR* B14)

This of course sets Kant's view on mathematics sharply apart from the Leibnizian view, according to which all necessary truth is ultimately reducible to the logical principle of identity or non-contradiction (Leibniz regarded these as equivalent). But the crucial point is that it is a mistake to think that the admitted fact of strict logical constraints on mathematics entails a reduction of mathematics to logic. Kant's view, on the contrary, is that mathematics can essentially include logical elements without in any way undermining its syntheticity. Now it has also been sometimes suggested by commentators that Kant is saying at B14 that only the *premises* and *conclusions* of mathematical reasoning are non-logical, while also holding that all the inferential transitions or steps of proof are of a purely logical nature.[40] So their idea is that while mathematics is indeed synthetic as regards its semantic content, its formal machinery of proof is purely logical. This I think is also a mistake, for reasons we will see in the next section.

Right now, the question on the table is whether Kant in the letter to Schultz in 1788 is intentionally or unintentionally backsliding towards some

---

[40] This view has been defended by L. W. Beck and Gottfried Martin. The opposite view—that for Kant pure intuition enters even into the inferential transitions of arithmetic proofs—has been defended by Russell, Hintikka, and Friedman. See Friedman, *Kant and the Exact Sciences*, pp. 80–95. As I indicated in n. 24, Friedman is also committed to the Russellian view that the weakness of Kant's logic is responsible for his doctrine that arithmetic is synthetic a priori. But it is of course perfectly consistent to hold that intuition enters even into the inferential transitions of arithmetic*al proofs and also that the weakness of Kant's logic is *not* responsible for his doctrine that arithmetic* is synthetic a priori.

sort of logicism about arithmetic*. And one reason for thinking that he is *not* backsliding, as we have just seen, is that in the B edition of the first *Critique*, published only a year before the letter to Schultz, he is explicitly committed to the idea that the presence of significant logical factors in mathematics is consistent with the denial of logicism. But the decisive reason for thinking that he is not has to do with his views on the role of logic in the semantic constitution of the concept NUMBER. In the letter to Schultz, Kant is saying, I think, that NUMBER does indeed have a purely logical source of representational content in our conceptual faculty, the understanding, but that this source of content does not exhaust the content of NUMBER.

So what, according to Kant, does NUMBER mean? Here is what he says explicitly in the first *Critique*:

No one can define the concept of a magnitude in general except by something like this: That it is the determination of a thing <u>through which it can be thought how many units are posited in it</u>. Only this how-many-times is grounded on successive repetition, thus on time and the synthesis of the homogeneous in it. (*CPR* A242/B300, underlining added)

And here is what I think he means by that remark, when we combine it with what he says in the letter to Schultz. Kant's view, it seems, is that NUMBER is necessarily partially based on the three "logical functions" of quantification in judgments:

Universal (e.g., *all* Fs are Gs),
Particular (e.g., *some* Fs are Gs),
Singular (e.g., *the* F is G, or *this* F is G) (*CPR* A70/B95).

The logical functions of quantification, in turn, correlate one-to-one with the three categories of quantity:

Totality
Plurality
Unity (*CPR* A80/B106).[41]

Now, in the Schematism, Kant says that "the pure image of all magnitudes (*quantorum*) ... for all objects of the senses ... is time" and that "the

---

[41] Kant sometimes reverses the terms "totality" and "unity"; but for a good defense of the claim that Kant's real intention is to put them in the order I have used in the text, see Longuenesse, *Kant and the Capacity to Judge*, pp. 248–9.

pure schema of magnitude (*quantitatis*), however, as a concept of the understanding, is number" (*CPR* A142/B182). As I understand it, what he means is that the concept NUMBER is what results if one takes the basic logical constants of quantity (all, some, the/this), maps them onto the corresponding metaphysical categories of quantity (totality, plurality, unity), and then systematically interprets those quantitative categories in terms of the pure intuition of time, as follows:

(1) The logical function of universality in judgments, corresponding to "all Fs," goes over into the infinite totality of successive moments of time, and so yields an exemplar or paradigm of the whole series of the natural numbers.
(2) The logical function of particularity in judgments, corresponding to "some Fs," goes over into any finite plurality of successive moments of time (i.e., a duration), and so yields exemplars or paradigms of any finite natural number.
(3) The logical function of singularity in judgments, corresponding to "the F" or "this F," goes over into any arbitrarily chosen single moment or unit of time, and so yields an exemplar or paradigm of the number 1.

Now for Kant all empirical magnitudes or quantities are finite or infinite (*CPR* A430/B458), discrete or continuous (*CPR* A526–7/B554–5), and extensive or intensive (*CPR* A162–3/B203–4, A165–1/B208–12).[42] And, as we have just seen, in the Schematism Kant tells us that all appearances, as magnitudes or quantities, fall under the schematized concept NUMBER (*CPR* A161–176/B202–218). So NUMBER is directly applicable to all sorts of empirical magnitudes by virtue of its construal in terms of the pure or formal intuition of time.

In other words, according to Kant, the concept NUMBER has a purely logical source of representational content, but that logical input does not exhaust its semantic content, since it also has a complementary non-logical source of its representational content, the pure or formal intuition of the infinite unidirectional successive time-series. So the concept NUMBER is *a partially logical*

---

[42] The extensive continuum has a magnitude equal to the natural numbers, and the intensive continuum has a magnitude equal to the real numbers. So, given Kant's conception of pure intuition, together with the schematized "mathematical" categories, it follows that the empirical world is both an extensive and intensive continuum. And in this way, it seems, Cantor's continuum hypothesis is determinately true in every experienceable world.

but not wholly logical concept: it represents numbers in purely logical terms, but these logical terms alone do not suffice to fix its meaning or objective validity adequately. Its meaning is adequately fixed, however, when we supplement its purely logical content by combining it with a certain *non-logical structure*. That is, when we represent natural numbers by using and specifying the concept NUMBER, we must also invoke a supplementary pure or formal intuition of the infinite unidirectional (= asymmetric or irreversible) successive time-series, which supplies the other fundamental semantic condition for the objective validity of NUMBER in particular and for arithmetic* more generally, by sharply constraining what will count as a model for the latter, and by securing the empirical applicability of the former.

If this interpretation of Kant's response to Schultz is correct, then it brings us back to Hacking's point in the second epigraph of the chapter, to the effect that the concept of a natural number cannot be categorically characterized in elementary logic. This is closely connected to the fact (originally discovered by Thoralf Skolem) that elementary arithmetic has non-standard models.[43] The Hacking-Skolem worry, then, is that by means of elementary logic alone we cannot determine just which of the many models of elementary arithmetic is the intended or standard model that is to be identified with the natural numbers.

Kant's view about the numbers, by contrast to that of any theory attempting to give a reduction of arithmetic to logic, is that something is a natural number if and only if it satisfies the purely logical categories of quantity and is isomorphic to some part of the infinite unidirectional successive time-series picked out by pure or formal intuition. Think of it this way: just as the President of USA is any person who plays *all the presidential roles* under the Constitution, so too a given natural number—say, two—is any entity that plays *all the two-roles* under the theory of arithmetic. Therefore, for Kant, a number concept is nothing more and nothing less than how we collect or colligate all Fs, or some Fs, or the/this F, in a way that formally-structurally mimics the unidirectional successive synthesis of moments in time. The concept "75," for example, is just how we collect or colligate whatever falls under the concept F (say, all the fingers on one hand including the thumb) in exactly the same way that we representationally generate just that many moments of time. And the concept "zero" is just how we collect

---

[43] See Hunter, *Metalogic*, pp. 202–5.

or colligate no Fs at all in exactly the same way that we representationally generate no moments of time by representing the specious present in which nothing has yet happened—the "beginning, the pure intuition = 0" (*CPR* A165/B208). More generally, the representational generation of numbers by counting is the logical representation of all objects, some objects, the/this object, or even no objects (which is represented in terms of negation and the particular quantifier), under some first-order (typically, empirical) concept C, taken together with the representation of time.

Number concepts, in other words, are *schematized concepts* of a unique kind. That is, they are concepts whose meaningful content is partially determined by a fundamental non-logical structure together with their positions or roles in that structure: the structure of total infinite unidirectional time, as delivered by pure or formal intuition. So the natural numbers are *essentially the fillers or players of a specific positions or roles in a logically-conceptually constrained pure or formal intuitional time-structure*, which is to say that Kant's theory of the numbers is a brilliantly original (and specifically non-platonic) version of *ante rem* mathematical structuralism.[44] So just as, in Kant's philosophy of matter, material things are essentially structure-dependent entities all of whose intrinsic structural properties are causal-dynamic laws of the system of attractive and repulsive forces in nature, so too, in Kant's philosophy of arithmetic, the natural numbers are essentially structure-dependent entities all of whose intrinsic structural properties are laws of the system of arithmetic*. But that is not to say that the natural numbers are really something *other* than the natural numbers. On the contrary, the numbers are what they are, and not some other things. Kant is *not* a reductionist about the natural numbers, precisely because his transcendental cognitive semantics is thoroughly non-reductive. Numbers are *sui generis* entities because they are fully determined by logical concepts with a *sui generis* semantic content, and this is the same as to say that the natural numbers are essentially the fillers or players of specific positions or roles in an empirically applicable and humanly graspable a priori time-structure

---

[44] See Shapiro, *Thinking about Mathematics*, ch. 10; and Shapiro, *Philosophy of Mathematics: Structure and Ontology*. Unfortunately it is hard to find a clear or widely-accepted statement of what is meant by saying that something (e.g., a universal) is *ante rem*. In any case, for me something is *ante rem* if and only if it is not uniquely located in spacetime and its existence does not logically require the existence of actual things. So, roughly speaking, for me something is *ante rem* if and only if it is abstract and neither *de re* nor *in re*. And Kant's logically-constrained pure intuitional representation of the infinite unidirectional time-series is *ante rem* in precisely this sense (*CPR* A30–6/B46–53, A291–2/B347–8).

under special logical-conceptual constraints, which is delivered to us solely by means of pure or formal intuition.

If my interpretation of Kant's response to Schultz is correct, then Kant is saying along with Hacking that pure logic on its own underdetermines the meaning of the concept NUMBER. But where Kant goes well beyond Hacking is by saying that *only* our pure or formal temporal intuition can do the further semantic job that logic fails to do on its own, and by saying that numbers are *sui generis* entities—therefore irreducible entities—with *sui generis* properties and relations. Arithmetic* is *about* the natural numbers and their formal features, and requires both pure logic and the pure or formal intuitional representation of time in order to be *about* such things. The natural numbers are the semantic values of numerical terms and among the semantic values of arithmetic*al propositions. But the natural numbers, in turn, are *natural* precisely because their special structuralist ontology is primitive and essentially bound up with *human nature*. So in an illuminating twist on Leopold Kronecker's famous quip about number theory to the effect that God made the integers and everything else was done by humans,[45] we can now say that for Kant *innate human sensibility made the natural numbers*, and everything else was done *by logic*. But pure general logic for Kant is an a priori cognitive science with categorically normative foundations:

Logic is a science of reason, not as to mere form but also as to intensional content; a science a priori of the the necessary laws of thought, not in regard to particular objects, however, but to all objects in general;—hence a science of the correct use of the understanding and of reason in general, not subjectively, however, i.e., not according to empirical (psychological) principles for how the understanding does think, but objectively, i.e., according to principles a priori for how it ought to think. (*JL* 9: 16)

Therefore, to the extent that arithmetic* is *also* pure logic, it is an a priori cognitive science of human sensibility *and* rational human discursivity, whose categorically normative foundations are directly inherited from its logical component. Pure logic tells us a priori how we ought to think, and arithmetic* tells us a priori how we ought to think *about numbers*. So Kant holds the thoroughly radical view that the concept NUMBER is not only an anthropocentric concept, but also a "thick" or *categorically normative* concept.

---

[45] Struik, *A Concise History of Mathematics*, p. 160.

6.4. Construction as Construal

No part of Kant's philosophy of arithmetic is a walk in the park. But I have been saving the trickiest bit of it for last. This is Kant's theory of mathematical "construction" (*Konstruktion*) in its particular application to arithmetic.

In his all-too-brief discussion of the nature of mathematics in chapter I, section 1 of the Transcendental Doctrine of Method, Kant distinguishes between two sorts of rational or a priori cognition: philosophical cognition and mathematical cognition. He had been concerned to draw this distinction sharply since the *Inquiry Concerning the Distinctness of the Principles of Natural Theology and Morality* of 1764, in order to explain what he regarded as a set of manifest differences in semantic and epistemic character between the two, despite their both falling into the realm of the a priori. In the *Inquiry* the distinction turned on a difference between two sorts of conceptual reasoning: philosophical reasoning is a priori analysis of metaphysical concepts, or a non-empirical advance from "given" metaphysical concepts to their decompositional parts; whereas mathematical reasoning is a priori synthesis, or the non-empirical "making" of new concepts by combining two given concepts. But this way of drawing the distinction has two important problems. First, on the side of philosophical cognition, it does not distinguish between the mere analysis of concepts and the specifically philosophical analysis of concepts; and therefore it does not show why the propositions of philosophy are synthetic a priori, not analytic. Second, on the side of mathematical cognition, it does not adequately discriminate between the making of new concepts by mere arbitrary decision or stipulation (*CPR* A729/B757), and specifically mathematical cognition. As a consequence, it threatens to make the distinction between analytic and synthetic a priori propositions wholly relative to the intentions of the judger,[46] since according to it every putatively synthetic a priori proposition can be reformulated as an analytic proposition whose predicate concept is contained in its subject concept by an act of sheer stipulation on the part of the judging subject.

So, in the first *Critique*, Kant thoroughly reworked the distinction between philosophical and mathematical cognition. For our purposes, we can leave aside the renovated notion of philosophical cognition. What is

---

[46] Hanna, *Kant and the Foundations of Analytic Philosophy*, pp. 191–2.

important for us is that the essence of mathematical cognition is now said to lie in "the construction of concepts":

[M]athematical cognition [is cognition from] from the **construction** of concepts. But to **construct** a concept means to exhibit (*darstellen*) a priori the intuition corresponding to it. For the construction of a concept, therefore, a **non-empirical** intuition is required, which consequently, as intuition, is an **individual** object, but that must nevertheless, as the construction of a concept (of a general representation), express in the representation universal validity for all possible intuitions that belong under the same concept. Thus I construct a triangle by exhibiting an object corresponding to this concept, either through mere imagination, in pure intuition, or on paper, in empirical intuition, but in both cases completely a priori, without having had to borrow the pattern for it from any experience. The individual drawn figure is empirical, and nevertheless serves to express the concept without damage to its universality, for in the case of this empirical intuition we have taken account only of the action (*Handlung*) of constructing the concept, to which many determinations, e.g., those of the magnitude of the sides and the angles, are entirely indifferent, and thus we have abstracted from these differences, which do not alter the concept of the triangle.... [M]athematical cognition considers the universal in the particular, indeed even in the individual, yet nonetheless a priori and by means of reason, so that just as this individual is determined under certain general conditions of construction, the object of the concept, to which this individual corresponds only as its schema, must likewise be thought as universally determined. (*CPR* A713–14/B741–2)

A few sentences later, Kant remarks that the form of mathematical cognition itself guarantees that it will pertain solely to quantities, because "only the concept of magnitudes can be constructed, i.e., exhibited a priori in intuition" (*CPR* A714/B742). Then following up on that, in the context of a notoriously puzzling passage on the nature of algebra,[47] Kant speaks of "constructions of magnitude in general (numbers)" (*CPR* A717/B745).

---

[47] I think that Lisa Shabel is correct in holding that "in a Kantian context 'algebra' cannot be taken simply to denote the arithmetic of indeterminate or variable numeric quantities but must be recognized as a method applied to the solution of arithmetic *and* geometric problems, resulting in a geometric construction of 'magnitude in general': a line segment expressing either a number, or the determinate size of a quantum"("Kant on the "Symbolic Construction" of Mathematical Concepts," p. 617). In other words, Kantian algebra is a general science of magnitude that comprehends both geometry and arithmetic and indeed requires the theoretical fusion of geometry and arithmetic. So, since this chapter focuses on Kant's theory of arithmetic, I will say nothing specifically about *symbolic* construction in Kantian algebra. Presumably, however, Kantian algebra could also be given a construction-as-construal reading according to which the exemplary schematic mental models for the concepts entering into algebraic propositions are either spatial, temporal, or spatiotemporal.

This obviously refers back to what Kant says in the Schematism about the pure concepts of magnitude or quantity, namely, that the concept NUMBER is what results from the schematization of the pure concepts of quantity and that the representation of time is the "pure image" or schema of all magnitudes for all objects of the senses in general (*CPR* A142–3/B182). In the Schematism, moreover, Kant directly ties the notion of a schema of a pure concept to the faculty of pure or productive imagination (CPR: A140–2/B180–1). These ideas are also carried beyond the first *Critique*. In the *Prolegomena* he says that "mathematics must first exhibit all its concepts in intuition, and pure mathematics in pure intuition, i.e., it must construct them" (*P* 4: 283). And, in a similar vein, in the letter to Schultz he speaks of "construction, a single counting up in an a priori intuition" and of "the construction of the concept of quantity" (*PC* 10: 556).

Three things are immediately clear from these texts and many other similar ones: (a) that mathematics requires the construction of concepts; (b) that mathematical construction of concepts is carried out by means of pure or formal intuition together with the pure imagination; and (c) that arithmetic in particular requires the construction of numerical concepts, that is, concepts of magnitudes. But that is where immediate clarity runs out. My goal is to understand (c), but obviously that is intelligible if and only if (a) and (b) are intelligible. So what we need to know are answers to these three questions, in sequence: (a*) what, generally speaking, is the construction of a concept?, (b*) how, specifically, does one construct a mathematical concept by means of pure intuition together with the pure imagination?, and (c*) what, precisely, does it mean to construct a numerical concept?

(1) *What, generally speaking, is the construction of a concept?* The German abstract noun *Konstruktion* and its associated verb *konstruieren*, just like the corresponding English terms "construction" and "construct", are ambiguous. On the one hand, they convey the notion of putting something together or building something new, by the assembly of diverse concrete or abstract materials, or by repeated operations on diverse concrete or abstract materials—as we might express in talking about "the construction of a house" or "the construction of a formal system." And on the other hand, they convey the notion of grammatical parsing or semantic interpretation, as we might express in saying "I chose to put a certain construction on that sentence" or "The judge constructs the law." Let

us call the first sense *construction as creation* and the second sense *construction as construal*. Given Kant's well-known interest in jurisprudence and his equally well-known fondness for using legal metaphors and analogies in metaphysical, epistemic, logical, and semantic contexts, it seems obvious that thinking of construction as construal would be as natural to him as thinking of construction as creation.

Nevertheless, I think it is almost universally assumed by readers of Kant that mathematical construction should be read as a some sort of creation of formal objects.[48] But Kant explicitly says that in mathematical construction it is *concepts* that are constructed by means of pure intuition. Now there is certainly a sense in which, for Kant, concepts are created from diverse cognitive materials by assembly or repeated mental operations, that is, by synthetic mental processes involving comparison, reflection, and abstraction. Kant calls this the "generation" of concepts (*JL* 9: 94–5). But the generation of concepts does not seem to be what Kant has in mind in the case of mathematics, since he makes no mention of this sort of mental activity in that context. Moreover, pure or formal intuition plays no special role in the generation of concepts.

For these reasons, I think, most readers of Kant typically tumble forward into a quick and unacknowledged interpretative slide from the notion of "the construction of concepts," to the notion of "the construction of objects falling under concepts," and then, taking into account the fact that Kant is talking about some sort of mental process involving pure intuition and pure imagination, over-hastily conclude that Kantian mathematical construction is the *mental creation of mathematical objects*.[49] This, for example, is precisely the sort of mathematical construction that is at work in Brouwer's intuitionism. According to Brouwer, natural numbers are generated by the infinitely interated application of formal operations to the conscious contents

---

[48] See, e.g., Hintikka, *Logic, and Language-Games, and Information: Kantian Themes in the Philosophy of Logic*; Kitcher, "Kant and the Foundations of Mathematics," pp. 42–50; Young, "Construction, Schematism, and Imagination"; and Young, "Kant on the Construction of Arithmetical Concepts."

[49] The mental creation of mathematical objects should be carefully distinguished from the mental creation of empirical objects. Kant officially holds that by constructing mathematical concepts a priori we also (partially) create empirical objects by determining basic elements of their form (*CPR* A723/B751). That follows directly from the thesis of strong transcendental idealism. But an empirical object is not a mathematical object, except insofar as mathematical concepts apply to empirical objects. My general point here is that for Kant there really are no mathematical objects—where such objects are taken to be ontically independent of number concepts—even though there are entities that are numbers. This apparent paradox is resolved when we recognize that numbers are essentially the fillers or players of positions or roles in a logically-conceptually constrained intuitional time-structure.

of the diachronic stream of an individual's mental states. So Brouwer posits an original (infinitist) creation of mathematical objects in inner sense.[50]

But why should we allow Brouwerian creationist intuitionism—with its psychologistic implications—to drive our interpretation of Kant? Although Kant is a cognitivist, he explicitly rejects psychologism of any sort.[51] More generally, if reading the construction of concepts as the mental creation of mathematical objects not only does violence to Kant's views, but also imports many of the problems of a philosophy of mathematics whose motivations and rationale are more or less foreign to Kant's, then why read it that way? Surely it is more charitable to Kant according to my explicit methodological principles of historico-philosophical interpretation, to try out the hypothesis that by the notion of a construction of mathematical concepts Kant means *the construal of mathematical concepts* and not the mental creation of mathematical objects. So let us do just that.

(2) How, specifically, does someone construct a mathematical concept by means of pure intuition together with the pure imagination? My working hypothesis is that, in general, to construct a mathematical concept for Kant is to parse or semantically interpret a numerical or geometric concept by means of pure intuition and pure imagination. We know from the crucial text at *CPR* A713–14/B741–2 that to construct a pure mathematical concept of the understanding is for the pure imagination to "exhibit" an instance of that concept in pure intuition. And we also know from the Schematism that for the pure imagination to exhibit an instance of a pure concept in pure intuition is to produce a schema of that concept. Because of its relative richness of detail, it is useful to quote from the Schematism at some length:

We will call this formal and pure condition of the sensibility, to which the use of the concept of the understanding is restricted, the **schema** of this concept of the understanding... The schema is in itself always only a product of the imagination; but since the synthesis of the latter has as its aim no individual intuition but rather only the unity in the determination of sensibility, the schema is to be distinguished from the image. Thus, if I place five points in a row, ..., this is an image of the number five. On the contrary, if I only think number in general, which could be five or a hundred, this thinking is more the representation of a method for

---

[50] See: n. 2; Brouwer, "Historical Background, Principles, and Methods of Intuitionism"; and Shapiro, *Thinking about Mathematics*, ch. 7.
[51] Hanna, *Kant and the Foundations of Analytic Philosophy*, pp. 54–65, and 154–159.

representing a multitude (i.e., a thousand) in accordance with a certain concept than the image itself, which in this case I could survey and compare with the concept only with difficulty. Now this representation of a general procedure of the imagination for providing a concept with its image is what I call the schema for this concept.

In fact it is not images of objects but schemata that ground our pure sensible concepts.... [T]he **image** (*Bild*) is a product of the empirical faculty of productive imagination, [but] the **schema** of sensible concepts (such as figures in space) is a product and as it were a monogram of pure a priori imagination, through which and in accordance with which the images first become possible ... The schema of a pure concept of the understanding ... is something that can never be brought to an image at all, but rather is only the pure synthesis, in accord with a rule of unity according to concepts in general, which the category expresses, and is a transcendental product of the imagination, which concerns the determination of inner sense in general, in accordance with conditions of its form (time). (*CPR* A140–2/B180–1)

So what does all this tell us? According to Kant, a mathematical concept is shown to be objectively valid or empirically meaningful just insofar as it can be supplied with a corresponding schema, in pure or formal intuition, by means of the pure imagination. This is the same as to construct that concept. But the schema is *not* itself an object in the strict or narrow sense (i.e., a *Gegenstand*), namely, an empirical substance or an object of experience; nor is it an empirical image of an empirical object. Instead, it is an object only in the loose or broad sense (i.e., an *Objekt*), which can include representational targets or intentional objects of all sorts.

More precisely, a schema is a *quasi-object* since it is no more than a sort of rule or method, pattern, or template, whose sole function is to illustrate the form and content of the relevant mathematical concept. Quasi-objects are both ontically incomplete (that is, they lack some properties required for objecthood in the strict and narrow sense) and partially indeterminate (that is, there exist properties for which it is neither true nor false that they apply to the quasi-object). Most precisely of all, a schema is a quasi-objective exemplary or paradigmatic instance of a concept, produced by the pure imagination, such that it encodes the relevant conceptual content or conceptual information in a specifically spatial or temporal format. In the terms of contemporary cognitive science, a schema is a "mental model."[52] The schema always bears spatial or temporal structure because it is generated with reference to pure

---

[52] See Johnson-Laird, *Mental Models*, p. 156. And for an application of the notion of mental models to Kant's theory of a priori knowledge, see ch. 7.

or formal intuition by the *synthesis speciosa* or figurative synthesis of the pure imagination (*CPR* B151), and therefore has a direct bearing on the possibility of sensory experience. So the act of construction does not create a mathematical object, but instead only construes a mathematical concept by imaginatively producing a pure spatial or temporal schematic exemplar of it (i.e., a mental model), which is only a quasi-object. The pure spatial or temporal schematic exemplar, in turn, exhibits or illustrates the content of the concept by providing a constraint on all possible models of any proposition or theory into which that concept enters, namely, that the model in question has to be at least isomorphic with the pure spatial or temporal exemplar that is used to construe that concept.

(3) *What, precisely, does it mean to construct a numerical concept?* To construct a mathematical concept, I have said, is to use the pure imagination to create a schema of that concept in pure intuition, hence to create a pure spatial or temporal mental model of it. This mental model encodes conceptual information in a spatial or temporal format. In the case of numerical concepts, the mental model is always temporal in character—not in the sense that it represents an event of some sort, but rather in the sense that it is itself a temporally formatted model of something that is among the natural numbers.

Unfortunately, Kant says very little indeed about how a schematized numerical concept enters into arithmetic*al propositions. In the controversial remarks about algebra, he says that algebra "exhibits every procedure (*Behandlung*) through which magnitude is generated and altered in accordance with certain rules in intuition" (*CPR* A717/B745), by which he clearly means that arithmetic* essentially includes operations as subtraction, addition, division, multiplication, exponentiation, extraction of roots, and so on. So the general idea seems to be that an arithmetic*al proposition is a logical complex consisting of schematized numerical concepts and some arithmetic*al operations on those concepts. In the Axioms of Intuition, Kant says explicitly that arithmetic*al propositions are "propositions of numerical relation" (*Zahlverhältnis*) or "numerical formulas" (*Zahlformeln*) (*CPR* A165/B206) that are neither general in form or content (but in fact singular) nor logically derivable from more general axioms. This is presumably because every atomic arithmetic*al proposition—say, "$7+5 = 12$" or "$3+4 = 7$"—expresses an operation on concepts of numbers, and each of

338 THE PRACTICAL FOUNDATIONS OF THE EXACT SCIENCES

those concepts can be construed in only one way, namely in terms of its own particular schema, in relation to the pure intuition of time:

> That 7+5=12 is not an analytic proposition. For I do not think the number 12 either in the representation of 7 nor in that of 5 nor in the representation of the combination (*Zusammensetzung*) of the two ... Although it is synthetic, however, it is still only a singular proposition. Insofar as it is only the synthesis of that which is homogeneous (of units) that is at issue here, the synthesis can take place only in a single way, even though the subsequent **use** of these numbers is general .... The number 7 [in the proposition 7+5=12] ... is possible in only a single way, and likewise the number 12, which is generated through the synthesis of the former with 5. Such propositions must therefore not be called axioms (for otherwise there would be infinitely many of them) but rather numerical formulas. (*CPR* A164–5/B205–6).

> In the problem, conjoin 3 and 4 in one number, the number 7 must arise not out of a decomposition of the constituent concepts by rather by means of a construction, that is, synthetically. This construction, a singular counting up in an a priori intuition, exhibits the concept of the conjunction of two numbers. (*PC* 10: 556)

This unusual doctrine of arithmetic*al propositions, based directly on Kant's idea that numerical concepts are constructed by means of the pure imaginative introduction of exemplary quasi-objects (schemata, mental models) that occur within the total logico-temporal structure of the numbers, has important implications for the question of whether inferential steps in arithmetic*al proofs are purely logical in character. Take the following simple arithmetic*al argument:

(1) 7+5 = 12
(2) 3+4 = 7
(3) Therefore, 3+4+5 = 12.

If Kant is right, then the logical substitution of "3 + 4" for " 7" under the extensional law of identity requires the constructions of the concepts THREE and FOUR, the operation-concept PLUS, and their synthesis. Arithmetic*al identity is not a purely logical relation.[53] So the inference-step of substitution

---

[53] It follows that arithmetic* equations cannot be entered into proofs as instances of the extensional law of identity. This presumably is why Kant thinks that arithmetic* truths must be entered into proofs as primitively true or unprovable premises (*indemonstrabilia*), premises that depend on no assumptions, and are logically based on the empty set of premises. Such premises cannot be properly speaking called "axioms" because they are not general in form or content (*CPR* A164/B205), although otherwise they function just like axioms in the sense that axioms are all primitively true or unprovable premises in

would not have been valid unless pure or formal temporal intuition and pure imagination had contributed representational content to the numerical concepts. This appears to be generally true of logical inferences in arithmetic*al proofs. Therefore logical inferences in arithmetic*al proofs require intuitional and imaginational supplementation and are not purely logical in character.[54]

As I see it, Kant is not asserting that arithmetic* is the pure science of time. Rather, as Hacking suggests, Kant is asserting a highly original two-part doctrine about the cognitive semantics of the concept NUMBER: (a) that the content of the concept NUMBER requires our pure or formal intuition of the sempiternal (or infinite unidirectional) series of successive moments of time as a non-logical necessary condition of that concept's objective representational content; and (b) that the content of the concept NUMBER equally requires the logical functions of quantity and their corresponding categories. If Kant is right about this, then arithmetic* is essentially the result of combining the formal ontology of our human intuitional representation of time with the conceptual resources of logic in Kant's sense. That his own conception of arithmetic comprehends at most the primitive recursive fragment of elementary arithmetic, and that Kant's own conception of logic comprehends at most the monadic fragment of elementary logic, are ultimately far less important than his deep insight into the essentially two-sided temporal/intuitional and logical/conceptual structure of the pure science of numbers. This dual structure is at once irreducibly anthropocentric and also strictly constrained by the a priori and categorically normative cognitive science of logic. So arithmetic* for Kant is not only an exact science, but also and even more fundamentally, *a human or moral science*. This perhaps surprising Kantian claim however does *not* mean that arithmetic* is

---

arithmetic*al arguments (CPR: A733/B761). In the *Foundations of Arithmetic*, pp. 5-6, Frege criticizes Kant for appealing to arithmetic*al indemonstrabilia, because he (I mean Frege) thinks that there must be only as many first principles as can be comprehended in a compact rational survey. Frege's criticism is odd for two reasons. First, Kant thinks that pure temporal intuition guarantees that arithmetic* will be cognitively accessible, thus satisfying the requirement of a compact rational survey—so Frege's worry depends entirely on the question-begging assumption that pure intuition is excluded by our rational mode of access to arithmetic truth. Second, Frege himself thinks that logical reasoning depends on unprovable logical laws, but has no way of showing that there are not infinitely many such logical indemonstrabilia.

[54] This too is perfectly consistent with Gödel-incompleteness and suggests the essentially Kantian thesis that there are true unprovable sentences in elementary arithmetic precisely because arithmetic*al proof is not purely logical, but on the contrary always requires intuitional and imaginational supplementation.

grounded on historical *Verstehen*, "interpretation," social practices, or individual psychologies. Instead it means that arithmetic*is partially constituted by the innate architecture of our human cognitive faculties, which in turn are sharply constrained in their rule-governed operations by our ultimately practical rational animal nature.

And this brings us back to the non-trivial metaphysical implications of Kant's weak transcendental idealism. We did *not* create the numbers. The numbers are *objectively real*, in the sense that their *sui generis* ontology is not dependent on our actual human existence: they were forever there before there were humans, and they will be forever there after the humans and other creatures minded like us are all dead and gone, precisely because numbers are determined by intrinsic formal-structural features of the larger natural world that envelops us. *But if rational human animals or persons had been metaphysically impossible, then there would have been no numbers.*

Otherwise put, the necessary possibility of persons is built into the natural world at its metaphysical foundations by virtue of its intrinsically containing some uninstantiated but instantiable mental properties. These properties are instantiated by animals minded like us whenever and wherever the world of matter reaches a sufficient level of causal-dynamic complexity. Charitably on Kant's behalf, and in order to bracket out any sort of explicit or even implicit teleological or so-called "intelligent design" argument on Kant's part, we can assume that this is all just a matter of sheer brute fact, and that the weak transcendental mentalistic architecture of nature did not come to be by any sort of *intentional* means. On the contrary, we can assume that the conscious intentionality of persons and other animals came to exist utterly unintentionally, and that nature is "intelligently designed" only in the sense that it has *an amazingly intricate and partially mentalistic structure that is irreducible to mere physical mechanism*. But the philosophical hypothesis of the sheer brute fact of the weak transcendental idealistic architecture of material nature smoothly explains why the larger natural world is manifestly *scientifically user-friendly*: it explains why *we* can actually do pure mathematics, and why the pure mathematics implicit in our fundamental physics actually *applies* to all of external nature. Indeed, it is now possible to see clearly how Kant's weak transcendental idealism is arguably a core element of the best overall explanation of all the basic properties of the exact science of pure mathematics, and, in turn, also of all the other sciences that involve or presuppose pure mathematics.

7

How Do We Know Necessary Truths?

The conception of the a priori points to two problems which are perennial in philosophy: the part played in knowledge by the mind itself, and the possibility of "necessary truth" or of knowledge "independent of experience." But traditional conceptions of the a priori have proved untenable. That the mind approaches the flux of immediacy with some godlike foreknowledge of principles which are legislative for experience, that there is any natural light of any innate ideas, it is no longer possible to believe.

C. I. Lewis[1]

Only that whose certainty is apodictic can be called authentic science (*eigentliche Wissenschaft*).

MFNS 4: 469

[T]he distrust of the "intuitional" basis of analytic philosophy... is rooted in nothing less than an imperfect understanding of scientific method.

Arthur Pap[2]

7.0. Introduction

How do we know necessary truths? The traditional answer is that we know them "independently of all sensory experience," or a priori. It is received wisdom within both the pre-Quinean and post-Quinean parts of the analytic tradition, however, that the only even minimally acceptable

---

[1] Lewis, "A Pragmatic Conception of the A Priori," p. 169.
[2] Pap, *Semantics and Necessary Truth*, p. 422.

candidates for explaining the a priori knowledge of necessary truth are platonism[3] and conventionalism.[4] Yet both platonism and conventionalism are subject to deep and familiar objections. The platonic perception of universals and other abstract objects is notoriously "mysterious";[5] and conventionalism is just as notoriously open to Quine's famous objections to the effect that it can neither (i) define logical truth without presupposing logic, nor (ii) coherently ground its fundamental distinction between the analytic and the synthetic.[6] Even worse, the double failure of platonism and conventionalism leads to an unhappy "forked" counter-reaction: either it is argued that our knowledge of necessity is primarily or wholly a posteriori, via the natural sciences (scientific essentialism); or else the very possibility of the human knowledge of necessity, be it a priori or a posteriori, is rejected (modal skepticism, for example, Quine himself).

From another point of view, however, the "*either* platonism or conventionalism, *or* scientific essentialism or modal skepticism" dilemma is a false one, since in my opinion it presupposes without sufficient justification the falsity of Kant's theory of a priori knowledge. This presupposition is nicely captured by one of the remarks made by C. I. Lewis—not altogether incidentally, one of Quine's teachers at Harvard—in the first epigraph for this chapter: "That the mind approaches the flux of immediacy with some godlike foreknowledge of principles which are legislative for experience... it is no longer possible to believe." Now it is easy enough to know what, for Lewis and the other philosophers in the Quinean and post-Quinean analytic tradition, automatically disqualifies Kant's doctrine from the start: the seeming impossibility of transcendental idealism.

Q: And can you please tell me again what transcendental idealism is?

A: It's funny you should ask me that. Because I was just about to say that Kant's *strong transcendental idealism* is the two-part doctrine to the effect (i) that the proper objects of human cognition are nothing but subjective sensory appearances or phenomena, not things-in-themselves or noumena (the idealism thesis), and (ii) that the human mind by means of various acts of a priori synthesis strictly

---

[3] See, e.g., Frege, "Thoughts"; Moore, "The Nature of Judgement"; and Russell, *The Problems of Philosophy*, pp. 46–68.
[4] See Wittgenstein, *Tractatus Logico-Philosophicus*, props. 6.122–6.123, p. 163; Ayer, *Language, Truth, and Logic*, pp. 71–87; and Carnap, "Meaning Postulates," pp. 222–9.
[5] Frege, "Logic [1897]," p. 145. See also Benacerraf, "Mathematical Truth"; and section 6.3.
[6] See Quine, "Truth by Convention,"; and Quine, "Two Dogmas of Empiricism."

imposes a set of non-empirical forms or structures upon original perceptual inputs from the world in order to generate its phenomenal objects (the transcendentalism thesis). In turn, the idealism thesis and the transcendentalism thesis jointly entail that empirical objects are token-identical with the contents of sensory representations, and type-identical with the a priori forms or structures innately specified in our cognitive faculties.

Now, from strong transcendental idealism and the plausible assumption that the human cognizer can be reflectively aware of her own cognitive capacities, we can immediately derive the basic Kantian epistemic doctrine that all knowledge of necessity is *self*-knowledge of the human mind's non-empirical formal-structural contributions to both the semantic contents and the external objects of its own cognition. This idea is captured in one of Kant's *Reflexionen* and most memorably in a famous slogan in the B edition Preface to the first *Critique*: "we can grasp only what we can make ourselves" (R 2398; 16: 345) and "reason has insight only into what it self-produces (*selbst... hervorbringt*) according to its own design (*nach ihrem Entwurfe*)" (CPR Bxiii).

So it seems that only the fear of transcendental idealism stands in the way of an adequate theory of a priori knowledge. To this someone might simply reply: "So just get over it, and become a transcendental idealist!" But on the other hand, and as with other serious bogeymen in life and in philosophy, there surely is *something* that needs to be dealt with here. For this reason I believe that the analytic philosopher's fear of transcendental idealism can at least be significantly allayed—if not altogether purged[7]—by showing that *strong* transcendental idealism, in the sense just spelled out, is logically *detachable* from the core of Kant's theory of a priori knowledge and to that extent inessential to it. So, as in chapter 6, my proposal is to reject any appeal to strong transcendental idealism,

---

[7] Finally getting over the fear of *all* forms of transcendental idealism, I think, would take a stronger, headier, and more Wittgensteinian sort of philosophical therapy. To use Tractarian language, the "metaphysical subject" must finally bring himself to see how "solipsism strictly carried out coincides with pure realism." See Wittgenstein, *Tractatus Logico-Philosophicus*, props. 5.64 and 6.43, pp. 153, 185. Less enigmatically put, the analytic philosopher in general and the scientific naturalist in particular must come to terms with the surprisingly disturbing fact *that mental properties are part of the fundamental constitution of the natural world*, as properties whose instantiation is necessarily possible with respect to the actual existence of space and time (as the necessary possibility of adequate conscious non-conceptual representations of space and time), and as properties that are really instantiated in all and only living organisms of sufficient neurobiological complexity (as causally efficacious conscious volitions). For the former, see section 6.1; for the latter, see sections 8.2–8.3.

both for the purposes of this chapter and the rest of the book. Nothing I will present as part of the core of Kant's theory of a priori knowledge will presuppose it. Correspondingly, however, I will also assume on Kant's behalf a minimal commitment to *weak* transcendental idealism, which says that the existence of space and time, and also the existence of all abstract objects, require the necessary possibility of *creatures minded like us*, that is, the necessary possibility of human cognizers who are capable of adequately representing space and time. The rationale for distinguishing between strong transcendental idealism, and also for attributing only weak transcendental idealism and not strong transcendental idealism to Kant under the charitable critical-interpretive methodology I have adopted, has already been worked out in section 6.1.

In order to go forward from here, I will also make two (I hope!) fairly uncontroversial assumptions that:

(1) There *actually are* some necessary truths (for example, in conceptual analysis, modal metaphysics, logic, or mathematics).
(2) Some individual actual human thinkers *already actually have known* some necessary truths, therefore that some individual actual or possible human thinkers *really can know* some necessary truths.

Let us call the conjunction of these two assumptions *minimal modal realism*. Minimal modal realism is so-called because it is consistent not only with all forms of modal realism (platonism, rationalism, essentialism, etc.), but also with every form of modal idealism (transcendental idealism, monadology, Carnap-style phenomenalism, absolute idealism, etc.), and even with modal anti-realism (conventionalism, holism, etc.).

How can minimal modal realism be justified? In adopting minimal modal realism as a starting point, I am excluding only two possible skeptical doctrines about strong modality and our knowledge of it:

(a) that although necessary truths do exist, they are all humanly unknowable (weak modal skepticism); and
(b) that there are no such things as necessary truths at all, and consequently there are neither any actual or possible human knowers of necessary truths nor any actual or possible human knowledge of necessary truths (strong modal skepticism, including modal eliminativism).

But it seems to me, just as I think it would also have seemed to Kant, that both (a) and (b) are excessively skeptical about rationality, given that the skeptic has to justify his modal skepticism rationally. How could a skeptic ever rationally justify any of his claims if he did not at least implicitly believe, for example, in the strongly modal notion of logical consequence?[8]

But here is another and interestingly different Kantian line of response to the modal skeptic.[9] For Kant, the very idea of strong modality *most deeply matters* to creatures minded like us. More precisely, according to Kant, in the psychological constitution of every rational human animal there exist at least two innate emotional dispositions towards natural necessity (including both analytic and synthetic necessary truth as well as natural law) and deontological necessity (the categorical "ought" and the moral law), each of which is also cognitively expressed in the form of categorically normative a priori principles of theoretical and practical reasoning. These innate emotional dispositions are partially constitutive of human rationality itself. For Kant, we would not be rational animals unless we were also innately disposed *to revere* the very idea of strong alethic and deontic modality, as an expression of the most fundamental need of all human persons for *self-transcendence*:

> Two things fill the mind with ever new and increasing admiration and reverence (*Ehrfucht*), the more often and steadily one reflects on them: *the starry heavens above me and the moral law within me.* I do not need to search for them as though they were veiled in obscurity or in the transcendental region beyond my horizon; I see them before me and connect them immediately with the consciousness of my existence. (*CrPR* 5:162)

Weak modal skepticism says, in effect, that these modal conations are ultimately nothing but—to use Sartre's striking phrase in *Being and Nothingness*—"useless passions." Kant anticipates this Sartrean mood in a closely related context: "reason, which is so desirous of this kind of [necessary a priori] cognitions, is more stimulated than satisfied by [what is merely

---

[8] See Hanna, *Rationality and Logic*, ch. 7.

[9] I am explicitly considering here only a *cognitive* modal skeptic, that is, a skeptic whose doubt consists in a negative propositional attitude towards the realistic modal theses of which he is skeptical. Strictly speaking there can also be a *non-cognitive* or purely practical skeptic whose skepticism consists merely in living a skeptical way of life or in affectively manifesting a Nietzschean "revaluation of values." But even so it is very hard to know what it would be like for a creature minded like us to live, move, or have its being in a world without, e.g., strong logical modality. See Hanna, *Rationality and Logic*, ch. 7.

contingent]" (*CPR* A1–2). In turn, strong modal skepticism says, in effect, that even though human rationality is fundamentally constituted by its strong modal aims, the very idea of strong modality is a pseudo-concept and therefore human reason is based on a myth. Kant also anticipates this mood in the special case of speculative pure theoretical reason: "human reason has the peculiar fate in one species of its cognitions that it is burdened by questions which it cannot dismiss ... but which it also cannot answer, since they transcend every capacity of human reason" (*CPR* Avii). But surely for rational human animals like us, it is an acceptable minimally modally realistic assumption that the non-cognitive core of human reason is neither a useless passion nor seriously self-deceived. To assume the contrary would be cognitively suicidal, not to mention the end of the world as we know it:

Proceeding skeptically nullifies all our effort, and it is an antilogical principle ... For if I bring cognition to the point where it nullifies itself, then it is as if we were to regard all human cognitions as nothing. (*VL* 24: 884)

Therefore, both weak and strong modal skepticism are rationally self-stultifying.

Granting minimal modal realism, then, what I want to do in this chapter is to work out Kant's highly original and highly important doctrine of an essential element of a priori knowledge: epistemic necessity, or what Kant calls "inner necessity" (*innern Notwendigkeit*) (*CPR* A1–2/B2). Epistemic necessity, we shall see, is the essential characteristic of the maximally strong propositional attitude that Kant calls "conviction" (*Überzeugung*) (*CPR* A820/B848). He argues that the ground of conviction—hence of epistemic necessity—is ultimately to be traced back to the mental act of "insight" (*Einsicht*). In turn, insight is the a priori mental act, state, or process of immediately (that is, non-inferentially) knowing a necessary truth: "what I have *insight* into, I must cognize as necessary and consequently as *a priori*" (R 1626; 16: 43). According to Kant, it is only the unique sort of subjective consciousness found in insight, or what he calls the "consciousness of necessity" (*Bewußtsein der Notwendigkeit*) (*JL* 9: 66), that explains the epistemic necessity found in conviction. So, in this way, his account of the strongest epistemic modality can be developed only through *the phenomenology of insight*.

In one sense, the phenomenology of insight is undoubtedly not original to Kant: in fact, Descartes and Leibniz very usefully initiated the study of it in the seventeenth century when they talked about the "clarity" and "distinctness" of rational intuitions.[10] Unhappily, however, the Cartesian and Leibnizian accounts of clarity and distinctness are notoriously unclear and indistinct. Also they both defended the really hopeless thesis that clarity and distinctness together supply an introspective *criterion* for truth (a criterion underwritten, of course, by a non-deceiving God). In sharp opposition to this, as we have already seen in chapter 5, Kant recognizes that truth is determined quite independently of all introspective access to our individual mental acts, states, or processes of *Fürwahrhalten*, "holding-to-be-true," or propositional affirmation. And this is precisely because the truth of a judgment depends instead on "upon the common ground, namely, upon the object" (*CPR* A821/B849), in the following dual sense:

(a) truth by (nominal) definition consists in an *Übereinstimmung*, "accordance," or correspondence between a judgment and its object (*CPR* A58/B52); and

(b) that:

> we lack a universally sufficient characteristic ... for correctly and infallibly distinguishing the subjective grounds of holding-to-be-true from the objective ones. (*BL* 24: 146–7)

In this way, "holding-to-be-true can be apodictic, without the cognition being objectively apodictic" (*R* 2479; 16: 388). Nevertheless, according to Kant, the attitude of conviction *by its very nature* includes the existence of a necessary truth: "objectively necessary holding-to-be-true, which, if it is at the same time subjectively necessary, is conviction" (*R* 2465; 16: 382). So this means that insight both logically requires and is partially constituted by something that is non-subjectively or objectively real (actually realized in space and time), that is, something given externally to and independently of the individual thinker. Put in contemporary terms, and odd as it may initially seem in view of his commitment to transcendental idealism, it follows that Kant is most definitely a *content externalist* about a priori belief and knowledge.

This point cannot be overemphasized: Kant is both a weak transcendental idealist *and* a content externalist. As noted in section 1.2, by "content

---

[10] See Descartes, "Rules for the Direction of the Mind," pp. 10–76, Descartes, "Principles of Philosophy," pp. 207–8; Descartes, "Meditations on First Philosophy," p. 24; and Leibniz, "Meditations on Knowledge, Truth, and Ideas."

externalism," I mean the doctrine that the representational content of at least some of our mental states is at least partially individuated or determined by the direct reference (or another direct relation) of those states to something existing outside the human mind in the worldly spatiotemporal environment, be it causal or otherwise physical, social, or historical.[11] We saw in chapter 1 that in the Refutation of Idealism Kant argues that "the consciousness of my existence is at the same time an immediate consciousness of the existence of other things outside me" (*CPR* B276). He is therefore a content externalist about introspective conscious awareness, and more generally about empirical apperception or self-consciousness. The deeper point I am making here is that not only does he defend content externalism for self-reporting conscious introspection, but he also defends it for *all* the other propositional attitudes, especially including the attitude that is essential for a priori knowledge, *conviction*. This is equally true for our cognition of analytic truths (*CPR* A6/B10), synthetic a priori truths in mathematics (*CPR* A27–8/B43–4, A36/B52–3, B293), and synthetic a priori principles of pure understanding (*CPR* B291). In each case, the content of the cognition is necessarily directly referred (or otherwise directly related) to the actual objectively real spatiotemporal world, via intuition, in at least a partially determining way.

After Putnam's and Burge's attacks on content individualism in the 1970s and 1980s, it became relatively commonplace to be a content externalist about empirical or a posteriori cognition and knowledge. But it remains unusual—perhaps even radically unusual—to be a content externalist about *a priori* cognition and knowledge. Given minimal modal realism, however, the Kantian step from empirical content externalism to apriorist content externalism seems both natural and smooth, since, as we shall see in the next four sections, the Kantian cognitive-semantic account of strong modality says precisely that alethic necessity and deontological necessity are constitutively *immanent* or embedded in at least some of our empirical mental representational contents, as a priori intrinsic structural properties of them.

So how is a priori insight possible on Kant's model if he eschews the Cartesian and Leibnizian "criterialist" approach and also defends nonempirical content externalism? In broad outline, here is how his answer

---

[11] For the classic defense of content externalism in the philosophy of mind see Burge, "Individualism and the Mental"; and for a defense of content externalism in epistemology, see Williamson, "Is Knowing a State of Mind?"

goes. According to Kant, when having an insight the rational human mind creates and manipulates a subsidiary formal-structural sensible image—a mental "model" or *Urbild*—of the semantic structures of its correlative necessary truth:

> I cannot have full insight into mere abstract cognitions in general unless I can portray and have insight into a case in *concreto*. (BL 24: 109)

"Portrayal" or *Darstellung*, I will argue later, is a *schematizing* function of the imagination. Kant's appeal to the schematic imagination thereby avoids the platonist's highly problematic appeal to a purely rational analogue of sense perception, for there is no need to assume any sort of mysterious causal contact between insight and its objects.[12] On the contrary, given Kant's theory of truth (see chapter 5), all we need to assume is the necessary possibility of an *actual precise isomorphism* between the propositional content of insight and its objects. Then: (i) the schematic imagination produces for itself a subsidiary mental model of the propositional content of insight; (ii) insight consists in consciously cognitively manipulating and thereby consciously apprehending this subsidiary mental model in a certain way;[13] and finally (iii) the correspondence-relation of actual isomorphism between the structure of the propositional content and the structure of the relevant parts of the external world independently guarantees the necessary truth of that very proposition.

Now this cognitive manipulation and conscious apprehension of a mental model or schema via the schematizing imagination—the best-known example of which occurs in a priori construction in mathematics (see section 6.4)—is a species of what Kant calls *theoretical technique* (CPJ 20: 200). Theoretical technique is active knowledge, and, in this case, more precisely it is the act of a priori *scientific knowing* (*Wissen*), insofar as it is governed by both hypothetical (instrumental) and categorical (non-instrumental) imperatives or norms. Kant's epistemic appeal to the theoretical technique of the a priori schematizing imagination thus avoids the conventionalist's problematic appeal to arbitrary decision or mere stipulation, which to be

---

[12] The general idea of a Kantian imagination-based strategy for avoiding the perils of platonism has been indicated in passing, although not actually developed, by Parsons in "Mathematical Intuition," p. 200.

[13] For a recent version of this idea in the context of cognitive psychology, see Johnson-Laird, *Mental Models*, pp. 2, 190, 407, 415. And see also Blachowicz, "Analog Representation Beyond Mental Imagery," pp. 78–83.

sure is practical-volitional and instrumental in character, yet absolutely unconstrained by categorical or non-instrumental norms. Conventionalism in effect says *that anything goes, so nothing really matters*. But Kant's view on the contrary says *that everything really matters, precisely because not everything goes*. So another way of describing the take-home message of this chapter, then, is to say that it provides a general framework for properly understanding Kant's famous modal-epistemological slogan, "reason has insight only into what it self-produces according to its own design," in terms of active a priori knowledge and pure practical reason, while rejecting any appeal to *strong* transcendental idealism, and while also effectively avoiding platonism, conventionalism, scientific essentialism, and modal skepticism alike.

This framework will be supplied in four stages. First, I explicate Kant's theory of epistemic necessity and epistemic apriority (section 7.1). Then, second, I exemplify that theory with a case study: his account of conceptual insight into simple analytic truths (section 7.2). Third, I make some remarks about the special role of insight in Kant's overall conception of a priori knowledge (section 7.3). And then, fourth and finally, I very briefly re-explicate the concept of insight in terms of his notion of theoretical technique (section 7.4).

7.1. Epistemic Necessity and Epistemic Apriority, Kant-Style

"Epistemic necessity" is a late twentieth- and early twenty-first-century philosophical term-of-art, used to draw a sharp distinction between a purely cognitive notion of necessity, and a purely semantic notion of it. According to this distinction, epistemic necessity concerns the strongest modality of affirmative attitudes, judgments, or beliefs about propositions, whereas logical or "metaphysical" necessity concerns only the strongest modality of propositional truth. As we saw in chapters 3 and 4, for Kripke, a proposition P is metaphysically necessary if and only if P is "strictly metaphysically necessary," that is, if and only if P is true in every logically possible world.[14] Unfortunately, however, although Kripke freely helps

---

[14] As I also pointed out in chs. 3 and 4, this formulation of Kripkean metaphysical necessity is slightly imprecise. Strictly speaking, a proposition P is metaphysically necessary if and only if either P is true in every logically possible world without qualification, or else P is true in every logically possible

himself to the contrastive notion of epistemic necessity, he never offers an explicit formulation of it.[15] On the other hand, Putnam does at least provide us with a formulation: a proposition P is epistemically necessary if and only if P is "rationally unrevisable."[16] But that is still too vague. Unrevisable in *what* sense? And according to *which* conception of rationality?

What I want to propose now is one Kantian way (and no doubt there are others) of making Putnam's vague notion more precise: A proposition P is epistemically necessary if and only if any mental act, state, or process which yields an understanding of P, suffices for belief in P. That is, P is epistemically necessary just insofar as anyone's comprehending the meaning of P, necessitates her belief in P. In the *Blomberg Logic*, Kant puts it like this:

> In the case of certainty it is not the truth of the thing we cognize that is necessary, but rather the holding-to-be-true ... This necessity of accepting the thing does not lie in the *objectum* itself, however, but instead in the subject. (BL 24: 229–30)

Corresponding respectively to metaphysical and epistemic necessity are metaphysical contingency or possibility, and epistemic contingency or possibility. A proposition P is metaphysically contingent if and only if P is true in some possible worlds and false in others; and P is metaphysically possible so long as P is not necessarily false. But on the other hand, I shall say, a proposition P is epistemically contingent if and only if some possible ways of understanding P lead to belief in P, and some other possible ways lead to a belief in P's denial; and P is epistemically possible just in case not all possible ways of understanding P lead to a belief in its denial. Since belief in epistemically contingent propositions is often sensitive to the special experiential conditions under which they are understood, it follows that epistemically contingent propositions are often believed or known a posteriori.[17]

To repeat, then: metaphysical necessity is the strongest modality of propositions (alethic modality), and epistemic necessity is the strongest

---

world in which a certain individual or kind exists; see Kripke, "Identity and Necessity," p. 164, and *Naming and Necessity*, pp. 38, 109–10, 125, 138. This disjunctive formulation is needed to account for the difference between analytic or logical necessity and synthetic or non-logical necessity (because the relevant individual or kind might not exist in every logically possible world), although Kripke does not explicitly put it in those terms.

[15] See, e.g., Kripke, *Naming and Necessity*, pp. 103–5, 123–5, 141–4, 150–3.
[16] See Putnam, "The Meaning of 'Meaning'," p. 233.
[17] Not all epistemic contingency is a posteriori, however, since it is possible to be unsure about propositions that are believed or known only a priori, e.g., Goldbach's conjecture that every even number greater than 2 is the sum of two primes. See Kripke, *Naming and Necessity*, 36–7.

modality of our attitudes towards or beliefs about propositions (doxic modality). So they are not definitionally equivalent notions. The question then arises whether they are extensionally equivalent notions. According to both Kripke and Putnam, and more generally according to scientific essentialism, some propositions are metaphysically necessary although epistemically contingent and a posteriori. Familiar examples of this are "Gold is the element with atomic number 79," and "Hesperus is Phosphorus." As I argued in section 3.3, however, there are good reasons for rejecting the very idea of the necessary a posteriori and for accepting Kant's claim that knowledge of the specific modal status of any necessary proposition is a priori: "if a proposition is thought along with its **necessity,** it is an *a priori* judgment" (*CPR* B3). So, if I am right on Kant's behalf, then contrary to scientific essentialism there are in reality no necessary a posteriori propositions—indeed, what essentialists have mislabeled as "necessary a posteriori propositions" all belong to Kant's class of "impure" a priori propositions (*CPR* B3).[18] In any case, I need not re-argue all those disputed points here, because even if one leaves the issue of the necessary a posteriori open, we can still easily show that metaphysical necessity and epistemic necessity are conceptually distinct by simply pointing out that they are not definitionally equivalent.

Let us focus now on epistemic necessity. Kripke does in fact explicitly hold that if a proposition is epistemically necessary then it is also a priori. But as in the case of epistemic necessity, he never offers us an explicit formulation of the crucial notion of epistemic apriority.[19] This is in sharp contrast to Kant.[20] What we must do is understand this text properly:

---

[18] This includes both analytic propositions such as "If Socrates is a bachelor, then Socrates is unmarried" and also synthetic a priori propositions such as "Every event has a cause" and "7 apples plus 5 apples equal 12 apples." By contrast, for Kant "Hesperus is Phosphorus" would be *strongly* synthetic a priori, that is, synthetically necessary and experience-independent under a special assumption about the existence and constitution of matter in the actual world and, correspondingly, about the actual set of natural causal-dynamic laws. See section 3.5.

[19] There is a difference between epistemic apriority and semantic apriority. Semantic apriority is how the meaning and/or truth-conditions of a proposition are underdetermined by its verification-conditions. Epistemic apriority, by contrast is how belief or knowledge is "independent of all sensory experience." In the text I focus solely on epistemic apriority.

[20] Kripke does draw a distinction between the meanings of the phrases "*can* be known a priori" and "*must* be known a priori" in order to account for cases in which someone learns, through experience, a proposition that can also be known independently of experience (*Naming and Necessity*, pp. 34–5). As I will argue immediately in the text, I take it to be an essential feature of Kant's doctrine that whatever is a priori is also actually acquired through (although not caused by) sensory experience. So Kripke's

We will understand by *a priori* cognitions, not those that occur independently of this or that experience, but rather those that occur *absolutely* independently of all experience. (*CPR* B2–3)

What is Kant saying here about a priori cognition? Taken at face value, it seems to say that it is possible for human cognizers to have belief or knowledge without ever having had any sensory experiences whatsoever. But that seems completely absurd, nonsense on stilts: how could there be *human* cognition without some inner or outer experiences? This absurdity is well brought out by Frege, who observes in a footnote appended to a brief discussion of pure rational knowledge that:

I do not mean in the least to deny that without sense impressions we should be as stupid as stones, and should know nothing either of numbers or anything else; but this psychological proposition is not of the slightest concern to us here.[21]

But of course "this psychological proposition" is of great concern to *us here*, for it implies the general impossibility of belief or knowledge that is not also in some fundamental respects sensory or experiential. Now Kant explicitly says that "there is no doubt that all our cognition begins with experience" (*CPR* B1), so it is obvious that Kant does not defend the absurd doctrine. What is he driving at then, when he says that cognition is a priori if and only if it is not independent of this or that experience but rather "*absolutely* independent of all experience"?

Here is an alternative analysis that both avoids the absurdity and also is smoothly consistent with Kant's own statements:

Epistemic Apriority: The belief in or knowledge of a proposition *P* is epistemically a priori if and only if there is *no particular set or specific sort* of sensory experiences that is either necessarily required or solely sufficient to believe or know *P*, even

---

distinction is not in opposition to Kant's view, although Kripke appears to believe that it is at odds with it—despite the fact that he also admits that "it would be still open to [Kant] to hold" that *a priori* knowledge can be merely possible and not necessitated (*Naming and Necessity*, pp. 159–60). But, even quite apart from the somewhat tangled issue of Kripke's interpretation of Kant, there are two sharp differences between Kant's and Kripke's conceptions of the a priori: (1) Kripke holds that it is possible to know a proposition without any sensory experience whatsoever (whereas for Kant every a priori proposition is also actually acquired through some sort of inner or outer experiential act, state, or process); and (2) Kripke holds that there are necessary a posteriori and contingent a priori truths (whereas Kant holds that if a proposition is knowable at all, then it is known a priori if and only if it is necessary).

[21] Frege, *The Foundations of Arithmetic*, p. 115, n. 2.

though every cognizer of P actually cognizes it through some mental state, act, or process involving sensory experiences.[22]

Thus all belief and knowledge involve sensory experiences: sensory experiences always trigger, or occasion, the operations of our cognitive faculties, and always accompany those operations as well. But in a priori belief or knowledge, every empirical mental state, act, or process by means of which a cognition of the proposition P actually occurs, *essentially underdetermines* belief in P or knowledge of P. Or in other words, whatever the empirical conditions, P could have been believed or known on the basis of a different set or sort of sensory experiences; and P is never believed or known solely because of that set or sort of sensory experiences. So apriority entails that any mental act, state, or process by which P is believed or known either contains content-elements, or invokes psychological capacities, which are neither reducible to nor strongly supervenient upon its sensory content—which is just to say that it contains non-sensible and *spontaneous* (that is, causally and temporally unprecedented, underdetermined by sensory inputs, creative, and self-guiding) elements or capacities.[23] Or, as Kant so crisply puts it: "although all our cognition begins **with** experience, yet it does not on that account all arise **from** experience" (*CPR* B1).

Now, as we have seen, if a proposition P is epistemically necessary, then it is such that to understand P is thereby necessarily also to believe P. If a proposition P is such that understanding P necessitates the cognizer's

---

[22] Harper, in "Kant on the A Priori and Material Necessity," p. 250, correctly points out that for Kant the empirical acquisition of a content is quite consistent with its being a priori. So too Philip Kitcher, in "A Priori Knowledge," pp. 3–10, works out an analysis of apriority that is in one respect—the idea that apriority is the underdetermination of a cognition that nevertheless has an empirical origin—quite similar to the one I have ascribed to Kant. Still, Kitcher's analysis does differ importantly from Kant's in three ways: (i) Kitcher adds a condition tying apriority necessarily to truth; (ii) he permits particular sets of experiences to operate as solely sufficient conditions of a priori cognition; and (iii) his analysis presupposes an explicitly naturalistic-psychologicistic framework of explanation.

[23] This spontaneity can be either rational (whether conceptual-logical, or volitional-autonomous) or proto-rational (whether intuitional-imaginational, or volitional-affective). It is a basic and striking feature of Kant's conception of the mind that cognitive spontaneity, or "the mind's power of self-producing (*selbst hervorzubringen*) representations" (*CPR* A51/B75), attaches not only to the understanding or the faculty for conceptualization, but also—and indeed fundamentally—to the imagination (*CPR* A78/B103, B152). McDowell in *Mind and World* has rightly drawn attention to the spontaneous component in all judgment and cognition; but Kant would sharply disagree with McDowell's attempt to reduce spontaneity to our conceptual or discursive capacities. Imaginational spontaneity necessarily includes a reference to intuition (*CPR* B151); and more generally the spontaneous production of schemata or mental models (*Urbilder*) or even mental images (*Bilder*) is not itself a function of conceptualization or discursivity. Kant's view is well supported by recent work in cognitive psychology; see Kosslyn, *Image and Brain: The Resolution of the Imagery Debate*.

believing P, then there is no act, state, or process of understanding P that will also disconfirm it for that cognizer. So no matter what set or sort of sensory conditions is combined with that act, state, or process of understanding P, it will still automatically yield belief in or knowledge of P. And, even if all the merely sensory components in the understanding of P tend towards its disconfirmation, nevertheless belief in or knowledge of P will still result from the non-sensory and spontaneous component of that understanding. Belief in or knowledge of P is necessarily underdetermined by all sets or sorts of sensory experiences. Therefore the epistemic necessity of P entails the apriority of belief in or knowledge of P.[24]

While all epistemically necessary propositions are cognized a priori, however, it is not the case that every proposition cognized a priori is epistemically necessary. As Kant puts it in one of his logic lectures:

Can't one also accept propositions *a priori* on belief? Yes. Mathematics is also of this kind. One believes mathematicians because it is not possible that they can err since they would hit upon false consequences at once. (DWL 24: 733)

In other words, a proposition can be epistemically contingent and still a priori. For example, right now I have an a priori belief in Gödel's first incompleteness theorem (which says, roughly, that not every true or valid sentence of arithmetic is provable from the laws of logic plus the Peano axioms of arithmetic), even though offhand I cannot quite remember all the steps of his ingenious proof. My understanding of Gödel's first theorem is such that the sensory experiences accompanying that understanding are neither necessarily required nor solely sufficient for my belief in it. Thus my belief is grounded on non-sensory, spontaneous components within my cognitive constitution. Also I rationally trust Gödel and lots of other brilliant logicians since 1931 not to have messed up. Hence I am

---

[24] It might be objected that propositions such as (R) "All red things are red things" will provide a counterexample to this account of apriority. For, so the objection goes, although (R) is clearly known a priori, the concept-term "red" cannot be understood or believed without sensory experiences of a certain kind; hence (R) cannot be understood or believed without sensory experiences of a certain kind, which violates the thesis of *Epistemic Apriority*. As I see it however, the error here lies in thinking that the concept-term "red" has an *semantically essential occurrence* in (R): in fact, (R) is a mere instantiation of the logically necessary truth "All Fs are Fs," into which *any* meaningful predicate-constant or concept-term could have been inserted. And obviously the mere empirical acquisition of the concept-term "red" will not automatically yield even an understanding of (R), much less belief in or knowledge of (R). So, although certain sensory experiences of red may still play a role in the ordinary acquisition of (R), they are neither necessarily required nor solely sufficient for believing or knowing (R).

presumptively warranted in believing Gödel's first incompleteness theorem. Yet, because I cannot run right through Gödel's proof at this moment, I am not rationally compelled to believe in the first incompleteness theorem merely by virtue my understanding it. For while I *partially* understand it and also believe it without being determined to do so by sensory content, I am still somewhat uncertain about its truth. For all I know, it could in fact be (necessarily) false. So just because every rationally unrevisable belief is experience-independent, it does not follow that every experience-independent belief is rationally unrevisable: some experience-independent beliefs are in fact still dubitable, revisable, or epistemically contingent.[25]

Another crucial feature of epistemic necessity is sometimes overlooked in contemporary discussions of it. Although the epistemic necessity of a proposition is not analytically equivalent with that proposition's metaphysical necessity, for Kant it is nevertheless the case that necessarily *if* a proposition is epistemically necessary (hence also a priori), *then* it is also necessarily true: "any cognition that is supposed to be certain *priori* proclaims that it wants to be held as intrinsically necessary (*schlechthinnotwendig*)" (*CPR* Axv).

Kripke has argued, to the contrary, for the existence of the "contingent *a priori*."[26] But those arguments, I believe, can be shown to be unsound. In section 3.3 I argued against the very idea of the necessary a posteriori by showing:

(a) that the mere fact that a necessary proposition (for example, "If Kant is a bachelor, then Kant is unmarried") is empirically learned

---

[25] See also Kripke, *Naming and Necessity*, p. 39. The case of a priori (or experience-independent) cognitions that are nevertheless dubitable or revisable allows for an important distinction between two kinds of rational intuitions: (1) those that provide a presumptive or provisional warrant for our belief by virtue of their non-sensory or purely rational component (let us call these "prima facie intuitions"); and (2) those that by contrast *necessitate* our belief by virtue of their non-sensory component (let us call these "actual intuitions," "authoritative intuitions," or "insights"). The difference, as I have indicated, seems to depend on the contrast between a merely partial or limited understanding of the proposition by means of the non-sensory, rational, spontaneous component on the one hand, and a complete or comprehensive understanding of it by means of the non-sensory, rational, spontaneous component on the other. More generally, I also think that there is an important analogy between the prima facie intuition vs. actual intuition pair, and Ross's famous distinction between "prima facie duties" and "actual duties" (in *The Right and the Good*). Much that is honorifically labeled "intuitions" by philosophers and non-philosophers alike falls into the category of mere prima facie intuitions; like prima facie duties, such intuitions often directly conflict with one another. Similar points are made in Tidman, "The Justification of A Priori Intuitions."

[26] Kripke, *Naming and Necessity*, pp. 54–7, 75–6.

and contains significant empirical content, does not entail that it is a posteriori; and

(b) that giving an a priori justification of one's belief in the *necessity* of a scientific essentialist identity proposition, by knowing a priori of propositions of that type that they are necessarily true if true at all, thereby confers apriority on the belief in its *truth*.

In order to argue effectively against the contingent a priori however, I would have to argue:

(a*) that the mere fact that a contingent proposition (for example, "Pigs can't fly" or "Cats don't grow on trees") is not empirically learned (everybody thinks she knows the truth of "Pigs can't fly" or "Cats don't grow on trees" as soon as she's understood it, and *never* bothers trying to verify it), does not entail that the proposition is a priori; and

(b*) that giving an a posteriori justification of one's belief in the *contingency* of a proposition (for example, showing that flying pigs are possible by putting actual pigs in zero gravity conditions, or showing that cats can grow on trees by successfully grafting actual cats onto trees) thereby automatically confers aposteriority on the belief in its *truth*.

This not the place to try to establish the falsity of Kripke's claim decisively. But we can now see that there are reasons to doubt it. And, in any case, I am far from being the only one who is skeptical about Kripke's conception of the contingent a priori.[27]

Nevertheless, even *granting* Kripke's questionable claim, we can see that he does not hold that it is possible that a proposition could be epistemically necessary and a priori, and yet false. For all of his examples involve cases in which the relevant proposition is supposed to be true-by-stipulation; so despite his objection to Kant, he still defends the weaker thesis that the epistemic necessity and apriority of P entails at least the actual truth of P. Needless to say, Kant would also hold that weaker thesis.

As a consequence of the deep connection between epistemic necessity and truth, however, it follows that the epistemic necessity of a proposition

---

[27] See, for example, Casullo, "Kripke on the A Priori and the Necessary," pp. 164–9; and Dummett, *Frege: The Philosophy of Language*, pp. 115–26.

is not the same as what is sometimes called "mere psychological certainty": the strong feeling of certainty about a proposition which can exist quite apart from any rational grounds for that feeling. Cartesian skepticism shows us that there exists a logical and conceptual possibility of a very powerful demon who causes in us "dreams of certainty," that is, mental acts, states, or processes that in all the relevant details superficially resemble the acts, states, or processes that are actually directed to epistemically necessary propositions, and that can even be introspectively indiscriminable from such acts or processes, but that do not correspond to the external world. Kant calls this the "*phaenomenon* of conviction" (*BL* 24: 146–7). Because of its relevantly superficial or phenomenally indiscriminable resemblance to genuine certainty, the *phaenomenon* of conviction can *seem* compelling and so lead to an actually false belief. But it is crucial to see that because it is only a psychological simulacrum or inauthentic counterpart of a mental act, state, or process that is authentically directed to a proposition that has epistemic necessity, it produces at best a mere psychological certainty. Now, strong feelings about a proposition are one thing and rational belief in it is quite another; hence the very same psychological *phaenomenon* can in some other possible worlds lead someone else (say, someone who is more self-critical about the epistemic import of her strong feelings) to doubt that proposition; hence the proposition in question is only epistemically contingent, not epistemically necessary. Otherwise put, the crucial distinction between mere psychological certainty and what Descartes calls "metaphysical certainty" or "absolute certainty"[28] is in fact merely a special interesting sub-case of the general distinction between epistemic contingency and epistemic necessity.

A central thesis of this chapter is that the label "epistemic necessity" is simply a newfangled term for an oldfangled idea that was developed by Kant as a crucial part of his theory of a priori knowledge. This is particularly evident in Kant's discussion of the propositional attitude of conviction or *Überzeugung*. He offers an explicit analysis of it in section 3 of the Canon of Pure Reason in the first *Critique*, "Of Opinion (*Meinen*), Scientific Knowing (*Wissen*), and **Belief** (*Glauben*)" (*CPR* A820–2/B848–50).[29] In

---

[28] See Descartes, "Discourse on the Method," p. 130; and Descartes, "Principles of Philosophy," pp. 289–90.
[29] There are also useful parallel discussion of these topics in the *Blomberg Logic* (*BL* 24: 142–53), the *Vienna Logic* (*VL* 24: 845–59), the *Dohna-Wundlacken Logic* (*DWL* 24: 730–5, 747), and the *Jäsche Logic*

this much-neglected text, Kant argues that conviction is a fundamental kind of holding-to-be-true, *Fürwahrhalten*, or assertoric propositional attitude. A proposition targeted by conviction is first and foremost subjectively sufficient, or such that it entails the subject's belief in that proposition. Everyone who rationally considers a proposition held-to-be-true by conviction must believe it: so the propositional object of a conviction is epistemically necessary. More than that, however, that proposition is also objectively sufficient, which is to say that (a) it is "valid for everyone merely as long as he has reason" and (b) it satisfies the "presumption" (*Vermutung*) that the proposition has "accordance (*Übereinstimmung*) with the object"—which is to say that the proposition corresponds to the actual world and is itself true. Conviction is therefore externalist in the sense described above.

Kant says that conviction is to be sharply contrasted with "persuasion" (*Überredung*) whose propositional object has a merely "private validity" and is therefore not such that every possible thinker must believe in that proposition. Mere persuasion can in some possible cases be subjectively mistaken for conviction "when the subject has taken something to be true merely as an appearance of his own mind." The "external touchstone" or operational test of the difference between conviction and mere persuasion—namely, "the possibility of communicating it and finding it to be valid for the reason of every human being to take it to be true," which is to say that the proposition is able to be universally intersubjectively believed—is displayed by means of a conceptual "experiment" (*Versuch*) by which I imagine many other possible thinkers like myself under the same epistemic conditions. If every possible thinker I can logically imagine under those possible conditions asserts the proposition, then I have conviction; if not every possible thinker I can logically imagine under those possible conditions asserts it, then it is merely persuasion, even if the experiential conditions superficially resemble those of conviction (as, for example, in a dream). Or, as Kant puts it even more simply—and in an almost Bayesian way—in the *Vienna Logic*: "I am convinced… when I would hold the thing to be true no matter what the risks; if I were to waver, then I would not really be convinced" (*VL* 24: 855).

(*JL* 9: 65–73). It should also be remembered that I am using the boldface version of "**belief**" and its cognates to indicate that the meaning of Kant's term *Glauben* differs subtly but crucially from that of the English word "belief" (see section 0.3).

Persuasion has two modes, "opinion and **belief**": in opining, the thinker can either assert the proposition or not assert the proposition; in **belief** (that is, subjectively sufficient but objectively insufficient non-inferential theoretical or practical commitment—see section 0.3), the thinker actually asserts the proposition, but it is not necessary that every possible thinker do so. Conviction, however, is essentially bound up with *scientific knowing*, which entails both subjective and objective sufficiency. In this way, conviction is essentially the subjective side of "certainty" (*Gewißheit*), which according to Kant is the same as the conscious recognition of the objective sufficiency of a proposition.

Now what more precisely is Kant's notion of certainty? To understand it, we need to focus on his pregnant remark that:

[Experience] tells us, to be sure, what is, but never that it must necessarily be thus and not otherwise. For that very reason it gives no true universality, and reason, which is so desirous of this kind of cognitions, is more stimulated than satisfied by it. Now such universal cognitions, which at the same time possess the character of inner necessity, must be clear and certain for themselves, independently of experience; hence one calls them *a priori* cognitions. (*CPR* A1–2)

And this should be put together with some texts taken from various versions of Kant's logic lectures, plus one from the *Reflexionen*:

Certainty is nothing but subjective necessity in the quality of judgments. (*BL* 24: 142)

Consciousness of a cognition ... through the sufficient ground of its truth is called *conviction*. This arises solely from reason's *consciousness* of the necessary in the cognition. (*BL* 24: 144)

To cognize the thing *from reason* (i.e., *a priori* through the understanding even if it is not given) from universal principles according to its grounds, is called *having insight* (*perspicere*). Hence to have insight *a priori* is to cognize not only that it is so ... but that it must be so. (*DWL* 24: 730–731)

Certain holding-to-be true, or *certainty*, is combined with consciousness of necessity (*Bewußtsein der Notwendigkeit*), while uncertain holding-to-be-true, or *uncertainty*, is combined with consciousness of the contingency or the possibility of the opposite. (*JL* 9: 66)

[R]ational certainty is distinguished from empirical certainty by the consciousness of necessity that is combined with it. (*JL* 9: 71)

We are rationally certain of that into which we would have had insight *a priori*. ...³⁰ (*JL* 9: 71)

Rational certainty is ineluctably bound up with the consciousness of the necessity of that which is taken to be true. (R 5645; 18:290)

Six factors come forward here. First, if a cognition has certainty then it is a priori. Second, the proposition cognized by that cognition has "true universality"—that is, strict universality—and is therefore necessarily true (*CPR* B4). Third, a certain cognition has "the character of inner necessity." As I have mentioned already, I take this to be Kant's way of saying that it is epistemically necessary. Fourth, all certainty involves "clarity." (More on this below.) Fifth, certainty involves a "consciousness of necessity" that distinguishes it sharply from any consciousness of contingency no matter how well-supported by empirical evidence or "empirically certain" it might be. (Later, we shall also see that this "consciousness of necessity" is equivalent to the *distinctness* of a mental state, act, or process.) Sixth and finally, certainty is essentially connected with insight. When we recall from the first *Critique* text at *CPR* A820–2/B848–50 that conviction is nothing but the conscious or subjectively experiential side of objective certainty, it should be obvious enough that Kantian conviction entails apriority, semantic necessity, epistemic necessity, clarity, the consciousness of necessity (distinctness), and the act of insight. Kantian conviction consists in the subject's *a priori, absolutely rationally compelling clear (and distinct) conscious experience of a necessary truth by means of insight*.

Now both conviction and insight have been almost entirely overlooked in recent and contemporary—in effect, post-C. I. Lewis—Anglo-American interpretive or critical discussions of Kant's theory of a priori knowledge.³¹ By contrast, I want to insist that they are essential to Kant's

---

³⁰ In the original text Kant adds the phrase "even without any experience" (*auch ohne alle Erfahrung*) to "*a priori*," which is puzzling because it seems to suggest that cognition could be acquired without any sensory experience whatsoever, which is not only absurd but also clearly *not* Kant's view. So I treat it simply as a redundant gloss on "*a priori*."

³¹ What explains this neglect? First, very few Kant commentators pay any attention at all to the Transcendental Doctrine of Method, far less to section 3 of the Canon of Pure Reason; nor do they pay much attention to Kant's logic lectures or to the logical *Reflexionen*. Second, and perhaps even more importantly, insight or rational intuition is by and large an off-limits topic for logical empiricists, and most recent and contemporary Anglo-American philosophers—whatever their current views—cut their philosophical teeth on logical empiricism. In this connection it is worth rereading Arthur Pap's trenchant remark in the second epigraph for this chapter.

theory, and also to any adequate theory of a priori knowledge; and what I want to emphasize especially in what follows is the considerable extent to which conviction or a priori knowledge for Kant is a function of the subject's mode of consciousness in the mental act, state, or process of insight. This emphasis on the relevant mode of consciousness in turn suggests the possibility of a special inquiry into the basic features of the subjective experience of necessary truth, or a phenomenology of insight.[32] An excellent example of such an inquiry can be found in Kant's account of the role of insight in conceptual analysis. What I will argue in the next section then, is that Kant's phenomenology of conceptual insight shows us that the epistemic necessity of a proposition targeted in conviction is in fact *grounded* in the inwardly conscious, imagination-generated structures of insight, just as propositional necessity is grounded in the logico-syntactico-semantic structures of concepts and other intrinsic constituents of propositions, and just as metaphysical necessity is grounded in the intrinsic constitution of the mind and/or the constitution of the world.

7.2. Insight in Conceptual Analysis

In the *Jäsche Logic*, Kant draws a crucial distinction between two aspects of cognition:

All our cognition has a *twofold* reference, *first* a reference to the object, *second* a reference to the subject. In the former respect it is referred to [mental] *representation* (*Vorstellung*), in the latter to *consciousness* (*Bewußtsein*), the universal condition of all cognition in general ... In every cognition we must distinguish [intensional content (*Inhalt*)],[33] i.e., the object, and form, the *manner in which we* cognize the object (*die*

---

[32] Few twentieth-century epistemologists have attempted to work out such a theory, for the Pap-style reasons indicated in n.31. An important exception is Husserl's theory of "self-evidence" (*Evidenz*) in *Logical Investigations*: "The self-evidence (*Evidenz*) of judgments resting on intuition (*Anschauung*) is rightly contested when such judgments intentionally transcend the content of the actual data of consciousness. They have true self-evidence when their intention rests on the content of the actual data of consciousness itself, and finds fulfilment in that content, just as it is." See Husserl, *Logical Investigations*, vol. 1, p. 143. In the Sixth Investigation, ch. 6 (*Logical Investigations*, vol. 2, pp. 773–802), Husserl works out this basic idea in much detail under the rubric of "categorial intuition." Another important exception is Parsons, "Mathematical Intuition." Parsons in turn explicitly notes his debt to Kant and Husserl ("Mathematical Intuition," pp. 197–200).

[33] In fact Kant uses 'matter (*Materie*)' here, which fails to heed the important distinction he makes in the first *Critique* between sensory matter and intensional content; so I have charitably assumed that he would have wanted to be understood as referring to intensional content.

Art, wie wir den Gegenstand erkennen). ... The difference in the form of the cognition rests on a condition that accompanies all cognition, on *consciousness*. (*JL* 9: 33)

On the one hand, then, and as I noted in section 2.2, for Kant all cognition inherently contains a reference to something or another, an X, which is broadly speaking an object (or, in the limit case of apperception, a reflexive and reflective reference to the conscious subject as an object), and this objective reference essentially depends upon the existence of a mental representation. The mental representation thereby has an intensional content (*Inhalt*)—a package of information about an object—whose cognitive-semantic function it is to refer to the object in a certain way. But on the other hand, cognition also has a reference to the subject, and this reference depends essentially upon consciousness. By the notions of the subject and of consciousness here, I think, Kant means to advert to inner sense (*CPR* A22/B37) or "my own existence as determined in time" (*CPR* B275). One is reflectively aware of one's own conscious states, acts, and processes through "empirical apperception" (*CPR* A107) or empirical self-consciousness.

As we saw in section 1.1, in the first *Critique* Kant does not analyze either consciousness or empirical apperception at any great length (nor, unfortunately, does he always carefully distinguish between them), but he does manage to tell us two important things about them:

(1) That in identifying consciousness with inner sense or the determination of my existence in time, he thereby identifies it with the subject's direct empirical intuition of the immediately apprehended and reproductively synthesized stream of purely subjective sensible contents (which includes feelings, endogenously-caused or exogenously-caused sensations, and mental images) under various modes of temporal ordering (*CPR* A22/B37, A100–2, A107, B152–5, A357–9, A361–3, B420, B422–3 n.).

And:

(2) That empirical apperception, as empirical self-consciousness, is importantly distinct from "pure" or "transcendental" apperception, the original synthetic unity of apperception, which is our innate a priori capacity for generating the self-representation "I think X" and functions as a prefix for any thought (*CPR* A106–8, B131–5, B139–40, B152–6, A346–8/B404–6, A355, A407–9).

One can best gloss these two points, I think, by saying:

(1*) That inner sense consciousness is a concrete, direct, first-order, immanently reflexive,[34] temporally-structured function of human sensibility, and not a conceptual, discursive, propositional, self-conscious, higher-order, and reflective mental function of human thought.

And:

(2*) That empirical self-consciousness or apperception is a conceptual, discursive, propositional, meta-conscious, higher-order, reflective, thoughtful representation of our inner sense consciousness, but still distinct from pure or transcendental apperception.

The *Jäsche Logic* text quoted two paragraphs above is misleading in one important way. In one of the portions I elided, Kant remarks in a parenthesis that "consciousness is really a [mental] representation that another [mental] representation is in me." We should not, I think, be fooled by Kant's use of a "representation-that" formulation here, which seems to suggest that primary consciousness is itself some sort of higher-order judgment or thought about its own mental representations.[35] Similarly, and only slightly less misleadingly in the *Dohna-Wundlacken Logic* he says that consciousness is "an *actio* in the mind... [a mental] representation of our [mental] representation" (*DWL* 24: 701). But in one of the *Reflexionen* and in the *Prolegomena* he gets his own view exactly right when he says:

(The inner sense) Consciousness is the intuition of its self. (R 5049; 18: 72)

[The ego] is nothing more than the feeling of an existence without the slightest concept and is only the representation of that to which all thinking stands in relation. (*P* 4: 334 n.)

That is to say, inner sense consciousness directly and intuitively picks out the subject's own act, state, or process of represent*ing*, which in turn contains that representing's intensional or objective representational

---

[34] I borrow this apt term from Frankfurt, "Identification and Wholeheartedness," p. 162. A similar idea can be found in Sartre's *Transcendence of the Ego*, pp. 41–60, under the rubrics of "unreflected," "non-positional," or "non-thetic" self-consciousness.

[35] For a contemporary version of the higher-order representation approach to consciousness, see, e.g., Rosenthal, 'Two Concepts of Consciousness." See also Carruthers, "Natural Theories of Consciousness."

content. Inner sense consciousness does not *refer* to its own intensional or objective content; rather, inner sense consciousness, as an immanent reflexivity, includes a "form" of the objective content, which expresses the "manner in which" or *the way* I cognize the object, over and above the *informational* features of intensional or objective content. What we want to know however, is just what such an immediately given subjectively conscious "form" or "manner-in-which" can be if it is neither identical to the intensional or objective representational content, nor refers to it.

Here is my interpretation of what Kant is saying. The Kantian distinction between cognition's "relation to an objective representation" and its "relation to consciousness," is, I think, very familiar to philosophical psychologists and philosophers of mind as the distinction between the *intentionality* of the rational human animal on the one hand, and on the other hand the rational human animal's capacity for *phenomenal consciousness*: "something that it is like to *be* that organism—something it is like *for* an organism."[36] Intentionality, which is the rational human animal's "directedness to objects" or the "aboutness" of its mental acts, states, and processes, is determined by intensional or objective mental representational content. As we know from section 2.2, this intensional content can be either on the one hand a descriptive, conceptual, discursive, propositional package of information about an object, or else on the other hand a non-conceptual, non-descriptive, intuitional, non-discursive, non-propositional package of information about the object.[37] In contrast to intentionality of either sort however, phenomenal consciousness is *subjective experience*, or the mind's capacity for first-order or immanent unreflective reflexive sensible awareness, or what Kant in the *Anthropology* calls "taking notice of oneself" (*das Bemerken*), as opposed to "observing oneself" (*Beobachten*) (*A* 7: 132) or empirical apperception. Otherwise put, whenever objective mental representation or intentionality is in play, phenomenal consciousness is the subject's capacity for directly getting non-conceptual, non-discursive, non-propositional, sensible information about her own mental states, acts, or processes of mental representation. This capacity for phenomenal consciousness is constantly being exercised so long as the experiencing subject

---

[36] Nagel, "What is it Like to be a Bat?," p. 166.

[37] As I argued in sections 2.3 and 2.4, this non-conceptual, non-discursive, non-propositional information is primarily *locational*: it situates the object in space, and tracks its movements and changing qualities over time, relative to the egocentric standpoint of the embodied conscious subject.

is relatively alert (that is, is not unconscious), and it is a necessary condition of intentionality, although it can in fact also occur even if the mind is not actually directed to any particular object, hence it is not a sufficient condition of intentionality.[38] What is crucial about phenomenal consciousness, however, is that its mental function is not to tell us what a mental representation is referred-to or about (which is the job of intensional or objective mental representational content), but rather it conveys in a sensible format the manner-in-which or the way-in-which the mind is directed to the object *via* its intensional or objective representational content. That is, the information delivered by phenomenal consciousness in an objective representational setting is essentially "adverbial" in character. In section 2.2, I called this adverbial informational feature of phenomenal consciousness its *representational character*.[39]

Again: according to Kant, in an objective representational setting, the phenomenal consciousness of inner sense is a non-conceptual, non-discursive, non-propositional, sensible, first-order, immanent, and unreflective but also reflexive or self-aware expression of the manner-in-which or way-in-which the subject perceives, conceives, or judges, rather than an expression of *what* the subject perceives, conceives, or judges. But if the intensional or objective representational contents of cognition are evaluated according to their basic semantic properties—namely, whether they are correct or incorrect, or true or false, of their objects—then according to what basic properties are representational characters, or manners-in-which or ways-in-which the subject represents objects, evaluated?

Kant's answer is that the subject can perceive, conceive, or judge either *clearly* or *obscurely* on the one hand, or (assuming clarity) *distinctly* or *indistinctly* on the other. He writes:

[C]onsciousness always has a degree, which can always be diminished. (*CPR* B414; see also *MFNS* 4: 542).

---

[38] That is, as I will argue shortly, for Kant there exist states of phenomenal consciousness which are intrinsically "obscure" or vague in that they lack sufficient internal differentiation or ordering of content to determine a represented object. Hence they are non-intentional. So, again, from the existence of "objectless" phenomenally conscious states it follows that, while consciousness is necessary for mental representation of objects or intentionality, it is not sufficient for intentionality; and it also follows that the mental representation of objects, or intentionality, is sufficient but not necessary for the existence of inner sense, or immanently reflexive conscious mind.

[39] Searle has recently rediscovered this Kantian notion under the rubric of "aspectual shape"; see Searle, *Rediscovery of the Mind*, p. 155.

Clarity (*Klarheit*) is not, as the logicians say, the consciousness of a [mental] representation; for a certain degree of consciousness, which, however, is not sufficient for recollection, must be met with even in some obscure [mental] representations, because without any consciousness we would make no distinction in the combination of obscure representations; yet we are capable of doing this with the characteristics of some concepts (such as those of right and equity, or those of a musician, who, when improvising, hits many notes at the same time). Rather a representation is clear if the consciousness in it is sufficient for **a consciousness of the discrimination** (*Bewusstsein des Unterschiedes*) between [this mental representation] and others. To be sure, if this consciousness suffices for a discrimination, but not for a consciousness of the discrimination, then the [mental] representation must still be called obscure (*dunkel*). So there are infinitely many degrees of consciousness down to its vanishing. (*CPR* B414–15 n.)

The difference in the form of the cognition rests on a condition that accompanies all cognition, on *consciousness*. If I am conscious of the representation, it is *clear*; if I am not conscious of it, *obscure*... All clear representations. Can now be distinguished in regard to *distinctness* (*Deutlichkeit*) and *indistinctness* (*Undeutlichkeit*). If we are conscious of the whole representation, but not of the whole manifold (*Mannigfaltigen*) that is contained in it, then the representation is indistinct. (*JL* 9: 33–4)

The clarity of a mental act or state therefore is not a semantic feature of an intensional or objective mental representation; rather, it is a representational character having essentially to do with attaining a certain threshold in the intensive magnitude or degree of phenomenal consciousness. The lower bound of the intensive magnitude of consciousness, the point at which it diminishes to zero, is the state of unconsciousness. By contrast, when phenomenal consciousness reaches a threshold of intensity at which it effectively presents and discriminates an intensional mental representational content, it is clear. If, despite the fact that the mind has an intensive magnitude greater than zero—that is, it is not unconscious, hence to some degree conscious—and contains intensional mental representational content, it nevertheless still fails to reach clarity, then it is an obscure consciousness.[40] In turn, if we assume that a mental act, state, or process is already clear, which is to say that it has a discriminable objective mental

---

[40] Kant briefly discusses obscure consciousness in the *Anthropology* (*A* 7: 135–7), where he likens it to the darker portions of an immense yet only partially illuminated map: the darker parts are "consciously there," but not within the focus of discriminating attention and hence are not objectively representational. Clear and distinct consciousness corresponds to the focal or illuminated areas, and to intentionality more generally.

representational content, then we can further distinguish between its being indistinct or distinct: if the content is presented to the mind without internal articulation of its parts, then it is indistinct; if it is presented along with an internal articulation of the parts of its "manifold," then it is distinct.

Now at this point it might be asked: If the clarity and distinctness of the representational character of phenomenal consciousness are not themselves semantic features of a cognition, then why should we be interested in them from the standpoint of epistemology? The answer lies in these texts:

A cognition is perfect (*vollkommen*) (1) as to quantity if it is universal; (2) as to quality if it is distinct; (3) as to relation if it is true; and finally (4) as to modality if it is *certain*. (JL 9: 38)

All distinguishing of the true from the false involves the cognition of inner sense, i.e., I must be and become conscious of what really lies in my concept, and what I think. Inner sense is often dull, and its horizon shrouded in fog, and it does not give us enough help. (BL 24: 87)

In other words, the clarity and distinctness of the representational character of phenomenal consciousness are necessary conditions of a "perfect cognition," or any cognition which has met the categorically normative standard of authentic knowledge. So what we *ought to do* epistemically, is to perfect our cognition until it yields authentic knowledge. Assuming that the propositional object of a given propositional attitude is indeed universally intersubjectively communicable[41] and objectively true, then adding clarity and distinctness to that attitude—thereby moving it from obscurity to lucidity—*suffices* for authentic knowledge and satisfies our epistemic obligation.

In the *Jäsche Logic*, Kant gives helpful examples of obscurity and clarity, and of indistinctness and distinctness, when discussing the non-conceptuality of sense perception (JL 9: 33–5), as we have already seen in section 2.2. If, for example, someone's visual field contains an array of contents that represent a bucolic country scene, but she altogether fails to notice a certain country house as a salient represented object within that

---

[41] It is clear from his discussion of the quantitative dimension of perfect cognition or authentic knowledge (JL 9: 40–9) that Kant does not require that all knowledge be expressible in universally-quantified propositions. What he requires instead is that there be "congruence of the limits of human cognition with the limits of the whole of human perfection in general" (JL 9: 41). In other words, the universality of a perfected cognition is universal intersubjective communicability, not logical universality.

field, then that is perceptual obscurity in relation to that particular object; if she notices that house as a salient object against the background of its bucolic environment, then that is perceptual clarity. If the house is perceptually registered and set apart from other objects in the perceptual field, but only as a slightly-bigger-than-mid-sized material object over there, which is not recognized *as* a house, or brought under the concept HOUSE, then that is perceptual clarity plus indistinctness. If the perceiver manages, finally, to recognize it *as* a house, then that is the same as to say that she perceptually isolates and identifies its windows, doors, roof, etc. This conscious unpacking of the internal visual manifold of the house-representation by means of the application of a concept is perceptual distinctness.

Perceptual distinctness or indistinctness, Kant describes as "sensible." So this is to be sharply distinguished from *intellectual* distinctness or indistinctness, which applies to strictly conceptual cognitions alone:

Distinctness itself can be of two sorts: *First, sensible:* This consists in the consciousness of the manifold in intuition ... *Secondly, intellectual; distinctness in concepts or distinctness of the understanding.* This rests on the decomposition (*Zergliederung*) of the concept in regard to the manifold that lies contained within it. Thus in the concept of *virtue* for example, are contained as characteristics (*Merkmale*) (1) the concept of freedom, (2) the concept of adherence to rules (to duty), (3) the concept of overpowering the force of inclinations, in case they oppose those rules. Now if we break up the concept of virtue into its individual constituent parts, we make it distinct for ourselves through this analysis (*Analyse*). By thus making it distinct, however, we add nothing to a concept; we only elucidate it. With distinctness, therefore, concepts are improved not as to [intensional content (*Inhalt*)]⁴² but only *as to form.* (*JL* 9: 35)

Kant's idea here is that the consciousness in a strictly conceptual cognition is distinct just in case the act or process of cognition fully represents and articulates the several sub-conceptual parts—the "characteristics" or *Merkmale*—that make up the internal manifold of that concept. This is also what he calls "logically complete distinctness":

[A]s for what concerns logical distinctness in particular, it is to be called *complete* distinctness insofar as all the characteristics which, taken together, make up the whole concept have come to clarity. (*JL* 9: 62)

⁴² See n.33.

Now, every concept according to Kant is individuated and defined by means of the ordered totality of necessary characteristics or sub-concepts constituting its conceptual manifold, and this totality is the "logical essence" or "conceptual essence" (*esse conceptus*) of that concept (*JL* 9: 61). The logical or conceptual essence of a given concept is best construed as a *conceptual microstructure*, in that the members of the manifold of characteristics are intrinsically ordered in relations of both "subordination" (species-inclusion, and especially the determinable-determinate relation) and "coordination" (non-redundant classification, and especially the determinate-exclusion relation) (*JL* 9: 59, 62, 95–9). For instance, the concept BACHELOR is subordinated to—that is, is *a determinate* under—the higher sub-concepts or characteristics—that is, *the determinables*—ADULT, UNMARRIED, and MALE, and in turn, ADULT, UNMARRIED, and MALE are each partially overlapping coordinates of one another, hence are not mutually exclusive determinates of some further determinable, like RED and GREEN in relation to COLOR. The total fine-grained non-overlapping conceptual microstructure of a given concept would be reflected in a strict analytic definition of that concept, were such a complete, "precise" or non-redundant analysis in fact humanly possible (*JL* 9: 142–5). The best we finite thinkers can do, however, is to carry out decompositions of a given concept up to just the point at which our awareness of the essential content of that concept becomes sufficiently distinct for our theoretical purposes: "a definition is a *sufficiently* distinct and precise concept" (*JL* 9: 140).

A concept is made clear, or discriminable, so long as its manifold is at least to some extent decomposed into its constituent characteristics.[43] By contrast, it is made distinct just insofar as either a sufficient part or (in the ideal case) all of its conceptual microstructure is represented to the mind via the conscious process of definitionally decomposing that concept. But to predicate a concept C_2 of another concept C_1, when C_2 is a decompositional part of C_1—hence C_2 is "contained in" C_1—suffices for analytic truth. This is because any attempt to deny that C_2 is predicable of C_1 leads directly to self-contradiction, and according to Kant a proposition is analytic if and only if its denial entails a logical or conceptual contradiction

---

[43] In contradistinction to definitional (complete, non-redundant or precise) decompositions, Kant calls decompositions which may be merely partial or redundant, "expositions" (*Erörterungen*) or "descriptions" (*Beschreibungen*) of a concept's microstructure (*JL* 9: 142–3).

(*CPR* A150–3/B189–93).⁴⁴ So to cognize a concept clearly or distinctly is implicitly to make an analytic judgment. All one then needs to do in order to make an analytic judgment *explicitly* is to describe its decompositional content out loud:

> If I say, for instance, "All bodies are extended," this is an analytic judgment. For I do not require to go beyond the concept which I combine with the word 'body'⁴⁵ in order to find extension bound up with it. To meet with this predicate, I have merely to decompose the concept, that is <u>to become conscious to myself of the manifold</u> which I always think in that concept. The judgment is therefore analytic ... [I]t would be absurd to found an analytic judgment on experience, since I do not need to go beyond my concept at all in order to formulate the judgment, and therefore need no testimony from experience for that. That a body is extended is a proposition that is established *a priori* and is not empirical. For before I go to experience, I have already have all the conditions for my judgment in the concept, from which I merely draw out the predicate in accordance with the principle of contradiction, and can thereby at the same time <u>become conscious of the necessity of the judgment</u>, which experience could never teach me. (*CPR* A7/B11–12, underlining added)

In this way the subjective act of analytic judging (*Urteilen*), by virtue of its representational character, can immanently reflexively trace the logico-semantic structure of the objective proposition (the *Urteil*). And in so doing, it becomes genuine analytic a priori insight. For it contains a phenomenal consciousness whose representational character, via the intellectual clarity and distinctness of that consciousness, reproduces or models the very same essential decompositional structure that makes the proposition analytically necessary. Thus the representational character of the phenomenal consciousness in analytic insight is isomorphic, and also directly experienced *as* isomorphic, with the underlying semantic form of an analytically necessary proposition.

Two important aspects of Kant's theory of conceptual analysis require further exposition.

The first has to do with the phenomenological structure of a representational phenomenal consciousness. What sort of representational character will constitute the isomorphic, or intellectually distinct, phenomenal

---

⁴⁴ For more details, see Hanna, *Kant and the Foundations of Analytic Philosophy*, ch. 3.

⁴⁵ The B edition reads "the body" (*dem Körper*) instead of the A edition's "the word 'body'" (*dem Wort Körper*). But the A edition version seems clearer and philosophically more informative.

consciousness in conceptual insight? Kant's explicit answer is that it an essentially *imaginational* character:

> We observe, that all concepts can be made distinct if one can make them comprehensible through images of the imagination (*Bilder der Imagination*). (*R* 1571; 16: 8–9)

These "images of the imagination" initially derive from the empirical "reproductive imagination," the innate faculty for generating empirical *Bilder* through immediate or short-term memory (*CPR* A100–1, B152). All cognition is automatically accompanied by empirical imagery, be it visual, auditory, tactile, olfactory, or whatever. And this imagery can be manipulated and reworked in order to provide quick and effective conceptual access to sensory manifolds, in the form of sensory icons or "empirical schemata" which summarize or gloss much intuitional information in simple patterned arrays or templates (*CPR* A140–2/B180–1).

In the simple case of purely conceptual cognition or thought however, this imagery is *linguistic* in nature:

> Our cognition has need of certain means, and this is language. (*VL* 24: 812)

> Language signifies thoughts (*Gedanken*) and, on the other hand, the means *par excellence* of signifying thoughts (*Gedankenbezeichnung*) is language, the most important way we have of understanding ourselves and others. Thinking is *talking* with ourselves...; so it is also *listening* to ourselves inwardly (by reproductive imagination). (*A* 7: 191)

According to Kant, then (i) linguistic meanings are thoughts, or mental representations essentially involving concepts, and (ii) all thinking is inner speech. Unfortunately, these brief texts taken from the *Vienna Logic* and the *Anthropology* exhaust virtually everything that Kant has to say explicitly about the philosophy of language in the entire corpus of his writings,[46]

---

[46] Kant was in fact sharply criticized by his contemporary Hamann for ignoring language; see Beiser, *The Fate of Reason*, p. 40. In my opinion, however, Kant does *not* "ignore language": instead, he ignores only what he regards as *secondary aspects* of language. Kant believes that there is a strict explanatory symmetry between discursive (conceptual, propositional, logical) cognition and the information-conveying aspects of language. Moreover his transcendental theory of discursive cognition *yields* a theory of language in the information-conveying sense. And because he is interested first and foremost in this aspect of language, he leaves aside the pragmatic and social dimensions of language. But nothing in his view excludes them. So to reject Kant's theory on that ground alone would be like rejecting Chomsky's linguistics because he focuses his attention primarily on "competence" rather than on "performance" or on "language games."

so it would be unwise to press too hard on them. Still, he does seem to state fairly unequivocally here that there is a strict symmetry between conceptual cognition (thought) and at least the information-conveying aspects of natural language. And he also does explicitly say that natural languages in turn all reflect—despite their superficially different specific grammars and lexicons—a single "universal grammar" which is innately stored as a generative capacity in the underlying transcendental logical and categorial structures of human cognition (*P* 4: 322–3) (*JL* 9: 11–13).

Focusing on thesis (ii), however, what is crucial for our purposes here is that Kant's theory of language, even in this highly elliptical form, does *not* entail the strong thesis that all thought or conceptual cognition occurs in a syntactically autonomous mental language or *lingua mentis*.[47] Rather, all it entails is that thoughts necessarily have *some* linguistic realization in consciousness, and more specifically all it entails is that thoughts are realized in such a way that their accompanying empirical mental imagery in phenomenal consciousness has a structure whose salient syntactical parts correspond one-to-one to the salient syntactical parts of physical linguistic inscriptions occurring in some natural language or another.[48] So in this sense Kant "thinks in"—that is, he experiences in the phenomenal consciousness which accompanies all of his thinking the reproductive linguistic imagery of—eighteenth-century German, just as the readers of this book are now "thinking in" twenty-first-century English.

On Kant's view, necessarily for any linguistically competent thinking subject, if by judging she unpacks the logico-semantic microstructure of an analytic proposition with insight, then she also thereby mentally manipulates her consciously-experienced reproductive linguistic imagery in order to bring about that insight, just as someone might mentally manipulate mental images of various shapes in everyday cognition (for example, in figuring out how to get the key in the keyhole) or under controlled conditions (for example, in the famous mental rotation experiments carried out by the psychologist Roger Shepard and his colleagues[49]). So in cognizing with insight

---

[47] For a contemporary version of the mental language thesis, see Fodor, *The Language of Thought*; but the general idea of "mentalese" (as Deborah Brown pointed out to me) goes at least as far back as Ockham.
[48] For a contemporary view very similar to Kant's, see Carruthers, *Language, Thought, and Consciousness*.
[49] See ch. 2, n.81.

the proposition (BE) "All bodies are extended," the conceptual thinker thereby inwardly consciously reproduces in empirical imagination some image-token or another having roughly the following shape or sign-design:

[(All x) (BODY x = <... etc. + EXTENDED x > pred EXTENDEDx)].[50]

Any image-token having the same ordered "multiplicity" is, quite obviously, an isomorph of the semantic containment-relation between subject-concept and predicate concept in the simple analytic proposition (BE), as Kant himself explicitly points out:

An example of an *analytic* proposition is, to everything *x*, to which the concept of body (*a* + *b*) belongs, belongs also *extension* (*b*). (*JL* 9: 111)

The crucial point here is that our everyday use of informative language not only denotes objects and connotes senses or concepts: it also, via the empirical imagination in its schematizing function, literally *portrays* logico-syntactico-semantic structures by self-producing mental models or *Urbilder* of them.[51] In effect, then, the essential source of the intellectual distinctness in conceptual insight is the basic psychological fact that the underlying logico-syntactico-semantic structure of an analytic proposition can be encoded in a single phenomenally conscious, empirically imaginational, mentally manipulable linguistic schema.

The second point concerns the relationship between the special sort of distinctness found in conceptual insight and the semantic content of

---

[50] This symbolism needs a little elaboration. The outermost square brackets enclose propositional content. "(All x) (... x ...)" is a universal quantifier. Concept-words are in capital letters. The bound variable "x" ranges over the comprehension (*Umfang*)—or possible-worlds extension—of the concept denoted by the concept-word to which it is appended. The identity-sign stands for identity of concepts. Wedge brackets enclose the decompositional content of a concept, the several constituent characteristics of which are joined by "+." And "pred" stands for an operator on concepts, which predicates the concept denoted by the concept-word on the right-hand side of "pred," of the concept denoted by the concept-word on the left-hand side of "pred." The whole expression thus means the proposition to the effect that the concept of being extended is predicated of the concept of being a body, which in turn contains (along with other sub-concepts) the concept of being extended as an intrinsic decompositional part. See also Hanna, *Kant and the Foundations of Analytic Philosophy*, ch. 3.

[51] This point is phenomenologically obvious by virtue of the fact that one can recognize, remember, and even recite words of a foreign language without having the slightest idea what they refer to or mean. And there is also empirical evidence in support of the thesis that language has an autonomous "displaying" function: recent work in cognitive psychology strongly indicates that the shape or design (whether it be the "look" or "sound" or even "feel") of language is processed quite differently—not only in the phenomenological and information-theoretic senses, but also in the neurobiological sense of involving a different localization in the brain—from the processing of its logico-semantic components. See Schacter, "Perceptual Representation Systems and Implicit Memory: Towards a Resolution of the Multiple Memory Systems Debate."

concepts. On Kant's view, the representational character of phenomenal consciousness in analytical insight varies quite *independently* of its conceptual content:

> A great part, perhaps the greatest part, of the business of our reason consists in *decompositions* of the concepts which we already have of objects. This [decomposition] affords us a multitude cognitions that, although they are nothing more than illuminations or clarifications (*Aufklärungen oder Erläuterungen*) of that which is already thought in our concepts (though still in a confused [*verworrene*[52]]way), are, at least as far as their form is concerned, treasured as if they were new insights (*neuen Einsichten*), though they do not extend the concepts we have in either matter (*Materie*) or intensional content (*Inhalte*), but only discriminate them (*aus einander setzen*) .... [T]his procedure does yield real *a priori* cognition. (*CPR* A5–6/B9)

As we have seen, intellectual distinctness consists in the consciously experienced isomorphism between an imaginational linguistic form (or empirical schema) that constitutes a judgmental cognition's representational character, and the conceptual microstructural form of the analytic proposition that is the logico-syntactico-semantic content of that cognition. But the analytic proposition possesses its overall logico-syntactico-semantic form—and its constituent concepts possess their several microstructures—quite independently of the character of the subject's phenomenally conscious schematic imagery. Distinctness is an adverb of the mental state, act, or process, not an adjective of the propositional or conceptual content. Or again, distinctness is an adverbial feature of the representational character of a "perfect cognition," hence a distinct grasp of the concept does not in any way change or affect the conceptual intensional content (*Inhalt*) of that concept: it only *fully reveals its conceptual microstructure to the subject*. To use Kant's apt analogy, grasping a concept distinctly by definitionally decomposing it is like the *my illumination* (*Aufhellung*) of a map, not like *my drawing* a map (*JL* 9: 64). In this way, conceptual insight has no *semantic* efficacy; instead, it only moves the consciousness of the conceptual analyst from obscurity to clarity to distinctness, by generating or manipulating a

---

[52] The term "confused" is borrowed by Kant from Leibnizian-Wolffian philosophy. In the *Jäsche Logic* (*JL* 9: 34–5), he remarks that confusion entails indistinctness although it is not precisely equivalent with indistinctness. And this is because cognitions involving *simple* mental representations (say, of this or that object given in intuition) are indistinct because those representations have no internal manifold, yet such cognitions are not confused. In this particular context, however, Kant is talking about the indistinct cognition of complex concepts.

376   THE PRACTICAL FOUNDATIONS OF THE EXACT SCIENCES

linguistic schema in empirical imagination. Nevertheless, the move from obscurity through clarity to distinctness involves a genuine changes in phenomenally conscious form: changes which, as Kant says, "are treasured as if they were new insights." So while an analytic proposition or definition does not supply new logico-syntactico-semantic information, a distinct conceptual or analytic consciousness supplies new phenomenological or noetic information.

So the overall Kantian picture of conceptual cognition is this: Conceptual microstructure and intensional content are static and architectural, while conceptual consciousness is dynamic and epistemically productive. This in turn means that the logico-syntactico-semantic architecture of a given concept can operate as a merely tacit element in the act, state, or process of conceptual thinking:

> Without doubt the concept of **right** that is used by common sense understanding (*gesunde Verstand*) contains the very same things that most subtle speculation can produce (*entwickeln*) out of it, only in common and practical use we are not conscious of these manifold representations in these thoughts. (*CPR* A43/B60–1)

Prior to its full acquisition by the understanding, then, the concept RIGHT is possessed obscurely. Perhaps I have simply picked up the word 'right' in conversation, casual reading, and other ordinary linguistic practices, and can even competently use it to some extent, but I cannot explicate its meaning. In this way I am, for the most part, an unreflective (and, sadly, often an almost "mindless") consumer of my language. Now obscure concept possession is the intellectual analogue of perceptual obscurity: so, just as it is possible to have perceptual content in one's perceptual field without bringing it into the focus of attention, so too it is possible to use words in such a way that one invokes concepts but does not properly understand them:

> [E]ven people who can speak and hear do not always understand themselves or others; and it is because their power of using signs is defective or because they use it incorrectly... that, especially in matters of reason, men who use the same language are poles apart in their concepts and only discover this accidentally, when each acts on the basis of his own concepts. (*A* 7: 193)

Concepts can thus be possessed "consumeristically" and "loosely." Indeed, even the fairly effective use of a concept in everyday and practical contexts does not alone suffice for clear consciousness of it. To have a clear conceptual consciousness of a concept is not only to be able to *apply it correctly* to some central cases but also to be able to *decompose its intensional content* to some extent. Furthermore, it is epistemically one thing to have a clear consciousness of some of the contents of a concept, and epistemically quite another to become distinctly aware of its conceptual essence or microstructure, which can then be recorded in an explicit analytic definition. But these important epistemological differences are not themselves differences in the microstructure or content of the concept being analyzed, which remains semantically identical under all dynamic variations in phenomenal discursive consciousness and its representational character:

> So far as the matter [of identical concepts] is concerned, I always think the same object, not always in the same way, however, but instead in a different way; namely, I represent distinctly in the definition what I previously represented [confusedly[53]] in the *definitum*; and every definition must accordingly be distinct. (BL 24: 265)

For Kant, then, conceptual analysis that has achieved intellectual clarity, and then distinctness via linguistic imagery in phenomenal discursive consciousness, is semantically trivial but noetically highly informative and spontaneous. In fact, it is a special kind of self-produced self-knowledge. The subject gradually becomes aware of the internal manifold and microstructures of her concepts, and makes them noetically salient, rather than treating them as mere intentionality-mediators yielding descriptive access to objects in the world. She actively learns more and more about her own conceptual repertoire.

In this way, Kant's theory of conceptual insight provides a rather crisp, satisfying epistemological solution to G. E. Moore's notorious "paradox of analysis."[54] The paradox is this. Since a correct analytic definition of a concept supplies a description that is completely semantically identical—or synonymous—with the term expressing the concept (as in "Bachelors are adult unmarried males"), it follows that a correct analysis necessarily reveals no new semantic information. So a conceptual analysis is correct only on condition that it is semantically trivial; but if on the other hand it is semantically non-trivial, it cannot be correct. Paradoxically, therefore, every analysis is either totally trivial or just false! From a Kantian point of view what is needed to solve the paradox is simply the sharp distinction between semantic and noetic information. For Kant, an analytic definition of a concept

---

[53] Kant uses the Latin term *confuse*.
[54] See Langford, "Moore's Notion of Analysis"; and Moore, "Analysis."

provides important, novel a priori noetic information for an epistemically active self-conscious conceptualizer; yet at the same time it safely preserves semantic identity or synonymy between definiens and definiendum.

If I am right so far, then Kant's theory of intellectual distinctness in conceptual analysis brings out a fundamental feature of all a priori knowledge: it essentially involves *cognitive dynamics*. Insofar as a priori knowledge implies insight, this insight consists in the thinking subject's cognitive capacity for actively bringing about an isomorphism between (a) an imaginational form immanent in the phenomenal consciousness (hence also constituting the representational character) of an act, state, or process of judging which is directed to a necessary truth, and (b) the underlying logico-syntactico-semantic form of that necessary truth. Even more precisely, insight consists in the thinking subject's capacity spontaneously to create and manipulate a semantic schema of the imagination. The schema is what reason "self-produces according to a design of its own." And the consciously-experienced isomorphism between the schema and the necessary propositional content is the same as the (clarity and) distinctness of the act of judging. In this way, the thinker's effective manipulation of her mental model of the proposition automatically generates that cognition's epistemic necessity and apriority.

This leads to an even more general point. For Kant an epistemically decisive justification (as opposed to a mere sufficient warrant for believing) is *not* a reason added extrinsically to a belief, which in turn merely happens to be true: *decisive justification is instead a direct appeal to the intrinsic psychological constitution of the act, state, or process of understanding a truth*. It is an *internal* epistemic reason based on the psychological constitution of the cognitive dynamics, not an *external* epistemic reason having nothing to do with psychological facts. On Kant's view then, a belief in a proposition P is decisively justified if and only if, assuming P is indeed intersubjectively communicable and true, the belief-act, belief-state, or belief-process has precisely the right kind of noetic structure:

Truth has objective characteristics; however <u>certainty, which in every case finds these objective characteristics, can have only subjective characteristics</u>, i.e., the agreement of cognition with itself. (R 3716; 17: 255, underlining added)

This in turn is equivalent to saying that authentic knowledge or science is the belief-act's, belief-state's, or belief-process's' joint possession of the four "perfections" (*Vollkommenheiten*) of cognition (*JL* 9: 40–73). Authentic knowledge or science is thus a categorically normative and cognitively ideal posture of the mind. Whether or not one accepts Kant's perfectionist internalist (about reasons, that is) normative epistemology,[55] it should at least be obvious that it is a very long way indeed from the all-too-familiar "analysis of knowledge" approach to our all-too-human epistemic condition.

7.3. The Role of Insight in A Priori Knowledge in General

Kant's conception of the role of insight in a priori knowledge in general is worked out in the Transcendental Doctrine of Method, section 1, "The Discipline of Pure Reason in its Dogmatic Employment" (*CPR* A712/B740).[56] The particular question on the table in the "Discipline" section is whether the epistemic method employed by mathematicians is the same as that employed by transcendental philosophers, or different. Kant's general interest in this question goes back at least as far as the so-called "Prize Essay" or "Inquiry Concerning the Distinctness of the Principles of Natural Theology and Morality" of 1764. But it is especially important and relevant for the Critical context we are interested in, because it explicitly raises the issue of whether there might be fundamentally different types of certainty within the total domain of a priori knowledge:

> It is ... very important for us to know whether the method of obtaining apodictic certainty that one calls **mathematical** in the latter science is identical with that by means of which one seeks the same certainty in philosophy. (*CPR* A712–13/B740–1)

Now Kant's answer to his own question is that the methods of getting certainty in transcendental philosophy and in mathematics are indeed

---

[55] It is interesting and illuminating to note that on the other hand Kant explicitly rejects any perfectionist *ethics*, because that would be heteronomous (*GMM* 4: 443–4). By contrast, the acceptable heteronomy of Kant's perfectionist epistemology lies precisely in its content externalism and the role of truth in a priori knowledge: the scientifically knowing subject must ultimately conform herself to the way the world is. Hence an act (in this case, of scientific knowing) can be categorically normative but still heteronomous.

[56] Fortunately this section has not been quite so badly neglected as section 3 of the Canon of Pure Reason: those interested in Kant's philosophy of mathematics, particularly Parsons and Friedman, have dealt with it extensively.

different: transcendental philosophy uses the method of reasoning from concepts, while mathematics uses the method of construction (*Konstruktion*) of concepts (*CPR* A712–20/B740–8). As we saw in section 6.4, mathematical construction involves the use of the productive imagination (*CPR* B152) and its schemata, in conjunction with pure intuition. Kant calls the mathematical certainty based on constructional insight "self-evidence" (*Evidenz*), and says that it is intrinsically clearer than discursive insight—in fact, self-evidence can never arise from discursive insight alone (*CPR* A734/762) (*JL* 9: 70–1). And this is due to the fact that while discursive insight uses merely empirical linguistic imagery, mathematical insight is based on non-empirical imagery derived directly from pure or formal intuition. So mathematics smoothly fits the imagination-based model of a priori knowledge I have ascribed to Kant, even though it contrasts in some non-trivial ways with discursive insight.

And this in turn leads Kant to a parallel distinction between the types of propositions known by means of these a priori methods: the propositions known by reasoning from concepts are "dogmata," and those known by construction of concepts are "mathemata" (*CPR* A736/B764). Both dogmata and mathemata are synthetic a priori propositions, and both have the same special type of necessity (they are true in all and only the humanly experienceable worlds). But while dogmata and mathemata do not differ semantically, still by virtue of the different types of insight required to know them, they do differ sharply as regards the type of epistemic necessity available in the correspondingly different methods of knowing them.

"Philosophical cognition" for Kant is discursive or conceptual reasoning. Nevertheless it should not be assumed that just because philosophy reasons "from" concepts, that its cognition always or even primarily takes the form of conceptual analyses. Rather, although it does indeed secondarily utilize direct decompositional insight into analytic propositions, its primary epistemic activity consists in carrying out what Kant calls "acroamatic" or philosophically discursive proofs (*CPR* A735/B763). All sound acroamatic proofs are transcendental deductions or "transcendental proofs" (*CPR* A782–94/B810–22). The most important kind of transcendental proof has the function of proving "transcendental propositions" or transcendental principles expressing schematized pure concepts of the understanding: for example, "Every event has a cause" (*CPR* A718–21/B746–9, A734–7/B762–5). And isolating this sort of proof points up the essential difference that Kant sees between knowledge of transcendental principles on the one hand, and knowledge of either analytic propositions or synthetic a priori truths in mathematics on the other hand. Explicating this difference, he writes:

> [T]hrough concepts of understanding... [pure reason] certainly erects secure principles, not directly from concepts, but rather always only indirectly through the relation of these concepts to something entirely contingent, namely **possible experience**; since if this (something as object of possible experience) is presupposed, then [these principles] are of course apodictically certain, but in themselves they cannot be cognized *a priori* directly at all. Thus no one can have fundamental insight (*einsehen*) into the proposition "Everything that happens has its cause" from these given concepts alone... But although it must be proved, it is called a **principle**, and not a **theorem** because it has the special property that it first makes possible its ground of proof, namely experience, and must always be presupposed in this. (*CPR* A736–7/B764–5)

Kant is saying here that while straightforward conceptual analysis can indeed give insight into analytic a priori truths "from these given concepts alone," a transcendental deduction by contrast cannot provide any comparable direct insight into synthetic a priori transcendental principles—even though it can indeed get indirect a priori knowledge of them by proving them. The reason for the essentially indirect character of philosophical cognition is that in order to give a sound transcendental proof of any transcendental principle we must:

(i) understand the logical meaning of a pure concept of the understanding;

(ii) employ the "limiting" notion of the possibility of experience (by means of an appeal to pure intuition) in order to restrict the scope of the application of that pure concept;

(iii) justify the application of that pure concept to objects of possible experience;

(iv) supply a pure intuitive temporal model or "transcendental schema" in order to show just how such an application of that pure concept can be carried out; and finally

(v) actually work out the application of that schematized pure concept to objects of possible experience by means of a universal synthetic a priori proposition.

To deduce a transcendental principle is therefore the same as to deduce a synthetic a priori proposition in the transcendental metaphysics of nature. But any such proof is obviously bound to be *far* from being immediately certain or self-evident.[57]

For my purposes here, the crucial point is this: the difference between the indirect a priori knowledge of transcendental principles, and the direct a priori knowledge of "mathemata," exemplifies a quite general and fundamental contrast between indirect or *proof-based* a priori knowledge, and direct or *insight-based* a priori knowledge.[58] Kant introduces the same distinction as a difference between "mediated certainty" and "unmediated certainty" (*JL* 9: 71). The difference turns on the logico-syntactico-semantic structure of the proposition cognized. Only a proposition (a) whose semantic structure is sufficiently simple, and (b) whose truth is wholly internal to that structure, is knowable with "immediate certainty" or is cognizable by means of insight. Good examples would be "$(P) \sim (P \& \sim P)$," "Bachelors are unmarried," and "$2 + 2 = 4$." Only propositions with these special logico-syntactico-semantic properties are such that merely by distinctly understanding them via imaginative schemata, we are *thereby* necessitated to believe them.

In this connection, it is worth strongly emphasizing that even necessary propositions accessible only to proof-based a priori knowledge have epistemic necessity. Nevertheless, the extent to which a merely provable necessary proposition is epistemically necessary is in direct proportion to the extent to which the several parts of its proof are separately accessible to insight. No merely provable necessary proposition is in the highest degree epistemically necessary, because it cannot be cognized as a simple

---

[57] Even the a priori knowledge gained through transcendental proof is based on the operations of the imagination. Here the pure productive imagination is combined with pure conceptualization in order to survey the semantic fit between synthetic propositional contents and humanly experienceable worlds. One isolates either invariant intuitional-spatiotemporal or conceptual-categorial structures across worlds by first running systematic imaginational searches for empirical counterexamples and then abstracting away from all search-domains that do not produce "hits." For example, "in regard to appearances in general one cannot remove time, though one can very well take the appearances away from time" (*CPR* A31/B46). This highly specialized conceptual function of the productive imagination is what Kant labels "so-called mother-wit" (*so genannten Mutterwitzes*) or, far more accurately, "productive wit" (*produktiven Witze*) (*CPR* A133/B172) (*A* 7: 220–7). See also Brook, "Kant's *A Priori* Methods for Recognizing Necessary Truths," pp. 247–52.

[58] In "Are There Synthetic A Priori Truths?," C. D. Broad introduces a similar distinction between "demonstrably *a priori*" and "intuitably *a priori*" necessary propositions. But the contrast between (i) a priori knowledge obtainable by means of proof and (ii) a priori knowledge obtainable by means of direct insight goes at least as far back as Descartes's basic distinction in the *Regulae* between "deduction" and "intuition"; see Descartes, "Rules for the Direction of the Mind," pp. 33–9.

self-contained logico-syntactico-semantic whole. It follows that the intensive magnitude of the conviction or certainty which characterizes a priori knowledge of a merely provable necessary proposition is always of a lesser degree than that which characterizes a priori knowledge of necessary propositions by insight. Moreover, while the several premises in a given proof may or may not each be known by insight (thereby increasing or decreasing the proof's overall degree of epistemic necessity), no actual proof has overall epistemic necessity unless each of its basic premises, and also each of its distinct deduction-links or inference-steps, is known by insight. And no knowledge by proof whatsoever is possible without a capacity for direct insight into the unprovable and absolutely fundamental logical principle of non-contradiction (*CPR* A151–2/B190–1) (*DWL* 24: 694). Therefore for Kant while not all a priori knowledge either is itself insight or else presupposes insight. In other words, insight is the primary mode of all a priori knowledge:

> All certainty is either *unmediated* or *mediated*, i.e., it either requires a proof or is not capable of and does not require any proof. Even if so much in our cognition is certain only mediately, i.e., through a proof, there must still be something *indemonstrable* or *immediately certain*, and the whole of our cognition must proceed from *immediately certain* propositions. (*JL* 9: 71)

7.4. Theoretical Technique: A Priori Knowledge in Action

The dynamic cognitive manipulation and conscious apprehension of a mental model or *Urbild* via the schematizing imagination in insight—the paradigmatic example of which occurs in a priori construction in geometry or arithmetic (see section 6.4), but which also occurs in conceptual insight (section 7.2), although with a lesser degree of certainty—is a species of what Kant in the First Introduction to the *Critique of the Power of Judgment* calls "theoretical technique" (*CPJ* 20: 200). There Kant argues that theoretical technique is a form of human practical action driven by hypothetical "imperatives of skill" (also called "technical imperatives"), rather than by hypothetical "imperatives of prudence" (also called "pragmatic imperatives") (*GMM* 4: 414–15) (*CPJ* 20: 197–201). What does he mean by this?

Here is one way of formulating it. If the general form of a hypothetical imperative of prudence (**HIP**) is

HIP: *If I desire X for the purposes of my happiness, then I should do Y*;

and if the general form of a hypothetical imperative of skill (**HIS**) is:

HIS: *If I want to produce X for the purposes of Y, then I should do Z*;

and if the specific form of a hypothetical imperative of theoretical skill (**HITS**) is:

HITS: *If I want to produce X for the purposes of A-type knowledge, then I should do Z*;

then it follows that the instantiated specific form of a hypothetical imperative of, for example, mathematical skill (**HIMS**) is:

HIMS: *If I want to produce conviction for the purposes of a priori mathematical knowledge, then I should construct mathematical concepts.*

But all forms of the hypothetical imperative—whether **HIP**, **HIS**, **HITS**, or **HIMS**—are a priori constrained by categorical imperatives of both logic and morality (*JL* 9: 11–13).[59] Therefore, for Kant a priori knowledge is in fact a priori *scientific knowing* (*Wissen*), which in turn is a form of a priori action.

The importance of Kant's idea here cannot be overemphasized. By construing a priori scientific knowing as a priori action, he thereby gets beyond the classical dichotomies between exact scientific theories and human practice, and between propositional cognition and volitional action. For according to the Kantian conception, a priori knowing in the exact sciences is not merely an instance of cognitive dynamics, it is also essentially a categorically normative achievement. More precisely, a priori scientific knowing expresses, to the extent that it is possible for finite cognizers, the ideal realization of our rational capacities. It is how we rationally perfect (*vervollkommnen*) our cognitive activity. Hence a priori scientific knowing alone constitutes "authentic science" (*eigentliche Wissenschaft*) (*MFNS* 4: 469).

To summarize. I have argued in this chapter that Kant's theory of our knowledge of necessary truth—a priori knowledge—can be traced back

---

[59] See also O'Neill, "Reason and Autonomy in *Grundlegung* III": and O'Neill, "Vindicating Reason."

directly to his theory of "inner" or epistemic necessity; that epistemic necessity is the same as the propositional attitude of "conviction"; that conviction is delivered by "insight"; that insight is based on schematic functions of the imagination; that this imagination-based theory of modal knowledge is well-exemplified in Kant's theory of simple conceptual analysis; that this theory offers a plausible phenomenological solution to the paradox of analysis; that this theory further points up the fact that for Kant decisive epistemic justification is at one externalistic and yet also grounded in the psychological constitution of certain mental acts, states, or processes (and thereby is internalist about epistemic reasons); that all a priori knowledge whatsoever either is insight or else presupposes insight; that Kant's epistemology is perfectionist and non-instrumentally normative; and finally that the role of the schematizing imagination in insight together with the notion of theoretical technique guarantees that a priori scientific knowing is a form of practical rational agency.

As I mentioned at the outset of the chapter, the main objection to Kant's theory of a priori knowledge has always been that his transcendental idealism is unacceptable. But the assumption of minimal modal realism together with weak transcendental idealism jointly obviate the need either to defend or to apologize for strong transcendental idealism. And when we consider that Kant's insight-based, imagination-based, dynamicist theory of a priori knowledge manages to avoid both the problems of platonism and conventionalism on the one hand, and the anti-apriorism of scientific essentialism and modal skepticism on the other, we can conclude that it arguably provides the beginnings of an adequate theory of our knowledge of necessary truths. C. I. Lewis is therefore wrong: it is *not* true—or at the very least it is not *obviously* true—that *all* "traditional conceptions of the *a priori* have proved untenable."

8

Where There's a Will There's a Way: Causation and Freedom

<u>Practical freedom can be proved through experience</u>. For it is not merely that which stimulates the senses, i.e., immediate affects them, that determines human choice, but we always have a capacity to overcome impressions on our sensory faculty of desire by representations of that which is useful or injurious even in a more remote way; but these considerations about that which in regard to our whole condition is desirable, i.e., good and useful, depend on reason. Hence this also yields laws that are imperatives, i.e., objective laws of freedom, and that say what ought to happen, even though it never does happen.... <u>We thus cognize practical freedom through experience, as one of the natural causes, namely a causality of reason in the determination of the will</u>.

CPR A802–3/B830–1, underlining added

THE HUMAN BEING AS A BEING IN THE WORLD, SELF-LIMITED THROUGH NATURE AND DUTY.

OP 21: 34

8.0. Introduction

In the three preceding chapters, I have explored three basic Kantian ways in which human practical reason is presupposed by the exact sciences: in relation to the nature of truth, in relation to the nature of mathematics, and in relation to the nature of a priori knowledge. The presuppositional relations here are both explanatory *and* ontological. The concept of truth makes no sense without the practical concept of sincerity or truthfulness, and the correspondence relation itself also requires the existence of a judger who is cognitively oriented and practically engaged with her surrounding world (see chapter 5). Correspondingly, the concept of number makes no sense without the pure human intuitional representation of time together with the categorically normative laws of pure logic, and neither numbers themselves nor arithmetic nor logic can exist unless weak transcendental idealism is true and rational human thinkers are necessarily possible (see chapter 6). And again, correspondingly, the concept of a priori knowledge makes no sense without the concept of the act of rational human insight, and necessary truths cannot exist unless it is necessarily possible for rational human thinkers to know those very truths a priori by means of the act of insight and theoretical technique (see chapter 7).

In this final chapter, I explore the same basic Kantian presuppositional links with respect to causation. Needless to say, both the concept and the fact of causation are at the explanatory and ontological core of the natural sciences. Kant's radical thesis is that both the concept and the fact of *naturally mechanized causation* not only permit but furthermore necessarily require both the concept and the fact of human *practical or intentional causation*—the causal efficacy and freedom of the human will—as a condition of their possibility. This is *not* however to say, like Nietzsche, that the cause-and-effect relation of classical Newtonian mechanistic physics is nothing but a psychological or conventional fiction:

> One should not wrongly reify "cause" and "effect" as the natural scientists do (and whoever, like them, now "naturalizes" in his thinking), according to the prevailing mechanical doltishness which makes the cause press and push until it "effects" its end; one should use "cause" and "effect" only as pure concepts, that is to say, as conventional fictions for the purpose of designation and communication—not for explanation.[1]

Kant's radical thesis, in sharp contrast to both what Nietzsche scornfully calls the "mechanical doltishness" of the scientific realist approach to causation on the one hand, and also to the causal anti-realism of the Nietzschean (or for that matter, Humean) approach on the other hand, is that naturally mechanized causation is perfectly objectively real, but also implicitly derivative in the sense that it both explanatorily and ontologically presupposes practical or intentional causation. For Kant, not only does the concept

---

[1] Nietzsche, *Beyond Good and Evil*, §21, p. 29.

of naturally mechanized causation ultimately make no sense without the concept of practical or intentional causation, but also the very fact of naturally mechanized causation would not have existed if human freedom of the will had not been necessarily possible.

Here is the outline of my reconstruction of Kant's argument for his radical thesis about causation. First, in section 8.1, I unpack the basics of Kant's metaphysics of causation, with special reference to the three Analogies of Experience and the Third Antinomy of Pure Reason. Second, in section 8.2, I zero in on the problem of free will and work out a new version of Kant's theory of freedom, which I call *the Embodied Agency Theory*. Third, in section 8.3, I explore some of the intimate Kantian links between freedom and nature, and develop a *biological* interpretation of the Embodied Agency Theory. And, finally, in section 8.4 I also argue that for Kant the irreversibility of time—"Time's Arrow"—entails a necessary connection between naturally mechanized causation and the possibility of human practical causation.

8.1. Kant's Metaphysics of Causation: Three Analogies and an Antinomy

In this section I explicate the elements of Kant's metaphysics of causation by offering a unified reading of the Analogies of Experience and the third Antinomy of Pure Reason. I need to emphasize that this is *not* intended as a full exposition of either the Analogies or the third Antinomy. That has already been done with great success—which is not to say with great philosophical agreement!—in the recent and contemporary Kant literature.[2] My goal is instead to isolate a single line of argumentation that captures what I take to be the essential and most philosophically defensible features of Kant's doctrine.

In some ways, the Analogies of Experience section is the most important part of the first *Critique*. This is because in it Kant offers, in effect, solutions to not one but *three* fundamental philosophical problems: Berkeley's problem of how to account for the objectivity of a material world made up entirely

---

[2] See, e.g., Allison, *Kant's Transcendental Idealism*, chs. 9–10, and 15; Guyer, *Kant and the Claims of Knowledge*, chs. 9–11 and 18; Melnick, *Kant's Analogies of Experience*; Strawson, *The Bounds of Sense*, part II, chapter iii; Van Cleve, *Problems from Kant*, chs. 8–9; and Watkins, *Kant and the Metaphysics of Causality*, chs. 3–5.

of wholly subjective sensory objects (a world of sensory ideas), Hume's problem about the continuity of object-identity over time, and Hume's problem about the objective validity of our idea of *causation* or necessary connection in nature.

Berkeley famously argues in *The Principles of Human Knowledge*[3] that the very idea of matter (the idea of a mind-independent cause of our sensory ideas) is logically and metaphysically impossible and that to be an object is nothing more than to be perceived by a thinking subject—to be an idea in a mind. And his speculative solution to the objectivity problem is that a divine mind imposes an order upon the totality of subjective sensory objects or ideas by systematically affecting us in sensibility.

Hume somewhat less famously but just as troublingly argues in *The Treatise of Human Nature*[4] that the continuity of object-identity over time cannot be either directly experienced through the senses or legitimately inferred from those experiences, and in fact is nothing but an irreal or fictional projection of the mind from the repeated association of similar experiences of distinct sensory objects in temporal succession.

And Hume most famously and most troublingly argues in the *Treatise* and again in the first *Enquiry Concerning Human Understanding*[5] (a) that the ideas of causally necessary connections we naturally ascribe to perceived objects are false and vacuous because of the logical contingency of all temporal connections immediately presented to us in sensory impressions (skepticism about causal necessity), and (b) that even if causally necessary connections can in some sense exist "secretly" behind mere sensory objects, they are totally unknowable by means of the senses (metaphysical agnosticism). His skeptical solution to the causal necessity problem is that we non-rationally form habits of mind in experiencing constantly conjoined sensory events, and unconsciously project our habitual expectations, in the form of a belief that a necessary connection exists between all events of those types, onto the sensory data (radical psychological empiricism).

Kant's transcendental solutions to these problems avoid both Berkeley's appeal to a transcendent being who causes our sensory ideas of an objective material world, and also Hume's skepticisms about object-identity and

---

[3] See Berkeley, *Treatise Concerning the Principles of Human Knowledge*, sections 1–33.
[4] See Hume, *Treatise of Human Nature*, book I, part IV, section vi.
[5] See Hume, *Treatise of Human Nature*, part III; and Hume, *Enquiry Concerning Human Understanding*, section 7.

causal necessity, as well as his radical psychological empiricism. Kant's solution, in a nutshell, is that creatures minded like us cannot represent an objective material world without also representing it: (i) as the persistent substrate of real changes in time; (ii) as self-identically enduring over time; (iii) as successively (diachronically) causally necessitated; and (iv) as constituted by a plurality of substances engaged in simultaneous (synchronic) mutual causal-dynamic interactions. Now, even if strong transcendental idealism is false but weak transcendental idealism is true, then it still follows that the objective material world actually *is* the persistent self-identically enduring substrate of changes in time, that it actually *is* diachronically causally necessitated, and that it also actually *is* constituted by a plurality of substances engaged in synchronic mutual causal-dynamic interactions.

Kant says that "[The Analogies'] principle is: Experience is possible only through the representation of a necessary connection of perceptions" (*CPR* B218). What he means is this. The three Analogies correspond to the three categories of relation (substance/attribute, cause/effect, community), which in turn correspond to the three relational forms of judgment (subject/predicate, hypothetical, disjunctive). We also know from Kant's theory of pure general logic that all the forms of judgment are law-governed or nomological, in that pure general logic "contains the absolutely necessary rules of thinking" (*CPR* A52/B76). The temporal schemata for the three categories are, respectively, duration (the continuing existence of a thing through time), succession (the asymmetric or unidirectional passage of events), and coexistence (simultaneity). Then the weak transcendental idealist application of all of these structures to the direct objects of our sensory perception constitutes an *objectively real, substantial, self-identically enduring, causally necessitated, causal-dynamic, law-governed material empirical world in time and space*. That is, broadly conceived as an equation: weak transcendental idealism + the categories (assuming their derivation from pure general logic) + schematization + sense perception = a spatiotemporal empirical world suitable for the application of both necessarily and empirically true causal law-propositions in physics. So much for Berkeley and Hume! Of course, as always, the devil is in the details. Let us now look at the finer grain of each of the three Analogies.

(1) *The First Analogy*

The first Analogy (*CPR* A182–9/B224–32). says: "In all changes of appearances substance persists, and its quantum is neither increased nor diminished in nature" (*CPR* B224). What does *that* mean? Think of it this way. Take the logical form of a subject/predicate proposition, and metaphysically interpret it by applying it to objects in general. The result is the notion of a substance (i.e., an independently existing thing that supports, or is the substrate for, properties), and its accidents or extrinsic properties (i.e., the contingent relational or non-relational properties of the substance). Now take the metaphysical notion and give it a temporal interpretation (that is, a schematization) in terms of duration. The result is the notion of something which exists "persistently" or enduringly through time, and is also the substrate for the various changes in extrinsic properties that occur through asymmetric successive time. But, what is the thing that exists persistently throughout all time and supports various changes in accidents or extrinsic properties?

Kant's initial answer is that the persistent thing must be *real time itself*, considered as a single structure:

All appearances are in time, in which, as substratum (as persistent form of inner intuition,) both **simultaneity** as well as **succession** can alone be represented. The time, therefore, in which all change of appearances is to be thought, lasts, and does not change; since it is that in which succession or simultaneity can be represented only as determinations of it. (*CPR* A181/B224–5)

But this initial answer is philosophically misleading, or at the very least a dialectical prolegomenon to Kant's central claim in the first Analogy. That is because real time as a single intrinsic structure is *not* an objective thing, and thus cannot be represented except as immanent in the changing or simultaneous perceivable material things in space that contain time intrinsically. So in turn, it is these changing or simultaneous perceivable things that presuppose a single substratum, namely, *physical matter as whole*:

Now time cannot be perceived by itself. Consequently it is in the objects of perception, i.e., the appearances, that the substratum must be encountered that represents time in general and in which all change or simultaneity can be perceived in apprehension through the relation of appearances to it. However, the substratum of everything real, i.e., everything that belongs to the existence of things, is **substance**, of which everything that belongs to existence can be thought only as a determination. Consequently that which persists, in relation to which alone all temporal relations of appearances can be determined, is substance in the appearance, i.e., the real in the appearance, which as the substratum of all change always remains the same. Since this, therefore, cannot change in existence, its quantum in nature can also be neither increased nor diminished. (*CPR* A181/B225)

Again, Kant is saying that necessarily every changing extrinsic or contingent property of appearances must be applied to, or predicated of, a single existing material stuff that self-identically endures through asymmetric successive time. An example of a changing extrinsic relational property of something X enduring through asymmetric successive time would be X's varying velocity, and an example of a changing extrinsic non-relational property would be X's varying shape. In any case, the single self-identical enduring material stuff persists through time by virtue of its *intrinsic* properties, which we know from chapters 1–4 and Kant's manifest realism more generally, to be exclusively intrinsic *structural* properties, or intrinsic relational properties based on spatiotemporal form.

In this connection, Kant quite puzzlingly however talks about two distinct levels of the material substrate of empirical nature: on the one hand, he talks of a single substratum that exists persistently throughout *all* time—One Big Substance, or the totality of matter; but on the other hand, he talks about a plurality of individual substances that exist persistently for a while, and then go out of existence—the particular material substances or bodies (*Körper*):

In all appearances that which persists is the object itself, i.e., the substance (*phenomenon*), but everything that changes or that can change belongs only to the way in which this substance or substances exists, thus to their determinations. (*CPR* A183–4/B227, underlining added)

In all appearances there is something that persists, of which that which changes is nothing but the determination of its existence. (*CPR* A184/B227, underlining added)

Substances (in appearance) are the substrata of all time-determinations. The arising of some of them and the perishing of others would itself remove the sole condition of the empirical unity of time, and the appearances would then be related to two different times, in which existence flowed side by side, which is absurd. For there is **only one** time, in which all different times must not be placed simultaneously but onlyone after another. (*CPR* A188–9/B231–2, underlining added)

These texts may seem to involve a serious confusion on Kant's part, in that his arguments from the need for a persistent substrate of change, together with the empirical unity of time, would seem to justify only the inference to the existence of One Big Substance, not to the existence of many distinct substances.[6] And this confusion may also seem to be compounded when, in the third Analogy, his account explicitly requires the existence of many distinct individual material substances or bodies, although this is a thesis he has not sufficiently proved in the first Analogy.[7]

These two apparently divergent Kantian perspectives on material substance can be smoothly reconciled, I think, by proposing that Kant's considered view on material substance is that the plurality of individual material substances or bodies are *real proper parts* of a *real integral structured whole*, which is the same as the One Big Substance, or material plenum, which in turn is the same as the totality of matter in space. More precisely, we can think of the spatial One Big Substance as "primary substance," and also think of the plurality of uniquely individuated material substances or bodies as a set of "secondary substances" whose existence and properties are also *nomologically strongly supervenient* on the existence and properties of the primary substance. We will remember that A-facts or higher-level facts about the instantiation of A-properties are strongly supervenient on B-facts or lower-level facts about the instantiation of B-properties if and only if:

(1) Necessarily if anything has a B-property, then it also has an A-property ("upwards determination").

And:

(2) Necessarily there can be no change in anything's A-properties without a corresponding change in its B-properties ("necessary covariation").

And *nomological* strong supervenience means that the "necessarily" in (1) and (2) is to be interpreted as *synthetic a priori necessity according to laws of nature*—that is, the "dynamic or material necessity" of the third Postulate of Empirical Thought (*CPR* A218/B266), which I briefly discussed in section 3.5 The primary substance, the totality of matter, is then preserved through the real but ontologically derivative coming-to-be and passing-away of the many "secondary substances" by virtue of the fact that particular material substances or bodies are nothing but special law-governed organizations of the totality

---

[6] See Strawson, "Kant on Substance."
[7] See Westphal, *Kant's Transcendental Proof of Realism*, ch. 4.

of matter at various temporal and spatial coordinates, all of which eventually "break up." But the "quantum" or total supply of matter persists and is permanently preserved as the condition of the empirical unity of time.

This mereological (i.e., part-whole theoretic) and supervenience-based interpretation of the first Analogy makes it possible for Kant to give a metaphysically elegant doctrine of alteration, or change. For a material substrate to alter is for that material substrate to have a succession of changing accidents or extrinsic properies instantiated in it, in a single or unique comprehensive time and a unique comprehensive space. Particular material bodies, the proper parts of the One Big Substance or material plenum, then come to be and pass away solely by virtue of the temporal succession of instantiated extrinsic properties in the material plenum. In this way, the individual material bodies we experience are nothing but particular temporal complexes consisting of the One Big Substance together with a proper subset of its changing accidents, at particular spatial places. Otherwise put, a particular material bodily object of experience is nothing but a particular necessarily rule-governed successive temporal sequence of spatially placed events—as it were, a constantly updated *Curriculum Vitae*—in the immensely long career of the spatiotemporal One Big Substance. Or, again, the unity of a particular secondary material substance or body is nothing more than a certain intrinsic *nomological orderliness* imposed on the asymmetric temporal succession of extrinsic properties applicable to the totality of matter or One Big Substance enduring self-identically through time in space.

In order to be correctly understood, Kant's two-leveled or dual-perspective metaphysics of material substance as persistence through nomologically ordered change over time must also be explicitly combined with his causal-dynamic structuralism, and his aether-theory. We will remember that, according to Kant's aether deduction in the *Opus postumum*, the actually existing totality of matter is identical to a structural aether consisting of a multiplicity of primitive attractive and repulsive forces standing in a multiplicity of law-governed causal-dynamic relations to one another (see section 4.2). This is the same as the spatialized One Big Substance or primary substance that persists as a single quantum or material plenum through all time. And we will also remember that according to Kant's causal-dynamic metaphysical structuralism, individual material substances or bodies are nothing but positions or roles in the total relational system of causal-dynamic forces in objectively real space and time (see sections 1.0, 3.2, and 4.2). These structure-dependent entities are the same as the individual nomologically-ordered successive temporal sequences of spatially placed events, the proper parts or secondary substances of the One Big Substance. This in turn clearly satisfies the nomological strong supervenience relation that obtains between the One Big Substance and its many proper parts or secondary substances. If individual material substances or bodies are nothing but positions in the total relational system of attractive and repulsive forces, then obviously it must be the case that:

(i) fixing all the properties of the One Big Substance sufficiently determines the existence and properties of the many individual material substances; and also
(ii) that there can be no change in any of the properties of the many individual material substances without a corresponding change in the properties of the One Big Substance.

So the overall preliminary picture we get from the First Analogy of Experience is a two-leveled structuralist and substantialist metaphysics of matter, as shown in Figure 8.1:

LEVEL 2 = THE LEVEL OF STRUCTURE-DEPENDENT ENTITIES
the temporarily-persisting substrates of changes in time = the proper integral parts of matter as a whole
= the plurality of individual material substances or bodies = the secondary substances
= the multiplicity of nomologically-ordered particular successive temporal sequences of spatially-placed events
= the many different positions or roles in the systematic totality of primitive attractive and repulsive forces
NOMOLOGICALLY STRONGLY SUPERVENES ON

⇕ ⇕ ⇕

LEVEL 1 = THE LEVEL OF THE STRUCTURE ITSELF
the permanently persisting substrate of changes in time = matter as a whole = the One Big Substance
= the primary substance = the material plenum in one comprehensive space and one comprehensive time
= the structural aether = the systematic totality of primitive attractive and repulsive forces

Figure 8.1. Kant's Metaphysics of Matter (Preliminary Version)

(2) The Second Analogy

The second Analogy of Experience (CPR A188–211/B232–56) is the most famous and important of the Analogies, mainly because it contains the pith and marrow of Kant's answer to Hume's skeptical analysis of our idea of causal necessity in nature. This philosophical salience has however led to the mistaken idea that the second Analogy is logically independent of the other two Analogies. On the contrary, in fact the second Analogy is both conceptually and metaphysically complementary to the other two Analogies, and can neither be understood in theory nor obtain in reality without them.

In any case, the second Analogy goes like this: "All alterations (*Veränderungen*) occur in accordance with the law of the connection of cause and effect" (*CPR*: B232). This principle clearly builds on the first Analogy. An alteration is "a way of existing that succeeds another way of existing of the very same object" (*CPR* A187/B230). That is, an alteration is an asymmetric temporal succession of at least two different instantiations of extrinsic properties in the same enduring material substance. Each such instantiation of an extrinsic property in a material substance at a particular time is what Kant calls a "state" (*Zustand*) of that material substance. Therefore "everything that is altered is **lasting**, and only its **state** changes" (*CPR* A187/B230). In a nutshell then, the Second Analogy is saying that the nomologically ordered asymmetric temporal succession of changing states of the persisting totality of matter is itself a set of law-governed necessary connections between earlier and later states such that each earlier state is nomologically sufficient for some corresponding later state, and each later state is nomologically necessitated by some earlier state. Here, also in a nutshell, is Kant's basic argument for that thesis.

(1) The concept of a cause analytically entails the concept of its effect.

(2) The general schematized concept of the cause-effect relation is the concept of a temporal relation of rule-governed necessitation or nomological sufficiency between a condition X and a conditioned entity Y that is determined by X (*CPR* B112, A144/B183).

(3) The first Analogy of Experience is true.

(4) The persistent totality of matter is thus the One Big Substance upon which a plurality of individual material substances nomologically strongly supervene. (From (3).)

(5) The One Big Substance can be self-identical and enduring only if it is presupposed that all its changing sensible states in asymmetric successive time are all necessarily connected in law-governed ways.

(6) Therefore, the set of law-governed necessary connections between asymmetric successive states of the persistent totality of matter is the same as the set of diachronic cause-effect relations actually obtaining in the empirical natural world. (From (1), (2), (4), and (5).)

Now I want to unpack Kant's reasoning in this basic argument in more detail.

According to Kant, we can think of each causally-ordered alteration between two asymmetrically successive states of matter as what I will call a *simple event* in the total career of matter. A simple event in my terminology is the same as what Kant calls an "occurrence" (*Begebenheit*) or "something that actually happens" (*etwas wirklich geschieht*) (CPR A201/B246). Then since distinct states of matter are individuated by their constituent properties and time-indices (for example, state $a = P_1$ instantiated at t_1, state $b = P_2$ instantiated at t_2, and state $c = P_3$ instantiated at t_3) whereas simple events (that is, occurrences or happenings) are individuated by their two constituent states together with the the law-governed necessary relation between them (thus simple event e_1 = state a's causal necessitation of state b, and simple event e_2 = state b's causal necessitation of state c), the second Analogy is also equivalently saying that the asymmetric temporal succession of changing states of the permanently persisting totality of matter is itself a set of necessary connections between earlier and later different simple events, such that each earlier simple event is nomologically sufficient for some later simple event, and each later simple event is nomologically necessitated by some earlier simple event, and that this set is the same as the set of diachronic cause-effect relations between simple events in the empirical natural world. Or, in other words, according to Kant in the second Analogy, every simple event both has a cause, which is another simple event, and also causes its own effect, which is yet another simple event.

From this thesis, together with the account of particular material substances that I developed in my interpretation of the first Analogy, it then directly follows that particular material substances in the empirical natural world are essentially particular successive temporal sequences of simple events (that is, occurrences or happenings) all related by nomological sufficiency or causation. In other words, for Kant a particular material substance is essentially what I will call a *complex event* (or *Ereignis*), which is a unified and causally-structured ordered set consisting of at least two simple events. It should be noted here for later discussion, however, that Kant's metaphysical analysis of complex events in terms of successively necessitated changing states of the One Big Substance, and then in turn his analysis of particular material substances or bodies in terms of complex events is not a *reductive* analysis, since it turns out that the material substances or bodies themselves have intrinsic structural properties, such as the property of standing in simultaneous or synchronic mutual causal-dynamic determination-relations with other particular material bodies, which are not possessed by any simple event alone or indeed by any mere irreversibly ordered n-tuple of simple events.

Why does Kant identify the concept of diachronic causation with the concept of the relation of nomological sufficiency obtaining either between asymmetrically successive states of a persisting material substance or else between asymmetrically successive simple events? This is because the category of cause/effect derives from the logical form of the hypothetical, which Kant understands as:

Logically or analytically necessarily (if P then Q),

which means that the antecedent is logically or analytically sufficient for the consequent under some law of logic. The cause/effect relationship, in other words, is *the nomological logical consequence relation as applied to objects in general*. But the temporal schema of the time-series restricts this relation to asymmetrically successive moments in time ("Time's Arrow") (*CPR* A144/B183, A215/B262). Hence the schematized category of cause/effect is *the nomological logical consequence relation as mapped either onto asymmetrically successive states of a persisting substance or else onto asymmetrically successive simple events*, which by virtue of its cognitive-semantic and metaphysical dependence on time makes it a non-logical or *synthetic* necessary connection, namely, a relation of *synthetic nomological sufficiency*.

Now it may seem that not every asymmetric temporal succession of either states or simple events bears this causal structure of synthetic nomological sufficiency, because some of them seem to be mere *coincidences*. Mere coincidences, if they really existed, would be randomly related successive changing states or simple events in material nature. Hence the existence of real coincidences would constitute breaks or ruptures in the otherwise seamless causal fabric of the material world. So Kant is obliged to deny the existence of real or objective coincidences, and to explain them away by making them purely psychological or subjective. I will come back to this important point shortly.

But in any case, even the existence of real coincidences would not undermine Kant's thesis that necessarily *whenever* either an earlier state of a substance or an earlier simple event is nomologically sufficient for a later state of a substance or a later simple event respectively, *then* the later one is the effect and the earlier one is the cause:

If, therefore, we experience something which happens, then we always presuppose that something else precedes it, which it follows in accordance with a rule. For without this I would not say of the object that it follows, since the mere sequence in my apprehension, if it is not, by means of a rule, determined in relation to something preceding, does not justify any sequence in the object. Therefore I always make my subjective synthesis (of apprehension) objective with respect to a rule in accordance with which the appearances in their sequence, i.e., as they occur, are determined through the preceding state, and only under this presupposition alone is the experience of something that happens possible. (CPR A195/B240)

States of material substances and their corresponding simple events, insofar as they are causally ordered in this way into the structure of a complex event (Kant's example is the successive positions of a boat floating downstream), are represented by *objective* orderings of perceptions; by contrast, states of material substances insofar as they are ordered in an arbitrary way, and do not constitute either simple or complex events, are represented by merely *subjective orderings* of perceptions (Kant's example is the succession of perceptions delivered by someone's gaze flitting over a house). The difference between the objective or necessitated ordering of perceptions and the merely subjective or "arbitrary" (*beliebig*) (CPR A193/B238) ordering of perceptions is also the difference between our fully determinate representations of a scientifically knowable objective material world given in outer sense and by means of judgments of experience, and the phenomenal stream of consciousness given in inner sense. The seeming existence of real coincidences in material nature can then be explained way as nothing but functions of the subjective or arbitrary ordering of perceptions in inner sense. For example, it may seem to me that the following three-part sequence of simple events is a real coincidence, or a tripartite coincidental complex event: I am sitting on a park bench and cross my legs; a child falls down nearby and scrapes her knee; and a jet airplane screams by overhead. But this seeming three-part coincidental complex event is no event at all, and in fact is nothing but the result of my own subjective or arbitrary ordering of perceptions, or what we would now call my capacities for *attention* and *interest*, which in turn are a function of *my perceptual consciousness* and *my desires*, not a function of the external material world.

This, however, raises another absolutely crucial point. The subjective or arbitrary ordering of perceptions described by Kant in the Second Analogy, and offered by him as a necessary element in his explanation of how we represent an objective material world, is a direct expression of a species of psychological freedom[8] I will call *the spontaneity of consciousness and desire*.[9] According to *the spontaneity of consciousness*, I can order the sequence of my sensory apprehensions of empirical objects at will—I can attend to them in any order, and even reverse them if I feel like it:

In the ... example of a house my perceptions could have begun at its rooftop and ended at the ground, but could also have begun below and ended above; likewise I could have apprehended the manifold of empirical intuition from the right or from the left. In the series of these perceptions there was therefore no determinate order that made it necessary when I had to begin in the apprehension in order to combine the manifold empirically. (*CPR* A192–3/B237–8)

But this spontaneity of consciousness is clearly also a function of "the power of choice" or *Willkür*: hence it is also *a spontaneity of desire*. Just as I can consciously perceptually attend to *this*, and then *that*, and then *that*, etc., at will, or even completely reverse the order of my sensory perceptions if I feel like it, so too I can consciously desire *this*, and then *that*, and then *that*, etc., at will, or even completely reverse the order or ranking of my preferences, if I feel like it. In other words, in the subjective ordering of my perceptions, I immediately feel myself to be under no sort of psychological or external compulsion whatsoever, whether cognitive or affective.

The spontaneity of consciousness and desire in the Second Analogy, while it is an *authentic species of psychological freedom*, is not however a sufficient condition of *real freedom of the will*, whether in the guise of "transcendental freedom" or "practical freedom" (see section 8.2). This is for two

---

[8] See also O'Neill, *Constructions of Reason*, pp. 62–3; and Sensen, "'No Causality without Freedom': Kant's Argument for Freedom in the 'Analogies of Experience'."

[9] See Hanna and Thompson, "Neurophenomenology and the the Spontaneity of Consciousness."

reasons. First, psychological spontaneity—as the unprecedented, sensibility-underdetermined, creative, and self-guiding features of a mental operation or act (see section 0.2)—is only a *non-absolute or relative* spontaneity, because the perceptions and other mental contents that can be ordered at will, are already empirically given to the conscious subject. Second, even granting the phenomenological fact of my first-person experience of willful control over the ordering of my perceptions, it is always logically and metaphysically possible that even though I immediately and vividly feel myself to be under no sort of internal or external compulsion, nevertheless in fact my attentive perceptual and desiderative focusings, my choices, and my actions are all either metaphysically necessitated by natural laws together with all the preceding or settled past empirical natural facts (*CPR* A549–53/B577–81),[10] or else they are all nothing but a mathematical output of aggregated empirical natural facts governed solely by probabilistic or statistical laws.[11] In other words, as Hume correctly saw—and as Harry Frankfurt has recently rediscovered—psychological freedom of the will is perfectly consistent with natural determinism and mechanism.[12]

Even so, the spontaneity of consciousness and desire is still a *necessary condition* of real freedom of the will, both transcendental and practical. For no human choice or act could actually constitute authentic freedom of the will if the subject at the same time consciously experienced himself as helplessly internally or externally compelled, constrained, manipulated, overwhelmed, or violated: I cannot actually *be free* without also *feeling unfettered and unforced* (which is not to say that this feeling is always or even usually self-consciously noticed). Suppose now that my will really *is* unfettered and unforced. This fact is called *negative freedom*. So what Kant is implicitly saying in drawing his distinction between the objective or necessitated ordering of perceptions on the one hand, and the subjective or arbitrary orderings of perceptions

---

[10] Suppose we deny that free will and determinism are metaphysically consistent, and also deny that free will exists, but *also* assert that determinism is true. This doctrine is then called "hard determinism." See Kane, *A Contemporary Introduction to Free Will*, ch. 7. See also section 8.2.

[11] Suppose we deny that free will and indeterminism are metaphysically consistent, deny that free will and determinism are consistent, and also deny that free will exists, but *also* assert that indeterminism is true. This doctrine is then what I will call "hard indeterminism." See section 8.2.

[12] See Hume, *Treatise of Human Nature*, book II, part III, sections i-ii; Frankfurt, "Alternate Possibilities and Moral Responsibility"; and Frankfurt, "Freedom of the Will and the Concept of a Person."

on the other, is that an objective or necessitated ordering of perceptions exists *only if* a certain form of psychological freedom—the consciousness of negative freedom—also exists, which in turn is a necessary condition of real transcendental and practical freedom of the will:

> I must therefore derive the **subjective sequence** of apprehension from the **objective sequence** of appearances, for otherwise the former would be entirely undetermined and no appearance would be distinguished from any other. The former alone proves nothing about the connection of the manifold in the object, because it is entirely arbitrary. (*CPR* A193/B238)

As we will see in sections 8.2 and 8.3, what corresponds to the subjective or arbitrary ordering of perceptions in apprehension is a set of "subjective subjects" (non-conscious or conscious mental states) and their correlated non-mechanical or *purposive* (teleological) causal-dynamic properties, neither of which can be accurately captured or known by classical or Newtonian physics.

Now back to the objective or necessitated orderings of perceptions. The material objects represented by means of such an ordering of perceptions are essentially complex events, which in turn are essentially causally-structured sequences of simple events (that is, occurrences or happenings), which in turn are essentially causally-structured sequences of states of material substances. Kant's analysis of simple events is therefore this:

Non-logically or synthetically necessarily for anything x and for any two distinct extrinsic properties P_1 and P_2, x is a simple event if and only if there exists an earlier state of x such that this state instantiates a property P_1 that is synthetically sufficient for a later state of x which in turn instantiates another quality P_2.

This is the same as to say that synthetically necessarily the earlier state of a simple event *causes* its later state as its *effect*. If you find quasi-formalizations helpful for clarity, then Kant's analysis of simple events looks like this:

Non-logically or synthetically necessarily (x) (P_1) (P_2) {x is a simple event ↔ (∃y) (∃z) [y and z are both states of x & y is prior to z & synthetically necessarily (P_1 is instantiated in y → P_2 is instantiated in z)]}

This analysis can then be easily extended to two-part complex events as follows:

Non-logically or synthetically necessarily (x) {x is a complex event ↔

(∃y) (∃z) [y and z are both distinct simple events contained in x & y occurs earlier than z & synthetically necessarily (y → z)]}

And similarly for complex events containing three simple events, and so on.

The upshot of Kant's account in the Second Analogy is thus that the individual material substances or bodily objects of experience are all essentially complex events. Complex events, in turn, are essentially causally-ordered sequences of at least two simple events, which in turn are essentially temporal sequences of nomologically necessarily connected time-indexed instantiations of extrinsic properties—that is, causally structured relations between states—predicated of the One Big Substance. Furthermore, these complex events all instantiate causal-dynamic laws of attractive and repulsive forces. Therefore for Kant the objective world of individual material substances or bodies is just the totality of causal-dynamic complex events, as analyzed above. This in turn is required as a presupposition of the self-identical and enduring One Big Substance, or material plenum, postulated in the first Analogy. And this solves both Berkeley's and Hume's problems in one fell swoop by proposing that both objectivity and causality in nature are the result of our weakly transcendentally ideally imposing the schematized second Analogy of Experience on the asymmetric successive temporal sequence of changing sensible states of the One Big Substance.

One of the theoretical advantages of this interpretation of the second Analogy is that it directly answers a famous criticism. In *The Bounds of Sense*, Peter Strawson says that Kant commits a "non sequitur of numbing grossness."[13] This is the inference from the premise that the *ordering* between two events is necessary (for example, at time t1 boat B is higher up the stream, and at later time t2 boat B is lower down the stream—but B could not have gone downstream without first being higher upstream) to the conclusion that the events follow each other necessarily (for example, that boat B's being further down the stream at t2 is a necessary consequence of its having been higher up the stream at t1). This inference from temporal irreversibility to temporal necessity is of course fallacious, just as Strawson insisted. But Kant is not inferring from temporal irreversibilty to temporal necessity. Instead, what Kant is saying is that what *constitutes* X's being an individual objective material substance is that X is a complex event which intrinsically contains not only temporal irreversibility but also temporal

[13] Strawson, *The Bounds of Sense*, p. 137.

necessity. So Kant's account, which postulates both temporal irreversibility and temporal necessity, is not a fallacious inference, rather it is nothing but the elaboration of a single metaphysical analysis.

Another theoretical advantage of this interpretation is that it forestalls a certain worry that could arise in connection with the third Analogy. The problem is that the third Analogy explicitly requires the existence of a plurality of material substances in mutual simultaneous causal-dynamic interaction: yet the second Analogy is explicitly all about alterations or changes in states of the One Big Substance, simple events, and complex events, not about substances. This, however, is a worry only if one assumes *an exclusivist and reductivist* approach to the ontology of simple and complex events, of the sort that one might find, for example, in Hume's account of causation. Exclusivism about event-ontology says *that nothing can both be event-like and substance-like*. And reductivism about event-ontology says *that there are no substances in reality but really only events*, even if we happen to use the language of substances for convenience and everyday purposes. Then, assuming that every event-ontology is both exclusivist and reductivist, either we would be forced to say either that the second Analogy and third Analogy are conceptually and logically inconsistent with one another, or that the second Analogy is really all about causal relations between substances, even though it seems to be about alterations of states, simple events, and complex events.[14]

As I interpret Kant, however, he is offering a theory according to which objects of experience are *both* material substances and *also* events, because a material substance just essentially *is* a complex event, that is, a unified and intrinsically causally-structured ordered set of at least two simple events. So that avoids exclusivism. And because material substances, as such, can have properties that are properties of neither simple nor complex events, as such then the ontology of material substances is irreducible to the ontology of simple and complex events. In short, the ontology of material substances is *nomologically superveniently emergent* from the ontology of simple and complex events, in the precise sense that the properties of material substances are nomologically strongly supervenient on but not identical with the properties of their constituent simple and complex events. And that avoids reductivism. In this way, one of Kant's most profound insights

---

[14] See Watkins, *Kant and the Metaphysics of Causality*, chs. 3–4.

in the second and third Analogies is that it is possible to combine an event-ontology consistently and non-reductively with a substance-ontology, and thus capture the core of truth in both empiricist and rationalist accounts of "what there is" in the natural world.

(3) *The Third Analogy*

It is an apparent consequence of Kant's metaphysical analysis of causation and objectivity in the second Analogy that causal relations hold exclusively between either earlier and later moments of events (in the case of simple events), or earlier and later simple events (in the case of complex events). But what about the many physical phenomena that are apparently both simultaneous and yet causal-dynamic, for example, the lacrosse ball now denting my pillow, the centripetal force of the earth's gravity now acting on me, the light now illuminating this room, the legs of the table now holding up the top of the table, and the pattern of iron filings which now exists in the presence of this magnet? Here is how Kant puts the problem:

> There is a reservation that must be raised. The principle of causal connection among appearances is, in our formula, limited to the succession of them, although in the use of this principle it turns out that it also applies to their co-occurrence (*Begleitung*), and cause and effect can be simultaneous. E.g., there is warmth in a room that is not to be encountered in the outside air. I look around for the cause, and find a heated stove. Now this, as the cause, is simultaneous with its effect, the warmth of the chamber; thus there is no succession in time between cause and effect, rather they are simultaneous, yet the law still holds. The majority of efficient causes in nature are simultaneous with their effects, and the temporal sequence of the latter is occasioned only by the fact that the cause cannot achieve its entire effect in one instant. But in the instant in which the effect first arises, it is always simultaneous with <u>the causality of its cause</u>, since if the cause had ceased to be an instant before then the effect would never have arisen. (*CPR* A202–3/B247–8, underlining added)

In the third Analogy (*CPR* A211–15/B256–62), Kant solves this problem to his own satisfaction in three steps.

The first step is to distinguish between the *order* of time and the *lapse* of time (*CPR* A203/B248). Where causation is concerned it is the former and not the latter that counts. There can still be temporal irreversibility and temporal necessitation even if two simple or complex events are simultaneous. For example, as I turn the corner in my car, the complex event consisting of the rotation of the steering wheel and the complex event consisting of the transverse movement of the tires against the pavement are simultaneous. But the temporal ordering between the two complex events runs irreversibly and necessarily from the steering wheel complex event to the tire movement complex event, in the sense that necessarily the tire movement event could not have happened *prior* to the steering wheel event. So to say that two events e_1 and e_2 have an irreversible temporal ordering is just to say that necessarily e_2 could not have happened prior to e_1. Or, in other words, necessarily e_2 occurs either simultaneously with e_1 or else after e_1.

The second step is Kant's extension of the general concept of the cause-effect relation from a *diachronic* temporal schematization to a *synchronic* temporal schematization. Or in other words, we can represent causal-dynamic relations not only as occurring over a temporal *succession*, but also as occurring *simultaneously* in time:

The schema of <u>the cause and of the causality of a thing in general is the real upon which, whenever it is posited, something else always follows. It therefore consists in the succession of the manifold insofar as it is subject to a rule</u>. The schema of community (reciprocity), or of <u>the reciprocal causality of substances with regard to their accidents</u>, is the simultaneity of the determinations of the one with those of the other, in accordance with a general rule. (*CPR* A144/B183–4, underlining added)

The crucial point to note here is that whereas one might have had the impression from looking at the Table of Categories, which includes the category of cause and effect as the second of the categories of relation, that for Kant there is only *one* type of causation, namely, diachronic or asymmetric successive causation, nevertheless and on the contrary for him there are in fact *two* types of causation: (1) diachronic or asymmetric successive causation; and (2) synchronic or simultaneous causation (= community or reciprocity). Each of these types exemplifies "the causality of the cause" or *intrinsic causal powers* with respect to the effect. In the case of diachronic causation, the direction of the realization of these intrinsic causal powers with respect to the effect is *one-way* or asymmetric in successive time, whereas in the case of synchronic causation, the direction of the realization of these intrinsic causal powers with respect to the effect is *two-way* or symmetric in simultaneous time.

The third and final step is to extend the general concept of the cause-effect relation from material complex events to material substances. As we have already seen, this step is automatically and smoothly mediated by Kant's thesis that the ontology of simple and complex events is the nomological strong supervenience base of the ontology of material substances.

These three steps together yield the third Analogy: "All substances, insofar as they can be perceived in space as simultaneous, are in thoroughgoing interaction" (*CPR*: B256). That is, co-temporal or simultaneous material substances stand to one another in necessary and mutual causal-dynamical relations. Given the third Analogy, we can now recognize that the relation of diachronic or successive causation as presented in the second Analogy was in fact all along implicitly *also* a relation of synchronic or simultaneous reciprocal causation between substances. Insofar as an earlier simple or complex event brings about a corresponding later simple or complex event with nomological sufficiency, there must also be a world of mutually interacting determinately-positioned substances existing in the present in order to guarantee the direct action of the causal powers of the earlier cause on the later effect. The action or causality of the earlier cause must always be simultaneous with the beginning of the later effect, which constantly emerges in the present until it fully exists and thereby further causally-dynamically mutually determines the extrinsic properties of all the other simultaneously existing substances, as time advances in the direction of its one-way arrow. Contrariwise, if the action or causality of the cause were to expire *prior* to the beginning of the effect—that is, expire in the past and not exist simultaneously in the present along with the beginning of the effect—then the effect would never get underway. In other words then, the diachronic event-causation of the second Analogy is both explanatorily and ontologically *complementary* with the synchronic substance-causation of the third Analogy.

This explanatory and ontological interplay of diachronic causation and synchronic causation also accounts for the existence of a world of substances in simultaneous mutual causal-dynamic interaction, since according to Kant the latter comes about precisely by means of diachronic causation. Thus, for example, the lacrosse ball that is now simultaneously denting my pillow came into existence because at an earlier time I carried the ball across the room and dropped it on the pillow. Similarly, the centripetal force of the earth's gravity now acting on me, the light now illuminating this room, the legs of the table now holding up the top of the table, and the pattern of iron filings which now exists in the presence of this magnet, all were originally caused by earlier complex events.

One last important point about the third Analogy. We will remember that in the second Analogy, Kant argued from the existence of two distinct ways of perceptually representing a world of material objects, together with weak transcendental idealism, to the conclusion that necessarily every alteration in a material substance is a diachronic causal relation between simple or complex events. Those two distinct ways of perceptually representing the world are the objective/necessitated ordering of perceptions and the subjective/arbitrary ordering of perceptions. And we will also remember that the existence of a subjective/arbitrary ordering of perceptions also implies the existence of a species of psychological freedom, the spontaneity of consciousness and desire. Now, in the third Analogy, Kant argues from the existence of *one* particular way of ordering perceptions, together with weak transcendental idealism, to the conclusion that all material substances, insofar as they are simultaneous, stand in causal-dynamic community with one another. This way of ordering perceptions is precisely the *reversibility* of the sequence of perceptions by which a conscious subject represents the influence by which two simultaneous substances mutually causally necessitate the extrinsic properties of one another:

Things are simultaneous insofar as they exist at one and the same time. But how does one cognize that they exist at one and the same time? If the order in the synthesis of the apprehension of the manifold is indifferent, i.e., if it can proceed from A through B, C, and D to E, but also conversely from E to A. For if they existed in time one after the other (in the order that begins with A and ends at E), then it would be impossible to begin the apprehension at the perception of E and proceed backward to A, since A would belong to past time, and thus can no longer be an object of apprehension ... In addition to the mere [simultaneous] existence [of A and B] there must be ... something through which A determines the position of B in time, and conversely also something by which B does the same for A, since only under this condition can those substances be empirically represented as **existing simultaneously**. Now only that determines the position of another in time which is the cause of it or its determinations. Thus each substance (since it can be a consequence only with regard to its determinations) must simultaneously contain the causality of the other, i.e., they must stand in dynamical community (immediately or mediately) if their simultaneity is to be cognized in any possible experience. But now everything in regard to objects of experience is necessary without which the experience of these objects would be impossible. Thus it is necessary for all substances in appearance, insofar as they are simultaneous, to stand in a thoroughgoing community of interaction with each other. (CPR A211–13/B258–60)

In other words, it is precisely the existence of *a subjective or arbitrary ordering of perceptions that is reversible at will* which (on the assumption of weak transcendental idealism) cognitively guarantees the existence of synchronic causation. To reuse Kant's example in the second Analogy, my gaze can flit freely over the surface of the house *only because* all parts of the house are in simultaneous causal-dynamic community. As a consequence, it also follows that the existence of a species of psychological freedom—the spontaneity of consciousness and desire—is built as a necessary condition into the existence of our representations of synchronic causation.

The Antinomy of Pure Reason follows the Paralogisms of Pure Reason and continues the job of transcendental dialectic, which is the logical diagnosis of our natural tendency to pure rational metaphysical illusion, or "transcendental illusion" (CPR A293/B349). According to Kant, we can never entirely remove a transcendental illusion, because the desire for self-transcendence is naturally innate in humans. In this respect, sadly, we *are* nothing but a "useless passion." Looking at the brighter side of things, however, we *can* come to terms with our natural tendency towards this transcendental illusion by exposing the metaphysical error that underlies it. To be sure, this will not make the transcendental illusion go away, just as in the case of perceptual "veridical illusions" (see section 2.2) we cannot help seeing the two lines as unequal in the Müller-Lyer diagram, or the rising moon as appearing larger closer to the horizon (CPR A297/B354), even when we know that it *is* a perceptual illusion and can explain how it happens. Still, the diagnosis and explanation of the Müller-Lyer illusion or the larger rising moon illusion tells us something important about our informationally encapsulated modular perceptual capacity; correspondingly, the diagnosis and explanation of transcendental illusion tells something about our innate rational capacity. But in this respect there are two crucial differences between the Paralogisms and Antinomies.

First, whereas the Paralogisms investigate transcendental illusion concerning the subject of cognition, the thinking *subject*, the Antinomies investigate transcendental illusion concerning the *object* of cognition: that is, the totality of appearances. We know from our study of the Analogies of Experience that this totality of appearances is equivalent to the empirical world, or material nature, and that material nature is a nomological diachronic and synchronic causal-dynamical system of simple and complex events or material substances in space and time, predicated of the One Big Substance, the self-identically enduring material plenum in space and time.

Second, whereas the Paralogisms expose a basic fallacy in metaphysical reasoning about the thinking subject (roughly, the fallacy that the fact of self-consciousness or apperception logically entails the existence of a Cartesian thinking substance), the Antinomies expose an antinomy or contradiction in our metaphysical reasoning about the object of cognition.

An antinomy, however, is no ordinary logical inconsistency or contradiction, which is a judgment that is, or entails, a judgment of the form "P and $\sim P$." An antinomy is, by contrast, a *paradox* or *hyper-contradiction*. A paradox or hyper-contradiction has the following criterial feature: on the assumption of the thesis, P, a contradiction can be derived; and on the assumption of the antithesis, $\sim P$, another contradiction can be derived. So an antinomy is a proposition whose truth entails a contradiction and whose falsity also entails a contradiction. Hence the defender of the thesis can "prove" his claim by a *reductio ad absurdum* argument on the antithesis, and the defender of the antithesis can also "prove" her claim by the same *reductio* strategy as applied to the thesis.

Kant believes that there are four basic forms of the Antinomy, corresponding to the four sets of categories: (1) quantity; (2) quality; (3) relation; and (4) modality. The first or quantitative Antinomy is about cosmological *extension*, and concerns the question as to whether the world is finite in time and space (thesis) or alternatively is infinite (antithesis). The second or qualitative Antinomy is about cosmological *composition*, and concerns the question as to whether the world is made of ultimate atoms (thesis), or alternatively is infinitely composite (antithesis). The third or relational Antinomy is about cosmological *causation*, and deals with the question as to whether nature includes some absolutely spontaneous causes or freedom (thesis), or alternatively includes no freedom whatsoever because it is completely determined by the causal-dynamic laws of nature (antithesis). And finally the fourth or modal Antinomy is about cosmological *origins*, and covers the question as to whether the world includes or has as its cause a necessary being (thesis), or alternatively whether it neither includes nor has as its cause a necessary being (antithesis).

For Kant, the logical clue to the solution of the several antinomies lies in the logical distinction between "contradictories" and "contraries." Both are forms of inconsistency. But whereas two contradictories cannot both be false and cannot both be true, so one of them *must* be true, two contraries cannot both be true, but both *can* be false. For example, "All *A*s are *B*s" and "Some *A*s are not *B*s" are contradictories, but "All *A*s are *B*s" and "No *A*s are *B*s" are contraries (that is, they can both be false if some *A*s are *B*s and some *A*s are not *B*s). In each form of the Antinomy, what we discover is that the thesis and antithesis are really contraries, not contradictories. For in each case we discover that both thesis and antithesis share a false presupposition: both sides falsely presuppose that there is no distinction between noumena and phenomena, hence both sides falsely presuppose that they must apply their principles to the *same* domain of substances or properties. But at least in principle they could still each consistently apply to *different* domains of substances or properties: that is, one side could apply *their* principles to noumena, and the other side could apply *their* principles to phenomena. Hence the Antinomy is not a genuine or insoluble paradox after all.

Kant's critical and diagnostic interest in the Antinomy is only methodologically skeptical: more profoundly, he wants to disclose, by a negative route, some a priori truths about the world or nature and about human reason. This is what he calls the "solution" to the Antinomy. The key to the solution of the Antinomy, not too surprisingly, is transcendental idealism, which, leaving aside for a moment the difference between its strong and weak versions, can be captured in this context by the sharp epistemic and metaphysical distinction between noumena and phenomena. In particular, the application of transcendental idealism to the third Antinomy, the antinomy of causation, yields a way in which both the thesis and antithesis can be *reinterpreted* as to come out jointly *true*, and thus mutually consistent or compatible (*CPR* A444–51/B472–9, A532–58/B560–86).

The third Antinomy is crucially constrained by two factors. First, whatever Kant has to say about solving this version of the Antinomy, the three Analogies of Experience ("in all change of appearances substance persists, and its quantum is neither increased nor diminished in nature," "all alterations occur in accordance with the law of the connection of cause and effect" and "all substances, insofar as they can be perceived in space as simultaneous, are in thoroughgoing interaction"), which tell us about the nature of causation in the natural empirical world of possible experience, must all come out true. Second, whatever Kant has to say about different types of causation, there must be a level of generality at which the concept of causation is univocal. In this connection we will remember that the schematized pure concept of causation for Kant is that something X (the cause) necessitates something else Y (its effect) in time according to a necessary rule or law. Or equivalently, to say that X causes its effect Y is to say that X is nomologically sufficient for Y in time.

But this schematized pure concept of causation allows for at least two distinct sub-concepts of causation. On the one hand, there is the concept of an *absolutely spontaneous* cause, and on the other hand there is the concept of a *naturally deterministic* cause. Strictly speaking, one could also postulate the notion of an *naturally indeterministic* cause, whose effects are brought about as the mathematical output of aggregated natural facts by means of probabilistic or statistical laws. Since the very idea of a systematic or nomological mathematical science of probability is a twentieth century invention,[15] Kant would simply have assumed, I think, that the very notion of chance, as non-nomological, logically excludes the notion of a cause.

In any case, the concept of an absolutely spontaneous cause depends on Kant's general notion of the spontaneity of a mental act or operation. As I noted in the Introduction (section 0.2), X is spontaneous if and only if X is a mental event that expresses some acts or operations of a creature, and X is:

(i) causally and temporally *unprecedented*, in that (ia) those specific sorts of act or operation have never actually happened before, and (ib) antecedent events do not provide fully sufficient conditions for the existence or effects of those acts or operations;

(ii) *underdetermined* by external sensory informational inputs, and also by prior desires, even though it may have been triggered by those very inputs or motivated by those very desires;

(iii) *creative* in the sense of being recursively constructive, or able to generate infinitely complex outputs from finite resources; and also

---

[15] See Hacking, *The Taming of Chance*.

(iv) *self-guiding*. (*CPR* A51/B75, B130, B132, B152, A445–7/B473–5)

Combining this with Kant's schematized pure concept of a cause as a nomologically sufficient condition for its effect in time, it follows that, according to him, X is an absolutely spontaneous cause of its effect Y if and only if:

(1) X is nomologically sufficient for Y in time.

And:

(2) X is a mental act or operation that is absolutely unprecedented, underdetermined by external sensory inputs and desires, creative, and self-guiding.

In turn, absolutely spontaneous causation is the same as *transcendental freedom*:

> By freedom in the cosmological sense... I understand the faculty of beginning a state **from itself** (*von selbst*), the causality of which does not in turn stand under another cause determining it in time in accordance with the law of nature. Freedom in this signification is a pure transcendental idea. (*CPR* A533/B561)

In other words, if I am transcendentally free, then I am the casually sufficient ground, source, or origin of what I choose or do. What I choose or do is ultimately *up to me*. It should thereby be noted here that the very idea of an absolutely spontaneous mental act or operation, and thus transcendental freedom, smoothly implies the existence of *a causally empowered substance* or *causally efficacious agent* which or who chooses or acts freely. Transcendental freedom for Kant, as applied to the human will, entails causally efficacious rational intentional agency or *personhood*. But as I will argue in the next two sections, it is possible to develop a metaphysically robust Kantian theory of causally efficacious rational intentional agency or personhood without thereby having to appeal to the substance-dualist "timeless agency" metaphysics of *agent-causation*.[16]

On the other hand, the concept of naturally deterministic causation is sharply distinct from the concept of transcendental freedom. According to Kant, X is a naturally deterministic cause of its effect Y if and only if:

(1) X is nomologically sufficient for Y in time.

---

[16] See: Chisholm, "Human Freedom and the Self"; and Watkins, *Kant and the Metaphysics of Causality*, ch. 5. Rational agents are substances. And the defenders of agent-causation adopt a noumenal substance-ontology. But this is not a problem for my Kantian analysis of causation, since as I have argued, Kant's non-reductive event-ontology is perfectly consistent with a phenomenal substance-ontology.

(2) The law under which X and Y both fall is a causal-dynamic natural law.
(3) X and Y are either simple events or complex events in asymmetric time.
(4) Y cannot precede X in time (hence either Y follows X in time or Y is simultaneous with X).
(5) X and Y are material substances or parts of material substances.
(6) X is itself the effect of an earlier cause $Z1$, which in turn is the effect of an earlier cause $Z2$, and so on indefinitely backwards in time.

And:

(7) From the existence of the causal-dynamic natural laws, together with the actual existence of all the simple or complex events prior to X, not only Y but also every other future simple or complex event in nature follows with metaphysical necessity from X.

In short, the causal metaphysical framework described by the three Analogies of Experience is precisely that of naturally deterministic causation.

So much for the conceptual stage-setting. We are now in a position to reconstruct the third Antinomy.

Thesis: "Causality in accordance with laws of nature is not the only one from which all the appearances of the world can be derived. It is also necessary to assume another causality through freedom in order to explain them" (*CPR* A444/B472).

In other words, naturally deterministic causation is not the only kind of causation and transcendental freedom therefore exists.

(1) Suppose that there is only naturally deterministic causation.
(2) If (1) is true, then every simple or complex event is necessitated according to a natural law by some earlier simple or complex event, and that earlier simple or complex event is in turn nomologically necessitated by an earlier one, and so on ad infinitum.
(3) But if (2) is true, then there is never a first beginning to the series of causes of a given simple or complex event, hence never a complete nomologically sufficient condition for that event. But that is absurd, since the very idea of a naturally deterministic cause is that it is the nomologically *sufficient* condition of the simple or complex event which is its effect.

(4) Therefore, by *reductio*, (1) is false, and we must assume the existence of an absolutely spontaneous cause, transcendental freedom, as the nomologically sufficient condition of every naturally deterministic causal series. **QED**.

Antithesis: "There is no freedom, but everything in the world happens solely in accordance with the laws of nature" (*CPR* A445/B473).

In other words, there is only naturally deterministic causation and transcendental freedom does not exist.

(1) Suppose that transcendental freedom exists.
(2) If (1) is true, then the nomologically sufficient condition of every naturally deterministic causal series itself has no cause.
(3) But if (2) is true, then transcendental freedom does not itself fall under any laws of nature and is a law unto itself (that is, a miracle) operating by totally uncognizable (that is, occult) means. But that is absurd, since the very idea of a cause is that it is the *nomologically* sufficient condition of the simple or complex event which is its effect.
(4) Therefore, by *reductio*, (1) is false: there is only naturally deterministic causation and transcendental freedom does not exist. **QED**.

What is Kant's solution for the third Antinomy? We will remember that according to Kant every Antinomy is diagnosed and (dis)solved by distinguishing sharply between noumena and phenomena. In this light, as I mentioned above, the shared error of Thesis and Antithesis in each case is that both fail to distinguish between noumena and phenomena and falsely assume that their principles apply to a single undifferentiated domain of substances or properties.

As I also mentioned above, the third Antinomy has a constructive reconciliation phase in which the recognition of the distinction between noumena and phenomena allows for a reinterpretation according to which the Thesis and the Antithesis both come out true. Why does Kant undertake this constructive reconciliation? One important reason is that Kant is assuming for the purposes of the third Antinomy that the three Analogies *are* true. And as I noted earlier, the concept of causation contained in the Analogies is equivalent to the concept of causation contained in the Antithesis, the concept of naturally deterministic causation. The only salient difference between the two presentations of that concept is that, in the Analogies, it is specifically restricted to the domain of phenomena or appearances, whereas in the Antithesis of the third Antinomy, at least initially, it is allowed to range ambiguously over the domains of phenomena and noumena alike.

But another even more important reason for the constructive reconciliation phase is that Kant thinks that morality is impossible without the concept of practical freedom, which is negatively defined as the ability to choose independently of all sensory impulses or empirical desires, and positively defined as autonomy or self-legislation according to the moral law or Categorical Imperative.[17] Now, as Kant argues in the third section of the *Grounding for the Metaphysics of Morals*, the concept of the moral law or Categorical Imperative reciprocally entails the concept of practical freedom or autonomy, and practical freedom presupposes transcendental freedom.[18] So, without a constructive reconciliation of the Thesis and Antithesis of the third Antinomy, morality itself would be undermined.

As I indicated earlier in this section, Kant's strategy for solving the third Antinomy is via his transcendental idealism, and in particular via the sharp distinction between noumena and phenomena, together with the idea of restricting the scopes of the Thesis and Antithesis alike. So he restricts the scope of the Thesis to the domain of noumena, and correspondingly he restricts the scope of the Antithesis to the domain of phenomena or appearances. Then it is formally consistent to claim that in the noumenal domain transcendental freedom exists and natural determinism does not exist, while in the phenomenal domain natural determinism exists and transcendental freedom does not exist. So both the Thesis and Antithesis can come out true, and thus transcendental freedom and natural determinism are both logically possible.

Are we there yet? Sadly, no. For this is precisely where the philosophical troubles start all over again, as we shall see in the next section.

8.2. The Problem of Free Will and Kant's Embodied Agency Theory

Determinism is the doctrine that the complete series of settled past events, together with the laws of nature (and perhaps also together with the

---
[17] See Allison, *Kant's Theory of Freedom*, ch. 3.
[18] Ibid. chs. 11–12.

irresistible causal and creative powers of an all-knowing deity[19]), logically or metaphysically necessitate all future events, including all the choosings and doings of intentional agents. Or in other words, determinism says that necessarily if *the same past* then *the same future*. By sharp contrast, *indeterminism* is the doctrine that determinism is false and that all connections between events, including the choosings and doings of intentional agents, are either strictly governed by mathematical laws of probability or else merely random. Or in other words, indeterminism says that necessarily if *the same past* then *possibly a different future*.

What is free will? Free will, at the very least, is an intentional agent's choosing or doing things without preventative constraints and without inner or outer compulsion (*negative freedom*), together with the ability to choose or do what she wants (*positive freedom*). Moreover, it also seems to be undeniably true that necessarily an intentional agent A can freely choose or do something X if and only if A is casually or morally responsible for X (*responsibility*). So a minimal definition of free will says that it is an intentional agent's choosing or doing things with negative freedom, positive freedom, and responsibility. Then the problem of free will is this: How can intentional agents choose or do things with negative freedom, positive freedom, and responsibility in a deterministic or indeterministic world? Or more starkly framed, the problem of free will is *how do I know that I am not just a deterministic or indeterministic puppet epiphenomenally dreaming that I am a real person?*

In turn, the standard positions on the metaphysics of freedom are these:

(1) *Hard Determinism*: Free will and determinism are metaphysically inconsistent, free will does not exist, and determinism is true.
(2) *Soft Determinism*: Free will and determinism are metaphysically consistent, free will exists, and determinism is true.
(3) *Libertarianism*: Free will and determinism are metaphysically inconsistent, free will exists, and determinism is false.[20]

---

[19] In what follows I will leave aside the special problems of divine determinism and focus exclusively on natural determinism.

[20] See Van Inwagen, *An Essay on Free Will*; and Van Inwagen, "Free Will Remains a Mystery." It is possible to deny that libertarianism entails violations of laws of nature. See, e.g., Lewis, "Are We Free to Break the Laws?" As we have seen in section 8.1, Kant holds that naturally determined causation is a relation of nomological sufficiency between diachronic or synchronic singular events or Events: so for Kant, as for Van Inwagen, agent-causal libertarianism entails violations of the causal laws.

Now, hard determinism and libertarianism are both *incompatibilist*: the incompatibilist says that freedom and determinism are *metaphysically inconsistent*. By contrast, soft determinism is *compatibilist*: the compatibilist says that freedom and determinism are *metaphysically consistent*. Then we can doubly reformulate the problem of free will as, on the one hand, the problem of whether hard determinism, soft determinism, or libertarianism is true, and, on the other hand, the problem of whether incompatibilism or compatibilism is true.

Even allowing for this double reformulation however, the reformulation is in at least two respects a *superficial* version of the free will problem.

First, there there are at least two other positions not included in the above schema: (4) *causal indeterminism*, and (5) *hard indeterminism*. As Peter van Inwagen has pointed out, indeterminism is as apt to undermine free will as determinism is.[21] Causal indeterminism, however, says that all relations between simple or complex events in nature are merely stochastic—that is, governed exclusively by probabilistic or statistical laws—and that freedom consists in an agent's causing something *by means of* some of these stochastic relations.[22] Causal indeterminism is a form of libertarianism which says that free will and indeterminism are metaphysically consistent. Hard indeterminism, by contrast, while it also says that all relations between singular events or complex events in nature are either stochastic or merely random, nevertheless denies that persons are capable of affecting any of these relations. Hard indeterminism, like hard determinism, says that free will and determinism are metaphysically inconsistent. But hand indeterminism *also* says that free will and indeterminism are metaphysically inconsistent. Hard determinism and hard indeterminism thus share a rejection of the possibility of free will and an affirmation of nature's thoroughly nomological, mathematical, physical, and scientifically comprehensible character. That is precisely what makes them both "hard." The disjunction consisting of *either* hard determinism *or* hard indeterminism is what I will call *natural mechanism*,[23] and it is in direct opposition to both compatibilism and libertarianism. Otherwise put, the thesis of natural mechanism says that the entire world—including living organisms, animals, human beings, and persons—operates according to non-teleological, mathematico-physical principles alone. This in turn

---

[21] van Inwagen, *An Essay on Free will*; and van Inwagen, "Free Will Remains a Mystery."
[22] See Kane, *A Contemporary Introduction to Free Will*, pp. 64–5.
[23] See also Bok, "Freedom and Practical Reason," pp. 130–1.

gives rise to a deeper version of the free-will problem: how can a person choose or act with negative freedom, positive freedom, and responsibility in a naturally mechanized world?

Second, and perhaps more surprisingly, there is in fact a third position in logical space between compatibilism and incompatibilism: I will dub this *post-compatibilism*. Post-compatibilism says that freedom and determinism are *metaphysically consistent* **and also** *metaphysically inconsistent*. This is not a paradox. Post-compatibilism is possible if there is more than one type of freedom, for in that case determinism can be metaphysically consistent with *one type* of freedom but also metaphysically inconsistent with *another type* of freedom.

In my opinion Kant's libertarian theory of freedom of the will is philosophically significant precisely because it is *neither* hard determinist, *nor* soft determinist, *nor* causal indeterminist, *nor* hard indeterminist, *nor* compatibilist, *nor* incompatibilist. To be sure, as I mentioned in passing above, because the Critical Philosophy pre-dates the invention of the science of logical and mathematical probability by at least a century, Kant would not have been able to recognize the possibility of purely stochastic determination relations between simple or complex events, and so would not have been able explicitly to consider the possibility of either causal indeterminism or hard indeterminism. Even so, various theses that Kant *does* hold do entail the denial of both of these forms of indeterminism. If we then add the claim that Kant's libertarian theory of free will, as distinct from all the standard alternative positions, is both metaphysically robust and arguably true, then his theory is not only philosophically significant, but also, well, *rules*.

How can Kant avoid all the standard positions on the metaphysics of free will, and still defend a metaphysically robust version of libertarianism? Here is how I think he does it. Because Kant asserts that *all* of the intentional acts of human persons are transcendentally free, and also that *some* of the transcendentally free intentional acts of human persons are practically free or autonomous, his view is not hard determinist. Because he asserts that all of the transcendentally and practically free intentional acts of human persons are (negatively) noumenal, and also that all of their empirical phenomenal *living* psychological and biological processes are non-mechanical, while at the same time asserting that all and only *inert* empirical phenomenal material beings are naturally mechanized, he is neither a soft determinist nor a compatibilist: although some beings in the natural world are transcendentally free and some beings in the natural world are naturally mechanized, there is no material substance X or person X such that X is both transcendentally free and naturally mechanized. Because he asserts that psychological freedom of the will is metaphysically consistent with natural determinism, he is not an incompatibilist. And, finally, because he asserts that psychologically free will, transcendentally free will, practically free will, and naturally deterministic causation all exist, and also that both transcendentally free and naturally deterministic causation involve actual realizations of causal powers—"the causality of the cause" (*CPR* A203/B248)—that constitute the diachronic or synchronic nomological sufficiency relations between simple or complex events, he is neither a causal indeterminist nor a hard indeterminist.

In short, I think that Kant's libertarian theory of freedom of the will is thoroughly *post-compatibilist*, for the following five basic reasons:

(1) Psychological freedom exists and is a necessary but not sufficient condition of transcendental freedom.
(2) Transcendental freedom exists and is a necessary but not sufficient condition of practical freedom.
(3) Practical freedom exists.
(4) Psychological freedom is metaphysically consistent with natural determinism.
(5) Transcendental freedom is metaphysically inconsistent with natural determinism.

One *other* reason for Kant's being neither a hard determinist, nor a soft determinist, nor a classical libertarian, nor a causal indeterminist, nor a hard indeterminist, nor an incompatibilist, nor a compatibilist, is the third Antinomy and its solution. As we have seen in section 8.1, Kant asserts there that classical metaphysical reasoning about free will and natural determinism leads to logical paradox. Now the Thesis of the third Antinomy expresses an *agent-causation* version of libertarianism, and the Antithesis expresses a version of hard determinism. Agent-causal libertarianism says that intentional actions are caused by positively noumenal substances called *agents*, and not by natural events. Agent-causes therefore either *violate natural laws* or else *causally overdetermine* the intentional acts they cause. (I will spell out the notion of causal overdetermination shortly.) Kant also of course asserts the truth of transcendental idealism and the sharp distinction between noumena and phenomena. Then, according to Kant's solution to the third Antinomy:

(1) When interpreted independently of transcendental idealism, both the Thesis and the Antithesis would be true if and only if false by a mutual *reductio* strategy, and so not only logically contradictory but also paradoxical. Hence:
(2) In order to avoid paradox, both the Thesis and Antithesis must be simply false and thereby logical contraries, not contradictories. But:
(3) When reinterpreted in terms of transcendental idealism, both Thesis and Antithesis come out harmlessly true, and logically consistent with each other.

So Kant can reject both agent-causal libertarianism and hard determinism alike without rational incoherence.

But given this solution to the third Antinomy, then the following question immediately arises: Is Kant in fact a soft determinist and therefore a compatibilist?[24] Answering that question decisively proves to be fairly difficult, however, because Kant's own positive theory of freedom, like his theory of noumena, contains a seemingly unresolvable dichotomy between two mutually exclusive and apparently globally exhaustive versions of the theory.[25] The purpose of the rest of this section is to sketch the outlines of a third (and, I will argue, both textually and independently defensible) Kantian theory of freedom of the will, *the Embodied Agency Theory*, which also turns out to be post-compatibilist.

In order to get a proper grip on Kant's post-compatibilist solution to the problem of free will however, we are going to need a metaphysical run-up.

---

[24] See, e.g., Hudson, *Kant's Compatibilism*; and Wood, "Kant's Compatibilism." Hudson construes Kant's theory of freedom as a version of soft determinism. Wood however argues for the deeper point that Kant is both a compatibilist *and* an incompatibilist, hence an "incompatibilistic compatibilism." This, I think, is fairly close to the truth. But it also assumes the Two Object or Two World Theory of the noumena vs. phenomena distinction; hence it disguises the possibility of what I call Kant's *post-compatibilism*. It should also be noted here that in more recent writings Wood has adapted a soft determinist, Two Standpoint reading of Kant's theory of freedom. See, e.g., Wood, *Kant's Ethical Thought*, pp. 180–2.

[25] One way out of the dichotomy would be to argue that Kant is in fact an agent-causal libertarian. This is the tack taken by Pereboom in "Kant on Transcendental Freedom." However, that's a hollow interpretive victory, since in effect it assimilates Kant's view to the un-reinterpreted Thesis of the third Antinomy.

So let us start with the noumena vs. phenomena distinction. Here the stubborn conjunctive fact is that some of Kant's texts clearly indicate the Two World or Two Object Theory, while other texts just as clearly indicate the Two Aspect or Two Standpoint Theory. Appearances or phenomena are thinkable but also observable, mind-dependent, spatiotemporal objects possessing both extrinsic and intrinsic-structural macrophysical properties but lacking "real essences," or unperceivable and unobservable mind-independent constitutive intrinsic non-relational properties. Noumenal objects are non-sensible objects. Some noumenal objects are, if they exist, causally efficacious with respect to our cognitive faculties, precisely because they would be identical with the real essences of causally efficacious material objects. These noumenal objects are *things-in-themselves* or positive noumena. Then the Two World or Two Object Theory says that that phenomena and things-in-themselves are two mutually exclusive classes of things. By contrast, the Two Aspect or Two Standpoint Theory says that that there is one and only one class of otherwise unspecified things, or perhaps one and only one class exclusively made up of phenomenal things, each of which is taken or believed by us to be phenomenal under one aspect or standpoint and also taken or believed by us to be noumenal under another aspect or standpoint.[26] It is crucial to note that because the Two Aspect or Two Standpoint Theory is all about how we take or have beliefs about things, and not a theory about the natures of those things, it is essentially an *epistemological* and not a metaphysical theory of the noumena vs. phenomena distinction.

The big problem with the Two World or Two Object Theory is the nature of the causal interaction between things-in-themselves, our cognitive faculties, and phenomenal objects. If things-in-themselves cause our objective perceptions of phenomena, then the empirically real causal relation between phenomenal material objects and our cognitive faculty is *causally overdetermined*. The thesis of causal overdetermination says that (i) there can be two ontologically distinct sufficient causes of the same event, and (ii) that there can be two complete and independent causal explanations of the same event. But, as Jaegwon Kim has compellingly argued, it seems entirely reasonable to hold that if there already exists a sufficient material

---

[26] See: Allison, *Kant's Transcendental Idealism*, p. 8; Allison, "Transcendental Idealism: The 'Two Aspect' View"; Paton, *Kant's Metaphysic of Experience*), vol. i., p. 61; and Prauss, *Kant und das Problem der Dinge an Sich*.

cause of some event, and if correspondingly a complete and independent material causal explanation of that same event also exists, then this cause and this causal explanation together necessarily *exclude* there being any other distinct cause or distinct causal explanation of the same event.[27] So causal overdetermination, although logically and metaphysically possible, is rationally unacceptable.

Correspondingly, the big problem with the Two Aspect or Two Standpoint Theory can be presented in the form of a dilemma. If on the one hand the Theory *were* interpreted metaphysically, then it would be obviously incoherent because it entails the existence of a class of otherwise unspecified objects, or perhaps made up exclusively of phenomenal objects, each of which instantiates two contradictory sets of intrinsic properties: phenomenal properties and positive noumenal properties. But then if on the other hand it is interpreted—as explicitly intended by its proponents—essentially epistemologically, in order to avoid the obvious contradiction between disjoint classes of intrinsic properties, then it simply does not do the philosophical work required of the noumena vs. phenomena distinction. It tells us only that there is one and only one class of otherwise unspecified objects, or perhaps of exclusively phenomenal objects, each of which is *taken or believed by us* to be things-in-themselves and also *taken or believed by us* to be phenomenal; but it neither explains why we perversely persist in ascribing contradictory intrinsic properties to the same objects, nor does it justify our beliefs in the objective correctness of those ascriptions.

Analogously and correspondingly, in the case of freedom of the will, the stubborn conjunctive fact also is that some texts clearly indicate the Timeless Agency Theory,[28] while other texts just as clearly indicate the Regulative Idea Theory.[29]

As we have seen, transcendental freedom is absolutely spontaneous causal efficacy or nomological sufficiency in successive or simultaneous time. If we ascribe transcendental freedom specifically to the will of a person, then it is the ability of a person to choose and act in an absolutely spontaneous way independently of all "alien causes," that is, independently of all pathological inner and unowned outer sources of nomologically sufficient compulsion

---

[27] See Kim, "Mechanism, Purpose, and Explanatory Exclusion."
[28] See Allison, *Kant's Theory of Freedom*, pp. 47–53; and Wood, "Kant's Compatibilism."
[29] See Allison, *Kant's Theory of Freedom*, ch. 13; and Nelkin, "Two Standpoints and the Belief in Freedom."

(*GMM* 4: 446). In other words, if a person is transcendentally free, then her choosings and doings are *up to her*. She herself is the nomologically sufficient ground, origin, or source of her choices and acts. Practical freedom presupposes but also exceeds transcendental freedom, in that practical freedom is the spontaneous causal efficacy of the will independently of all alien causes and also *all sensible impulses* (empirical desires):

> It is this **transcendental** idea of freedom on which the practical concept of freedom is grounded.... **Freedom in the practical sense** is the independence of the power of choice (*Willkür*) from **necessitation** by impulses of sensibility. For a power of choice is **sensible** insofar as it is pathologically affected (through moving-causes of sensibility); it is called an animal power of choice (*arbitrium brutum*) if it can be **pathologically necessitated**. The human power of choice is indeed an *arbitrium sensitivum*, yet not *brutum*, but *liberum*, because sensibility does not render its action necessary, but in the human being there is a faculty of determining oneself from oneself, independently of necessitation by sensible impulses. (*CPR* A534/B562)

This is merely a negative characterization of practical freedom, however. As positively characterized, it also involves the capacity for *self-legislation* in conformity with the moral law or Categorical Imperative. Or in other words, positively characterized practical freedom is the same as *autonomy* (*GMM* 4: 440–1, 446–63).

The Timeless Agency Theory adopts the Two World or Two Object Theory of the noumena vs. phenomena distinction and asserts that a noumenal subject is autonomous in that it has absolutely spontaneous causal efficacy or nomological sufficiency of the self-legislating positively noumenal will apart from all alien causes and all sensible impulses, in conformity with the Categorical Imperative, by causing *from outside of time and space* phenomenal human behavioral movements (in outer sense) and psychological processes (in inner sense) that are themselves independently necessarily causally determined by natural laws plus antecedent simple events and complex events. By contrast, the Regulative Idea Theory adopts the Two Aspect or Two Standpoint Theory of the noumena vs. phenomena distinction and says that we are required by our practical reason to take or believe ourselves to be acting morally only under the rational idea of own practical freedom or autonomy.

One big problem with the Timeless Agency Theory is that if all phenomenal events are independently necessarily determined by natural laws together with antecedent events, then the noumenal causality of the will implies a rationally unacceptable causal overdetermination of phenomenal human behavioral movements in outer sense and psychological processes in inner sense. Another big problem is that timeless agency does not place moral responsibility where we would pre-theoretically want it, namely in the empirical acts we know ourselves to perform.

Correspondingly, the big problem with the Regulative Idea Theory is that even if it is true, it simply does not do the philosophical work required of the noumenal causation vs. phenomenal causation distinction, because it does not entail the *actual or real existence* of freedom of the will but rather entails only at best our *belief* in its actual or real existence, which is not only ontologically deflationary but also, arguably, does not even rationally justify that belief.

Now let us briefly step back again in order to push forward presently. One obvious way around the seemingly unresolvable dichotomy between the two versions of Kant's theory of the noumena vs. phenomena distinction would be to find, with appropriate textual grounding, a third Kantian theory of the distinction which avoids the problems of the Two World or Two Object Theory and the Two Aspect or Two Standpoint Theory alike. Here is one such proposal, briefly described earlier in section 4.0, which I call *the Two Concept*[30] *or Two Property Theory*:

(1) Although we have thinkable concepts of noumenal things-in-themselves, and therefore such entities are logically possible, nevertheless because we do not have *objectively valid* concepts of such entities, it follows that we can neither empirically meaningfully assert nor empirically meaningfully deny their existence, hence we must remain *consistently agnostic* about them, and *methodologically eliminate* them for the purposes of objectively valid metaphysics and epistemology.

(2) We do have objectively valid concepts of phenomenal things, hence only macrophysical empirical phenomenal things can be objectively validly asserted to really exist and be causally efficacious with respect to our cognitive faculties.

(3) Corresponding to our distinct noumenal concepts and phenomenal concepts are ontologically distinct noumenal properties and

---

[30] See Hanna, *Kant and the Foundations of Analytic Philosophy*, pp. 110–13.

phenomenal properties. Since things-in-themselves are methodologically eliminable, then although positive noumenal properties are in principle instantiable, we do not know whether they are uninstantiated or instantiated. But both negative noumenal properties and phenomenal properties are not only instantiable but also actually instantiated, and directly known or felt by us to be instantiated.

To summarize, then, the Two Concept or Two Property Theory of the noumena vs. phenomena distinction that I am proposing is *ontologically monistic* about phenomena (which is to say that we can empirically meaningfully assert the existence of one and only class of entities, namely empirically real phenomenal entities, whether empirical objects or empirical subjects), *consistently agnostic and methodologically eliminativist* about things-in-themselves (which is to say that things-in-themselves are thinkable but uncognizable and unknowable, and therefore otiose for objectively valid metaphysics and epistemology), *conceptually dualistic* (which is to say that there are distinct and mutually irreducible noumenal and phenomenal concepts), *property dualistic* (which is to say that there are distinct and mutually irreducible noumenal and phenomenal properties), and finally also *metaphysically economical* (which is to say that only negatively noumenal properties and phenomenal properties are actually instantiated, in both cases by empirically real phenomenal entities).

I do not deny that the Two Concept or Two Property Theory is controversial! Not only does it deviate from the familiar Two World Theory vs. Two Standpoint Theory framework of contemporary Kant scholarship, but also there are some Kantian texts that will not support it without some squinting and tweaking. If pressed, I suppose I *could* squint and tweak them. In point of fact however, given the charitable interpretive strategy I have explicitly adopted—see section 0.2—this is not necessary. My twofold claim is only (a) that the Two Concept or Two Property Theory accurately captures a doctrine that is found explicitly in some, even if not all, of Kant's texts, and (b) that the Two Concept or Two Property Theory has at least three extremely important theoretical virtues.

What are those virtues? First, it avoids the problems of the Two World or Two Object Theory because it avoids both interactionist substance dualism and the rationally unacceptable doctrine of causal overdetermination.

Second, it also avoids the problems of the Two Aspect or Two Standpoint Theory because, even though it shares with that theory a commitment to the existence of only one world, it also avoids the Two Aspect Theory's dilemma of interpreting that commitment *either* as a metaphysical commitment to an ontology-bloating "neutral," non-phenomenal, non-noumenal kind of object *or* as an epistemological commitment to pointlessly ascribing contradictory sets of intrinsic properties to the same objects. The last and perhaps most important theoretical virtue is that the Two Concept or Two Property Theory underwrites a *third* Kantian theory of free will. So the leading claim of this section, now fully explicitly formulated, is that a third Kantian theory of freedom can be found which presupposes the Two Concept or Two Property Theory, which avoids the problems of the Timeless Agency Theory and Regulative Idea Theory alike, and which also appears to be independently defensible. As I have mentioned, I call this third theory *Kant's Embodied Agency Theory*.

Kant's Embodied Agency Theory can be encapsulated in the following seven theses:

(1) Rational human animals are autonomous in that these animals have absolutely spontaneous causal efficacy or nomological sufficiency of the *negatively but not positively noumenal self-legislating will* in conformity with the Categorical Imperative.

(2) The negatively noumenal self-legislating will is nothing but *the empirical phenomenal will of a rational human animal insofar as it also has some non-sensible properties which are also absolutely spontaneously causally efficacious or nomologically sufficient* (for example, the property of trying to do the right thing even when it is counter-prudential, or against the animal's best interests or strongest desire, to do the right thing).

(3) This absolute spontaneous causal efficacy or nomological sufficiency of the negatively noumenal self-legislating will occurs *inside of* space and time, with respect to empirical phenomenal behavioral movements and psychological processes of the rational human animal's own living body that are *not naturally mechanistically caused but instead naturally practically or intentionally caused*, because they have been nomologically sufficiently uniquely causally determined by means of the rational human animal's lower-level volitional capacity for *animal choice*.

(4) Animal choice guarantees *animal freedom* in accordance with what I will call *the Principle of Pathologically-Conditioned Alternative Possibilities*, which says that that *if the animal had spontaneously desired otherwise, then it would have willed or done otherwise*:

> a faculty of choice...is merely **animal** (*arbitrium brutum*) which cannot be determined other than through sensible impulses, i.e., **pathologically** (*CPR* A802/B830)

(5) Although satisfying the Principle of Pathologically-Conditioned Alternative Possibilities guarantees animal freedom, and although animal choice is such that if animal freedom is lacking then a rational human animal cannot be morally responsible for his or her actions,[31] nevertheless animal freedom constitutes only a necessary condition of both transcendental freedom and also practical freedom or autonomy alike, and not a sufficient condition of either of them,[32] since on its own animal freedom is only *psychological freedom*, that is, the non-absolute or relative spontaneity of consciousness and desire presupposed by the Second and Third Analogies of Experience (see section 8.1):

> If these determining representations [i.e., instincts or motives] themselves have the ground of their existence in time and, more particularly, in the antecedent state and these again in a preceding state, and so on ... ; and if they are without exception internal; and if they do not have mechanical causality but a psychological causality through representations instead of through bodily movements: they are nonetheless determining grounds of the causality of a being insofar as his existence is determinable in time ....[33] Thus these conceptions do indeed imply psychological freedom (if one wishes to use this

---

[31] According to the Embodied Agency Theory, a necessary and sufficient condition of moral responsibility for an intentional act is psychological freedom, plus transcendental freedom, plus the *capacity* for autonomy. Hence the rational human animal can at once be psychologically and transcendentally free, not actually acting autonomously, and still be morally responsible for doing the wrong thing.

[32] This means that Kant would reject Frankfurt's influential intuition that moral responsibility is *generally* consistent with determinism (i.e., soft determinism). See Frankfurt, "Alternate Possibilities and Moral Responsibility," pp. 1–10. But there is nevertheless a crucial affinity between Kant's theory of psychologically and transcendentally free animal choice in rational humans and Frankfurt's notion of freedom of the will as a rational animal's ability to determine its effective first-order desires by means of what he calls "second-order volitions." See n. 47.

[33] I have elided the sentence "As such, this being is under necessitating conditions of past time which are no longer in his power when he acts," for it misleadingly suggests that animal choice is unfree. It's true that the animal could do otherwise only if it antecedently *desired* to do otherwise. But animal

word for a merely internal concatenation of representations in the mind), but nonetheless they also imply natural necessity,³⁴ leaving no room for transcendental freedom³⁵ which must be thought of as independence from everything empirical³⁶ and hence from nature generally, whether regarded as an object of inner sense merely in time or also as an object of outer sense in both space and time. (*CPrR* 5: 97, underlining added)

(6) The absolute spontaneous causal efficacy or nomological sufficiency of the negatively noumenal transcendentally free will is a psychologically free animal choice which *also* brings about a *natural causal singularity*, and thereby implies the existence of a *non-mechanistic and new or "one-off" causal-dynamic law of nature*.

(7) The absolute spontaneous causal efficacy or nomological sufficiency of the negatively noumenal autonomous will is nothing other than a transcendentally free animal choice *which also satisfies the Categorical Imperative*.

One sentence below the text cited just above, Kant also famously writes:

[A]ll necessity of events in time according to natural law can be called the "mechanism of nature," even though it is not to be supposed that things which are subject to it must really be material machines. Here reference is made only to the necessity of the connection of events in a temporal series as they develop according to natural law, whether the subject in which this development occurs be called *automaton materiale* when the machinery is impelled by matter, or, with Leibniz. *automaton spirituale* when it is impelled by representations. And if the freedom of our will were nothing else than the latter, i.e., psychological and comparative and not at the same time transcendental or absolute, it would in essence be no

---

choice fully exemplifies the *control* of an animal over its own bodily movements and also the animal's *guidance* of its own bodily movements, via its own desires. See Frankfurt, "The Problem of Action."

[34] I'm interpreting "they also imply natural necessity" to mean the same as "they also entail the existence of natural necessity in the causal-dynamic relations between the parts of inert mechanical matter in the external world", in precisely the same way in which, in the second Analogy, the subjective and arbitrary ordering of perceptions in inner sense entails the existence of an objective and necessitated ordering of perceptions in outer sense.

[35] I'm interpreting the phrase "leaving no room for transcendental freedom" to mean the same as "is not in and of itself sufficient for transcendental freedom" and thereby leaving open the possibility that the psychological freedom of free animal choice is a necessary condition of transcendental and practical freedom alike.

[36] I'm interpreting the phrase "independence from everything empirical" to mean the same as "underdetermination by everything empirical" or "non-supervenience on everything empirical" but *not* to mean the same as "the necessary exclusion of everything empirical".

430 THE PRACTICAL FOUNDATIONS OF THE EXACT SCIENCES

better than the freedom of a turnspit, which <u>when once wound up also carries its motions from itself</u>. (*CPrR* 5: 97, translation slightly modified, underlining added)

There are four important things to notice about these two fascinating texts from the second *Critique*.

First, according to Kant's notion of "mechanical causality," every simple or complex event in the world of inert material substances is the product of wholly impersonal, nomological, non-teleological, mathematically-describable, physical causal forces. This is the same as the world of "objects of experience" governed by the three Analogies of Experience: "experience reveals only the law of appearances and consequently <u>the mechanism of nature</u>, the direct opposite of freedom" (*CPrR* 5: 29, underlining added). As I have mentioned twice already, Kant's Critical Philosophy pre-dates the invention of the science of probability, so he could not explicitly allow for the possibility of statistical natural laws and stochastic determinative relations between simple or complex events. But supposing counterfactually for a moment that he *had* allowed for stochastic determination between simple or complex events, then all material objects of experience whose operations fall exclusively under probabilistic or statistical laws would *also* have counted for him as naturally mechanized.

Second, and correspondingly, according to Kant's conception of natural mechanism, there are three fundamental differences between an *automaton materiale* and an *automaton spirituale*.

(1) Whereas the material automaton is an "objective object" of outer experience, represented by an objectively valid outer "judgment of experience" (*Erfahrungsurteil*) in a naturally mechanistic physics, by contrast the spiritual or psychological automaton is not even a "subjective object" of inner sense, represented by an objectively valid self-reporting judgment of empirical apperception in a soft determinist empirical psychology,[37] but rather only a "subjective subject" of immanently reflexive phenomenal consciousness in inner sense, represented by a merely subjectively valid so-called[38] "judgment

---

[37] See Sturm, "Kant on Empirical Psychology: How Not to Investigate the Human Mind"; and Sturm, "A Matter of Character: Hume and Kant on the Causation and Empirical Explanation of Actions."

[38] "So-called" because it has only *subjective validity* and therefore cannot strictly speaking *be* a judgment by the definition given in the B Deduction, which defines a judgment as an *objectively valid* truth-evaluable representation (*CPR* B141–2). One way out of this puzzle is to identify the subjectively

of perception" (*Wahrnehmungsurteil*) (*P* 4: 298–9). Thus the psychological automaton is essentially *not* governed by the general deterministic laws of natural mechanism.

(2) The spiritual or psychological automaton is driven by conscious desires, whereas the material automaton is not. Hence, the material automaton is essentially a mere machine made of inert matter, whereas the psychological automaton is essentially *not* a mere inert machine, precisely because it is driven internally by conation and intention, and more generally by what the pre-Critical Kant would have called "living forces," in opposition to "dead" or inertial forces. As we saw in the second and third Analogies, moreover, the subjective or arbitrary ordering of perceptions entails the non-absolute or relative spontaneity of consciousness and desire. The psychological automaton is thus a *living and relatively spontaneous teleological natural process*, not a mere inert machine.

(3) The psychological domain for Kant is naturally governed by unique non-deterministic, non-mechanistic or teleological "psycho-psycho" laws of introspective empirical psychology, but *neither* intrinsically structured by the deterministic mechanistic causal laws of inert matter *nor* nomologically strongly supervenient on inert matter, and thus *not* naturally determined by any deterministic psycho-*physical* laws. So the psychological automaton is not really an "automaton" in the strict sense at all, precisely because unlike the material automaton it is neither naturally mechanized in itself nor strictly determined by what is naturally mechanized.

Third, the rational human will is free in the psychological sense and *at the same time* transcendentally free. Its freedom would be the "freedom of a turn-spit" only if it were *merely* psychologically free and not *also* transcendentally free. For, as Hume and Frankfurt have both clearly seen, psychological freedom is metaphysically consistent with natural determinism.

Fourth, my *automaton spirituale*, once it has been wound up by my desires, "carries out its motions from itself (*von selbst*)." Similarly, in transcendental freedom, my will "begin[s] a series of occurrences entirely from itself (*ganz von selbst*)" (*CPR* A534/B562). Transcendental freedom is different

---

valid *Wahrnehmungsurteil* with the immanently reflexive character of inner sense, and then distinguish it sharply from objectively valid judgments of empirical apperception; see section 1.1.

from psychological freedom. But my *own* personal capacity for willing is immediately manifest in *both* cases. In both cases, my *choices* are up to me.

Noting all this, we can then see that the Embodied Agency Theory avoids the problems of the Timeless Agency Theory because the causally efficacious powers of the noumenal will are *inside* phenomenal space and time, and not *outside* it, hence there is no rationally unacceptable causal overdetermination. In order for this to be possible however, it also thereby follows that some causally-dynamically efficacious processes in space and time—namely, the purposive bodily movements and psychological processes of the living organismic bodies of rational human animals—are non-mechanistic and (either relatively or absolutely) spontaneous.

So too, the Embodied Agency Theory avoids the problems of the Regulative Idea Theory because, while the Embodied Agency Theory *also* accepts the thesis that that we are required by our practical reason to take or believe ourselves to be acting morally only under the rational idea of own autonomy, this is explicitly combined with an assertion of the actual existence and phenomenal causal efficacy or nomological sufficiency of rational human freedom, whether as transcendentally free animal choice or as autonomy.

One element of the Embodied Agency Theory that needs to be further elaborated is the negatively but not positively noumenal character of autonomy. As I pointed out in section 4.0, negative noumena for Kant are entities that possess some intrinsic non-sensible properties, whether non-relational or relational, and whether mind-dependent or mind-independent. Positive noumena, by sharp contrast, are entities which, if they actually existed, would be constituted by a set of intrinsic non-relational, non-sensible, mind-independent properties (*CPR* B308–9). Positive noumena are things-in-themselves. All positive noumena are also negative noumena, but not all negative noumena are positive noumena. The crucial difference between negative noumena and positive noumena is that negative noumena can also include *empirical phenomenal spatiotemporal entities insofar as they possess some non-sensible intrinsic structural properties*, whereas positive noumena essentially exclude all sensible and empirical properties, as well as all intrinsic relational properties, and also all mind-dependent properties, and must therefore be *transcendent entities* (*CPR* A296/B352–3). Otherwise put, negative noumena are things that can have some intrinsic properties that are a priori or non-reducible to all other empirical things or facts, but this is perfectly consistent with their also possessing sensible and empirical, relational, weakly transcendentally ideal intrinsic properties and therefore also perfectly consistent with their actual existence in objectively real space and time. Still otherwise put, the acceptance of the existence of negative noumena commits one only to the truth of *property dualism about the difference between the noumenal and the phenomenal*, and correspondingly to *the distinct existences of both noumenal properties and phenomenal properties, even though all substances are still real empirical phenomenal material things*, whereas the acceptance of the existence of positive noumena would also commit one to the truth of *substance dualism about the difference between the noumenal and the phenomenal*, and correspondingly to *the existence of unknowable things-in-themselves*.

Now according to the Embodied Agency Theory, framed against the backdrop of the Two Concept or Two Property Theory of the noumena vs. phenomena distinction, Kantian noumenal subjects or persons are all *negatively* noumenal but *not* positively noumenal, in the sense that Kantian persons are nothing but empirical selves or conscious living human animals who also exemplify some real intrinsic structural non-sensible properties, such as willing the Categorical Imperative. Therefore the basic ontology behind Kant's third theory of freedom of the will is *ontological phenomenal monism together with property dualism*: the same individual living human animal has both an intrinsic structural "empirical character" (the empirical self or person) which is motivated by conscious desires and psychologically free, and also an intrinsic structural "intelligible character" (the negatively noumenal self or person) which is both transcendentally free and also capable of practical freedom or autonomy (*CPR* A538–69/B566–8).

*Are we there yet **now**?* Again sadly, we are not. Even if someone were to accept the Two Concept or Two Property Theory of the noumena vs. phenomena distinction, *and* also were to accept the notion of a negative noumenon as I have spelled it out, it would not of course follow that she accepted the rest of the Embodied Agency Theory. So we still have a few miles to go before we sleep.

8.3. Freedom and Nature

It may seem, on the face of it, that there should be no direct connection between the rational human animal's absolutely spontaneous, autonomous will and its existence in physical nature. But in fact Kant explicitly asserts otherwise:

> Now although there is an incalculable gulf fixed between the domain of the concept of nature, as the sensible, and the domain of the concept of freedom, as the supersensible ... yet the latter **should** have an influence on the former, namely the concept of freedom should make the end that is imposed by its laws real in the in the sensible world; and nature must consequently also be able to be conceived in such a way that the lawfulness of its form is at least in agreement with the possibility of the ends that are to be realized in it in accordance with the laws of freedom. (*CPJ* 5: 176)

I will now reconstruct Kant's reasoning for this surprising conclusion, and, in so doing, argue that Kant's Embodied Agency Theory is in fact *a biological theory of freedom of the will*.

In the two Introductions and the second half of the *Critique of the Power of Judgment*, Kant argues that the concepts LIFE and ORGANISM, and in particular the concept of a "natural purpose" (*Naturzweck*) or living organism, are not ordinary empirical concepts of matter, and that they invoke a type of causation which cannot be known in classical Newtonian mechanistic physics:

> For a body to be judged as a natural purpose in itself and in accordance with its internal possibility, it is required that its parts reciprocally produce each other, as far as both their form and their combination is concerned, and thus produce a whole out of their own causality, the concept of which, conversely is in turn the cause (in a being that would possess the causality according to concepts appropriate for such a product) of it in accordance with a principle; consequently the connection of **efficient causes** could at the same time be judged as an **effect though final causes**. In such a product of nature each part is conceived as if it exists only **through** all the others, thus as if existing **for the sake of the others** and **on account** of the whole, i.e., as an instrument (organ), which is, however, not sufficient (for it could also be an instrument of art, and thus represented as possible at all only as a purpose); rather it must be thought of as an organ that **produces** the other parts (consequently each produces the others reciprocally), which cannot be the case in any instrument of art, but only of nature, which provides all the matter for instruments (even those of art): only then and on that account can such a product, as an **organized** and **self-organizing** being, be called a **natural purpose**. (*CPJ* 5: 373–4)

Strictly speaking, the organization of nature is ... not analogous with any causality that we know. (*CPJ* 5: 375)

Because the causality of living organisms is scientifically unknowable, the basic concepts of biology are merely "regulative" or "hypothetical" concepts of reason, that is, heuristic and logical-fictional concepts for the unification and promotion of natural scientific inquiry (*CPJ* 5: 369–415; see also *CPR* A642–7/B670–5).[39] But it does not follow that organismic life (in particular, the organismic life of my own animal body) cannot be directly cognized by *non-conceptual, non-propositional* means. According to Kant in the first Part of the third *Critique*, the feelings of pleasure and pain, and of the bodily affects and proprioceptive emotions more generally, constitute "the feeling of life" (*CPJ* 5: 204, 278), or the feeling of embodied vitality. Furthermore, there is an intrinsic connection between the affective-emotional psychological life of my mind and the biological life of my own body:

[L]ife is the subjective condition of all our possible experience. (*P* 4: 335)

Life without the feeling of the corporeal organ is merely consciousness of one's existence, but not a feeling of well- or ill-being, i.e., the promotion or inhibition of the powers of life; <u>because the mind for itself is entirely life (the principle of life itself)</u>, and hindrances and promotions must be sought outside it, though in the human being himself, hence in combination with his body. (*CPJ* 5: 278, underlining added)

This Kantian thesis, as I understand it, means that biological life is literally identical to non-conscious or conscious mind. So non-conceptual phenomenal affective-emotional consciousness in inner sense entails embodied biological life: conscious beings are necessarily also living organisms.

This is a crucially important point. The semantic and epistemic constraints Kant places on teleological judgments about distal material objects in space in the context of biological science—namely, that such judgments are always "regulative" and not "constitutive"—do *not* in fact apply to the human conscious experience of embodiment, which is essentially

---

[39] See Ginsborg, "Kant on Understanding Organisms as Natural Purposes"; Guyer, *Kant's System of Nature and Freedom*, chs. 5 and 13; and Kreines, "The Inexplicability of Kant's *Naturzweck*: Kant on Teleology, Explanation, and Biology."

intuitional, and affective-emotional in character, and *not* conceptual, propositional, or judgmental. So there is an important Kantian distinction to be drawn between teleological *judgments* (which are neither directly referential nor existentially committed, because they are essentially based on concepts and regulative) and teleological *intuitions* (which are both directly referential and also existentially committed). According to Kant, then, I have *teleological inner sense intuitions* of my own biological life. In this way, even if teleological *judgments* are only regulative, I can still have a non-conceptual, non-propositional, teleological *phenomenology* that is constitutive. If so, then for Kant there *are* real biological facts in nature. It is just that I cannot *scientifically know* them. But I can still truly *feel* at least some of them, precisely by feeling my own embodied animal life. As we know from section 1.2, according to Kant my living embodiment is metaphysically necessarily entailed by my capacity for empirical apperception. And, as we know from section 2.2, according to Kant feelings are also non-conceptual.

This in turn raises an important general issue issue about how the psychological and biological properties of human animals are cognized or known in the sciences. Kant has notoriously high standards for something's qualifying as a science. Not only must a science involve a systematic organization of objective facts or objective phenomena of some sort, it must also be strongly nomological in the sense that it expresses necessary a priori laws (*MFNS* 4: 468). Sciences in this sense, in turn, can include either "constitutive" (that is, existentially committed without conditions, and assertoric) principles or else "regulative" (that is, at best hypothetically existentially committed, logical-fictional, and non-assertoric) principles. But perhaps most importantly, a science can be a naturally mechanized or *physical* science—that is, an exact science of material nature—only if its phenomena and its laws are mathematically describable (*MFNS* 4: 470). But, as we saw in section 6.2, Kant's notion of mathematics is significantly narrower than our contemporary notion. So we must assume that mathematical describability for Kant is equivalent at best to *arithmetic*al analyzability*, that is, analyzability in terms of primitive recursive arithmetic or PRA, the quantifier-free theory of the natural numbers and the primitive recursive functions over the natural numbers, including the successor function, addition, multiplication, exponentiation, and so on.

As we have just seen, Kant regards biology as a merely regulative non-mechanistic "life science" that supplements Newtonian mechanistic mathematical physics with the teleological concept of a natural purpose or living organism (*CPJ* 5: 369–415). But at the same time Kant regards this biological supplementation of physics as *explanatorily necessary*. And that is because biology provides concepts of natural phenomena that are themselves explanatorily irreducible to mechanistic concepts:

> It is quite certain that we can never adequately come to know the organized beings and their internal possibility in accordance with merely mechanical principles of nature, let alone explain them; and this is indeed so certain that we can boldly say that it would be absurd for humans ever to make such an attempt or to hope that there might yet arise a Newton who could make comprehensible even the generation of a blade of grass according to natural laws. (*CPJ* 5: 400)

In contemporary terms, this means that according to Kant, biology adds the notion of what I will call *natural causal singularities*, and correspondingly the concept of the non-linear non-equilibrium thermodynamics (also known as "complex systems dynamics") of self-organizing systems,[40] to the familiar classical notions of mechanistic natural causal regularities and the linear equilibrium dynamics of inertial physical systems. The general mathematical theory of complex dynamic systems is called "dynamical systems theory" or DST. Strictly speaking, DST is metaphysically neutral, and consistent with both determinism and indeterminism. But DST is *also* perfectly consistent with saying that there are natural systems of interacting proper parts or elements whose actual behaviors over time can be neither digitally computed or nomologically predicted due to random exchanges of causal information, energy, and matter with the surrounding environment, and which exemplify *dynamically emergent* causally efficacious properties that are neither reducible to nor strongly supervenient on the intrinsic non-relational properties of the elements of the system.[41] One direct implication of DST when it is interpreted as a theory of dynamic emergence is the existence of natural causal singularities. More explicitly, *X* is a natural causal singularity if and only if:

---

[40] See Haken, *Principles of Brain Functioning: A Synergetic Approach to Brain Activity, Behavior, and Cognition*; Juarrero, *Dynamics in Action*; Kelso, *Dynamic Patterns*; Port and Van Gelder (eds.), *Mind as Motion: Explorations in the Dynamics of Cognition*; Thelen and Smith, *A Dynamic Systems Approach to the Development of Cognition and Action*; Varela, *Principles of Biological Autonomy*; and Weber and Varela, "Life After Kant: Natural Purposes and the Autopoietic Foundations of Biological Individuality." The notion of self-organization used by contemporary theorists of complex systems dynamics is slightly broader than Kant's, in that it includes non-living complex systems as well, e.g., the rolling hexagonal "Bénard cells" that appear as water is heated. Kantian self-organizing systems are all holistically causally integrated or "autopoietic," such that the whole and the parts mutually produce each other.

[41] See, e.g., Silberstein and McGeever, "The Search for Ontological Emergence."

438 THE PRACTICAL FOUNDATIONS OF THE EXACT SCIENCES

(i) X is an event-type;
(ii) X is tokened exactly once in this actual world $W@$;
(iii) it is (synthetically) metaphysically impossible for any token of X to occur more than once in $W@$, in the sense that even in any other (synthetically) metaphysically possible world accessible from $W@$ in which absolutely all of the antecedent conditions of the occurrent token of X are replicated, then another token of X still would *not* also occur; and
(iv) X, *as an event-type*, (synthetically) metaphysically necessarily requires the passage of elapsed actual time in $W@$ in order to exist.[42]

For example, according to the accounts provided by contemporary cosmological physics, it follows from (i)–(iv) that the Big Bang and black holes are natural causal singularities.[43] Other good candidates for being natural causal singularities are living organisms and the intentional body movements of animals.

But contemporary cosmological physicists have not, it seems, noticed the striking metaphysical implications of natural causal singularities for the problem of free will. Think of it this way. The determinist says: necessarily whenever the *same past* obtains, then the *same future* will occur. Contrariwise, the indeterminist says: necessarily whenever the *same past* obtains, then *different possible futures* might have occurred. But if there are natural causal singularities, then:

(1) Determinism is false, because according to (iii) above, even though the same past of X obtains in some other (synthetically) metaphysically possible world accessible from this actual world, the same future will *not* occur. But:
(2) Indeterminism is *also* false, because according to the conjunction of (ii) and (iii) above, there is (synthetically) necessarily one and only one way that X will *ever* occur, even given the very same past.

What undermines both determinism and indeterminism alike is the fact that the brute actual passage of real time is an intrinsic feature of a natural causal singularity: causal singularities are *radically existential and essentially temporal*. By contrast, neither determinism nor indeterminism takes actual

---

[42] Thus X is "essentially indexical." See Perry, "The Problem of the Essential Indexical."
[43] See, e.g., Hawking, *A Brief History of Time*.

temporal passage seriously. The present for them is nothing but a logical and mechanical *conditionalizing conduit* between the antecedent past and the consequent future.

Now for our current purposes what is crucial is not the fact that the Big Bang and black holes are natural causal singularities, but rather that for Kant the biological, psychological, and rational processes of human animals *also* constitute natural causal singularities. They are, as it were, *little bangs*. They are fully causal-dynamical, yet they are also *underdetermined by mechanical laws of nature* and *nomologically unique*: via the brute actual passage of time, together with their special organismic and (in the case of conscious psychological processes) spontaneous operations, they bring into existence one-time-only or "one-off" causal-dynamical laws of biological, psychological, and rational activity, which can then be added to the existing repertoire of empirical natural causal-dynamic laws.

There is therefore for Kant an irreducible explanatory gap between biology and classical or Newtonian physics, which is the same as the contemporary explanatory gap between the non-linear, non-equilibrium, non-mechanistic dynamics of self-organizing living organismic systems on the one hand, and the classical linear, equilibrium, mechanistic dynamics of inertial physical systems on the other hand. So for Kant all biological facts are explanatorily irreducible—and, if any biological facts can be shown to exist in actuality, also *ontologically* irreducible—to the mechanistic facts of classical or Newtonian physics.[44] But we consciously possess the feeling of biological life occurring in our own bodies via our teleological inner sense intuitions, and thus at least some biological facts actually exist. Therefore there can never be a Newton of the actual biological life of the human animal body in both an explanatory and an ontological sense.

In view of these points, Kant must then regard empirical psychology as *a constitutive and nomological yet nevertheless non-mechanistic and non-deterministic "life science" of the mind*. Even though psychology contains unique "psycho-psycho" laws which strictly govern the phenomenological facts of inner sense[45]—which, we now recognize, must also be actual

---

[44] See Ginsborg, "Two Kinds of Mechanical Inexplicability in Kant and Aristotle."

[45] For Kant, laws do not have to be semantically insensitive to contextual conditions or mentalistic facts in order to be necessary and strict, since they can of course be non-logically or synthetically necessary, that is, restrictedly necessary. See Hanna, *Kant and the Foundations of Analytic Philosophy*, ch. 5. Fodor also calls such psychological laws "ceteris paribus laws": see his "Making Mind Matter More."

biological facts—nevertheless mental phenomena cannot be *arithmetic*ally analyzed* because their merely subjective temporal ordering in inner sense is "entirely arbitrary" (*ganz beliebig*) (*CPR* A193/B238) according to the relatively spontaneous desires and purposes of the rational human animal or person. That is, the radical open-endedness of possible orderings in inner sense means that the set of all mental phenomena cannot be put into a one-to-one correspondence with the set of natural numbers, or reconstructed as computable functions of PRA. But Kant's conception of mathematics, together with the Axioms of Intuition and the Anticipations of Perception—that is, the *mathematical* synthetic a priori principles of pure understanding (*CPR* A160–2/B199–201)—and the Analogies of Experience, show that for him mechanistic natural determinism requires the simple primitive recursive arithmetization of causal processes in time. Thus for Kant psychological laws *cannot be deterministic*:[46]

> The empirical doctrine of the soul must always remain ... removed ... from the rank of what may be called a natural science proper. <u>This is because mathematics is inapplicable to the phenomena of the inner sense and their laws</u> .... It can, therefore, never become anything more than <u>a historical (and, as such, as much as possible) systematic natural doctrine of the inner sense, i.e., a natural description of the soul, but not a science of the soul.</u> (*MFNS* 4: 471, underlining added)

Furthermore, since mental life entails biological life, it follows directly from Kant's thesis that there can never be a Newton of biological life, that there can also never be a Newton of the human mind. So, again, our psychological life, especially including our "power of choice" or *Willkür*, cannot be naturally mechanized.[47]

How does all this apply to Kant's Embodied Agency Theory of freedom of the will? The answer is that according to the biological interpretation of the Theory that I have been developing, even if all the inert parts of material nature, as metaphysically described by the Analogies of Experience, fall under the naturally deterministic and mechanistic causal-dynamic laws of physics, nevertheless the existence of these naturally mechanized parts of nature is fully consistent with the instantiation of an irreducibly different

---

Where Kant and Fodor would disagree is that, for Kant, these synthetically necessary psychological laws are wholly particular and *one-time-only* or "one-off," not general.

[46] See also Lucas, *The Freedom of the Will*, chs. 24–30; and Lucas, "Minds, Machines, and Gödel."
[47] See also Westphal, *Kant's Transcendental Proof of Realism*, pp. 229–43.

set of properties in the rational human animal. This is a set of negatively noumenal properties whose precise pattern of instantiations constitutes both that animal's power of choice or *arbitrium brutum* and also its transcendental and practical freedom of the will, or its autonomy (*CPR* 532–58/560–86), and brings dynamically emergent natural causal singularities of rational animal movement into existence.

In order to see this point, we must look more closely at Kant's account of the structure of the human will:

> The capacity for desiring in accordance with concepts, insofar as the ground determining it to action lies within itself and not in its object, is called the capacity for *doing or refraining from doing as one pleases*. Insofar as it is joined with one's consciousness of the capacity to bring about one's object by one's action it is called *the capacity for choice* (*Willkür*); if it is not joined with this consciousness its act is called a *wish*. The capacity for desire whose inner determining ground, hence even what pleases it, lies within the subject's reason, is called the *will* (*Wille*). The will is therefore the capacity for desire considered not so much in relation to action (as the capacity for choice is) but rather in relation to the ground determining choice to action. The will, strictly speaking, has no determining ground; insofar as it can determine the capacity for choice, it is instead practical reason itself. Insofar as reason can determine the capacity for desire in general, not only *choice* but mere *wish* can be included under the will. The choice which can be determined by *pure reason* is called free choice. That which can be determined only by *inclination* (sensible impulse, *stimulus*) would be animal choice (*arbitrium brutum*). Human choice, however, is a capacity for choice that can indeed be *affected* but not *determined* by impulses, and is therefore of itself (apart from an acquired aptitude of reason) not pure but still can be determined to action by pure will. *Freedom* of choice is this independence from being *determined* by sensible impulses; this is the negative concept of freedom. The positive concept of freedom is that of the capacity of pure reason to be itself practical. But this is not possible except by the subjection of the maxim of every action to the condition of its qualifying as universal law. (*MM* 6: 213–14)

According to Kant, *Willkür*, or the power of choice, is the power of intentional causation by means of effective first-order desires, that is, first-order desires that can or do move us all the way to action.[48] By contrast *Wille*, or the will, is a higher-order volitional power of self-legislation, which operates by means of recognizing either instrumental or non-instrumental

---

[48] See Frankfurt, "Freedom of the Will and the Concept of a Person."

reasons for the determination of choice. To act on the basis of *Willkür* is to move our animal bodies by means of our effective first-order desires. This can of course occur in a Humean way by means of instrumental reasoning according to the hypothetical imperative. Since instrumental reasoning is itself a form of self-legislation, it involves what we might call the "impure" *Wille*. To act on the basis of the pure *Wille*, however, is to constrain and determine our *Willkür* by recognizing the Categorical Imperative, which, as recognized, provides a universal overriding non-instrumental reason for action, and can causally trigger an innate higher-order emotional disposition existing in all human persons (also known as *respect* or *Achtung*) to desire to have non-egoistic and morally correct effective first-order desires.[49] So to act on the basis of pure *Wille* is to do the right thing as determined by our own pure practical reason, no matter what the external and psychological antecedents, and no matter what the consequences.

It is possible to act freely and rationally on the basis of our power of choice or *Willkür* and also on the basis of our impure *Wille*, even when we are acting selfishly or otherwise merely instrumentally on the basis of ordinary desires and reasoning according to the hypothetical imperative—provided of course that we are not compelled to do so by any overwhelming external or unowned forces or pathological inner mechanisms ("alien causes") or by overwhelming empirical desires ("sensible impulses"). As long as we satisfy the Principle of Pathologically-Conditioned Alternative Possibilities, then we are psychologically free, or equivalently we have animal freedom of the will. But in order for us to be *morally responsible*, psychological freedom must then also be combined with *occurrent* transcendental freedom, that is, the actual production of natural causal singularities by means of the will, together with the *capacity* for practical freedom or autonomy. Nevertheless, in doing an act for purely instrumental reasons in a psychologically and transcendentally free way, we are also at the same time acting either immorally or at least without moral worth, since as acting merely instrumentally we are not conforming ourselves to the moral law. Suppose however that this *Willkür*-driven action happens also to be consistent with the moral law. This action may have instrumental *moral value*. It nevertheless remains an

---

[49] This Kantian thesis entails what A. W. Moore aptly calls "conative objectivism"—the doctrine that we innately possess non-egoistic emotional dispositions–which in turn solves the classical problem of how the Categorical Imperative can motivate the empirical will. See Moore, *Noble in Reason, Infinite in Faculty*; and Hanna and Moore, "Reason, Freedom, and Kant: An Exchange."

act without *moral worth* if it is also counterfactually true that, given the same set of external and psychological antecedents, and the fact that it had been in our selfish or even benevolent interest to do something morally wrong, then we would have gone ahead and done the wrong thing.

Here, however, is the crucial point. Even when we are acting wrongly or merely without moral worth, it remains true that our *capacity* for acting practically freely and autonomously on the basis of pure *Wille* is undiminished, despite the fact that we have not adequately realized that capacity in that context. *Only a being with an undiminished capacity for practical freedom and autonomy can act immorally or in a way that is without moral worth.* Hence we remain morally responsible even for things that we have done non-autonomously, provided that we have also done them psychologically freely, transcendentally freely, instrumentally, and rationally via our power of choice or *Willkür* and our impure *Wille*. This is because the capacity for pure *Wille* counterfactually guarantees that even if, given the same set of external and psychological antecedents, together with the fact that it had been in our selfish or even benevolent interest to do something morally wrong, nevertheless we *still* could have gone ahead and done the right thing instead of the wrong or morally worthless thing we actually did, by recognizing the Categorical Imperative as an overriding non-instrumental reason for action.

The fact that our recognition of the Categorical Imperative can causally trigger our innate higher-order emotional disposition for feeling respect, which can then in turn determine our first-order effective desires and thereby our *Willkür*, is also what Kant calls *the fact of reason* (*Faktum der Vernunft*):

> The consciousness of this fundamental law [of pure practical reason, which says: so act that the maxim of your will could always hold at the same time as a principle of universal law giving] may be called a fact of reason, since one cannot ferret it out from antecedent data of reason, such as the consciousness of freedom (for this is not antecedently given), and since it forces itself upon us as a synthetic proposition a priori based on no pure or empirical intuition ... In order to regard this law without any misinterpretation as given, one must note that it is not an empirical fact, but the sole fact of pure reason, which by it proclaims itself as originating law. (CPrR 5: 31)

For example, someone intentionally raises her arm in order to stop a crime just because she feels in her heart and mind that it is the morally right thing to do, even though she thereby risks her own life, and even though she desperately wants to avoid getting involved. *This is a "little bang" with a big moral attitude.* It does not happen very often, but (I think!) it happens at least sometimes.

If so, then according to Kant this sort of intentional act is possible only because there are some macrophysical *non-mechanistic, non-deterministic* causal-dynamical processes—more specifically: (i) the non-linear, non-equilibrium causal-dynamic biological processes of my own living organismic body, including its purposive bodily movements; and also (ii) the non-arithmetic★ally analyzable spontaneous psychological processes in my inner sense, driven by mental representations and desires—that are perfectly logically and metaphysically consistent with the existence of mechanistic deterministic physical laws that apply to material events and facts *other than* those specifically involved in either the biological processes of my animal body or the psychological processes in my inner sense.

This in turn implies that to say that the practical-rational properties of a human animal are inconsistent with natural causal laws—"experience reveals only the law of appearances and consequently the mechanism of nature, the direct opposite of freedom" (*CPrR* 5: 29)—is fundamentally only to say that the intrinsic practical-rational properties of a human animal in action are inconsistent with the coexistence of intrinsic *naturally mechanistic* causal laws in the very same animal. Indeed, in the second *Critique* Kant explicitly asserts that rational personhood (*Persönlichkeit*) itself is just:

freedom and independence from the mechanism of nature regarded as a capacity of a being subject to special laws (pure practical laws given by its own reason). (*CPrR* 5: 87, underlining added).

Therefore the difference between naturally mechanistic, deterministic causal laws (with which categorically normative moral laws of human action are inconsistent, when applied to one and the same intentional act of choice), and *non-mechanistic, non-deterministic one-time-only or "one-off" laws* of natural causal singularities (with which categorically normative moral laws are perfectly consistent, since both transcendental freedom and practical freedom alike require underdetermination of animal choice by naturally mechanistic laws) is the metaphysical core of Kant's Embodied Agency Theory of freedom.

So far in this section, I have been developing a biological interpretation of Kant's Embodied Agency Theory of freedom of the will. What we need to see now more explicitly is just how this interpretation entails the denial of both compatibilism and incompatibilism—that is, how it entails that Kant's Embodied Agency Theory is *neither* compatibilist *nor* incompatibilist, precisely because it is *post-compatibilist*.

Consider first compatibilism. Compatibilism says that freedom of the will and determinism are metaphysically consistent. On the biological interpretation of Kant's Embodied Agency Theory of freedom, compatibilism is false. This is because according to this interpretation, all causation bottoms out in event-causation, and there are no simple or complex events that are at once transcendentally free and naturally mechanized. And since all individual material substances and agents are complex events, there are also no individual material substances or agents that are at once transcendentally free and naturally mechanized. All the conscious animals and in particular the rational animals and their actions are both alive and have spontaneity of consciousness and desire, and are not naturally mechanized, hence also not determined.

Consider now incompatibilism. Incompatibilism says that freedom of the will and natural mechanism are metaphysically inconsistent. On the biological interpretation of Kant's Embodied Agency Theory of freedom, incompatibilism is also false. This is for two reasons. First, psychological freedom of the will, although it is a necessary condition of transcendental freedom, is metaphysically consistent with natural determinism. Second, according to the Embodied Agency Theory there are parts of of the natural world that are naturally mechanized, and parts of the natural world that are not naturally mechanized. Living organisms, for example, are not naturally mechanized. As Kant puts it, there could never be a biological Newton who could explain the generation of even a single blade of grass. Most relevantly, conscious animals and in particular rational animals are not naturally mechanized. They are alive and psychologically spontaneous, which is to say that they have animal freedom of the will. So transcendental freedom and natural determinism can both exist in the same natural world. In this way, the metaphysical consistency of psychological freedom with natural determinism, plus the thesis that there is *a strong continuity between biological life and the relative or absolute spontaneity of the human will*,[50] when it is combined with a non-reductive approach to biological explanation and biological facts, entails the denial of incompatibilism.

[50] See Godfrey-Smith, *Complexity and the Function of Mind in Nature*.

In other words, Kant is a post-compatibilist. Kant's post-compatibilism combines Hume's deep insight about the metaphysical consistency of psychological freedom and natural determinism with the uniquely Kantian notions of transcendental freedom and practical freedom as interpreted according to the Embodied Agency Theory.

Here is a crucial consequence of Kant's post-compatibilism. It does not follow from the fact that something is transcendentally free, that it *violates* the mechanistic laws of nature. We can do only those things that are *permitted* by the mechanistic laws. But at the same time the mechanistic laws themselves together with the settled facts *do not compel or necessitate* our intentional actions, even if what merely happens to us—as opposed to what we will or do—still *contingently conforms* to the mechanistic laws. In a precisely similar way, in a moral context, as Kant points out, we can morally do only those acts that are permitted by the moral law (universalizability) But at the same time the moral law itself does not necessitate our intentional actions (*ought* does not entail *is*), even if what merely happens to us (as opposed to what we will or do) still contingently conforms to the moral law. It is also true that for Kant we can actually will or do things that only contingently conform to the moral law, when we have done them for reasons other than the moral law itself. So that leaves the threefold distinction between: (i) something's being *permitted by the law*; (ii) something's being *compelled or necessitated by the law*; and (iii) something's *contingently conforming to the law*, perfectly intact.

Indeed, this is the crucial Kantian distinction that is needed for resolving the free will *vs.* natural mechanism problem: what is *permitted* by the mechanistic laws is not the same as what is *compelled or necessitated* by the mechanistic laws, nor is it the same as what *contingently conforms* to the mechanistic laws. We are transcendentally free precisely when, by means of our own spontaneous conscious desires, we will and do things in a naturally causally singular way that is permitted by the mechanistic laws but not compelled or necessitated by them, even if what merely happens to us—as opposed to what we will or do—still contingently conforms to the mechanistic laws of nature. Therefore, even if everything in nature must heed the general mechanistic laws in the minimal extrinsic sense that they cannot violate those laws—for example, I cannot violate Newton's Laws of Motion, or run faster than the speed of light, etc.—it does not follow that everything in nature is *compelled or necessitated* by those laws. In particular, other things being equal, the mechanistic laws do not determine precisely how I will move my arm right now. They minimally extrinsically constrain it but do not determine it. Analogously, just because I heed the moral law in the sense that I do not violate it, it does not follow that the moral law compels or necessitates my actions. As long as my maxim universalizes, the moral law minimally extrinsically constrains my actions but does not determine them. So, by the same token, living systems operate consistently with the laws of natural mechanism, and are minimally extrinsically constrained by them, but are not determined by them.

Or, put in another way, just because everything in material nature is governed (in a minimal extrinsic sense) by mechanistic laws, and is indeed partially explicable by reference to those laws, it does not follow that everything is *completely* or even *essentially* explicable by the mechanistic laws *alone*. Only *some* things contain the mechanistic laws as intrinsic structural properties—inert material bodies—and living systems are not amongst those things, even if the living systems are still minimally extrinsically constrained by mechanistic laws. Living systems themselves contain the unique non-deterministic laws of non-linear non-equilibrium self-organizing organismic causal-dynamic action as intrinsic structural properties. So there cannot be a Newton of a blade of grass; consequently there cannot be a Newton of my living body or *Leib*; hence there cannot be a Newton of my human mind; and therefore there cannot be a Newton of my freely willed intentional actions either.

This point is intimately connected to Kant's idea, developed in the First Introduction to the *Critique of the Power of Judgment*, that there is an explanatory and ontological gap between what in the first *Critique* he had called the "transcendental affinity" of nature, its transcendentally nomological character, and its "empirical affinity," its empirically nomological character (*CPJ* 20: 208–11; see also *CPR* A122–8, B163–5).[51] And this in turn is intimately connected to the problem of "empirical laws" (see section 3.5). More specifically, Kant is committed to the thesis that even allowing for the existence of universal, transcendental laws of nature, and also for the existence of general mechanistic laws of nature, it does not automatically follow that that there are specific empirical laws of nature "all the way down." Indeed, nature might still be lawless and chaotic in its particular

---

[51] See Westphal, *Kant's Transcendental Proof of Realism*, ch. 3.

448 THE PRACTICAL FOUNDATIONS OF THE EXACT SCIENCES

empirical details. The assumption that nature is pervasively nomological is a regulative, but not constitutive principle of natural science. Neither the universal transcendental laws nor the general mechanistic causal laws of nature determine the specific behaviors and natures of all material objects. And in particular they do not determine the specific behaviors and natures of non-animal organisms, non-rational human or non-human animals, or rational human animals.

So what can actually close the gap? The answer offered by the biological interpretation of the Embodied Agency Theory is that transcendentally free rational animal choices produce natural causal singularities, along with their novel one-time-only or "one-off" laws, and thereby *freely complete nature*. Transcendentally free agents thus create new unique empirical causal-dynamic laws of nature that fall under, and are *permitted by*, but are not *compelled or necessitated by*, the general mechanistic laws of nature. This in turn is closely connected to Kant's notion of the mental power of *artistic genius*, which is such:

> that it cannot itself describe or indicate scientifically (*wissenschaftlich*) how it brings its product into being, but rather that it gives the rule to **nature**, and hence the author of a product that he owes to his genius does not scientifically know himself how the ideas for it came to him, and also does not have it in his power to think up such things...according to plan, and to communicate to others precepts that would put them in a position to produce similar products. (*CPJ* 5: 308)

In short, as transcendentally free rational animals with embodied wills, we enrich and ramify the causal-dynamic nomological structure of material nature by being *authors* of its most specific empirical laws. In this way not only do we make a causal difference, we also freely *make* nature, in part and on an appropriately human scale. As finite and radically evil, we are most certainly not gods. But we *are* small-time creators. And how much more power over nature could we really want?

But what then is nature? According to Kant's view, as we have seen, nature contains nothing but material spatialized simple events, complex events, and substances, which in turn are nothing but positions or roles in the total causal-dynamic structure of attractive and repulsive forces. Yet some of these material beings are not naturally mechanized bodies but are in fact biologically alive and thereby instantiate some non-mechanical non-deterministic intrinsic structural properties, and in particular the property of being conscious and rational. To put yet another twist on Royce's salty definition of idealism ("the world and the heavens, and the stars are all *real*, but not so *damned* real"), for Kant the natural world is everywhere *material or physical*, but not so *damned* physical.

The fundamental aim of Kant's later or "post-Critical" philosophy in the third *Critique* and the *Opus postumum* is to work out a detailed conception of the deep connection between freedom and nature. According to this conception, biological life and (non-conscious or conscious) mind, including the spontaneous human will, are one and the same. Furthermore, as we learn in the final metaphysics of matter that Kant develops in the *Transition*-project in the *Opus postumum*, both biological properties and volitional properties are intrinsic structural properties of material substances that are themselves structuralistically metaphysically grounded in a cosmologically fundamental "stuff" or *aether*, which consists in a nomological system of primitive causal-dynamic forces (see chapters 1–4), some of whose systematic functions or roles are really manifest as mechanical inert material bodies, and some of whose systematic functions or roles are really manifest as natural purposes or organisms (*OP* 21: 206–33). In one draft of the *Transition*-project, Kant puts it this way:

The primitive-moving forces of matter are the dynamic forces. <u>The mechanical are only derivative</u>. (*OP* 22: 241, underlining added)

So, refining the preliminary picture I sketched of Kant's metaphysics of matter in section 8.1, Figure 8.2 shows what the finished picture looks like.

LEVEL 2 = THE LEVEL OF STRUCTURE-DEPENDENT ENTITIES
                                                non-conscious or conscious minds
                                                DYNAMICALLY EMERGE FROM
                                                       ↕ ↕ ↕
inert material bodies = mechanical systems   *organisms = living systems*
NOMOLOGICALLY STRONGLY SUPERVENE ON   DYNAMICALLY EMERGE FROM
        ↕ ↕ ↕   ↕ ↕ ↕
LEVEL 1 = THE LEVEL OF THE STRUCTURE ITSELF
the structural aether = the systematic totality of primitive attractive and repulsive forces

Figure 8.2. Kant's Metaphysics of Matter (Refined Version)

Now, for Kant, by virtue of his doctrine of the strong continuity of life and mind, consciousness is continuous with organismic life in suitably complex, suitably intrinsically structured animals. Some of those animals are rational human animals or human persons. So the natural world contains, in addition to natural mechanisms and biological/mental facts, a further set of purely rational intrinsic structural properties, which together with the basic biological/mental facts, jointly constitute human persons and their living, embodied, spontaneous wills. In this way, Kant is what I will call *a causal-dynamic neutral monist*, for whom the cosmologically fundamental structural aether, as a total system of primitive causal-dynamic moving forces, is *not* itself inert. But this is not hylozoism either: the aether is *not* itself alive (*CPJ* 5: 392). Mechanical bodies and organisms alike are emergent *functions*, or causal-dynamic *roles*, of the structural aether. So, in a metaphysical structuralist sense, the causal-dynamic aether remains strictly ontologically neutral as between the actual existence of living thinking systems and the actual existence of inert mechanical systems. Indeed, from this emergentist causal-dynamic metaphysical structuralist point of view, the fact that inert mechanical systems came into being earlier in cosmological time than the living thinking systems did is a relatively superficial or trivial fact, without any deep metaphysical significance.

8.4. Freedom, Causation, and Time's Arrow

By way of conclusion, I want to point up one more deeply important implication of Kant's Embodied Agency Theory. This implication links the concept and fact of an embodied autonomous will with the concept and fact of a naturally mechanized causal process. Naturally mechanized causation, as we have seen in our discussion of the three Analogies of Experience in section 8.1, requires the asymmetric directionality or irreversibility of time. The irreversibility of time, in turn, is a logically contingent feature of time. The proposition that time runs backwards is not analytically false. Indeed, the naturally mechanized linear equilibrium causal-dynamic processes of classical Newtonian mechanics clearly can run *either* forwards *or* backwards.[52] What then synthetically necessarily guarantees that time actually runs forwards and not backwards?

The Kantian answer I propose invokes a direct connection between the naturally mechanized causation described in the Analogies of Experience,

---

[52] This phenomenon is also known as "time reversal invariance." See Savitt, "Introduction" to *Time's Arrows Today*, pp. 12–18.

and irreversible non-linear non-equilibrium thermodynamic processes, especially including biological processes.[53] For Kant, who identifies life with (non-conscious or conscious) mind, all mental processes are also biological processes, and all biological processes are also mental processes. So the relative or absolute spontaneity of the embodied human will is *also* an irreversible non-linear non-equilibrium thermodynamic process. The basic thought here is that the actual existence of an asymmetric "Time's Arrow" in naturally mechanized causation requires the necessary possibility of free rational human action, whether as transcendentally free animal choice or as practical freedom or autonomy, in order to fix its forward direction as a non-logically or synthetically necessary metaphysical fact. Time synthetically necessarily runs forwards because human action is inherently future-oriented or teleological, and because human action, as necessarily embodied in space and representationally framed by the forms of intuition, necessarily also occurs in time. *Therefore, for Kant, the forward direction of time is metaphysically fixed by the teleological notion of a purpose, both natural and practical.*

Given the centrality of the notion of causation to the very idea of a naturally mechanized material world, this in turn leads to the Kantian modal-metaphysical conclusion that there there cannot be a scientifically knowable naturally mechanized material world in which human value, human action, human freedom, and human morality are impossible. This actual naturally mechanized material world is to that extent *our* world, for better or worse. And if my biological interpretation of Kant's Embodied Agency Theory of free will is correct, then in addition to the naturally mechanized parts of the natural world, there also actually exist other causally efficacious parts of this world that are *not* naturally mechanized—even if they are minimally extrinsically constrained by the mechanistic laws—including all the living organisms, all the non-rational animals whether human or non-human, and all the free acts of rational human animals and other embodied persons. This constitutes the last part of Kant's thesis of the primacy of human nature, and it fully fleshes out his liberal naturalism. *Welcome home.*

---

[53] See Nicolis and Prigogine, *Self-Organization in Nonequilibrium Systems*; and Prigogine, *Being and Becoming: Time and Complexity in the Physical Sciences*.

Bibliography

Abbott, B., "A Note on the Nature of 'Water'," *Mind* 106 (1997): 311–19.
Abela, P., *Kant's Empirical Realism* (Oxford: Oxford University Press, 2002).
Agazzi, E., and Pauri, M. (eds.), *The Reality of the Unobservable* (Dordrecht: Kluwer, 2000).
Allison, H., "Causality and Causal Laws in Kant: A Critique of Michael Friedman," in P. Parrini (ed.), *Kant and Contemporary Epistemology* (Netherlands: Kluwer, 1994), 291–307.
―― *Kant's Transcendental Idealism* (New Haven, CT: Yale University Press, 1983).
―― *Kant's Theory of Freedom* (Cambridge: Cambridge University Press, 1990).
―― "Transcendental Idealism: The 'Two Aspect' View," in B. Den Ouden (ed.) *New Essays on Kant* (New York: Peter Lang, 1987), 156–76.
Almog, J., et al. (eds.), *Themes from Kaplan* (New York: Oxford University Press, 1989).
Ameriks, K., *Interpreting Kant's Critiques* (Oxford: Oxford University Press, 2003).
―― *Kant and the Fate of Autonomy* (Cambridge: Cambridge University Press, 2000).
―― "Kant on Science and Common Knowledge," in Watkins (ed.), *Kant and the Sciences* (2001), 31–52.
Anderson, E., "Kant, Natural Kind Terms, and Scientific Essentialism," *History of Philosophy Quarterly* 11 (1994): 355–73.
Aquila, R., "Kant's Phenomenalism," *Idealistic Studies* 5 (1975): 108–26.
―― *Representational Mind* (Bloomington, IN: Indiana University Press, 1983).
Aristotle, *De Anima, Books II and III*, trans. D. W. Hamlyn (Oxford: Clarendon/Oxford University Press, 1968).
Austin, J. L., *How to Do Things with Words* (2nd edn., Cambridge: Harvard University Press, 1962).
―― "Truth," in J. L. Austin, *Philosophical Papers* (3rd edn., Oxford: Oxford University Press, 1979), 117–33.
Ayer, A. J., *Language, Truth, and Logic* (2nd edn., New York: Dover, 1952).
Ayers, M., "Locke vs. Aristotle on Natural Kinds," *Journal of Philosophy* 78 (1981): 247–72.
Baker, L. R., "Why Constitution is Not Identity," *Journal of Philosophy* 94 (1997): 599–621.
Baldwin, T., "The Identity Theory of Truth," *Mind* 100 (1991): 35–52.

Bealer, G., "The Philosophical Limits of Scientific Essentialism," *Philosophical Perspectives* 1 (1987): 289–365.

Beiser, F., *The Fate of Reason* (Cambridge, MA: Harvard University Press, 1987).

Benacerraf, P. "Frege: The Last Logicist", in P. French et al. (eds.), *The Foundations of Analytic Philosophy*, Midwest Studies in Philosophy VI (Minneapolis, MN: University of Minnesota Press, 1981), 17–35.

—— "Mathematical Truth," *Journal of Philosophy* 70 (1972): 661–79.

—— "What Mathematical Truth Could Not Be—I," in A. Morton and S. Stich (eds.), *Benacerraf and his Critics* (Oxford: Blackwell, 1996), 9–59.

—— "What Numbers Could Not Be," *Philosophical Review* 74 (1965): 47–73.

Bennett, J., *Locke, Berkeley, Hume: Central Themes* (Oxford: Oxford University Press, 1971).

Berkeley, G., *Treatise Concerning the Principles of Human Knowledge* (Indianapolis, IN: Hackett, 1982).

Bermúdez, J., "Nonconceptual Mental Content," *Stanford Encyclopedia of Philosophy (Spring 2003 Edition)*, E. N. Zalta (ed.), URL=<http://plato.stanford.edu/archives/fall2004/entries/content-nonconceptual/>.

—— *The Paradox of Self-Consciousness* (Cambridge, MA: MIT Press, 1998).

—— *Thinking without Words* (New York: Oxford University Press, 2003).

—— et al. (eds.), *The Body and the Self* (Cambridge, MA: MIT Press, 1995).

—— and Macpherson, F., "Nonconceptual Content and the Nature of Perceptual Experience," *Electronic Journal of Analytical Philosophy* 8 (1998) URL=<http://ejap.louisiana.edu/EJAP/1998/bermmacp98.html>.

Blachowicz, J., "Analog Representation Beyond Mental Imagery," *Journal of Philosophy* 94 (1997): 55–84.

Blackburn, S., *Ruling Passions: A Theory of Practical Reasoning* (Oxford: Clarendon/Oxford University Press, 1998).

Bloor, D., *Knowledge and Social Imagery* (2nd edn., Chicago, IL: University of Chicago Press, 1991).

Boghossian, P., and Peacocke. C. (eds.), *New Essays on the A Priori* (Oxford: Clarendon/Oxford University Press, 2000).

Bok, H., "Freedom and Practical Reason," in Watson (ed.), *Free Will*, 130–66.

Boolos, G., and Jeffrey, R., *Computability and Logic* (3rd edn., Cambridge: Cambridge University Press, 1989).

Brewer, B., *Perception and Reason* (Oxford: Oxford University Press, 1999).

—— "Self-Location and Agency," *Mind* 101 (1992): 17–34.

Brittan, G., *Kant's Theory of Science* (Princeton: Princeton University Press, 1978).

Broad, C. D., "Are There Synthetic A Priori Truths?," *Proceedings of the Aristotelian Society* 15 (1936): 102–7.

Brook, A., *Kant and the Mind* (Cambridge: Cambridge University Press, 1994).

—— "Kant's *A Priori* Methods for Recognizing Necessary Truths," in P Hanson and B. Hunter (eds.), *Return of the A Priori* (Calgary, AL: University of Alberta Press, 1992), 215–52.

Brouwer, L. E. J., "Historical Background, Principles, and Methods of Intuitionism," *South African Journal of Science* 49 (1952): 139–46.

—— "Intuitionism and Formalism," in P. Benacerraf and H. Putnam (eds.) *Philosophy of Mathematics* (Englewood Cliffs, NJ: Prentice Hall, 1964), 66–77.

Brown, J., "Natural Kind Terms and Recognitional Capacities," *Mind* 107 (1998): 275–302.

Brueckner, A., "Brains in a Vat," *Journal of Philosophy* 83 (1986): 148–67.

—— "Transcendental Arguments from Content Externalism," in Stern (ed.), *Transcendental Arguments: Problems and Prospects*, 229–50.

—— "Trying to Get Outside Your Own Skin," *Philosophical Topics* 23 (1995): 79–111.

Buchdahl, G., "The Conception of Lawlikeness in Kant's Philosophy of Science," in L. W. Beck (ed.), *Kant's Theory of Knowledge* (Dordrecht: D. Reidel, 1974), 128–50.

—— *Kant and the Dynamics of Reason* (Oxford: Blackwell, 1992).

—— *Metaphysics and the Philosophy of Science* (Cambridge, MA: MIT Press, 1969).

Burge, T., "Cartesian Error and the Objectivity of Perception," in P. Pettit and J. McDowell (eds.), *Subject, Thought, and Context* (Oxford: Oxford University Press, 1986), 117–36.

—— "Individualism and the Mental," in P. A. French et al. (eds.), *Studies in Metaphysics*, Midwest Studies in Philosophy IV (Minneapolis, MN: University of Minnesota Press, 1979), 73–122.

—— "Reason and the First Person," in C. Wright et al. (eds.), *Knowing Our Own Minds* (Oxford: Clarendon/Oxford University Press, 1998), 243–70.

Butts, R. (ed.), *Kant's Philosophy of Physical Science* (Dordrecht: D. Reidel, 1986).

Campbell, J., *Past, Space and Self* (Cambridge, MA: MIT Press, 1994).

—— "The Structure of Time in Autobiographical Memory," *European Journal of Philosophy* 5 (1997): 105–18.

Carnap, R., *Der Raum. Ein Beitrag zur Wissenschaftslehre* (Berlin: Reuther and Reichard, 1922).

—— *Meaning and Necessity* (2nd edn., Chicago, IL: University of Chicago Press, 1956).

—— "Meaning Postulates," in Carnap, *Meaning and Necessity*, 222–9.

Carrier, M. "Kant's Mechanical Determinations of Matter in the *Metaphysical Foundations of Natural Science*," in Watkins (ed.), *Kant and the Sciences*, 117–35.

Carruthers, P., *Language, Thought, and Consciousness* (Cambridge: Cambridge University Press, 1996).

—— "Natural Theories of Consciousness," *European Journal of Philosophy* 6 (1998): 203–220.

Cartwright, R., "Some Remarks on Essentialism," *Journal of Philosophy* 65 (1968): 615–26.

Cassam, Q., "Inner Sense, Body Sense, and Kant's Refutation of Idealism," *European Journal of Philosophy*, 1 (1993): 111–27.

Casullo, A., "Kripke on the A Priori and the Necessary," in Moser (ed.), *A Priori Knowledge*, 161–9.

Caygill, H., *A Kant Dictionary* (Oxford: Blackwell, 1995).

Chalmers, D., *The Conscious Mind* (New York: Oxford University Press, 1996).

—— "The *Matrix* as Metaphysics," in *Philosophy and the Matrix* (2003) URL= <https://whatisthematrix.warnerbros.com/rl_cmp/phi.html>.

Chisholm, R., "Human Freedom and the Self," in Watson (ed.), *Free Will*, 26–37.

—— *Perceiving* (Ithaca, NY: Cornell University Press, 1957).

Chomsky, N., "Language and Nature," *Mind* 104 (1995): 1–61.

Churchland, P. M., *Matter and Consciousness* (Cambridge, MA: MIT Press, 1985).

—— *Scientific Realism and the Plasticity of Mind* (Cambridge: Cambridge University Press, 1979).

Clark, A., *Being There: Putting Brain, Body, and World Together Again* (Cambridge, MA: MIT Press, 1997).

—— "Embodiment and the Philosophy of Mind," in A. O'Hear (ed.), *Current Issues in the Philosophy of Mind* (Cambridge: Cambridge University Press, 1998), pp. 35–51.

—— *Mindware* (New York: Oxford University Press, 2001).

—— "Time and Mind," *Journal of Philosophy* 95 (1998): 354–76.

Clarke, T., "The Legacy of Skepticism," *Journal of Philosophy* 69 (1972): 754–69.

Collins, A., *Possible Experience: Understanding Kant's Critique of Pure Reason* (Berkeley: University of California Press, 1999).

Compton-Burnett, I., *Manservant and Maidservant* (New York: Oxford University Press, 1983).

Crane, T., "The Nonconceptual Content of Experience," in T. Crane (ed.), *The Contents of Experience: Essays on Perception* (Cambridge: Cambridge University Press, 1992), 136–57.

Damasio, A., *The Feeling of What Happens: Body and Emotion in the Making of Consciousness* (San Diego, CA: Harcourt, 1999).
Danto, A., "Naturalism," in Edwards (ed.), *The Encyclopedia of Philosophy*, vol. 5, 448–50.
Darnton, R., *The Great Cat Massacre* (New York: Basic Books, 1984).
Davidson, D., "The Folly of Trying to Define Truth," *Journal of Philosophy* 3 (1996): 263–8.
—— "The Structure and Content of Truth," *Journal of Philosophy* 87 (1990): 279–328.
Davies, M. K., and Humberstone, L., "Two Notions of Necessity," *Philosophical Studies* 38 (1980): 1–30.
Denyer, N., "Pure Second-Order Logic," *Notre Dame Journal of Formal Logic* 33 (1992): 220–4.
Derden, J., "A Different Conception of Scientific Realism: The Case for the Missing Explananda," *Journal of Philosophy* 100 (2003): 243–67.
De Rose, K., "Solving the Skeptical Problem," *Philosophical Review* 104 (1995): 1–52.
Descartes, R., "Discourse on the Method," in Descartes, *The Philosophical Writings of Descartes*, vol. I, 111–51.
—— "Meditations on First Philosophy," in Descartes, *The Philosophical Writings of Descartes*, vol. II, 3–62.
—— *The Philosophical Writings of Descartes*, trans. John Cottingham et al. (3 vols.; Cambridge: Cambridge University Press, 1984).
—— "Principles of Philosophy," in Descartes, *The Philosophical Writings of Descartes*, vol. I, 177–291.
—— "Rules for the Direction of the Mind," in Descartes, *The Philosophical Writings of Descartes*, vol. I, 7–78.
Dicker, G., *Kant's Theory of Knowledge* (New York: Oxford University Press, 2004).
Donnellan, K., "Kripke and Putnam on Natural Kind Terms," in C. Ginet and S. Shoemaker (eds.), *Knowledge and Mind* (New York: Oxford University Press, 1983), 84–104.
—— "Necessity and Criteria," *Journal of Philosophy* 59 (1962): 647–58.
—— "Reference and Definite Descriptions," in A. P. Martinich (ed.), *The Philosophy of Language* (2nd edn., New York: Oxford University Press, 1990), pp. 235–47.
Dretske, F., "Conscious Experience," *Mind* 102 (1993): 263–83.
—— *Knowledge and the Flow of Information* (Cambridge, MA: MIT Press, 1981).
—— *Seeing and Knowing* (Chicago, IL: University of Chicago Press, 1969).

Dummett, M., *Frege: The Philosophy of Language* (2nd edn., Cambridge, MA: Harvard University Press, 1981).
—— *Origins of Analytical Philosophy* (Cambridge, MA: Harvard University Press, 1993).
—— "Realism (1963)," in Dummett, *Truth and Other Enigmas* (London: Duckworth, 1978).
—— "Realism (1982)," in Dummett, *The Seas of Language*, 230–76.
—— "Realism and Anti-Realism," in Dummett, *The Seas of Language*, 462–78.
—— *The Seas of Language* (Oxford: Oxford University Press, 1993).
—— *Truth and Other Enigmas* (London: Duckworth, 1978).
Eddington, A., *The Nature of the Physical World* (Cambridge: Cambridge University Press, 1929).
Edwards, J., *Substance, Force, and the Possibility of Knowledge: On Kant's Philosophy of Material Nature* (Berkeley: University of California Press, 2000).
Edwards, P. (ed.), *Encyclopedia of Philosophy* (8 vols., New York: Macmillian, 1967).
Eilan, N., "Objectivity and the Perspective of Consciousness," *European Journal of Philosophy* 5 (1997): 235–50.
Ellis, B., *Scientific Essentialism* (Cambridge: Cambridge University Press, 2001).
Evans, G., *The Varieties of Reference* (Oxford: Oxford University Press, 1982).
Falkenstein, L., "Kant's Account of Sensation," *Canadian Journal of Philosophy* 20 (1990): 63–88.
Farrell, R., "Metaphysical Necessity is not Logical Necessity," *Philosophical Studies* 39 (1981): 141–53.
Feyerabend, p. *Against Method* (London: Verso, 1978).
Field, H., "The Deflationary Conception of Truth," in G. MacDonald and C. Wright (eds.), *Fact, Science, and Value* (Oxford: Blackwell, 1986), 55–117.
Fodor, J., *The Language of Thought* (Cambridge, MA: Harvard University Press, 1975).
—— "Making Mind Matter More," in J. Fodor, *A Theory of Content and Other Essays* (Cambridge: MIT Press, 1990), 137–59.
—— *The Modularity of Mind* (Cambridge: MIT Press, 1983).
—— "Special Sciences, or The Disunity of Science as a Working Hypothesis," *Synthese* 28 (1974): 97–115.
Fogelin, R., "Quine's Limited Naturalism," *Journal of Philosophy* 94 (1997): 543–63.
Förster, E., *Kant's Final Synthesis* (Cambridge, MA: Harvard University Press, 2000).
—— (ed.), *Kant's Transcendental Deductions* (Stanford, CA: Stanford University Press, 1989).
Frankfurt, H., "Alternate Possibilities and Moral Responsibility," in Frankfurt, *The Importance of What We Care About*, 1–10.

―― "Freedom of the Will and the Concept of a Person," in Frankfurt, *The Importance of What We Care About*, 11–25.

―― "Identification and Wholeheartedness," in Frankfurt, *The Importance of What We Care About*, 159–76.

―― *The Importance of What We Care About* (Cambridge: Cambridge University Press, 1988).

―― "The Problem of Action," in Frankfurt, *The Importance of What We Care About*, 69–79.

Frege, G., *Basic Laws of Arithmetic*, trans. M. Furth (Berkeley, CA: University of California Press, 1964).

―― *Collected Papers on Mathematics, Logic, and Philosophy*, trans. M. Black et al. (Oxford: Blackwell, 1984).

―― *The Foundations of Arithmetic*, trans. J. L. Austin (2nd edn., Evanston: Northwestern University Press, 1980).

―― "Logic [1897]," in Frege, *Posthumous Writings*, 126–51.

―― "On Sense and Meaning," in Frege, *Collected Papers on Mathematics, Logic, and Philosophy*, 137–56.

―― *Posthumous Writings*, trans. H. Hermes et al. (Chicago, IL: University of Chicago Press, 1979).

―― "Thoughts," in Frege, *Collected Papers on Mathematics, Logic, and Philosophy*, 351–72.

Friedman, M., "Causal Laws and the Foundations of Natural Science," in P. Guyer (ed.) *The Cambridge Companion to Kant*, 161–99.

―― *Dynamics of Reason* (Stanford: CSLI, 2001).

―― *Kant and the Exact Sciences* (Cambridge, MA: Harvard University Press, 1992).

―― "Matter and Motion in the *Metaphysical Foundations* and the First *Critique*: The Empirical Concept of Matter and the Categories," in Watkins (ed.), *Kant and the Sciences* (2001), 53–69.

―― *A Parting of the Ways: Carnap, Cassirer, and Heidegger* (La Salle, IL: Open Court, 2000).

―― "Philosophical Naturalism," *Proceedings and Addresses of the American Philosophical Association* 71 (1997): 7–21.

―― *Reconsidering Logical Positivism* (Cambridge: Cambridge University Press, 1999).

―― "Transcendental Philosophy and A Priori Knowledge: A Neo-Kantian Perspective," in Boghossian and Peacocke (eds.), *New Essays on the A Priori*, 367–83.

Gaukroger, S., *Descartes: An Intellectual Biography* (Oxford: Clarendon/Oxford University Press, 1995).

Gay, P., *The Enlightenment* (2 vols.; New York: Knopf, 1966–69).

Gendler, T., and Hawthorne, J. (eds.), *Conceivability and Possibility* (New York: Oxford University Press, 2002).

Ginsborg, "Kant on Understanding Organisms as Natural Purposes," in Watkins (ed.), *Kant and the Sciences*, 231–58.

—— "Two Kinds of Mechanical Inexplicability in Kant and Aristotle," *Journal of the History of Philosophy* 42 (2004): 33–65.

Ginzburg, C., *The Cheese and the Worms: The Cosmos of a Sixteenth Century Miller*, trans. J. and A. Tedaschi (Harmondsworth, Middlesex: Penguin 1982).

Godfrey-Smith, P., *Complexity and the Function of Mind in Nature* (Cambridge: Cambridge University Press, 1996).

Grene, M., and Depew, D., *The Philosophy of Biology* (Cambridge: Cambridge University Press, 2004).

Grice, H. P., "The Causal Theory of Perception," in P. Grice, *Studies in the Way of Words* (Cambridge, MA: Harvard University Press, 1989), 224–47.

Griffin, D. R., *Animal Minds* (2nd edn., Chicago, IL: University of Chicago Press, 2001).

Grimm, P., "Kant's Argument for Radical Evil," *European Journal of Philosophy* 10 (2002): 160–77.

Gunther, Y. (ed.), *Essays on Nonconceptual Content* (Cambridge, MA: MIT Press, 2003).

Guyer, P. (ed), *The Cambridge Companion to Kant* (Cambridge: Cambridge University Press, 1992).

—— *Kant and the Claims of Knowledge* (Cambridge: Cambridge University Press, 1987).

—— *Kant and the Experience of Freedom* (Cambridge: Cambridge University Press, 1993).

—— "Kant's Conception of Empirical Law," *Proceedings of the Aristotelian Society*, Supp. Vol. 63 (1990): 221–42. Also in Guyer, *Kant's System of Nature and Freedom*, ch. 2.

—— "Kant on Common Sense and Skepticism," *Kantian Review* 7 (2003): 1–37.

—— *Kant's System of Nature and Freedom* (Oxford: Oxford University Press, 2005).

Hacker, P., *Appearance and Reality* (Oxford: Blackwell, 1987).

—— *Wittgenstein's Place in Twentieth-Century Analytic Philosophy* (Oxford: Blackwell, 1996).

Hacking, I., *The Taming of Chance* (Cambridge: Cambridge University Press, 1990).

—— "What is Logic?," *Journal of Philosophy* 86 (1979): 285–319.

Haken, H., *Principles of Brain Functioning: A Synergetic Approach to Brain Activity, Behavior, and Cognition* (Berlin: Springer, 1996).

Hale, B., *Abstract Objects* (Oxford: Blackwell, 1987).

Hale, B., and Wright, C. (eds.), *A Companion to the Philosophy of Language* (Oxford: Blackwell, 1997).

Hall, B., "A Reconstruction of Kant's Ether Deduction in *Übergang* 11," *British Journal of the History of Philosophy* (forthcoming).

―― "Understanding *Convolut* 10 of Kant's *Opus Postumum*," *Proceedings of the 10th International Kant Congress* (forthcoming).

Hanna, R., "Direct Reference, Direct Perception, and the Cognitive Theory of Demonstratives," *Pacific Philosophical Quarterly* 74 (1993): 96–117.

―― *Kant and the Foundations of Analytic Philosophy* (Oxford: Clarendon/Oxford University Press, 2001).

―― "Kant, Causation, and Freedom: A Critical Notice of *Kant and the Metaphysics of Causality*," *Canadian Journal of Philosophy*, 36 (2006):281–306.

―― "Kant in Twentieth-Century Philosophy," in D. Moran (ed.), *Routledge Companion to Twentieth-Century Philosophy* (London: Routledge, forthcoming).

―― "Kant, Wittgenstein, and the Fate of Analysis," M. Beaney (ed.), *The Analytic Turn* (London: Routledge, forthcoming).

―― "Kant's Theory of Judgment," *The Stanford Encyclopedia of Philosophy (Fall 2004 Edition)*, Edward N. Zalta (ed.), URL=<http://plato.stanford.edu/archives/fall2004/entries/kant-judgment/>.

―― *Rationality and Logic* (Cambridge, MA: MIT, 2006).

―― "The Trouble with Truth in Kant's Theory of Meaning," *History of Philosophy Quarterly* 10 (1993): 1–20.

Hanna, R., and Moore, A. W., "Reason, Freedom, and Kant: An Exchange," *Kantian Review* (forthcoming).

Hanna, R., and Thompson, E., "The Mind–Body–Body Problem," *Theoria et Historia Scientiarum* 7 (2003): 24–44.

―― "Neurophenomenology and the Spontaneity of Consciousness," in E. Thompson (ed.), *The Problem of Consciousness* (Calgary, AL: University of Alberta Press, 2005), 133–62.

Harper, W., "Kant on the A Priori and Material Necessity," in Butts (ed.), *Kant's Philosophy of Physical Science*, 239–72.

Hatfield, G., *The Natural and the Normative* (Cambridge, MA: MIT Press, 1990).

Hawking, S., *A Brief History of Time* (New York: Bantam, 1988).

Hawkins, J., and Allen, R. (eds.), *Oxford Encyclopedic English Dictionary* (Oxford: Oxford University Press, 1991).

Hazen, A. P., "Logic and Analyticity," in A. C. Varzi (ed.), *The Nature of Logic* (Stanford, CA: CSLI, 1999), 79–110.

Heal, J., "Indexical Predicates and Their Uses," *Mind* 106 (1997): 619–40.

Heck, R., "Nonconceptual Content and the 'Space of Reasons'," *Philosophical Review* 109 (2000): 483–523.

Hill, T., *Human Welfare and Moral Worth* (Oxford: Clarendon/Oxford University Press, 2002).

Hintikka, J., *Logic, and Language-Games, and Information: Kantian Themes in the Philosophy of Logic* (Oxford: Clarendon/Oxford University Press, 1973).

Hirst, R., "Primary and Secondary Qualities," in Edwards (ed.), *Encyclopedia of Philosophy*, vol. 6, 455–7.

Holden, T., *The Architecture of Matter: Galileo to Kant* (Oxford: Clarendon/Oxford University Press, 2004).

Hudson, H., *Kant's Compatibilism* (Ithaca, NY: Cornell University Press, 1990).

Hughes, G. E., and Cresswell, M., *An Introduction to Modal Logic* (London: Methuen, 1968).

Humberstone, L., "Intrinsic/Extrinsic," *Synthese* 108 (1996): 205–67.

Hume, D., *Enquiry Concerning Human Understanding* (Indianapolis, IN: Hackett, 1977).

—— *Treatise of Human Nature* (Oxford: Clarendon/Oxford University Press, 1978).

Hunter, G., *Metalogic* (Berkeley, CA: University of California Press, 1971).

Hurley, S., *Consciousness in Action* (Cambridge, MA: Harvard University Press, 1998).

Hurley, S., and Noë, A., "Neural Plasticity and Consiousness," *Biology and Philosophy* 18 (2003): 131–68.

Husserl, E., *The Crisis of European Sciences and Transcendental Phenomenology*, trans. D. Carr (Evanston, IL: Northwestern University Press, 1970).

—— *Logical Investigations*, trans. J.N. Findlay (2 vols.; London: Routledge and Kegan Paul, 1970).

Hylton, P., "The Nature of the Proposition and the Revolt against Idealism," in R. Rorty (ed.), *Philosophy and its History* (Cambridge: Cambridge University Press, 1984), 375–97.

Jackendoff, R., *Consciousness and the Computational Mind* (Cambridge: MIT Press, 1987).

—— "Unconscious Information in Language and Psychodynamics," in R. Jackendoff, *Languages of the Mind* (Cambridge, MA: MIT Press, 1992), 83–98.

Jackson, F., "Finding the Mind in the Natural World," in D. Chalmers (ed), *Philosophy of Mind: Classical and Contemporary Readings* (New York: Oxford University Press, 2002), 162–9.

—— *From Metaphysics to Ethics: A Defense of Conceptual Analysis* (Oxford: Oxford University Press, 1998).

James, W., *Principles of Psychology* (2 vols.; New York: Dover, 1950).
Johnsen, B., "How to Read 'Epistemology Naturalized'," *Journal of Philosophy* 102 (2005): 78–93.
Johnson-Laird, P., *Mental Models* (Cambridge, MA: Harvard University Press, 1983).
Johnston, M., *The Manifest* (unpublished MS), excerpts posted online at URL=<http://consc.net/online.html>.
—— "Manifest Kinds," *Journal of Philosophy* 94 (1997): 564–83.
Juarrero, A., *Dynamics in Action* (Cambridge, MA: MIT Press, 1999).
Kain, P., "Kant's Conception of Human Moral Status," (unpublished MS).
Kalderon, M. E., "The Transparency of Truth," *Mind* 106 (1977): 475–97.
Kane, R., *A Contemporary Introduction to Free Will* (Oxford: Oxford University Press, 2005).
Kaplan, D., "Afterthoughts," in Almog et al. (eds.), *Themes from Kaplan*, 565–614.
—— "Demonstratives: An Essay on the Semantics, Logic, Metaphysics, and Epistemology of Demonstratives and Other Indexicals," in Almog et al. (eds.), *Themes from Kaplan*, 481–563.
Kelly, S., "The Non-conceptual Content of Perceptual Experience: Situation Dependence and Fineness of Grain," *Philosophy and Phenomenological Research* 62 (2001): 601–8.
—— "What Makes Perceptual Content Non-conceptual?," *Electronic Journal of Analytical Philosophy* 8 (1998) URL=<http://ejap.louisiana.edu/EJAP/1998/kelly98.html>.
Kelso, J. S., *Dynamic Patterns* (Cambridge: MIT Press, 1995).
Kemp Smith, N., *Commentary to Kant's "Critique of Pure Reason"* (2nd edn., Atlantic Highlands, NJ: Humanities Press 1992).
Kim, J., "Epiphenomenal and Supervenient Causation," in Kim, *Supervenience and Mind*, 92–108.
—— "Mechanism, Purpose, and Explanatory Exclusion," in Kim, *Supervenience and Mind*, 237–64.
—— "Multiple Realization and the Metaphysics of Reduction," in Kim, *Supervenience and Mind*, 309–35.
—— *Supervenience and Mind* (Cambridge: Cambridge University Press, 1993).
Kitcher, Patricia, *Kant's Transcendental Psychology* (Oxford: Oxford University Press, 1990).
Kitcher, Philip, "A Priori Knowledge," *Philosophical Review* 89 (1980): 3–23.
—— "Apriority and Necessity," in Moser (ed.), *A Priori Knowledge*, 190–207.
—— "Kant and the Foundations of Mathematics," *Philosophical Review* 84 (1975): 23–50.

Kitcher, Philip, "Real Realism: The Galilean Strategy," *Philosophical Review* 110 (2001): 151–97.
Kleist, H., *Über das Marionettentheater* (Wiesbaden: Insel-Verlag, 1980).
Köhnke, K., *The Rise of Neo-Kantianism*, trans. R. J. Hollingdale (Cambridge: Cambridge University Press, 1991).
Kosslyn, S., *Image and Brain: The Resolution of the Imagery Debate* (Cambridge, MA: MIT Press, 1994).
Kreines, J., "The Inexplicability of Kant's *Naturzweck*: Kant on Teleology, Explanation, and Biology," *Archiv für Geschichte der Philosophie* 3 (forthcoming).
Kripke, S., "Identity and Necessity," in A. W. Moore (ed.), *Meaning and Reference* (Oxford: Oxford University Press, 1993), 162–91.
―― *Naming and Necessity* (2nd edn., Cambridge, MA: Harvard University Press, 1982).
―― "Semantical Considerations on Modal Logic," in L. Linsky (ed.), *Reference and Modality* (Oxford: Oxford University Press, 1971), 63–72.
―― *Wittgenstein on Rules and Private Language* (Cambridge, MA: Harvard University Press, 1982).
Kroon, F., and Nola, E., "Kant, Kripke, and Gold," *Kant-Studien* 78 (1987): 442–58.
Kuklick, B., *The Rise of American Philosophy* (New Haven, CT: Yale University Press, 1977).
Langford, C. H., "Moore's Notion of Analysis," in P. Schilpp (ed.), *The Philosophy of G. E. Moore* (New York: Tudor, 1952), 321–42.
Langton, R., *Kantian Humility* (Oxford: Oxford University Press, 1998).
Langton, R., and Lewis., D., "Defining 'Intrinsic'," *Philosophy and Phenomenological Research* 58 (1998): 333–45.
LaPorte, J., "Chemical Kind Term Reference and the Discovery of Essence," *Noûs* 30 (1996): 112–32.
―― "Living Water," *Mind* 107 (1998): 451–5.
Latour, B. *Science in Action* (Cambridge, MA: Harvard University Press, 1987).
Laywine, A., *Kant's Early Metaphysics and the Origins of the Critical Philosophy* (Atascadero, CA: Ridgeview, 1993).
Levine, J., "Materialism and Qualia: The Explanatory Gap," *Pacific Philosophical Quarterly* 64 (1983): 354–61.
Leibniz, G. W., "Meditations on Knowledge, Truth, and Ideas," in Leibniz, *G. W. Leibniz: Philosophical Essays*, trans. R. Ariew and D. Garber (Indiana: Hackett, 1989), 23–7.
Leplin, J. (ed.), *Scientific Realism* (Berkeley: University of California Press, 1984).

Lewis, C. I., *Mind and the World Order* (New York: Dover, 1956).
——— "A Pragmatic Conception of the A Priori," *Journal of Philosophy* 20 (1923): 169–77. Also in Moser (ed.), *A Priori Knowledge*, 15–25.
Lewis, D., "Are We Free to Break the Laws?," in Watson (ed.), *Free Will*, 122–9.
Locke, J., *Essay Concerning Human Understanding* (Oxford: Oxford University Press, 1975).
Longuenesse, B., *Kant and the Capacity to Judge* (Princeton, NJ: Princeton University Press, 1998).
Lucas, J. R., *The Freedom of the Will* (Oxford: Clarendon/Oxford University Press, 1970).
——— "Minds, Machines, and Gödel," *Philosophy* 36 (1961): 112–27.
Maddy, P., "Naturalism and the A Priori," in Boghossian and Peacocke (eds.), *New Essays on the A Priori*, 92–116.
Marcus, R., "Modalities and Intensional Languages," *Synthese* 13 (1961): 303–22.
Martin, M. G. F., "Perception, Concepts, and Memory," *Philosophical Review* 101 (1992): 745–63.
McDowell, J., "Having the World in View: Sellars, Kant, and Intentionality," *Journal of Philosophy* 95 (1998): 431–91.
——— "On the Sense and Reference of a Proper Name," *Mind* 86 (1977): 159–85.
——— *Mind and World* (Cambridge, MA: Harvard University Press, 1994).
——— "Two Sorts of Naturalism," in R. Hursthouse et al. (eds.), *Virtues and Reasons* (Oxford: Clarendon/Oxford University Press, 1995), 149–79.
McGinn, C., *Mental Content* (Oxford: Blackwell, 1989).
——— *The Subjective View* (Oxford: Oxford University Press, 1983).
Meerbote, R., "Kant on the Nondeterminate Character of Human Actions," in W. Harper and R. Meerbote (eds.), *Kant on Causality, Freedom, and Objectivity* (Minneapolis, MN: University of Minnesota Press, 1984), 138–63.
——— "Kant's Refutation of Problematic Material Idealism," in B. den Ouden (ed.), *New Essays on Kant* (New York: Peter Lang, 1987), 112–38.
Mellor, D. H., "Natural Kinds," *British Journal for the History of Philosophy of Science* 28 (1977): 299–312.
Melnick, A., *Kant's Analogies of Experience* (Chicago, IL: University of Chicago Press, 1973).
Merleau-Ponty, M., *Phenomenology of Perception*, trans. C. Smith (London: Routledge and Kegan Paul, 1962).
Monk, R., *Bertrand Russell* (London: Jonathan Cape, 1996).
Montague, R., "Logical Necessity, Physical Necessity, Ethics, and Quantifiers," *Inquiry* 4 (1960): 259–269.
Moore, A. W. (ed.), *Meaning and Reference* (Oxford: Oxford University Press, 1993).

Moore, A. W. (ed.), *Noble in Reason, Infinite in Faculty: Themes and Variations in Kant's Moral and Religious Philosophy* (London: Routledge, 2003).

Moore, G. E., "Analysis," in Schilpp (ed.), *The Philosophy of G. E. Moore*, 660–7.

——— "Certainty," in Moore, *G. E. Moore: Selected Writings*, 171–96.

——— "A Defense of Common Sense," in Moore, *G. E. Moore: Selected Writings*, 106–33.

——— *G. E. Moore: Selected Writings* (London: Routledge, 1993).

——— "The Nature of Judgement," in Moore, *G. E. Moore: Selected Writings*, 1–19.

——— "Proof of an External World," in Moore, *G. E. Moore: Selected Writings*, 147–70.

——— "The Refutation of Idealism," in Moore, *G. E. Moore: Selected Writings*, 23–44.

——— "Truth and Falsity," in Moore, *G. E. Moore: Selected Writings*, 20–22.

Moser, P. K., (ed.), *A Priori Knowledge* (Oxford: Oxford University Press, 1987).

Musil, R., *The Man without Qualities*, trans. E. Wilkins and E. Kaiser (London: Martin Secker and Warburg, 1954).

Nagel, T., "Brain Bisection and the Unity of Consciousness," in Nagel, *Mortal Questions*, 147–64.

——— *Mortal Questions* (Cambridge: Cambridge University Press, 1979).

——— *The View from Nowhere* (New York: Oxford University Press, 1986).

——— "What is it Like to be a Bat?," in Nagel, *Mortal Questions*, 165–80.

Neiman, S., *The Unity of Reason* (New York: Oxford University Press, 1994).

Nelkin, D., "Two Standpoints and the Belief in Freedom," *Journal of Philosophy* 97 (2000): 564–76.

Nicolis, G., and Prigogine, I., *Self-Organization in Nonequilibrium Systems* (New York: Wiley, 1977).

Nietzsche, F., *Beyond Good and Evil*, trans. W. Kaufmann (New York: Vintage, 1966).

——— *The Portable Nietzsche*, trans. W. Kaufmann (Harmondsworth, MIDDX: Penguin, 1976).

Olson, E., *The Human Animal* (Oxford: Oxford University Press, 1977).

O'Neill (Nell), O., *Acting on Principle* (New York: Columbia University Press, 1975).

——— *Constructions of Reason* (Cambridge: Cambridge University Press, 1989).

——— "Reason and Autonomy in *Grundlegung* III," in O'Neil, *Constructions of Reason*, 51–65.

——— "Vindicating Reason," in Guyer (ed.), *The Cambridge Companion to Kant*, 280–308.

O'Shaughnessy, B., *The Will* (2 vols; Cambridge: Cambridge University Press, 1980).

Pap, A., *Semantics and Necessary Truth* (New Haven, CT: Yale University Press, 1958).
Papineau, D., *Philosophical Naturalism* (Oxford: Blackwell, 1993).
Parsons, C., "Arithmetic and the Categories," in Posy (ed.), *Kant's Philosophy of Mathematics*, 135–58.
Parsons, C., "Kant's Philosophy of Arithmetic," in C. Parsons, *Mathematics in Philosophy*. (New York: Cornell University Press, 1983), 110–149.
―― "Mathematical Intuition," *Proceedings of the Aristotelian Society*, 80 (1979–80), 145–68.
―― "Mathematics, Foundations of," in Edwards (ed.), *Encyclopedia of Philosophy*, vol. 5, 188–213.
Paton, H. J., *Kant's Metaphysic of Experience* (2 vols.; London: George Allen & Unwin, 1970).
Peacocke, C., "Does Perception Have a Nonconceptual Content?," *Journal of Philosophy* 98 (2001): 239–64.
―― "Nonconceptual Content Defended," *Philosophy and Phenomenological Research* 58 (1998): 381–8.
―― *A Study of Concepts* (Cambridge, MA: MIT Press, 1992).
Pereboom, D., "Kant on Transcendental Freedom," *Philosophy and Phenomenological Research* (forthcoming).
Perry, J., "Indexicals and Demonstratives," in Hale and Wright (eds.), *A Companion to the Philosophy of Language*, 586–612.
―― "The Problem of the Essential Indexical," *Noûs* 13 (1979): 3–21.
Pitcher, G. (ed.), *Truth* (Englewood Cliffs: Prentice Hall, 1964).
Plaass, P., *Kant's Theory of Natural Science*, trans. A. Miller and M. Miller (Dordrecht: Kluwer, 1994).
Port, R., and Van Gelder (eds.), *Mind as Motion: Explorations in the Dynamics of Cognition* (Cambridge, MA: MIT Press, 1995).
Posy, C., (ed.), *Kant's Philosophy of Mathematics* (Dordrecht: Kluwer, 1992).
Potter, M., *Reason's Nearest Kin* (Oxford: Oxford University Press, 2000).
Prauss, G., *Kant und das Problem der Dinge an Sich* (Bonn: Bouvier, 1974).
―― "Zum Wahreitsproblem bei Kant," *Kant-Studien* 60 (1969): 166–82.
Prigogine, I., *Being and Becoming: Time and Complexity in the Physical Sciences* (New York: W. H. Freeman, 1980).
Putnam, H., "Explanation and Reference," in Putnam, *Mind, Language, and Reality: Philosophical Papers, Volume 2*, 196–214.
Putnam, H., "Is Semantics Possible?," in Putnam, *Mind, Language, and Reality: Philosophical Papers, Volume 2*, 139–52.
―― "Is Water Necessarily H2O?," in J. B. Conant (ed.), *Realism with a Human Face* (Cambridge, Mass.: Harvard University Press, 1990) 54–79.

Putnam, H., *The Many Faces of Realism* (La Salle, IL: Open Court, 1987).
—— "Meaning and Reference," *Journal of Philosophy*, 70 (1973), 699–711.
—— "The Meaning of 'Meaning'," in Putnam, *Mind, Language, and Reality: Philosophical Papers, Volume 2*, 215–71.
—— *Mind, Language, and Reality: Philosophical Papers, Volume 2* (Cambridge: Cambridge University Press, 1975).
—— *Reason, Truth, and History* (Cambridge: Cambridge University Press, 1981).
Quine, W. V. O., *From a Logical Point of View* (2nd edn., New York: Harper and Rowe, 1963).
—— "Epistemology Naturalized," in Quine, *Ontological Relativity*, 69–90.
—— *From Stimulus to Science* (Cambridge: Harvard University Press, 1995).
—— "Natural Kinds," in *Ontological Relativity*, 114–38.
—— *Ontological Relativity* (New York: Columbia University Press, 1969).
—— "On What There Is," in Quine, *From a Logical Point of View*, 1–19.
—— "Reference and Modality," in Quine, *From a Logical Point of View*, 139–59.
—— "Two Dogmas of Empiricism," in Quine, *From a Logical Point of View*, 20–46.
—— "Truth by Convention,' in Quine, *The Ways of Paradox*, 77–106.
—— *The Ways of Paradox* (2nd edn., Cambridge, Mass.: Harvard University Press, 1976).
—— *Word and Object* (Cambridge, Mass.: MIT Press, 1960).
Ramsey, F., "Facts and Propositions," in Pitcher (ed.), *Truth*, 16–17.
Reichenbach, H., *The Theory of Relativity and A Priori Knowledge*, trans. M. Reichenbach (Berkeley: University of California Press, 1965).
Reid, T., *Essays on the Intellectual Powers of Man* (Cambridge, MA: MIT Press, 1969).
Rescher, N., *Kant and the Reach of Reason* (Cambridge: Cambridge University Press, 2000).
Richardson, A., *Carnap's Construction of the World* (Cambridge: Cambridge University Press, 1998).
Rockwell, T., *Neither Brain Nor Ghost: A Nondualist Alternative to the Mind–Brain Identity Theory* (Cambridge, Mass.: MIT Press, 2005).
Rorty, R., *Philosophy and the Mirror of Nature* (Princeton, NJ: Princeton University Press, 1979)
Rosenberg, G., *A Place for Consciousness* (New York: Oxford University Press, 2004).
Rosenthal, D., "Two Concepts of Consciousness," *Philosophical Studies* 94 (1986): 329–59.
Ross, W. D., *The Right and the Good* (Oxford: Oxford University Press, 1930).
Royce, J., *The Letters of Josiah Royce* (Chicago, IL: University of Chicago Press, 1970).

Russell, B., *An Essay on the Foundations of Geometry* (Cambridge: Cambridge University Press, 1897).
—— "Knowledge by Acquaintance and Knowledge by Description," in B. Russell, *Mysticism and Logic* (Totowa, NJ: Barnes and Noble, 1976), 152–67.
—— *Logic and Knowledge* (London: Unwin and Hyman, 1956).
—— "Mathematical Logic as Based on the Theory of Types," in Russell, *Logic and Knowledge*, 59–102.
—— "The Philosophy of Logical Atomism," in Russell, *Logic and Knowledge*, 177–281.
—— *Principles of Mathematics* (2nd edn, New york: W. W. Norton, 1996).
—— *The Problems of Philosophy* (Oxford: Oxford University Press, 1967).
Rynasiewicz, R., "Absolute versus Relational Space-Time: An Outmoded Debate?," *Journal of Philosophy* 93 (1996): 279–306.
Salmon, N., "How *Not* to Derive Essentialism from the Theory of Reference, *Journal of Philosophy* 76 (1979): 703–25.
—— *Reference and Essence* (Princeton, NJ: Princeton University Press, 1981).
Sanford, D., "Determinates vs. Determinables," *Stanford Encyclopedia of Philosophy* (Summer 2002 Edition), E. Zalta (ed.), URL=<http://plato.stanford.edu/archives/sum2002/entries/determinate-determinables/>.
Sartre, J. P., *Being and Nothingness*, trans. H. Barnes (New York: Washington Square, 1966).
—— *The Transcendence of the Ego: An Existentialist Theory of Consciousness*, trans. F. Williams and R. Kirkpatrick (New York: Farrar, Strauss, and Geroux, 1987).
Savitt, S., "Introduction," in S. Savitt (ed.), *Time's Arrows Today* (Cambridge: Cambridge University Press, 1995), 1–19.
Schacter, D. L., "Perceptual Representation Systems and Implicit Memory: Towards a Resolution of the Multiple Memory Systems Debate," *Annals of the New York Academy of Sciences* 608 (1990): 543–71.
Schilpp, P. (ed.), *The Philosophy of G. E. Moore* (New York: Tudor, 1952).
Schönfeld, M., *The Philosophy of the Young Kant: The Precritical Project* (New York: Oxford University Press, 2000).
Schorske, C., *Fin de Siècle Vienna* (New York: Knopf, 1979).
Schwyzer, H., "Subjectivity in Descartes and Kant," *Philosophical Quarterly* 47 (1997): 342–57.
Searle, J., *Rediscovery of the Mind* (Cambridge, MA: MIT Press, 1992).
Sedgwick, S., "McDowell's Hegelianism," *European Journal of Philosophy* 5 (1997): 21–38.
Sedivy, S., "Must Conceptually Informed Perceptual Experience Involve Nonconceptual Content?," *Canadian Journal of Philosophy* 26 (1996): 413–31.

Sellars, W., "Empiricism and the Philosophy of Mind," in Sellars, *Science, Perception, and Reality*, 127–96.
―― *Kant and Pre-Kantian Themes* (Atascadero, CA: Ridgeview, 2003).
―― *Kant's Transcendental Metaphysics* (Atascadero, CA: Ridgeview, 2003).
―― "Philosophy and the Scientific Image of Man," in Sellars, *Science, Perception and Reality*, 1–40.
―― *Science and Metaphysics: Variations on Kantian Themes* (London: Routledge and Kegan Paul, 1968).
―― *Science, Perception and Reality* (New York: Humanities Press, 1963).
―― "Some Remarks on Kant's Theory of Experience," *Journal of Philosophy* 64 (1967): 633–647.
Sensen, O., '"No Causality without Freedom': Kant's Argument for Freedom in the 'Analogies of Experience'," (unpublished MS).
Shabel, L. "Kant on the 'Symbolic Construction' of Mathematical Concepts," *Studies in the History and Philosophy of Science* 29 (1998): 589–621.
Shapiro, S., "Induction and Indefinite Extensibility: The Gödel Sentence is True, But Did Someone Change the Subject?," *Mind* 107 (1998): 597–624.
―― *Philosophy of Mathematics: Structure and Ontology* (New York: Oxford University Press, 1997).
―― *Thinking about Mathematics* (Oxford: Oxford University Press, 2000).
Shepard, R., "The Mental Image," *American Psychologist* 33 (1978): 125–37.
Shephard, R., and Chipman, S., "Second Order Isomorphisms of Internal Representations: Shapes of States," *Cognitive Psychology* 1 (1970): 1–17.
Shephard, R., and Cooper, L., *Mental Images and their Transformations* (Cambridge, MA: MIT Press, 1982).
Shephard, R., and Metzler, J. "Mental Rotation of Three-Dimensional Objects," *Science* 171 (1971): 701–3.
Shoemaker, S., "Self-Knowledge and 'Inner Sense'." *Philosophy and Phenomenological Research*, 54 (1994), 249–314.
―― "Self-Reference and Self-Awareness," *Journal of Philosophy*, 65 (1968), 555–67.
Sidelle, A., *Necessity, Essence, and Individuation* (Ithaca, NY: Cornell University Press, 1989).
Sidelle, A., "Rigidity, Ontology, and Semantic Structure," *Journal of Philosophy* 89 (1992), 410–30.
Silberstein, M., and McGeever, J., "The Search for Ontological Emergence," *Philosophical Quarterly* 49 (1999): 182–200.
Skolem, T., "The foundations of elementary arithmetic established by means of the recursive mode of thought, without the use of apparent variables ranging

over infinite domains," in J. v. Heijenoort (ed.), *From Frege to Gödel* (Cambridge, MA: Harvard University Press), 302–33.

Smart, J. J. C., *Philosophy and Scientific Realism* (London: Routledge & Kegan Paul, 1963).

Smith, E., "Concepts and Thought," in R. Sternberg and E. Smith (eds.), *Psychology of Human Thought* (Cambridge: Cambridge University Press, 1988), 19–49.

Smith, E., and Medin, D., *Categories and Concepts* (Cambridge, MA: Harvard University Press, 1981).

Smith, N. K., *Commentary to Kant's 'Critique of Pure Reason'* (Atlantic Highlands, NJ: Humanities Press, 1992).

Soames, S., *Philosophical Analysis in the Twentieth Century* (2 vols.; Princeton, NJ: Princeton University Press, 2003).

Speaks, J., "Is There a Problem about Nonconceptual Content?," *Philosophical Review* (forthcoming).

Stalnaker, R., "What Might Nonconceptual Content Be?," in E. Villanueva (ed.), *Concepts* (Atascadero, CA: Ridgeview, 1998), 339–52.

Stanley, J., "Names and Rigid Designation," in Hale and Wright (eds.), *A Companion to the Philosophy of Language*, 555–85.

Stapp, H., *Mind, Matter, and Quantum Mechanics* (Munich: Springer, 1993).

Stern, R., (ed), *Transcendental Arguments: Problems and Prospects* (Oxford: Oxford University Press, 1999).

―― *Transcendental Arguments and Skepticism* (Oxford: Oxford University Press, 2000).

Sternberg, R. J., *Cognitive Psychology* (Fort Worth, TX: Harcourt Brace, 1996).

Stoljar, D., "Physicalism and the Necessary a Posteriori," *Journal of Philosophy* 97 (2000): 33–54.

Strawson, P. F., *The Bounds of Sense: An Essay on Kant's Critique of Pure Reason* (London: Methuen, 1966).

―― "Imagination and Perception," in R. Walker (ed.), *Kanton Pure Reason* (Oxford: Oxford University Press, 1982).

―― *Individuals* (London: Methuen, 1959).

―― *Introduction to Logical Theory* (London: Methuen, 1952).

Strawson, P. F., "Kant on Substance," in P. F. Strawson, *Entity and Identity* (Oxford: Clarendon/Oxford University Press, 1997), 268–79.

―― "On Referring," *Mind* 59 (1950): 320–44.

Stroud, B., "The Charm of Naturalism," *Proceedings and Addresses of the American Philosophical Association*, 70 (1996), 43–55.

―― "Kant and Skepticism," in M. Burnyeat (ed.), *The Skeptical Tradition*, (Berkeley, CA: University of California Press, 1983), 413–34.

Stroud, B., "Kantian Argument, Conceptual Capacities, and Invulnerability," in P. Parini (ed.), *Kant and Contemporary Epistemology*, (The Netherlands: Kluwer, 1994), 151–231.

―― *The Significance of Philosophical Skepticism* (Oxford: Oxford University Press, 1984).

Struik, D. J., *A Concise History of Mathematics*. New York: Dover, 1967).

Sturm, T. "Kant on Empirical Psychology: How Not to Investigate the Human Mind," in Watkins (ed.), *Kant and the Sciences*, 163–84.

―― "A Matter of Character: Hume and Kant on the Causation and Empirical Explanation of Actions," (unpublished MS).

Tarski, A., "The Concept of Truth in Formalized Languages," in A. Tarski, *Logic, Semantics, and Metamathematics* (Oxford: Clarendon, 1956), 152–278.

―― "The Semantic Conception of Truth and the Foundations of Semantics," *Philosophy and Phenomenological Research* 5 (1943–1944): 341–75.

Tennant, N., "On the Necessary Existence of Numbers," *Noûs* (1997): 31: 307–36.

Thelen, E., and Smith, L., *A Dynamic Systems Approach to the Development of Cognition and Action* (Cambridge, MA: MIT Press, 1994).

Thompson, E., *Mind in Life* (Cambridge, MA: Harvard University Press, forthcoming).

Thompson, E., and Varela, F., "Radical Embodiment: Neural Dynamics and Consciousness," *Trends in Cognitive Sciences* 5 (2001): 418–25.

Tidman, P., "The Justification of A Priori Intuitions," *Philosophy and Phenomenological Research* 56 (1990): 543–71.

Tooley, M., "Causation: Reductionism vs. Realism," in E. Sosa and M. Tooley (eds.), *Causation* (Oxford: Oxford University Press, 1993), 172–92.

Toulmin, S., and Janik, A., *Wittgenstein's Vienna* (New York: Simon and Schuster, 1973).

Trevarthyn, C., "Analysis of Cerebral Activities that Generate and Regulate Consciousness in Commissurotomy Patients," in S. Diamond et al(eds.), *Hemishphere Function in the Human Brain* (London: Elek Science, 1974), 235–61.

―― and R. W. Sperry, "Perceptual Unity of the Ambient Visual Field in Human Commissurotomy Patients," *Brain* 96 (1973): 547–70

Troelstra, A. S., and Dalen, D. V., *Constructivism in Mathematics: An Introduction*, vol. 1 (Amsterdam: North Holland, 1988).

Tye, M., *Consciousness, Color, and Content* (Cambridge, MA: MIT Press, 2000).

―― *Ten Problems of Consciousness* (Cambridge: MIT Press, 1995).

Vaihinger, H., *Commentar zu Kants Kritik der reinen Vernuft* (2 vols., Stuttgart: W. Spemann, 1881–92).

Van Cleve, J., *Problems from Kant* (New York: Oxford University Press, 1999).

Van Fraassen, B., *The Scientific Image* (Oxford: Clarendon/Oxford University Press, 1980).
van Inwagen, P., *An Essay on Free Will* (Oxford: Clarendon/Oxford University Press, 1983).
—— "Free Will Remains a Mystery," in R. Kane (ed.)., *The Oxford Handbook of Free Will* (Oxford: Oxford University Press, 2002), 158–77.
Varela, F., *Principles of Biological Autonomy* (New York: Elsevier/North-Holland, 1979).
—— Thompson, E., and Rosch, E.,, *The Embodied Mind* (Cambridge, MA: MIT Press, 1996).
Vogel, J., "The Problem of Self-Knowledge in Kant's 'Refutation of Idealism': Two Recent Views," *Philosophy and Phenomenological Research* 53 (1993): 875–87.
Walker, R., "Kant's Conception of Empirical Law," *Proceedings of the Aristotelian Society*, Supp. Vol. 63 (1990): 243–58.
Warren, D., "Kant's Dynamics," in Watkins (ed.), *Kant and the Sciences*, 93–116.
—— *Reality and Impenetrability in Kant's Philosophy of Nature* (New York: Routledge, 2001).
Watkins, E. (ed.), *Kant and the Sciences* (New York: Oxford University Press, 2001).
—— *Kant and the Metaphysics of Causality* (Cambridge: Cambridge University Press, 2005).
—— "Kant's Justification of the Laws of Mechanics," in Watkins (ed.), *Kant and the Sciences*, 136–59.
Watson, G., (ed)., *Free Will* (2nd edn., Oxford: Oxford University Press, 2003).
Weber, A., and Varela, F., "Life After Kant: Natural Purposes and the Autopoietic Foundations of Biological Individuality," *Phenomenology and the Cognitive Sciences*, 1 (2002): 97–125.
Weiskrantz, L., *Blindsight* (Oxford: Oxford University Press, 1986).
Westphal, K., *Kant's Transcendental Proof of Realism* (Cambridge: Cambridge University Press, 2005).
Whyte, "Boscovich, Roger Joseph," in Edwards (ed.), *Encyclopedia of Philosophy*, vol. 1, 350–2.
Wiggins, D., "Putnam's Doctrine of Natural Kind Words and Frege's Doctrines of Sense, Reference, and Extension: Can They Cohere?," in Moore (ed.), *Meaning and Reference*, 197–207.
Williams, B., *Descartes: The Project of Pure Inquiry* (Atlantic Highlands: Humanities Press, 1978), 65–8.
—— "Internal and External Reasons," in B. Williams, *Moral Luck* (Cambridge: Cambridge University Press, 1981), 101–13.

Williams, B., *Truth and Truthfulness* (Princeton, NJ: Priceton University Press, 2002).

Williamson, T., "Is Knowing a State of Mind?," *Mind* 104 (1995): 533–65.

Wilson, C., *The Invisible World: Early Modern Philosophy and the Invention of the Microscope* (Princeton, NJ: Princeton University Press, 1995).

Wilson, M., "The 'Phenomenalisms' of Kant and Berkeley," in A. Wood (ed.), *Self and Nature in Kant's Philosophy*, 157–73.

Wisdom, J., *Problems of Mind and Matter* (Cambridge: Cambridge University Press, 1963).

Wittgenstein, L., *Culture and Value*, trans. P. Winch (Chicago, IL: University of Chicago Press, 1980).

—— *On Certainty*, trans. D. Paul and G. E. M. Anscombe (New York: Harper and Row, 1969).

—— *Philosophical Investigations*, trans. G. E. M. Anscombe (New York: Macmillan, 1953).

—— *Remarks on the Foundations of Mathematics*, trans. G. E. M. Abscombe (2nd edn., Cambridge: MIT Press, 1983).

—— *Tractatus Logico-Philosophicus*, trans. C. K. Ogden (London: Routledge and Kegan Paul, 1981).

Wood, A., "Kant's Compatibilism," in Wood (ed.), *Self and Nature in Kant's Philosophy*, 73–101.

—— *Kant's Ethical Thought* (Cambridge: Cambridge University Press, 1999).

—— (ed.), *Self and Nature in Kant's Philosophy* (Ithaca, NY: Cornell University Press, 1984).

Wright, C., *Frege's Conception of Numbers as Objects* (Aberdeen: Aberdeen University Press, 1983).

Yablo, S., "Is Conceivability a Guide to Possibility?," *Philosophy and Phenomenological Research* 53 (1993): 1–42.

Young, J. M. "Construction, Schematism, and Imagination," in Posy (ed.), *Kant's Philosophy of Mathematics*, 159–75.

—— "Kant on the Construction of Arithmetical Concepts," *Kant-Studien* 73 (1982): 17–46.

Zemach, E., "Putnam's Theory on the Reference of Substance Terms," *Journal of Philosophy* 73 (1976): 116–27.

Index

Abela, P. 89
absolute conception of the world 245
accordance (*Übereinstimmung*), *see*
 correspondence (*Übereinstimmung*)
 theory of truth, Kant's
acroamatic proofs 380
action-at-a-distance 235
aether 46, 55n37, 149, 179, 189n71,
 235–6, 394–5, 449, 450
affection 20, 136–8, 165
affinity of laws of nature 447
Allison, H. 301
Alterations (*Veränderungen*) 396, 404, 412,
 see also events, Kant's theory of
Ameriks, K. ix, 14n34
Analogies of Experience, the 182, 410–11,
 414–15, 428, 430–1, 440, 450
 first Analogy 54–5, 62, 390–7, 403
 second Analogy 395–405, 407–9,
 429n34
 third Analogy 393, 404–9
analyticity, *see* propositions, analytic
analytic-synthetic distinction 28, 236n58,
 304, 312, 314–16, 331, 342
analytic phenomenalism, *see* constructive
 empiricism
analytic tradition 3–5, 8, 10, 16–17, 250,
 341–3
animalism 75n66
Anthropic Principle 138n85
anthropocentrism, Kant's 7, 20, 22–3,
 49–50, 98, 138n85, 189, 236,
 240, 251, 269, 271, 288n4, 305,
 330, 339
anti-microphysicalism, Kant's 160–86,
 220–46
Antinomy of Pure Reason, third 411–16,
 420–1
antinomy, definition 410
Antinomy of Essentialism 177–81
anti-realism 33, 200n13, 224, 251, 270–1,
 344, 387
anti-supernaturalism 10–11
appearances, Kant's theory of 1, 1n2,
 14–15, 19, 21–3, 41–2, 48, 102,
 112, 118, 120–4, 163–4, 169–70,
 182–3
 authentic or objective
 (*Erscheinungen*) 199–200, 210,
 227–8, 230–3, 236–7, 242, 245, 280
 mere or subjective (*Schein*) 33, 96, 200,
 225
apperception 21, 104–5, 245n65
 empirical 52–3, 55, 58–62, 76, 104
 original synthetic unity of 53, 123–4
 pure or transcendental 53, 104, 234
a priori –a posteriori distinction 3
a priori judgments
 analytic, *see* propositions, analytic
 impure or "empirically infected" 176–7
 pure 176
 synthetic, *see* propositions, synthetic a
 priori
a priori knowledge, *see* knowledge, Kant's
 theory of
a priori science 1–2, 146
a priori truth viii, 3, *see also* propositions,
 analytic; propositions, synthetic a
 priori
apriority
 epistemic 144, 352–4
 and necessity 151, 171, 177, 191, 193n8
 semantic 144, 352n19
Aristotle 15, 154
arithmetic, Kant's theory of 312–30
Arnold, J. 240
aspect blindness 105–6, 108
atomless gunk 179
atoms 46, 148–9
automaton materiale 429, 431
automaton spirituale 429, 431–2
autonomy 27–8, 419, 424, 441–4

Basic Laws of Arithmetic (Frege) 266, 323
beauty, aesthetic experience of 109–110
Being and Nothingness (Sartre) 345
belief (*Glaube*) 39, 320–1, 358–60
Benacerraf, P. 323
Berkeley, G. 9, 51, 136, 224–5, 227,
 388–90, 403

Bermúdez, J. 85
Big Bang, the 181n61, 300, 438–9
biology 153n30, 433–7, 439–40
bivalence, law or principle of 259n13– 260
blind intuitions 97–9
blindsight 99
Blomberg Logic (Kant) 351
body (*Körper*), concept of 74–9, 216–19, see also Bohr-Rutherford conception of matter; embodied minds; matter
Bohr–Rutherford conception of matter 150, 167, 171, 235–6, 245
Boscovich, R. 148, 232
Bounds of Sense (Strawson) 90, 115
Boyle, R. 9, 154n31, 223–5, 227
Brentano. F. 6n18, 85, 95n35, 97
Brouwer, L. E. J. 287, 334–5
Buffalo Trace Kentucky Straight Bourbon Whiskey 26
Burge, T. 38, 39n5, 40, 348

Candide (Voltaire) 15
Cantorian transfinite numbers 197
capacities, mental (*Fähigkeiten*), see faculties or powers (*Vermögen*), mental
Carnap, R. 4, 6, 10, 344
Categorical Imperative, the 25–7, 427, 443–4
causal laws, Kant's theory of 181–90, 444–50
  Armstrong-Dretske-Tooley (ADT) conception of 189
  Humean conception of 189
causal theory of perception, the 138
causation, Kant's theory of 387–416, 433–51
certainty 358, 360
Chalmers, D. 176n57, 185n65, 193n8, 223n49
characteristics (*Merkmale*) 209–10, 369–71
Chisholm, R. 16
Churchland, Paul. 223n49
clarity vs. obscurity 94, 366–9
classification problem about non-conceptual content, the 91, see also non-conceptual content, Kant's theory of
cognition (*Erkenntnis*) 17–18, 96–100, see also knowledge, Kant's theory of

cognitive-semantic interpretation of Kant 3, 115, 303
cognitive world atlas 269
coincidences 388–9, see also causation, Kant's theory of
Commentar zu Kants Kritik der reinen Vernunft (Vaihinger) 301
commissurotomy 129, 132
community, dynamic, see Analogies of Experience, third Analogy
compatibilism 401, 418–21, 431
compositional plasticity 219
Compton-Burnett, I. 1, 14
conceivability vs. possibility 185n67, 318n28
concept BACHELOR, the 113, 173–4
concept possession 112–14
concept problem about non-conceptual content, the 88, 91–2, see also non-conceptual content, Kant's theory of
concept RIGHT, the 113, 376
conceptualism 87–91, see also non-conceptual content, Kant's theory of
concepts (*Begriffe*) 91–2, 112–14, 203–20, 369–79
conceptual analysis and natural science 202–3, 244–5
consciousness 52–3, 363–7, 399–402, 409
construction of concepts 331–9, see also concepts; mathematical knowledge
constructive empiricism 6
content, mental 68–79, 81–132, 203–20, 347–8
contextualism in epistemology 38
continuum, extensive vs. intensive 327
contraries vs. contradictories 411
conventionalism 350
convention-T 272, 276
conviction (*Überzeugung*) 346–8, 358–62, see also knowledge, Kant's theory of
Copernican Revolution, Kant's 34, 49
Corman, R. 241
corpuscularianism 148, 154 n31, 163, 232
correspondence (*Übereinstimmung*) theory of truth, Kant's 252–3, 262–72, 283, 303, see also truth
country house example 108
Crane, T. 85
Crisis of European Sciences (Husserl) 8

INDEX 477

Davidson, D. 249, 254–5, 276–7
deferred ostensive reference 137
definitions 210–11, 260–2, 312, 370
de re senses 89
Descartes, R. 9, 37–40, 50–1, 65, 80, 94, 134, 269, 276
Descriptivism 89
desire (*Begierde*) 19, 23, 26–8, 400–1, 441–4
determinable-determinate 204, 370
determinations (*Bestimmungen*) 195, 210, 218, 230–1, 233, 238, *see also* properties, Kant's theory of
determine (*bestimmen*) 54, 118
determinism 8n32, 17, 401, 416–21, 431, 437–8, 440, 445–6, *see also* natural mechanism
direct perceptual realism, Kant's 44–50, 78, 132–9
"Directions in Space" (Kant) 71–3, 76, 79
directly referential terms 100–1, 213–7
disembodiment 79
disenchantment of nature 13, 277
distinctness 113, 134, 369
Dogma of Semantic Overdetermination 83, 90, 141
dogmata 380
Dohna-Wundlacken Logic (Kant) 277
dream problem, the 64–8
Dretske, F. 45, 85, 87
Dummett, M. 270–1, 283
dynamic emergence 437
dynamic field theory, Einstein's 236
Dynamical Systems Theory (DST) 437
dynamics, cognitive 378–9, 383–5

Eddington, A. 162, 235
Einstein, A. 10, 34, 167, 235–6
eliminativism
 about mental properties 223n49
 about things-in-themselves 15, 198, 245, *see also* things-in-themselves
 about truth 286n50
 modal 344–5
 scientific realist 12, 83
Embodied Agency Theory, Kant's 416–50
embodied minds 74, 75n67, 79, *see also* Embodied Agency Theory, Kant's
empirical realism thesis, the 29

empirical realism, Kant's theory of 68–80, 220–46, *see also* transcendental idealism, Kant's theory of
empty concepts 97–9
encapsulation, informational 106
Enlightenment, the 9, 9n24
Enquiry concerning Human Understanding (Hume) 389
epistemological scientism 11–12
epistemology, highly demanding 38
Escher, M. C. 70
Essay concerning Human Understanding (Locke) 160
essential indexicality 214
essentialism
 Kantian 147–51, 181–87, 203–42
 Lockean 160–1, 166
 scientific 143–7, 151–60
Evans, G. 81, 85, 87–90, 92, 115
events, Kant's theory of 388–409, *see also* time, Kant's theory of
exact sciences and human sciences 1, 242–6, 330, 339–40, 384
excluded middle, law or principle of 259
"exists" as a predicate 274n36
explanatory gap between the microphysical and macrophysical worlds, the 162–8

fact of reason, the (*Faktum der Vernunft*) 443
faculties or powers, mental (*Vermögen*) 17–18
feelings 103, 109
fictionalism 245
Fischer, K. 301
Fodor, J. 106, 108
folk science 156
forces, dynamic 45–48
form of representation (*Form der Abbildung, Form der Darstellung*) 266
Formula of Autonomy (FA) 25
Formula of Humanity as an End-in-Itself (FHE) 25
Formula of the Kingdom of Ends (FKE) 25
Formula of Universal Law (FUL) 18, 25, 441, 443
Foundations of Arithmetic (Frege) 3
Frankfurt, H. 52n34, 401, 428n32, 429n33, 431, 441n48
freedom of the will, Kant's theory of 23, 411–16, 419–50

478 INDEX

Frege, G. vii, 2n4, 3, 10, 11n29, 89n18, 263n14, 266–7, 213, 323–4, 339n53, 353
Frege–Russell logic 2n4
Frege–Russell semantics 89
Friedman, M. 4–5, 147n15, 321, 325n40, 369n56
FUBAR 50

Galileo 9
generality constraint, Evans's 92
genius, artistic 448
geography and its philosophical relevance for Kant 56–7, 268–70
God 11, 19, 31, 40, 49,, 146n12, 197, 238, 299, 309–10, 330, 347
Gödel, K. 174
Gödel's incompleteness theorems 274, 312–13, 339n54, 355–6
Guyer, P. ix, 60n44

Hacking, I. 287, 323n37, 328, 330, 339
Harries, K. 246n69
Harvard 6n19, 10, 287, 342
Hazen, A. 2n6, 313
Heck, R. 85, 108
Heisenberg-Schrödinger quantum mechanics 236
"Hesperus is Phosphorus" (HP) 175, 352
Hobbes, T. 9
holding-to-be-true (*Fürwahrhalten*) 247
holism 344
Hume, D. 9–10, 24, 28, 56, 62n48, 155n32, 189, 223–5, 227, 229, 387, 390, 395, 401, 431, 442, 446
Hume's principle 312
Hurley, S. 85
Husserl, E. 8, 85, 95n35, 97
hypothetical imperatives 26, 383–4, 442

"I think", *see* apperception
idealism 21–2, 41–4, 50–1, 168–70, 194–8, 296–7, 299–312, 342–4, 422, 425–7, *see also* idealism thesis, the; transcendental idealism, Kant's theory of
idealism thesis, the 20–1, 41, 289, 309, 342–3
ideas of reason 245
identity, law of 318n28
identity, nature of 145n11

Identity of Indiscernibles and Indiscernibility of Identicals 205n28
illusion of contingency 147n15
illusion of necessity 222
illusions 77n72, 106–7, 409
imagination, faculty of (*Einbildungskraft*) 20–2, 64–7, 86, 96, 110–11, 333–7, 354n23, 372–6
Inaugural Dissertation, Kant's 117
Incompatibilism 418–19, 445
incongruent counterparts 119–20
The Incredible Shrinking Man (1957) 240
indeterminism 418
indexical predicates 202, 217
inference-to-the-best-explanation 164
inner sense 39, 47, 52–5, 58–64, 69, 71, 73–5, 80, 363–8, 399–402
insight, Kant's theory of 346–50, 361–83, *see also* knowledge, Kant's theory of
intellectual intuition 19, 57, 62, 101, 168, 194, 238
intentionality 365
intrinsicness of space and time, the 21, 42–3, 290, 307
Intuitionism 287, 334–5
intuition (*Anschauung*), Kant's theory of 97–103, 115–26, *see also* non-conceptual content, Kant's theory of
irreversibility, temporal 130–1, 328, 388, 398, 403–6, 450–1
isomorphism 264, 265–9, 273, 277–9, 281, 283, 286, 294, 371, 375, 378

James, W. 53
judgments, *see* propositions
judgments of taste 109

Kant and the Metaphysics of Causality (Watkins) ix
Kant's Transcendental Proof of Realism (Westphal) ix
Kelly, S. 85
Kemp Smith, N. 320
Kim, J. 422
Kitcher, Philip 320
Kleist, H. 13–14, 226
knowledge, Kant's theory of
 a priori 144, 171–5, 177, 184, 188, 210, 331–40, 350–85
 perceptual 132–6

scientific 160–71, 220–46, *see also* scientific knowing (*Wissen*)
Kripke, S. 144, 144n7, 145, 146n12, 147n13, 151–2, 154n31–32, 163, 173–6, 186n68, 191–2, 193n8, 219, 221n42, 223n49, 350n14, 352, 352n39, 356–7
Kroon, F. 207

language 372–3
laws of nature 181–90, 436–40, 446–50
Leibniz, G. W. 9, 15, 29n47, 49n29, 55n38, 72–3, 102, 116, 127, 145, 146n12, 152, 154n31, 187, 189, 193, 193n8, 197, 205n28, 276, 346–8, 375n52, 429
Lewis, C. I. 10, 95n24, 304, 341–2, 361, 385
Lewis, D. 29n47
Liar Paradox, the 272
life
  concept of 202, 434
  feeling of 109, 435–6
  living gold 220–1
  and matter (hylozoism) 221, 221n43
  and organisms 203, 435
  vs. mechanism 439–40, 449
  and mind 88n7, 109, 131, 435–6, 449, 451
  strong continuity between life, mind, and the power of choice 445, 449, 451
linguistic division of labor 153
lived experience (*Erlebnis*) 75n67, 96
location, unique 126
Locke, J. 9–10, 38, 39n5, 86, 94–5, 116, 140, 150n21, 154n31, 160–1, 163–4, 166, 197, 212, 223–5, 227, 229, 239
logic 2, 2n4, 3, 11, 11n29, 13, 18, 21, 23, 256, 274–5, 287, 299, 302, 304, 312–14, 318n28, 321, 322n33, 323n37, 324–30, 338–9
logicism 2n5, 10, 11n29
Logische Aufbau der Welt (Carnap) 6
logocentric predicament, the 12n29

magnetoscopes 241
Manicheanism 24
Der Mann ohne Eigenschaften (Musil) 227
"The Man with the Blue Guitar" (Stevens) vii

The Man with X-Ray Eyes (1963) 241
manifest image, the 8, 14, 90, 102n43, 141, 249–50
manifest realism, Kant's viii, 44, 47–9, 141, 142–3, 150, 152, 170, 220–42, 250, 271, 311, 340, 392, 449
maps 111, 114, 267, 267n25, 268–9, 271, 286, 367n40, 375
Martin, M. G. F. 85, 136
mathemata 380
mathematical knowledge 331–40, 379–83
The Matrix (1999) 66–8
matter, Kant's theory of 48, 148–51, 177–81, 220–42, 388–409, 433–50
McDowell, J. 8, 88–9, 90, 99, 107–8, 132–3
"The Meaning of 'Meaning'" (Putnam) 152
meaning and necessity 3
Meditations on First Philosophy (Descartes) 50
Meinong, A. 6n19, 85, 95n35, 97
memory 20, 45n20, 52, 56, 372, 374n51
mental images (*Bilder*) 96, 110, 354n23, 363, 373
mental language (*lingua mentis*) 373
mental models 143, 332n47, 336n52, 338, 354n23, 374
mental rotation of images 134–5
meta-consciousness 53–4, 125
meta-languages 274
Metaphysical Foundations of Natural Science (Kant) 46, 55n27, 179, 217n39, 234
methodological principles for interpreting Kant 7
Michelson-Morley experiments 235
microscopes 161, 199–200, 239, 241
Mill, J. S. 10
Mind and World (McDowell) 88, 132
Möbius strip 131
modal actualism, Kant's 304
modal conceptualism, Kant's 304
modal dualism, Kant's viii, 23, 193n8, 236n58, 304
modal skepticism 342, 344–6, 350, 385, *see also* skepticism
modal status of a proposition 172, 174–6, 184, 190, 352
mode of presentation (*Art des Gegebenseins*) 267
monadology 344
Moore, A. W. ix

Moore, G. E. 31n51, 37–8, 39n5, 40, 60–1, 80, 263n14, 268n37, 377
moral law, see Categorical Imperative, the
Müller–Lyer illusion, see illusions
Musil, R. 13
Myth of the Given, the 83, 88–9

Nagel, T. 53, 76, 132n74
Naming and Necessity (Kripke) 152
natural causal singularities 429, 438–9
natural kind terms
    Kant's theory of 203–20
    manifest 158–9, 170
    scientific 157–9, 174, 176
natural kinds, see essentialism
natural mechanism 8n32, 15, 418, 430–1, 445–7, 450
natural purpose (*Naturzweck*) 434
naturalism
    Humean 16
    liberal 16n35, 16–17, 310–11, 451
    reductive or scientific 4–8, 10, 10n28, 11n29, 13–17, 30, 33, 90, 151, 193n8, 250, 310
necessary possibility 304, see also necessity and possibility
necessity and possibility
    alethic 146n12, 147n15, 171–7, 184–5, 193n8, 304, 315–16, 318–19
    epistemic 350–83
    phenomenological 133–6, 362–83
neglected alternative 289, 301–4
Neiman, S. 30n48, 284
neo-Hegelianism 10
neo-Kantianism 4–6, 10, 89, 310
neo-logicism 312–13
neural plasticity 67
neutral monism, Kant's 450
Newton, I. 2, 15, 46, 75n61, 148, 167, 184, 187n69, 202n36, 232, 234, 387, 402, 434, 436–7, 439–40, 445–7, 450
Newton's laws of motion 184, 446
Nietzsche, F. 1, 34, 148, 285n47, 387
Noble in Reason, Infinite in Faculty (Moore) ix
Nola, R. 207
non-conceptual content, Kant's theory of 94–132, 295–6, 314, 322, 435–6, see also intuition, Kant's theory of
non-conceptualism, see non-conceptual content, Kant's theory of

non-conscious mind 74n64, 84, 84n7, 94, 94n33, 402, 435, 449, 451
non-contradiction, law or principle of 18, 126, 259, 260n13, 325, 383
non-epistemic perception 45, 45n17–18, 77n72, 78, 82–3, 132, 138, see also non-conceptual content, Kant's theory of
non-Euclidean geometry 130–1, 235–6
"non sequitur of numbing grossness" 403–4
normativity
    epistemic 135–6, 383–5
    moral 18–19, 23–9
    logico-mathematical 339–41
"nothing but" 167
nothing is hidden 142–3
noumena 14–15, 21, 41, 43, 141–2, 165–8, 194–9, 238, 240, 242
noumenoscopes 239
noumenotrons 199
numbers, Kant's theory of 320–31

objective reality
    in Descartes 94
    in Kant, see representation, objective reality of
O'Neill, O. 30n48, 284
ontological argument, the 197
Opus postumum (Kant) 47, 55n37, 189n71, 234, 394, 449
order of time vs. lapse of time 405
orientation
    cognitive 267–70
    spatial 71–80
Origins of Analytical Philosophy (Dummett) 89
Our Knowledge of the External World (Russell) 6
outer sense 63–80, see also intuition, Kant's theory of
overdetermination, causal 422–3
ox and barnyard gate example 105

Pap, A. 341, 361n31, 362n32
paradox, see antinomy
paradox of analysis 377–8
Paralogisms of Pure Reason 409–10
Parsons, C. 318n28, 322, 349n12, 362n32, 379n56
Paton, H. J. 301

pattern-matching activities 134–5
Peacocke, C. 85, 108, 115
Peano, G. 312–14, 319n29
Peirce, C. S. 10
perception, *see* direct perceptual realism, Kant's; non-conceptual content, Kant's theory of
perfectionist epistemology, Kant's 379, 379n55
permitted by law vs. compelled by law vs. contingently conforming to law 446
persistence of substance(s), *see* Analogies of Experience, first Analogy
personal identity 56, 62n48, 75n66
personalism 17
personhood 413
phenomena, *see* appearances, Kant's theory of
phenomenal character of consciousness 94, 124n66
"Philosophy and the Scientific Image of Man" (Sellars) 8n33, 250
Philosophy and Scientific Realism (Smart) 49, 72
Physical Monadology (Kant) 46, 148
physicalist metaphysics 11–12, 16
physics, *see* matter, Kant's theory of
platonism 17, 288, 342, 344, 349n12, 323, 329, 350, 385
Pontius Pilate 252, 286n50
positivism 6, 10, 273
possibility, *see* necessity and possibility
post-analytic philosophy 5
post-compatibilism, Kant's 420
post-Critical Philosophy, Kant's 449
postures of the mind 94
power of choice (*Willkür*) 18, 25–6, 28, 32, 400, 424, 440–3
practical foundations of the exact sciences thesis 1, 16–17, 30, 32, 252
pre-established harmony 309
preformation-system of pure reason 309
primacy of the exact sciences thesis 30
primacy of human nature thesis 29–30, 32–3
primary qualities vs. secondary qualities 9, 154n31, 222–5
Primordial Cheese 179–180, 181n61
Principia Mathematica (Whitehead and Russell) 266

Principles of Human Knowledge (Berkeley) 389
Principles of Mathematics (Russell) 34
Principle of Pathologically-Conditioned Alternative Possibilities 428
productive wit 382n57
properties, Kant's theory of 29, 29n47, 47–8, 127–9, 142, 153–6, 177–81, 192–201, 222–42
property dualism vs substance dualism 196
propositions
  analytic 3, 219–22, 243, 244n65, 277–9, 331, 338, 345, 348, 350, 351n14, 371, 374–6, 380–1
  synthetic a priori vii, 22–3, 181–7, 190, 211, 312–20
proto-rationality 87, 105, 125
prototypes and stereotypes 207, 207n27–8, 209n29, 213n32
psychologism 266n22, 335
purposiveness *see* teleology
Putnam, H. 8, 38, 39n5, 40, 67n58, 141n2, 144, 144n7, 145, 146n12, 147n13, 147n15, 151–2, 155n32, 160n41, 163n45, 182n62, 207n27–28, 219, 221n42, 348, 351–2

qualities 230–1
quantum entanglement 235
quantum field theory 235
quasi-objects 336
Quine, W. V. O. vii, 3, 10, 11n28, 17, 198, 244n65, 323n37, 341–2
quus-world 319n29

radical empiricist epistemology 11, 13
radical evil 24
radically finite world 318
radically unrecursive world 318–19
rational anthropology viii
rational intuitions 256n25, *see also* insight, Kant's theory of
real essences 166
realism 29, 33, 43–50, 79–80, 82, 132–9, 141–3, 194–202, 220–42, 264–75, 283
reason 18–19, 30, 251–2, 441–4
reasons 26, 30–1, 132–3, 135–6, 379, 385, 442, 446
reflection 94n33, 95
Reflexionen (Kant) 50n31, 52, 66, 69, 72, 117, 343, 360, 361n31, 364

Refutation of Idealism, Kant's 50–68
"The Refutation of Idealism" (Moore) 60
Reid, T. 30n49, 38, 39n5, 40n8
Reinhold, K. L. 166
Regulative Idea Theory, the 423, 425
Relativity, Einstein's theory of 10, 34, 235–6
representational character 94–5, 113, 366–8, 371, 375, 377–8
representations (*Vorstellungen*) 94–7
Rescher, N. 26–7, 442–3
respect (*Achtung*) 26–7, 442–3
The Right and the Good (Ross) 356n25
rigid designators 152–3, 156–7, 157n37, 158–9, 160n41, 192
Rorty, R. 83
Ross, W. D. 356n25
Royce, J. 287, 289, 449
Russell, B. 2n5, 4, 6, 10, 11n29, 34, 55n38, 85, 89, 96n35, 259n12, 266–7, 272, 302, 312, 313n24, 325n40
Russell's Paradox 312

Sartre, J.-P. 60–1, 345, 364n44
"savage" examples 106–8
scenario content 115
schematism 20, 110–11, 114, 135, 267n25, 274n36, 326–7, 329, 332–3, 335–8, 349, 354n23, 372, 374–6, 378, 380–3, 385
Schopenhauer, A. 186n68
schpace and schtime 302
Schultz, J. 324–6, 328, 330, 333
scientific image, the 8, 14, 90, 141, 161, 246, 249–50, *see also* manifest image, the
scientific knowing (*Wissen*) 30
second-order volitions 428n32, 442
seeing is believing 134
Seeing and Knowing (Dretske) 87
Seinfeld 113
self-consciousness, *see* apperception; meta-consciousness
self-organizing systems 407
self-orientation 76
self-transcendence 345
Sellars, W. viii, 4, 6n19, 8–11, 17, 30, 83, 89–90, 99, 107, 140, 161, 164, 167n49, 249, 268n27
semantic paradoxes 272; see also Liar Paradox, the

sensation (*Empfindung*) 19, 56–7, 59–60, 87, 95–6, 109, 118, 134
Shepard, R. 373
skepticism 37–41, 50–1, 344–6
Skolem, T. 273n35, 314n26, 328
Smart, J. J. C. 49
social construction of knowledge 244
Solaris (1972) 187
solipsism 38
space, Kant's theory of 21, 42–3, 63–80, 290, 307 *see also* intuition, Kant's theory of; time, Kant's theory of
space of reasons 107
spontaneity 18, 28, 30n47, 66, 354n33, 400–1, 408–9, 412, 428, 431, 445, 451
Stevens, W. vii
Strawson, P. ix, 17, 60n44, 75n68, 80n77, 86n12, 90, 115, 257
Stroud, B. 41
structuralism
  causal-dynamic metaphysical, Kant's 148–50, 200, 234, 394
  mathematical, Kant's 329
substance(s), Kant's theory of, *see* Analogies of Experience
substance dualism, *see* property dualism vs. substance dualism
sufficient reason, law or principle of 259
supervenience 12–13, 16, 72, 123, 144–5, 155–6, 179, 180–1, 201, 223n49, 291–2, 295, 320, 354, 393–6, 404, 407, 429n36, 431, 437, 449
synopsis of the sensory manifold 86
synthesis speciosa 22, 86, 110, 337
synthetic a priori propositions, *see* propositions, synthetic a priori

tactile-visual substitution systems 67
Tarkovsky, A. 187
Tarski, A. 11n29, 254, 272–4, 276–7, 283, 303, 323
teleology 4n11, 8, 15, 140, 153n30, 402, 418, 430–1, 435–6, 439, 451, *see also* life
temporal irreversibility vs. temporal necessitation 405–6
theoretical reason, *see* reason, theoretical
theoretical technique 349
thick Euclidean structure vs. thin Euclidean structure 120
things-in-themselves, *see* noumena

Thoughts on the True Estimation of Living Forces (Kant)
Three Alternatives Argument, the 42, 289, 301
time, Kant's theory of 21, 42–3, 290, 307, 388–409, 450, *see also* intuition, Kant's theory of; space, Kant's theory of
Timeless Agency Theory, the 423–5
Time's Arrow 130, 398, 450–1
"to be is to be the value of a bound variable" 198
togetherness (of concepts and intuitions) principle, the 97–8
Tractatus Logico-Philosophicus (Wittgenstein) 265
transcendence of the ego, the Transcendental Aesthetic
transcendental deductions 63, 98, 115–18, 124, 126
transcendental idealism, Kant's 21–2, 41–4, 168–70, 194–8, 289, 292, 296–7, 299–312, 342–4, 422, 425–7
transcendental proofs 380
transcendentalism thesis, the 20–1, 41, 43, 289, 343
Transition from the Metaphysical Foundations of Natural Science to Physics (Kant) 46
transparency of consciousness 60
Treatise of Human Nature (Hume) 389
Trendelenberg, F. A. 301
Trevarthyn, C. 129
truth
 criteria of 275–82
 definition of 255–75
 types of 277–82, *see also* propositions
truthfulness 284–6
truth-value gaps 258
truth with a human face 270
Twin Earth 147n13, 155n32, 180
Two Aspect or Two Standpoint Theory, the 5n17, 422–7
Two Concept or Two Property Theory, the 5n17, 426–7, 433
two dimensional modal semantics 193n8, 426–7, 433
"Two Dogmas of Empiricism" (Quine) 3

Two Images Problem, the 8–10, 14–15, 29, 34, 40n8, 84, 90, 140, 151, 249–50, 288, 311
Two Object or Two World Theory, the 5n17, 421n24, 422, 424–6
two tables, Eddington's 162–3
Tye, M. 61n47, 85

Über das Marionettentheater (Kleist) 13
unity problem about non-conceptual content, the 9, 126–32, *see also* non-conceptual content, Kant's theory of
universal grammar 373
unperceived unperceivables 238, 245, *see also* noumena
useless passions 345
user-friendly facts 33, 50, 267–8, 271, 286, 340

Vaihinger, H. 301
van Inwagen, P. 418
Varieties of Reference (Evans) 89n18
vicious circle principle, the 272
Vienna Circle, the 6
volition, *see also* desire (*Begierde*); power of choice (*Willkür*); will (*Wille*)
Voltaire 15

Watkins, E. ix
Westphal, K. ix
"What is Orientation in Thinking?" (Kant) 268–9
Whitehead, A. N. 2n5, 11n29, 266, 302
will (*Wille*) 14–15, 17, 19, 22–7, 29n47, 30, 32–3, 386, 388, 427–34, 442–4
Williams, B. 245, 255, 285n47
Wisdom, J. 252n4, 253
Wittgenstein, L. 1, 8, 13–14, 31, 40n8, 60n43, 79n76, 96, 105, 142–3, 186n68, 226, 249, 254, 264n16, 265–7, 343n17
Wolff, C. 116, 375n52
"The World as I Found It" 269

XXX-constitution 178–80